Automation, Collaboration, & E-Services 14

Series Editor

Shimon Y. Nof, *PRISM Center, and School of Industrial Engineering, Grissom Hall, Purdue University, West Lafayette, IN, USA*

The Automation, Collaboration, & E-Services series (ACES) publishes new developments and advances in the fields of Automation, collaboration and e-services; rapidly and informally but with a high quality. It captures the scientific and engineering theories and techniques addressing challenges of the megatrends of automation, and collaboration. These trends, defining the scope of the ACES Series, are evident with collaborative automation: collaborating systems, wireless communication, Internetworking, multi-agent systems, sensor networks, cyber-physical collaborative systems, interactive-collaborative devices, and social robotics – all enabled by collaborative e-Services. Within the scope of the series are monographs, lecture notes, selected contributions from specialized conferences and workshops.

Chin-Yin Huang · Sang Won Yoon
Editors

Systems Collaboration and Integration

See Past and Future Research through the PRISM Center

Springer

Editors
Chin-Yin Huang
Industrial Engineering and Enterprise Information
Tunghai University
Taichung, Taiwan

Sang Won Yoon
Department of Systems Science and Industrial Engineering
State University of New York at Binghamton
New York, NY, USA

ISSN 2193-472X ISSN 2193-4738 (electronic)
Automation, Collaboration, & E-Services
ISBN 978-3-031-44372-5 ISBN 978-3-031-44373-2 (eBook)
https://doi.org/10.1007/978-3-031-44373-2

This Springer imprint is published by the registered company Springer Nature Switzerland AG
The registered company address is: Gewerbestrasse 11, 6330 Cham, Switzerland

Paper in this product is recyclable.

Foreword

When Shimon Nof first approached me to pen a foreword for this book in celebration of PRISM'S 30th Anniversary, I felt equally honored as I did inspired. Being able to witness Shimon's leadership and vision firsthand throughout the founding of the PRISM Center 30 years ago to today has been one of the more remarkable experiences in my professional career. Yet, while this volume was crafted to celebrate PRISM's conclusion of this milestone year; it also serves as both a call to action and case study in articulating the ever-going continuance of innovation within the systems collaboration and integration field. The PRISM Center operates as a home base for thinkers and topic fields that share the drive and delight for positively moving the needle of today's challenges while eagerly embodying the notion that satisfaction comes not just from what is reached, but from the perpetual quest for what can be.

In the editors, Professor Chin-Yin Huang from Tunghai University, Taiwan; and Professor Sangwon Yoon from Binghamton University, New York, we have highly accomplished curators for a collection that so proficiently exemplifies the field. Both are Purdue Industrial Engineering graduates, former PhD Senior Researchers of the PRISM Center, and continue in their professional roles as active leaders of the Purdue Global Research Network (PGRN). In addition, Sangwon was an outstanding researcher assisting on a project applying systems integration and collaboration to profoundly impact sales and supply chain under the umbrella of my former responsibilities.

Throughout my professional experience with Fortune 500 global corporations, the ideals espoused by the PRISM Center offer an intellectual backbone for seeking to refine systems and improve integrations. From personal experience implementing in real-time the exchange of key metrics between handheld devices and remote computer analytics for dramatically improving immediate customer experience, to thought-provoking examples in the book such as: Collaborative Requirement System using Matrix and AI Approach, and Crop Plants Stress Monitoring with Bayesian Network Inference in Cyber-Physical Robotic System. The PRISM Center's orientation illustrates the importance of development and improvement of systems integrations and collaboration across numerous disciplines and applications. PRISM also displays each year how its collective scholars, innovators, and implementors build upon yesterday's thought products to create and explore new frontiers not only for tomorrow, but for today. This work is remarkable, inspiring, and endlessly promising.

To the reader, it is my best wish that you engage with these chapters through the lens of appreciating that success is a function of how you surround yourself, a belief that has served me well throughout my life. Such is the case with PRISM as the successful people surrounding its work nurture the core principle that innovation is not a next step, but rather a north star. A constant guide that is always needed, brightly inviting, and certainly never captured into a closed grip.

To my great friend and colleague Shimon, thank you for your decades of passionate commitment to collaboration, innovation, and exploration. Without your curiosity, reverence and rigor for learning, none of this would be possible.

And lastly, to The PRISM Center, its alumni, current cohorts, and future explorers, thank you for all your work in furthering the field of computer, human, and machine interactions. Your contributions have and will continue to have an enormous impact as systems integration and collaboration is a vital framework for progress and growth.

May all be inspired by the works within this book and continue to innovate and collaborate in the future. Seek and enjoy.

August 2023

Juan Ernesto de Bedout
Purdue University BSIE'67, MSIE'67, HD'13
Retired President Latin American Operations, Kimberly-Clark (and proudly, the first Distinguished PRISM Center Scholar)
Atlanta, Georgia

The original version of the book has been revised: The affiliation of the author Chin-Yin Huang has been updated. A correction to this book can be found https://doi.org/10.1007/978-3-031-44373-2_27

Foreword

Robotics and AI for manufacturing have become one of the most important research areas in both academia and industry today. They were not always the hottest or most obvious topics in the years past. We are proud that Purdue, thanks to the pioneering preeminence of PRISM, has been a leading institution over the last four decades in this area.

As a Boilermaker, reading the first chapter reviewing the luminous history and many achievements of PRISM makes me immensely proud of the roles that Purdue faculty, students, staff and collaborators have played in the evolution of what is now called Industry 4.0.

The co-editors of this valued book and long-time leaders of PRISM, Chin-Yin Huang and Sang Won Yoon, have assembled a very impressive collection of the latest research results. Readers will enjoy perusing through the wide-ranging choice of topics in these chapters. I also hope this book will become a series when a new edition is published each decade of the institute.

In the meantime, Purdue University, across many academic units and research labs, will continue to prioritize the advancement of research at the interface between the virtual/digital and the physical, between bytes and atoms, between what we code and what we touch. And central Indiana, through partnership with a growing number of industry ecosystems, will build out a "hard tech corridor" that deploys some of the fundamental research advances as reported in this delightful volume.

August 2023

Mung Chiang
President

Roscoe H. George Distinguished Professor
of Electrical and Computer Engineering
Purdue University

Contents

Future Challenges in Systems Collaboration and Integration

About the Authors

Mohammad Sa'eed Alattar is a master's student in the Department of Systems Science and Industrial Engineering, State University of New York at Binghamton, Binghamton, NY, USA. His research interests are modeling and simulation of large-scale complex systems and applying machine learning methodologies and data mining in manufacturing systems.

Prof. Avital Bechar is a senior research scientist and the director of the Institute of Agriculture Engineering (IAE) at the Agriculture Research Organization (ARO), Volcani Institute in Israel. He was an appointed adjunct professor in the school of IE at Purdue University, USA, in the years 2011–2012. He holds a B.Sc. degree in Aerospace Engineering and a M.Sc. in Agricultural Engineering, both from the Technion, Israel, and a Ph.D. in Industrial Engineering from Ben-Gurion University, Israel, on agricultural robotics and human–robot integrated systems. He is the founder and the head of the Agricultural Robotics Lab at IAE, where he is conducting fundamental and applied research in robotics for agriculture, human–robot collaborative systems, sensor technologies, and developing new concepts and approaches for the operation and development of agricultural robots and proximal sensing. He is the author of more than 100 articles in peer-reviewed scientific publications. He authored several chapters, edited a book on robotics and precision agriculture and has led and participated in more than 50 local and international research projects. He is the former chairman of the Israeli Society of Agricultural Engineering (ISAE), a co-founder of the Israeli Robotics Association (IROB), an IEEE senior member and a member of the IEEE Robotics and Automation and IEEE SMC Societies, a member of the IEEE Technical Committee on Agricultural Robotics, and a member of the CIGR Section V committee (Systems management).

Sigal Berman is an associate professor in the Department of Industrial Engineering and Management at Ben-Gurion University of the Negev, Beer-Sheva. She received a B.Sc. in Electrical and Computer Engineering, from the Technion, an M.Sc. in Electrical and Computer Engineering, and a Ph.D. in Industrial Engineering, both from Ben-Gurion University. She leads the Intelligent Systems Engineering Laboratory (ISEL) where her research focuses on the analysis and engineering of intelligent systems capable of dexterous motion. She develops data-driven models for the synthesis of robotic motion and the analysis of human motion.

John Biechele-Speziale is a Ph.D. student in Industrial Engineering at Purdue University in the Mario Ventresca group. His research interests fall within complexity research, machine learning, optimization, and applying methods to improve models and their interpretability.

Tim Blackwell is a senior lecturer at Goldsmiths, University of London. He has BSc, MSc, and PhD degrees in physics and computer science. His main research interests are swarm intelligence and live algorithms (computer–human improvised music). He leads AI, neural networks, and algorithms modules at Goldsmiths and runs an evening class on The Multiverse. Apart from swarm intelligence and live algorithms, his output includes computer music, perceptions of music complexity, and digital art. His current research interests are algorithm parameter tuning and well-structured benchmarks.

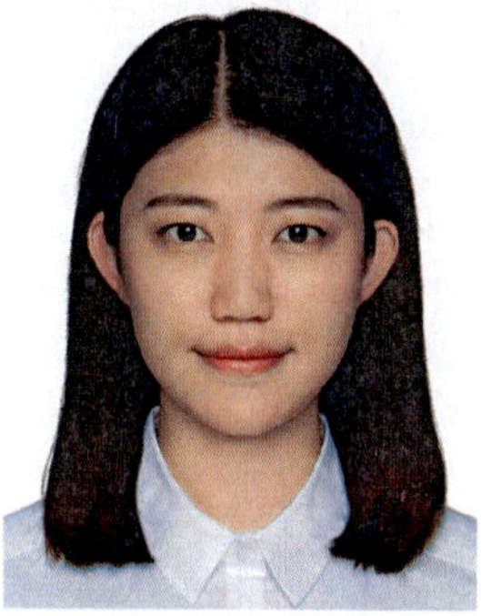

Nieqing Cao is currently pursuing a Ph.D. degree with the Department of Systems Science and Industrial Engineering, the State University of New York at Binghamton, Binghamton, NY, USA. Her research interests include modeling and simulation of large-scale complex systems, large-scale data predictive modeling, and systems optimization in manufacturing.

José A. Ceroni graduated as an industrial engineer from Pontifical Catholic University of Valparaiso, Chile 1988, and received his Master of Science in 1996 and PhD in 1999 in Industrial Engineering from Purdue University, Indiana, USA. His research interests include collaborative production and control, industrial robotics systems, collaborative robotics agents, and collaborative control in logistics systems. He is a member of the Board of the International Federation for Production Research and a member of IFAC and IEEE.

Xin W. Chen is a professor in the School of Engineering at Southern Illinois University Edwardsville. He received the MS and PhD degrees in Industrial Engineering from Purdue University and BS degree in Mechanical Engineering from Shanghai Jiao Tong University. His research interests cover several related topics in the area of conflict and error prognostics and prevention, production/service optimization, and decision analysis. He is the author of the book "Network Science Models for Data Analytics Automation" in Springer ACES book series.

Cristian Ciurea is a professor at the Department of Economic Informatics and Cybernetics from Bucharest University of Economic Studies. He is also the head of department. He has graduated the Faculty of Economic Cybernetics, Statistics and Informatics from the Bucharest University of Economic Studies in 2007. He has a master's in Informatics Project Management (2010) and a PhD in Economic Informatics (2011) from the Bucharest University of Economic Studies. He has a solid background in computer science and is interested in collaborative systems related issues. Other fields of interest include intelligent systems, software metrics, data structures, object-oriented programming, windows applications programming, mobile devices programming, and testing process automation for software quality assurance.

Vincent G. Duffy is a professor of Industrial Engineering and Agricultural & Biological Engineering at Purdue University. He has served as a faculty member at Purdue since 2005 and is a fellow of the UK Ergonomics Society (CIEHF), Chartered Institute of Ergonomics and Human Factors in the UK.

Puwadol Oak Dusadeerungsikul received Ph.D. in Industrial Engineering from Purdue University. He also received M.S. in Industrial Engineering from Georgia Institute of Technology. Currently, he is a lecturer at the Department of Industrial Engineering, Chulalongkorn University, Thailand. His research focuses on applications of operations research in agriculture, health care, and logistic systems.

Florin Gheorghe Filip was born in 1947 in Bucharest, Romania. He graduated and received his PhD in Automation at *Politehnica* University of Bucharest in 1970 and 1982, respectively. He was elected as a corresponding member of the Romanian Academy in 1991 and became a full member in 1999. During 2000–2010, he was the vice president of the Academy. In 2010, he was the elected president of the *Information Science and Technology* section of the Academy (re-elected in 2015, and 2019). He was the managing director of the *National Institute for R&D in Informatics*-ICI, Bucharest (1991–1997). He is an honorary member of the *Romanian Academy of Technical Sciences* and *Academy of Sciences of Moldova*. His main scientific interests include optimization and control of large-scale complex systems, decision support systems (DSS), technology management and foresight, and IT applications in the cultural sector. He authored/co-authored over 350 papers published in international journals and contributed volumes. He is also the author/co-author of thirteen monographs published in Romanian, English, and French and the editor/co-editor of 30 volumes of contributions.

Chin-Yin Huang is a professor of Industrial Engineering and Enterprise Information and the dean of Engineering College at Tunghai University, Taiwan. He had a Ph.D. degree from Purdue University, USA. His research interests include healthcare management, clinical data analysis, manufacturing process optimization, and intelligent manufacturing. He is currently the vice president of International Foundation for Production Research. He also serves as the chairman in the Asia Pacific Region. He is a board member of Asia Pacific Industrial Engineering and Management Society. He has been serving as the professional consultants and advisors in government, universities, hospitals, and manufacturing sectors. His publications appear in International Journal of Production Research, International Journal of Production Economics, Computers in Industry, Computers and Industrial Engineering, Robotics and Computer-Integrated Manufacturing, Epilepsy Research, Production Engineering, Engineering Computations, etc. He also co-authored chapters for Handbooks of Industrial Engineering, Handbook of Industrial Robotics, and Handbook of Automation. His has co-authored two books in Industrial Engineering and Management published in Taiwan.

Yu (Chelsea) Jin is an assistant professor of Department of Systems Science and Industrial Engineering, State University of New York at Binghamton, Binghamton, NY, USA. Her research focuses on sensing and analytics, optimization, and simulation for advanced manufacturing and service applications. She also served as a reviewer for IISE Transactions, ASME Journal of Manufacturing Science and Engineering, IEEE Transactions, Journal of Intelligent Manufacturing, etc.

Yuta Kitano was a master's student in the Management Science and Social Informatics Program, Department of Informatics, Graduate School of Informatics and Engineering, the University of Electro-Communications (UEC), Tokyo, Japan. He received his BS and MS from UEC in 2019 and 2021, respectively. Additionally, he stayed the UK Nottingham University Business School from September to November in 2019 hosted by Prof. Kim Hua Tan. His research interests are text, innovation, and Twitter analysis.

Dr. Pisut Koomsap is currently an associate professor in the Department of Industrial Systems Engineering at the Asian Institute of Technology. He received his doctoral degree in Industrial Engineering from the Pennsylvania State University, USA. His current research focuses on customer-oriented manufacturing covering a broad spectrum from concept development for designing products, service, and experience for better serving customers to developing technologies such as 3D printing and mosaic tiling automation to support the design concepts. In terms of education research, he is currently working with his doctoral student to develop a learning experience-focused course design and development for better learning for students. He is also the project coordinator and a researcher for two Erasmus+ Capacity Building in Higher Education projects. The first one is Curriculum Development of master's Degree Program in Industrial Engineering for Thailand Sustainable Smart Industry (MSIE 4.0). Another one is Reinforcing Non-University Sector at the Tertiary Level in Engineering and Technology to Support Thailand Sustainable Smart Industry (ReCap 4.0). Besides, he was an ISE department head and a chair for the Academic Development Review Committee that processes institute-wide academic matters, including reviewing curricula and new academic proposals.

Soongeol Kwon is currently an assistant professor in the Department of Industrial Engineering at Yonsei University. His research has mainly focused on developing decision-making methodologies using mathematical optimization models and machine learning techniques for demand-side management with energy storage and renewable energy. His research interests include energy storage control policy, data centers server provisioning, building energy management, and energy-aware scheduling.

Steven J. Landry, Ph.D. is a professor and the Peter and Angela Dal Pezzo chair and the department head in the Harold & Inge Marcus Department of Industrial and Manufacturing Engineering at the Pennsylvania State University. He was previously a faculty member, the associate head, and the interim head in the School of Industrial Engineering at Purdue University, with a courtesy appointment in the School of Aeronautics and Astronautics. He has conducted research and published in air transportation systems engineering and human factors and has taught undergraduate and graduate courses in human factors, statistics, and industrial engineering. Prior to joining the faculty at Purdue, he was an aeronautics engineer for NASA at the Ames Research Center. He was also previously a C-141B aircraft commander, an instructor, and a flight examiner with the U.S. Air Force with over 2,500 heavy jet flying hours.

Seokcheon Lee received his B.S. and M.S. degrees in Industrial Engineering from Seoul National University (Seoul, South Korea) in 1991 and 1993, respectively, and his Ph.D. degree in Industrial Engineering from the Pennsylvania State University (PA, USA) in 2005. He is currently a professor in the School of Industrial Engineering, Purdue University (West Lafayette, IN, USA). His research focuses on production scheduling and logistics.

Dr. Mark R. Lehto is a professor in the School of Industrial Engineering, at Purdue University, and also the co-chair of the Interdisciplinary Graduate Program in Human Factors. He received his M.S.I.E. from Purdue in 1981 followed by his Ph.D. from the University of Michigan. His research interests include human factors, text mining, safety engineering, and hazard communication and decision support systems. He has taught and developed several different undergraduate and graduate courses, within the School of Industrial Engineering, including classes on safety engineering, engineering economics, industrial ergonomics, and work design. He has also served as the committee chair or a committee member for over 200 graduate students. In 2008, he published a well-received textbook entitled *An Introduction to Human Factors and Ergonomics for Engineers* that synthesizes his years of experience in the field of human factors, in teaching, research, and engagement with outside organizations.

Dr. Yuanyuan Li received a master's degree (2019) and a Ph.D. degree (2022) from the Department of Systems Science and Industrial Engineering, Binghamton University, Binghamton, NY, USA. Her research interests include large-scale data predictive modeling, machine learning, data analytics in healthcare and manufacturing, modeling and simulation, and dynamic system optimization.

Dr. Yu-Ju Lin received both his Ph.D. and M.Sc. in Electrical Engineering from National Cheng Kung University (Taiwan) and his bachelor's degree in Electrical Engineering from Fu Jen Catholic University (Taiwan). In 2008, he began working at National Cheng Kung University as a doctoral researcher in the Department of Electrical Engineering, where he is leading a team of researchers in organic semiconductors and flexible electronics. In 2011, he joined United Microelectronics Corporation (UMC) at their Advanced Technology Development (ATD) as the chief engineer and a research scientist for the semiconductor process. Since 2017, he joined Tunghai University (Taiwan) as an assistant professor in Department of Industrial Engineering and Enterprise Information. His research areas are on Internet of Things, semiconductor manufacturing process, and mechatronics.

Ramses V. Martinez, Ph.D. received his Ph.D. degree in Physics and Materials Science from the Spanish National Research Council (CSIC) in 2009. He then worked as a postdoctoral researcher in the Chemistry and Chemical Biology laboratory of Prof. George M. Whitesides at Harvard University. Currently, he is an assistant professor at the Schools of Industrial and Biomedical Engineering, Purdue University. The interests of his research group, *The FlexiLab*, include biosensing, flexible electronics, wearables, smart textiles, and soft robotics.

Masayuki Matsui is an emeritus professor at The University of Electro-Communications, Tokyo, and a faculty of Kanagawa University, Yokohama, Japan. He received his BS and MS degrees in Industrial Engineering from Hiroshima University and his DEng in research on conveyor-serviced production systems from the Tokyo Institute of Technology, Japan. He was a visiting scholar at UC Berkeley and Purdue University from 1996 to 1997.

His recent research interests include production/queueing model, stochastic management, 3M&I-artifacts science, and nature vs. artifacts body. He has authored over 500 technical papers and 20 books, including many Springer books. At academics, he is the former president of the Japan Industrial Management Association (JIMA) and is now a board member of the International Foundation for Production Research (IFPR). Recently, he was honored with the PRISM award (Distinguished PRISM Center Scholar Award) at Purdue University in 2021.

His main inventions are queueing laws, pair matrix/map Matsui's logic, and so on. He served as the editor/director of the JIMA journal (2000–2005), was honored with the JIMA Prize and Award in 2005, has been an emeritus member of JIMA since 2016, and is currently a professional member of IISE. He is listed in Nihon-Shinshiroku (Who's Who in Japan), Marquis Who's Who in Science and Engineering, and in the World (USA).

Shogo Matsuno is an assistant professor in Faculty of Informatics in the Gunma University. He received the Ph.D. degree in Engineering from Graduate School of Information Science and Engineering, University of Electro-Communications, in 2017. From 2017 to 2021, he was a researcher at Hotlink, Inc. Since 2021, he has been in the current position. His research interests include biological signal processing, sensory information processing, and human interface. In recent years, he has been interested in contextual awareness services and human–agent interaction using social media and sensor data. He is a member of the Institute of Electrical Engineers of Japan, Information Processing Society of Japan, Institute of Electronics, Information and Communication Engineers, IEEE, and ACM.

Mohsen Moghaddam is an assistant professor of Mechanical and Industrial Engineering and an affiliated faculty with Khoury College of Computer Sciences at Northeastern University, Boston. He received his PhD from Purdue University and served the GE-Purdue Partnership in Research and Innovation in Advanced Manufacturing as a postdoctoral associate prior to joining Northeastern in 2018. His focus is on building novel methods, tools, and technologies at the intersection of AI/ML, XR, and human-centered computing to augment the interaction of people with machines in industrial settings. He collaborates with over several faculty and research groups across Northeastern University with a diverse set of convergent disciplines ranging from learning sciences and psychology to social sciences, economics, business, and design. He is the co-author of dozens of peer-reviewed journal articles and chapters and has served as a guest editor for the ASME Journal of Mechanical Design. His research is sponsored by the National Science Foundation, Department of Defense, Northeastern University, and industry.

Tomohiro Nakada is an associate professor in the English Communication Department, Faculty of Foreign Studies at Bunkyo Gakuin University, Tokyo, Japan. He received his Doctor of Engineering degree in Artificial Intelligence from the University of Electro-Communications. He was an assistant professor of Department of Electrical Engineering at Matsue College of Technology from 2010 to 2012. He worked on public transportation bidding systems and services evaluations as a researcher at the Institute of Transportation Economics from 2012 to 2014 and as a researcher and a senior researcher at the Policy Research Institute for Land, Infrastructure, Transport and Tourism, Ministry of Land, Infrastructure, Transport and Tourism from 2014 to 2018. He was an associate professor in the Department of Informatics and Electronics at the Daiichi Institute of Technology from 2018 to 2022. He is a member of the IEEE, the Institute of Electronics, Information and Communication Engineers in Japan (IEICE), and Japan Industrial Management Association (JIMA).

Dr. Gaurav Nanda is an assistant professor in the School of Engineering Technology at Purdue University. He obtained his Ph.D. in Industrial Engineering from Purdue University and his bachelor's and master's from Indian Institute of Technology (IIT) Kharagpur, India. He works on research problems involving applied machine learning, text mining, and intelligent decision support systems with applications in Industry 4.0, business intelligence, safety, health care, and STEM education. He teaches courses on supply chain, operations management, artificial intelligence, warehouse management, and lean and sustainable systems.

Win P. V. Nguyen received the B.S. and Ph.D. in Industrial Engineering from Purdue University, Indiana, USA. Currently, he is a collegiate assistant professor at the Grado's Department of Industrial and Systems Engineering at Virginia Tech. His research interests include cyber-physical systems, Industry 4.0, and disruption propagation response. He previously worked as an industrial engineer in Vietnam.

Shimon Y. Nof, Ph.D., D.H.C., is a professor of Industrial Engineering and the director of the NSF- and industry-supported PRISM Center (*Production, Robotics and Integration Software for Manufacturing & Management*; established 1991) linked with PGRN (*PRISM Global Research Network*) at Purdue University. His expertise is in cybernetics, automation, and robotics; a pioneer of robot ergonomics, collaborative control theory, and hubs for collaborative intelligence; a co-inventor of six cyber automation patents; the former president, the secretary general, and the current board member of International Foundation of Production Research (IFPR); the former chair of International Federation of Automatic Control (IFAC) Coordinating Committee "Manufacturing & Logistics Systems"; also active in IFIP, IISE, and INFORMS; a fellow of IFPR and of IISE; a recipient of John Engelberger Medal for Robotics Education; the editor of Springer Book Series ACES (*Automation, Collaboration, & E-Services*); the author/co-author of over 200 refereed journal articles; 400 conference papers; and seventeen books, including: *Handbook of Industrial Robotics* 1st and 2nd editions; *International Encyclopedia of Robotics and Automation*; *Information and Collaboration Models of Integration*; *Industrial Assembly*; *Springer Handbook of Automation 1st and 2nd editions; Revolutionizing Collaboration Through e-Work, e-Business, and e-Service*; *Best Matching Theory & Applications*; and *Dynamic Lines of Collaboration – Disruption Handling & Control.*

Hettiarachchige Don Kavitha Perera received his BS in Mechanical Engineering and MS in Industrial Engineering from Purdue University, West Lafayette, Indiana. He is currently pursuing a PhD in Industrial Engineering at Purdue University, with a focus on human-integrated energy harvesting devices.

William Raymer is a Ph.D. candidate at Purdue University in Industrial Engineering. William is from Mapleton Utah raised by Danny and Marilyn Raymer. He earned a master's degree in industrial engineering in 2022 from Purdue University. William graduated with his master's degree by being nominated as the commencement responder for the Purdue University commencement ceremony. William earned his bachelor's degree in mathematics and statistics composite from Utah State University in 2021. As an undergrad, he enrolled into the Air Force Reserve Officer Corps which allowed him to commission as a space force officer upon completing his undergraduate studies.

Ran Shneor is a Ph.D. student in the Department of Industrial Engineering and Management at Ben-Gurion University of the Negev, Israel. He investigates automatic robotic assembly planning of industrial products containing deformable objects. He served in various leadership and engineering roles in the Israeli Air Force. He holds M.Sc. and B.Sc. in Industrial Engineering and Management, both from Ben-Gurion University of the Negev.

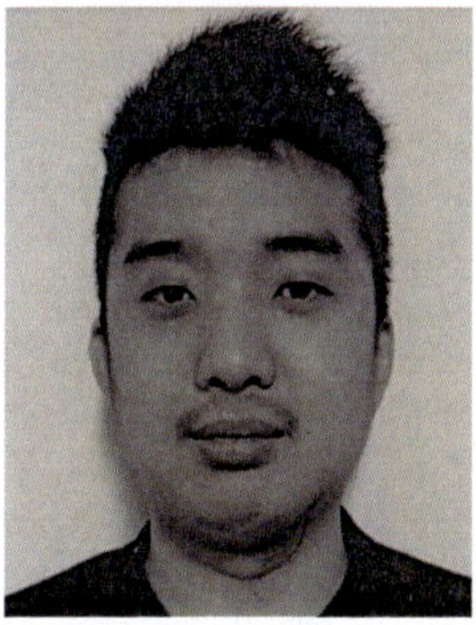

Dr. Hendri Sutrisno received the PhD and M.B.A. degrees in Industrial Management from National Taiwan University of Science and Technology (NTUST) in 2021 and 2018, respectively, and B.Eng. degree in Industrial Engineering from Petra Christian University (PCU), Indonesia, in 2016. He is a postdoctoral fellow with the Institute of Statistical Science, Academia Sinica (ISSAS), since 2021. His research interests include time series analysis, metaheuristic optimization, production research, and machine learning.

Chih-Fan Tan received degrees of B.S. and M.S. from Tnghai University, Taiwan. She also obtained a M.S. degree from Asian Institute of Technology, Thailand. Currently, she is an engineer in Taiwan Semiconductor Manufacturing Company. Her research interests are in production planning and control for manufacturing and services with AI approaches.

Dr. Kim Hua Tan is a professor of Operations and Innovation Management in the UK Nottingham University Business School. He is also an associate dean of Research and Knowledge Exchange. Prior to this, he was a research fellow and a teaching assistant at Centre for Strategy and Performance, University of Cambridge. He spent many years in industry, holding various executive positions before joining academia in 1999. His current research interests are accelerated innovation, lean management, operations strategy, sustainable operations, and supply chain management. He has spoken on these subjects across the globe, including China, Taiwan, Japan, Latin America, Europe, and other locales. He has consulted many Fortune 500 companies and appointed as Our Common Future Fellow by the Volkswagen Foundation in 2009.

Itshak Tkach is a senior scientist and a visiting professor at Goldsmiths, University of London. He holds a PhD in Industrial Engineering focused on AI and swarm intelligence, an MSc in intelligent systems, and a BSc in Mechanical Engineering, all from Ben-Gurion University of the Negev, Israel. He is the co-author of the Distributed Heterogeneous Multi Sensor Task Allocation Systems book and of many Journal articles and conference papers on AI, control, and robotics. He is a board member and a reviewer of several leading Journals.

Prof. Agostino Villa is a former faculty professor of Politecnico di Torino since November 1, 2020, with the official title of Full Professor of Technologies and Production Systems. He is a member, the past president, and a fellow of international scientific institutions among which IFPR, IFAC, IFIP, and a member of the editorial committees of international journals. June 2007, he founded the Association "European Virtual Institute on Innovation in Industrial Supply Chains and Logistic Networks" in cooperation with other European University Professors. In 2010, he was the co-founder of the Non-profit Association Kiron "Studies on the Communication and Organization Mediation". In 2017, he has been the co-founder of the "PMInnova Program", an agreement between Politecnico di Torino and Banca di Asti, to supporting the development and innovation of small-mid enterprises. Coordinator and scientifically responsible of about 15 European and national research projects, he is the author of 10 books and of more than 240 papers on international scientific journals and conferences.

Dr. Daehan Won received a B.S. (2008) and M.S. (2010) in industrial engineering from the Korea Advanced Institute of Science and Technology, Daejeon, S. Korea, and a Ph.D. (2016) in industrial and systems engineering from the University of Washington, Seattle, WA. In 2016, he joined the Department of Systems Science and Industrial Engineering, Binghamton University, SUNY, and is currently an assistant professor. His research interests primarily concern large-scale data (*aka.* big data) analysis via mathematical programming and developing new computationally efficient algorithms. He has published more than sixty peer-review journal articles and conference proceedings, including Journal on Computing, Scientific Reports, IEEE Intelligent Systems, etc.

Dr. Wenzhuo Wu is the Ravi and Eleanor Talwar Rising Star associate professor in the School of Industrial Engineering at Purdue University. He received his Ph.D. from Georgia Institute of Technology in Materials Science and Engineering in 2013. His research interests encompass the design, manufacturing, and integration of nanomaterials for applications in energy, electronics, optoelectronics, quantum devices, and wearable sensors. He was a recipient of the Oak Ridge Associated Universities Ralph E. Powe Junior Faculty Enhancement Award, the IOP Semiconductor Science and Technology Best Early Career Research, the Society of Manufacturing Engineers Barbara M. Fossum Outstanding Young Manufacturing Engineer Award, ARO Young Investigator Award, NSF Early CAREER Award, the TMS Functional Materials Division Young Leaders Professional Development Award, and an elected fellow of the Royal Society of Chemistry (FRSC) and the Royal Society of Arts (FRSA).

Tetsuo Yamada is a professor in the Management Science and Social Informatics Program, Department of Informatics, Graduate School of Informatics and Engineering, the University of Electro-Communications (UEC), Tokyo, Japan. He received his B.S., M.S., and Dr. Eng. in research on Designs for Assembly Line Systems from UEC. He was an assistant and associate professor from 2007 to 2011 in the Department of Environmental and Information Studies, Tokyo City University (Old name: Musashi Institute of Technology). Additionally, he was a visiting senior research scientist at the Department of Mechanical and Industrial Engineering, College of Engineering, Northeastern University, USA, from 2013 to 2014. He was a pre-examiner of a PhD thesis at Aalto University in Finland in 2018. He is a council member of the Japan Industrial Management Association and the Society of Plant Engineers Japan and a member of Institute of Industrial and Systems Engineers, Operations Research Society of Japan, Scheduling Society of Japan, the Institute of Life Cycle Assessment, Japan. His research interests are carbon neutral and closed-loop supply chains, renewable energy management, machine learning application, ERP systems as well as disassembly/assembly systems, remanufacturing, healthcare systems engineering, work-life balance, and e-Learning. He is the recipient of the Outstanding Paper Award at the 23rd International Conference on Production Research (ICPR-23), Manila, Philippines, in 2015 and of the appreciation in his service at Northeast Decision Sciences Institute (NEDSI) 2017 conference, Springfield, MA, USA, in 2017.

Dr. Chao-Lung Yang received a B.S. degree in Mechanical Engineering and an M.S. degree in Automation Control from National Taiwan University of Science and Technology, Taiwan, in 1996 and 1998, respectively. He also received the M.S.I.E. and Ph.D. degree in Industrial Engineering from Purdue University, West Lafayette, in 2004 and 2009, respectively. He is currently with the Department of Industrial Management, National Taiwan University of Science and Technology, Taiwan, as a professor. His research area is data mining, machine learning, big data analytics, metaheuristic algorithm, and human action recognition. He has developed a data streaming analytics framework by applying deep learning models such as CNN and LSTM and metaheuristic algorithms to quickly detect the process shift and classify the product detects. Recently, he works in the computer vision domain to develop machine learning models to recognize human action, particularly in manufacturing operations. He is a member of INFORMS, IEEE, and ASME.

Sang Won Yoon received his Ph.D. from the School of Industrial Engineering, Purdue University, West Lafayette, IN, USA, in 2009. He is currently a professor with the Department of Systems Science and Industrial Engineering, Binghamton University, Binghamton, NY, USA. He directs the Complex Systems Design and Analysis Laboratory and is a faculty member with the Watson Institute for Systems Excellence, Binghamton. He has published over 150 internationally renowned journals and conference proceedings.

Constantin-Bălă Zamfirescu obtained his PhD degree in Automation from the Politehnica University Bucharest in 2007. Since 1998, he is with the Computer Science and Electrical Engineering Department of the "Lucian Blaga" University of Sibiu where he is currently leading as a full professor the INCON research center, part of the regional European Digital Innovation Hub. He also worked as an invited researcher in Austria, Spain, Belgium, and Germany, participating as the principal investigator in many international projects. He is a member of IFAC TC 5.4 "Large scale and complex systems" and IEEE TC on Computational Collective Intelligence. He is trained in foresight methods by UNIDO. Current research interest includes multi-agent systems, cyber-physical social systems, and group decision support systems.

Zhenxuan Zhang received his M.S. (2017) in industrial engineering from the Department of Systems Science and Industrial Engineering, Binghamton University, Binghamton, New York, where he is currently pursuing a Ph.D. degree. His research interests include large-scale data predictive modeling, machine learning, dynamic system optimization in smart manufacturing, data analytics in smart manufacturing, modeling, and simulation.

Vision, Historical Perspectives, and Progress

Brief History of the PRISM Center and the PRISM Global Research Network (PGRN)

Chin-Yin Huang[1(✉)], Sang Won Yoon[2], and Shimon Y. Nof[3]

[1] Industrial Engineering and Enterprise Information, Tunghai University, Taichung, Taiwan
huangcy@thu.edu.tw

[2] Thomas J. Watson College of Engineering and Applied Science, State University of New York at Binghamton, Binghamton, NY 13902, USA
yoons@binghamton.edu

[3] Purdue University, 610 Purdue Mall, West Lafayette, IN 47907, USA
nof@purdue.edu

Abstract. The PRISM Center at Purdue University and the PRISM Global Research Network worldwide engage with both theoretical and applied research projects; and applied, industry- and government-oriented R&D projects. Such rich combination enables the participating students and scholars to be exposed to real scientific, technological, business, and society's challenges. It also enables responsible and responsive learning and feedback, better testing and validation. It also enables faster and competitive impact on application and implementation of research innovations, outcomes, and deliverables. Mostly, it enables the PRISM and PGRN members (fondly called "PRISMers") to learn from and encourage each other, and strengthen and enjoy our PRISM Family worldwide. The purpose of this chapter is to (1) review these activities since 1980, through the projects, their sponsors, and the creative people involved with them; and (2) provide some perspective on the significance, progress, accomplishments, and future visions based on them.

Keywords: Collaborative Automation · Collaborative Control Theory (CCT) · Cyber-Collaborative Protocols · HUB-CI (HUB for Collaborative Intelligence) · Task Administration Protocols (TAPs)

1 Background

The PRISM (Production, Robotics, and Integration Software for Manufacturing & Management) Program was established formally as PRISM Lab at Purdue University in 1991. It began about ten years earlier, with a series of projects on computerized manufacturing, production, and industrial robotics. The relatively new term robotics meant even then not just the robot machines, but the integrated automation of intelligent systems of humans, machines, computers, and robots.

C.-Y. Huang and S. W. Yoon (Eds.): ICPR1 2021, ACES 14, pp. 3–50, 2023.
https://doi.org/10.1007/978-3-031-44373-2_1

The purpose of the PRISM program and Lab was then and still is today to investigate and educate how to apply information and communication technologies to most effectively improve the performance of industrial systems, particularly in areas of computer-supported integration and collaboration. Support for PRISM has been obtained from NSF, other government agencies, and companies. Hundreds of graduate and undergraduate research students, and many international scholars have participated directly in hundreds of projects completed so far. Thousands of students and many companies have benefitted from its lab experiments, demonstrations, projects, presentations and publications; and through its activities, innovations, and deliverables.

From the outset, we recognized that:

- Production, service, supply, logistics systems and networks, and in general, productive work systems need augmentation tools for complementing (not replacing) human workers and organizations, for the design, operations, management and control of those work systems; and that
- There are opportunities for developing such augmentation by robotics and automation based on computing, communication, optimization, and AI (artificial intelligence).

Why? Automation, manufacturing and supply plants and networks, production and service facilities and networks all have five common characteristics:

1. They have a large number of participants (human and human-designed) with numerous interconnections, interactions, and dependencies; & those lead to complexity;
2. They all seek to be lean and efficient, have less redundancy of resources and efforts → Vulnerability
3. They have shared nodes that interconnect several networks → Overlap & interdependencies
4. Capacity and throughput are influenced by the systems' structure, the networks' topology, and the way protocols and procedures are designed, implemented, and utilized.
5. Their Inter-networked resources operate in normal modes and must be prepared for disrupted modes, since disruption effects can spread through those networks, reducing their ability to fulfill their goals and objectives.

Hence, they must be augmented as described above.

- Over the past forty years, PRISM's focus has shifted from local collaboration, such as multi-robot system, team-based facility design, and human-robot interaction, to distributed collaboration where collaborators (local and remote; machines, sensors, robots, computers, individuals, teams, and enterprises) have a higher degree of autonomy with self-objectives and even competing objectives.
- During the first two decades (1980–2000), PRISM research focused primarily on e-Work: The collaborative, computer supported activities and communication-supported operations in distributed organizations of humans and/or robots or autonomous software agent systems, such as agent-based manufacturing, CIM workflow and enterprise modeling, and middleware. The premise was, and still is that without well-designed and effective e-Work, the potential of electronic work, such as digital manufacturing, e-Supply, and e-Commerce, cannot be fully materialized.

Therefore, the foundations of CCT, the collaborative control theory, were developed, including its design principles, tools and methods that were investigated, discovered, implemented, and validated. They were developed as a major contributor to systems' augmentation, through systems' collaboration and integration.

- In the past two decades (2000–2022), with the further advent of computing and automation, AI, robotics, and cyber science and technologies, PRISM activities leveraged the learning and discoveries from the previous decades to integrate cyber-physical systems, collaborative intelligence, and collaborative robotics and automation.
- In addition, the PRISM lab, and later PRISM Center has collaborated with other research labs and centers, at Purdue and elsewhere, nationally and worldwide, with many industries, and with different disciplines, including Agriculture, AAE, ABE, CE, CEE, ChE, CS, ECE, IE, ME, NE, Management and Economics, Polytech, and Statistics.
- Since the formal establishment of the PRISM Lab in 1991, three anniversary celebrations through international Symposia and Reunions were organized:

    - In 2001, the Symposium was organized on Purdue Campus, and resulted with a CD-ROM Proceedings. By the recommendation of Purdue Vice President for Research, the decision to formalize the PRISM Center was accepted by the attendees at the conclusion of the Symposium.
    - In 2011, the Symposium took place in Stuttgart, Germany, during the International Conference on Production Research-21. It was decided by the attendees to formalize the many fruitful international activities of the PRISM Center under the PRISM Global Research Network, PGRN. Another outcome -- another CD-ROM proceedings.
    - In 2021, the Symposium was organized in Taichung, Taiwan, during the International Conference on Production Research-26. This time it resulted in the recommendation to write and publish this book, by the many co-authors who are all affiliated with PRISM and PGRN.

- The goals of those Symposia and Reunions was to enable researchers and industrial sponsors, who have been active with PRISM and PGRN over the years, to reflect on the research, development, and education effectiveness so far, and discuss and exchange thoughts on the future agenda for PRISM activities. Specifically:

    - Revisit the PRISM Program's vision, goals, objectives, and achievements by previous members (alumni), current members and affiliates;
    - Introduce the recent development of the PRISM Program to alumni and affiliates;
    - Share the research experiences and reflect on current research and education directions of the PRISM Program;
    - Brainstorm by symposium participants to prioritize research strategies and directions for the next decade, to serve both the academic and industrial communities (Fig. 1).

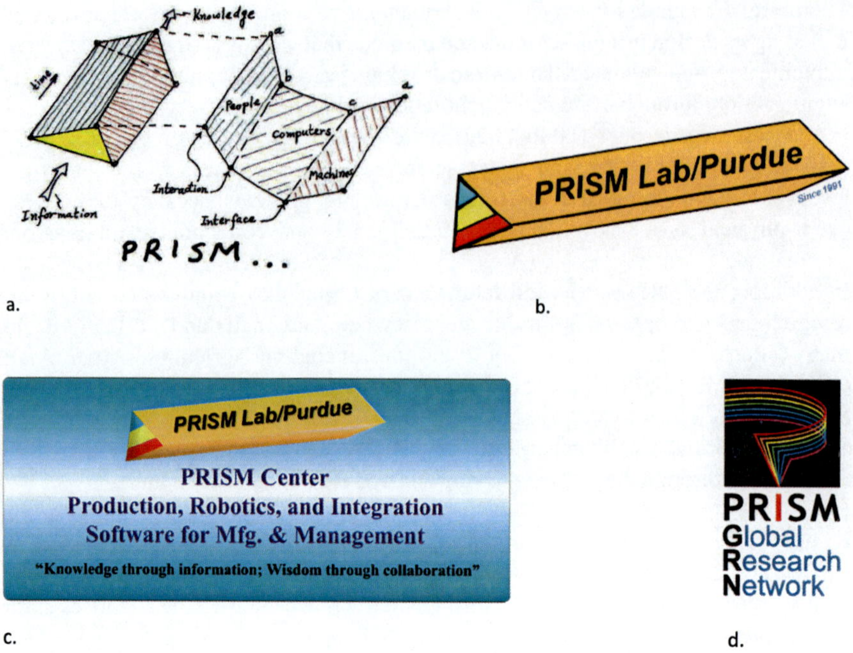

Fig. 1. The PRISM and PGRN Logos: a. The original logo; b. PRISM Lab logo; c. PRISM Center logo; d. The PGRN logo.

In summary: The vision and mission of the PRISM Center and PGRN has been to foster innovation and creativity in the Center's scope by all participants affiliated with our Center, and inspire both current and emerging leaders and pioneers of industry, scholarship, and service to flourish.

2 What We Have Accomplished

The following tables (Table 1, 2 and 3) and figures describe in some detail the specific activities discussed above (Figs. 2, 3, 14, 15, 16, 17, 18, 19, 20, 21, 22, 23, 24, 25, 26, 27, 28, 29, 30 and 31).

Table 1. Pioneers of PRISM Lab -- Projects on Production, Robotics, and Integration Software for Manufacturing and Management (Years: 1980–1990)

Pioneers of PRISM Lab	PRISM Lab Project (Samples)	Company/Sponsor	Sample References	Pioneered Topic/Key Significant Impact
Andrew B. Whinston, Krannert School of Business; Economics	1. AI & Mfg 2. MOS, Mfg. O.S 3. DSS for CIM	IBM; CIDMAC (Computer Integrated Design, Mfg., & Automation Center); NATO	Bullers et al., 1980 [1] Nof et al., 1980 [2] De et al., 1985 [3] Balakrishnan et al. 1994 [4]	1. Integrating AI and mfg. Systems 2. Automating mfg. by Operating Systems 3. Enabling decision support in computer integrated mfg
Richard P. Paul, Electrical Engineering	Analyzing human vs. robot task performance	General Motors; NSF	Paul and Nof, 1979 [5, 6] Paul, 1981 [7] Paul et al., 1983 [8]	1. Foundations of robot work analysis 2. Robot motion modeling
Colin L. Moodie Industrial Engineering	1. Flow control and facility planning 2. Knowledge bases for automated production	NSF; Purdue and State of Indiana Technology Assistance Program; NATO	Moodie et al., 1981 [9] Nof and Moodie, 1983 [10] Ben-Arieh et al., 1985 [11] Weber and Moodie, 1989 [12]	1. Knowledge bases and flow control models for computer integrated production systems 2. Integrated production design and control
Yukio Hasegawa, System Science Institute, Waseda University, Japan	Industrial robots' standardization and economic justification	Hitachi; General Motors; NSF	Hasegawa and Yonemoto, 1984 [13] Hasegawa, 1985a [14], 1985b [15] Bijl et al., 1986 [16]	1. Industrial robotics systems analysis and integration 2. Economic justification of robot systems 3. System science and integration
Wilbur L. Meier, Industrial Engineering	Production Systems and real-time control	NSF; Fischerwerke & Co	Nof et al., 1979 [17]; 1980 [18]	Physical simulators of robotic mfg. For education and training
Gavriel Salvendy, Industrial Engineering	Human factors and industrial robots	NSF; General Motors	Nof et al., 1980 [19] Salvendy and Nof, 1984 [20] Salvendy, 1985 [21]	1. Human factors in robotics 2. Human-computer interaction

(continued)

Table 1. (*continued*)

Pioneers of PRISM Lab	PRISM Lab Project (Samples)	Company/Sponsor	Sample References	Pioneered Topic/Key Significant Impact
Arnold Sweet Industrial Engineering	Reliability; Time series analysis and stochastic modeling	NSF; industry	Sweet, 1987 [22], 2001 [23], 2002 [24], 2007 [25]; Sweet and Tu, 2007 [26]	1. Collection workflow modeling 2. Best matching
Masaaki Yamamoto, Industrial Engineering, Hosei University, Japan	MOS (mfg. O.S) and scheduling theory	NSF; Factrol	Yamamoto, 1981 [27] Yamamoto and Nof, 1982 [28]; 1985 [29]	Adaptive scheduling and self-learning scheduling
Guy Doumeingts, University of Bordeaux I, France	1. CIM architectures 2. GRAI models of advanced manufacturing	European Union grants	Doumeingts,1985 [30] Doumeingts et al., 1987 [31] Doumeingts, 1989 [32]	Architectures for integrating production and manufacturing in enterprises
Hans J. Bullinger, Fraunhofer Institute, Germany	1. Work automation 2. Robot-integrated production	Fraunhofer grants	Bullinger and Lentes, 1982 [33], Bullinger et al., 1986 [34], Bullinger and Ziegler, 1999 [35] Nof et al., 1989 [36], Bullinger and Wagner, 1994 [37]	1. Human-automation interaction 2. Robot integrated production lines
Hank F. Grant, Factrol, Inc	Adaptive scheduling	Factrol; NSF	Grant. 1989 [38], Grant et al., 1989 [39], Grant and Nof, 1989 [40] Nof and Grant, 1991 [41]	Adaptive/predictive scheduling theory and real-time algorithms
Shimon Y. Nof, Industrial Engineering	1. Robot ergonomics and multi-robot work 2. Integrated production knowledge-based modeling and control	NSF; CIDMAC (Computer Integrated Design, Mfg., & Automation Center); NATO; US Forest Service	Nof, 1980 [42, 43]; 1982 [44] Lechtman and Nof, 1983 [45], Nof, 1985 [46] Fisher and Nof, 1987 [47] Nof et al., 1989 [36]	1. Simulation and prediction models of robotic work and robot ergonomics 2. Application of computers, AI, and cyber technologies for mfg. And production systems management 3. CCT, Collaborative Control Theory

(*continued*)

Table 1. (*continued*)

Pioneers of PRISM Lab	PRISM Lab Project (Samples)	Company/Sponsor	Sample References	Pioneered Topic/Key Significant Impact
Other key participants during this period/Topic (1980–1990): **Michael P. Deisenroth**, (Purdue IE/Realtime manufacturing control); **Zvi Drezner** (School of Business, University of California/Optimization of robot work and assembly plans); **R.H. Hollier** (UMIST Manchester, UK/Automated Guided Vehicle Systems); **George R. Karlan** (Purdue College of Education/Robotics to aid the disabled); **Jean Claude Latombe** (Stanford University/Robot path planning); **Benoit Montreuil** (Universite Laval, Canada/Knowledge representation and layout design); **Colleen L. Philips** (W. Michigan University/Human-computers interaction); **Jack W. Posey**, Purdue TAP/Industry projects); **Tibor Vamos** (Hungarian Academy of Sciences, Hungary/Computer science); **Richard H. Weston** (Loughborough University of Technology, UK/Frameworks for systems integration); **Daniel E. Whitney** (MIT & Draper Labs/Assembly automation)				
Other key impacts of the PRISM Lab, PRISM Center and PGRN projects, participants and affiliates during this period are with six books: ***Handbook of Industrial Robotics*** (1985); 2nd edition (1999), John Wiley and Sons; Translated to Russian (1989;1992) ***Robotics and Material Flow**, Elsevier Science Publishers, 1986*; ***International Encyclopedia of Robotics and Automation*** (1988), John Wiley and Sons; **Advanced Information Technologies for Industrial and Material Flow Systems**, NATO Series F: Computer and Systems Science, Vol. 53, Springer, 1989 ***Concise International Encyclopedia of Robotics, Applications and Automation*** (1990), John Wiley and Sons;				

Fig. 2. M.P. Deisenroth and S.Y. Nof observing a pioneering production and robotics physical simulator under minicomputer control at the MGL IE Labs, 1981.

Fig. 3. J.F. Engelberger, "the Father of Industrial Robots" and S.Y. Nof discuss the Handbook of Industrial Robotics at a robotics and automation conference, 1984

Table 2. PRISM Lab, PRISM Center, and PGRN projects at IE Robot Labs (samples)

Robots	At Purdue since	Robots' Source*	Sample Project Topics	Researcher, Lab Project Date, Where They Worked Later	References	Pioneered Topics & Key Significant Impacts
T^3, large articulated robot with dual gripper (Fig. 4, 5)	1982	Cincinnati Milacron Corp		**Hanan Lechtman**, 1983 International Harvester Co	Nof and Lechtman, 1982 [48]; Lechtman and Nof, 1983 [45]	1. Robot Time and Motion method (RTM) 2. Robotic facility simulators
Dual Cybotech 6-axis large articulated Robots (Fig. 6, 7)	1983	Cybotech Corp. (at the Civil Engineering high ceiling Lab)	Activity Controller for Multiple Robot Operation	**Oded Z. Maimon**, 1983 MIT and Tel Aviv University, Israel **Shimon Y. Nof**, 1985	Maimon and Nof, 1984 [49], 1985 [50], 1986 [51] Nof, 1985 [46]	1. MOS for multi-robots 2. Collaborative Robots
IBM SCARA 7545 robot (Fig. 8, 9)	1984	IBM Corp., by Arturo A. Rodriguez, former IE574 student; Tecnomatix (ROBCAD)	Robotic assembly workstation (T^3 and IBM 7545); Multi-robot integration Computer integrated and robotic product assembly and testing	**Venkat N. Rajan,** 1990, i2 Technologies **Keyvan Esfarjani**, 1995 Intel	Rajan and Nof, 1990 [52]; 1996a [53], b [54]; 1999 [55]; Nof and Rajan, 1993 [56] Esfarjani and Nof, 1998 [57]	1. Robotic cells optimization 2. CCT protocol: Collaborative Requirements Planning (CRP) 3. CCT Resource Sharing Task Administration Protocol (TOP TAP)

(*continued*)

Table 2. *(continued)*

Robots	At Purdue since	Robots' Source*	Sample Project Topics	Researcher, Lab Project Date, Where They Worked Later	References	Pioneered Topics & Key Significant Impacts
Mitsubishi Movemaster Small articulated Robot (Fig. 10)	1990	Mitsubishi (borrowed from Purdue Polytech MET Lab)	IRD, Interactive Robotic Device to aid severely disable children, and monitor learning and performance progress	**Neal S. Widmer**, 1990 Purdue Polytechnic	Widmer, Karlan, and Nof, 1992 [58]; Widmer and Nof, 1992 [59]	1. Interactive robotic devices 2. Learning and performance progress monitoring devices
IBM RS-1 Gantry Robot with gripper sensors	1992	IBM Corp., PHD Inc	Small part assembly Assembly with vision-based conveyor tracking and two-robot collaboration Assembly tasks, inspection and test	**Robert M. Remski**, 1993 Microelectronics Mfg. Co **Venkat Rajan**, 1993 i^2 Technologies **Naraye P. Williams**, 1995 i^2 Technologies	Remski and Nof, 1993 [60]; Nof and Rajan, 1993 [56] Williams et al., 2002 [61]; 2003 [62]	1. Sensor-based robot grippers for quality inspection 2. CCT protocol for Collaboration Requirements Planning 3. TestLAN protocols for networked sharing of computer-Integrated Testers
Two Adept 4-axis high-speed SCARA Cobra 600 and 800 Robots (Fig. 11)	2005	Adept Technologies, Inc	Assembly operations error prediction and diagnostics, prediction, prevention	**Xin W. Chen**, 2006 University of Southern Illinois	Chen and Nof, 2007 [63]	5 Patents on IPDN, Integrated error and conflict prevention and detection networks, algorithms and protocols
Numerous robots in several Purdue labs and PGRN Labs		Fanuc LR Mate 200i (at the IE ISAT lab) (Fig. 12)	Collaborative telerobotics	**Hao Zhong**, 2013 Climate Corp	Zhong, Wachs, and Nof, 2014 [64]	Collaborative telerobotics under hub for collaborative intelligence
		(At the Volcani Institute Agricultural Robotics Lab, Israel (Fig. 13)	Robotic food security collaborative project by Purdue University, University of Maryland, and Volcani Institute: Research on remote manipulation of human--robot cart – robot manipulator-- sensors for monitoring and early detection of crop-plants' stress, and precision preventive treatment. (The ARS, agricultural robotics system project.)	**Praditya Ajidarma** 2017; Bandung University, Indonesia **Ashwin Nair** 2018; John Deere **Oak Puwadol Dusadeerungsikul** 2019 Chula University, Thailand **Maitreya Sreeram** 2019; Decision Analytics; **Win P.V. Nguyen** 2021; Virginia Tech University • Additional researchers at Volcani and at UMD	Guo et al., 2018 [65] Nair et al., 2019 [66] Wang et al., 2019 [67] Sreeram and Nof, 2021 [68] Ajidarma and Nof, 2021 [69] Dusadeeru-ngsikul and Nof, 2019a [70], 2020 [71], 2021a [72] Nguyen et al., 2020 Bechar, 2021 [73]	1. Cyber-physical agricultural robotic system 2. Cyber collaborative TAPs (task administration protocols) for optimizing and harmonizing integrated humans-robots-sensors systems and operations 3. Effective machine learning of multi-spectral vision ag. Sensors

* All located at the IE/PRISM Robot Labs in Grissom Hall and MGL, unless indicated otherwise.

Fig. 4. The T^3 robot with dual gripper in assembly experiments at Purdue IE MGL/PRISM Lab,1981

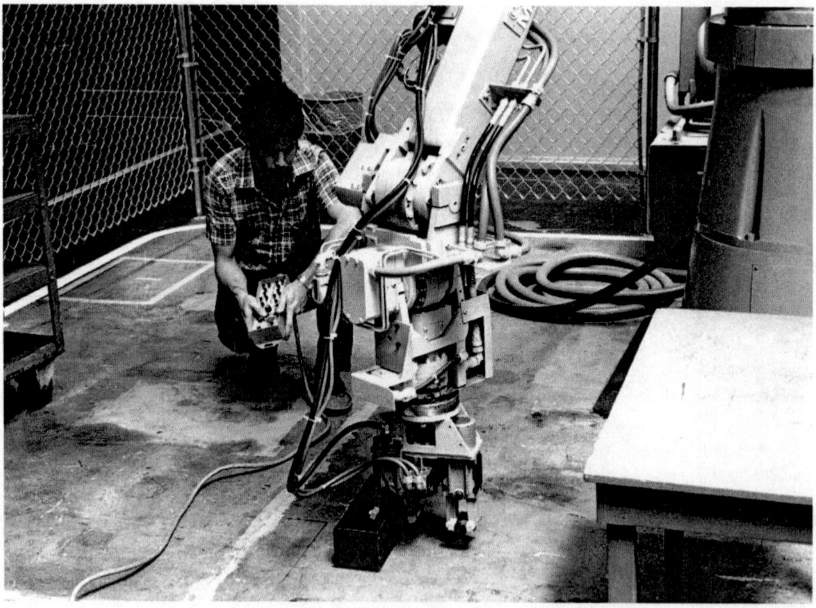

Fig. 5. T^3 RTM experiments, H. Lechtman, 1981

Fig. 6. Dual Cybotech robots set up for collaborative robot experiments 1982

Fig. 7. Multi-robot collaboration experiments, O. Maimon, 1982

Fig. 8. Multi-robot (T3 and IBM 7545) collaboration and machine vision experiments of assembly with conveyor tracking, V. Rajan and W. Eubanks, 1990

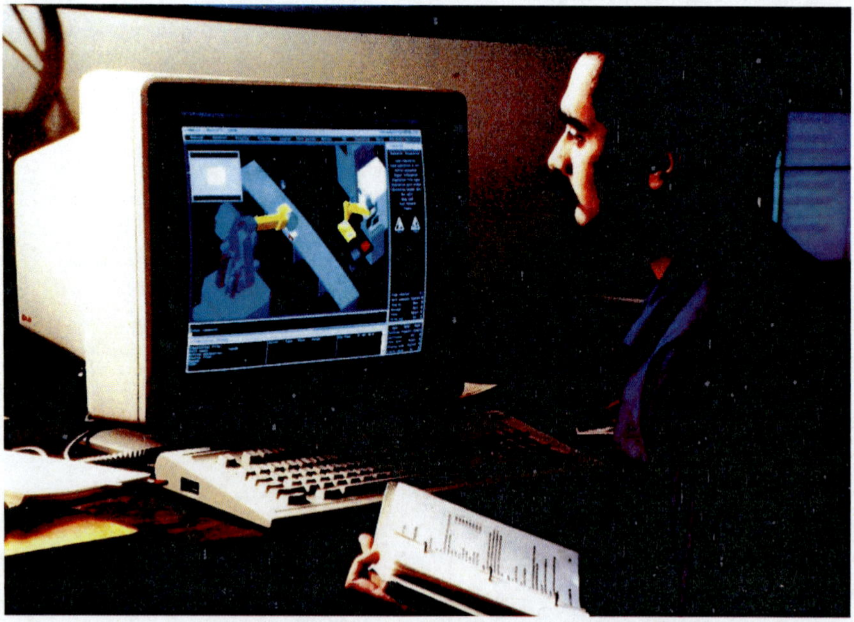

Fig. 9. RobCAD workstation showing T^3 and IBM 7545 Robots with machine vision, integrated for collaborative assembly, Venkat Rajan, 1991

a.

b.

Fig. 10. The IRD experiments, 1990
a. Neal Widmer programming the Mitsubishi Movemaster robot in the IRD2 system for the school experiments. (The IRD1 prototype was developed and tested in the PRISM Lab with the IBM RS-1.)
b. The child learns by activating with the IRD2 toy-playing actions not possible for the child: Note the special back and seating support required to stabilize the child during the experiment. Note the child interacts and activates the robot to execute his/her desired play-response by using a fist, or elbow, or head-mounted horn, depending on the individual needs. Following an instruction by the child, the slinky toy was selected from a fixture; the robot now operates the slinky toy under the interactive commands selected and activated by the child during the learning experiment. Another toy (clown with two horns) is set up in the fixture, in case the child decides to select and activate it next. All interactions by the child are through the large instrumented touch-sensitive board, which is covered with visual symbols for the child to select which toy to apply next, and which actions are requested (audio, spatial, etc.)

Fig. 11. Two collaborating Adept Cobra robots (800 and 600) at the PRISM Lab in MGL, 2006

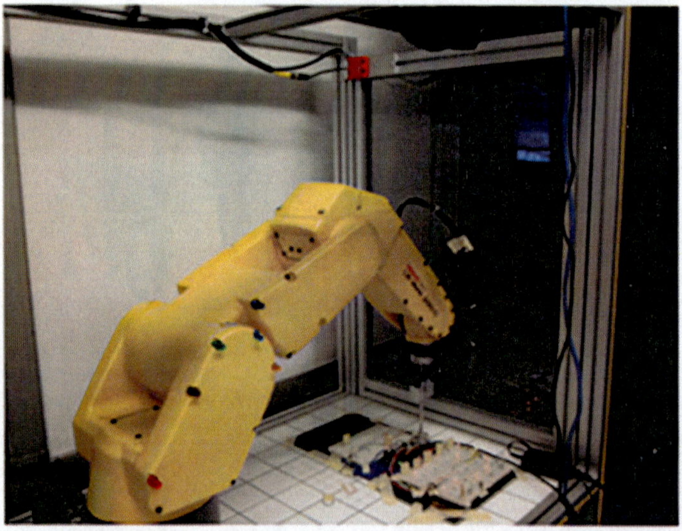

Fig. 12. The collaborative telerobot prototyping cell integrated with a HUB-CI and Fanuc robot at the ISAT Lab (Zhong et al., 2013)

Fig. 13. ARS system at a greenhouse experiment in Volcani Institute, Israel. The mobile robot cart (developed at Volcani's Agricultural Robotics Lab) with a mounted manipulator (UR5 Universal Robot) and instrumented with monitoring sensors is controlled remotely from Purdue PRISM Lab through a HUB-CI with (1) collaborative control protocols, routing and search optimization algorithms; and (2) sensors' data analysis and machine learning algorithms developed by University of Maryland Professor Yang Tao's Bio-Imaging and Machine Vision Lab. On-line local and remote farmers, and a knowledge-base developed by plant biologists are also integrated in the ARS system. (Photo courtesy of Professor Avital Bechar.)

Table 3. Pioneers of PRISM Center and PGRN, PRISM Global Research Network -- Projects on Production, Robotics, and Integration Software for Manufacturing and Management (Years: 1991 --)

Pioneers of PRISM Center & PGRN	PRISM Lab Project (Samples)	Company/Sponsor	Sample References	Pioneered Topic/Key Significant Impact
Aditya P. Mathur Computer Science	Software and Information integration, assurance and security	SERC; CERIAS	Mathur, 1990 [74]; 1991 [75] Wiegner and Nof, 1993 [76]	A model for integrated workflows information assurance
Ray E. Eberts, Industrial Engineering	Computerized Tools for collaborative work integration	NSF; Alcoa Foundation	Eberts, 1994 [77] Eberts and Nof, 1993 [78]; 1995 [79]	Cognitive interfaces for collaborative human and robot workers
Ferdinand F. Leimkuhler, Industrial Engineering	Students' externships in design and manufacturing	NSF; Alcoa; Allison; Rolls-Royce; Delphi Manufacturing Co.; Kirby Risk	Leimkuhler et al., 1996 [80]	Students' industry projects on design and implementation of e-Work and e-Mfg. Systems

(continued)

Table 3. (*continued*)

Pioneers of PRISM Center & PGRN	PRISM Lab Project (Samples)	Company/Sponsor	Sample References	Pioneered Topic/Key Significant Impact
Agostino Villa*, Politecnico di Torino, Italy	Manufacturing systems modeling, management & control	International Foundation for Production Research (IFPR); International Federation of Automatic Control (IFAC)	Villa et al., 1994 [81] Brandimarte and Villa, 1999 [82] Huang and Nof, 1999 [83]	Models for collaborative innovation networks, and their implementation with SME organizations
Yael Edan*, ABC Robotics, Ben-Gurion University, Israel	1.Robot motion economy and sensor economy 2. Fault-tolerant monitoring for supply networks	Industry grants	Edan and Nof, 1995 [84]; 1996 [85]; 2000 [86] Tkach and Edan, 2020 [87]; Tkach et al., 2017 [88]; 2018 [89]	1. Robot motion economy, and sensor economy principles; 2. System collaboration protocols for Supply network security
Masayuki Matsui*, University of Electro-communications, Japan	1. Production enterprise models of communica-tion, coordination. And collaboration 2. Effectiveness analysis of cooperation and collaboration	Industry and Japan Ministry of Education grants	Matsui et al., 1997 [90] Ceroni et al., 1999 [91] Yamada and Matsui, 2003 [92] Fujii et al., 2005 [93] Matsui et al., 2003 [94] Yoon et al., 2011 [95]	1. Advancement in production models with coordination, cooperation, and collaboration considerations; 2. Design of effective systems' collaboration
Moshe Yerushalmy, MBE Simulations, Israel	Serious games for learning production/supply systems' integration and collaboration	Industry grants	Chen et al., 2001 Vastag and Yerushalmy, 2009 [96] Yoon et al., 2011 [95]	Applying serious games as research tools for integration and collaboration
Kinya Tamaki, Aoyama Gaukin University, Japan	1. Assembly systems 2. e-Learning		Warnecke et al., 1992 [97] Tamaki and Nof, 1991 [98] Ishii and Tamaki, 2009 [99]; 2023 [100] Tamaki and Goda, 2009 [101]	Programs of e-Learning of production and manufacturing planning, management and control
Doug Mansfield*, Kirby Risk Manufacturing & Service Center	Assembly and testing automation integration and Task Administration Protocols (TAPs)	Kirby Risk; TAP; NSF	Williams et al., 2002 [61]; 2003 [62]	TestLAN TAPs applications and validation case studies of electrical wiring and control panels
Mark R. Lehto Industrial Engineering	1. Safety and reliability of robotics and automation 2. Decision making automation	Industry; NSF	Clark and Lehto, 1999 [102] Lehto et al., 2009 [103] Proctor et al. 2011 [104] Nanda et al., 2014 [105]	1. Safety of work with automation 2. Integrated knowledge-bases for decision support in automation

(*continued*)

Table 3. *(continued)*

Pioneers of PRISM Center & PGRN	PRISM Lab Project (Samples)	Company/Sponsor	Sample References	Pioneered Topic/Key Significant Impact
Pioneers of PGRN (2001 --)				
Juan Ernesto deBedout*, Kimberly Clark, Latin America	CCT applications integration in a global supply chain of consumer goods	Kimberly Clark Latin America	Nof, 2004 [106], 2007 [107] Nof et al., 2005 [108] Velasquez and Nof, 2009 [109], Yoon and Nof, 2010 [110], 2011 [111, 112] Reyes Levalle et al., 2013 [113] Scavarda et al., 2015 [114]; 2017 [115, 116] Seok and Nof, 2018 [117]	Implementation, validation of theory and business value in a global supply chain
Florin G. Filip* Romanian Academy, Romania	1. Modeling and design of complex systems 2. Collaborative decision support	Various	Nof et al., 2005 [118] Filip and Leiviska, 2009 [119] Nof and Filip, 2010 [120] Seok et al., 2012 [121] Zhong et al., 2014 [122]	Applications of integration and CCT, collaborative control theory in distributed decision support
Carlos E. Pereira Universidade Federal do Rio Grande do Sul, Brazil	Integrating Control, automation, and robotics in industry	Various	Pereira and Neumann, 2009 [123] Nof et al., 2008 [124] Morel et al., 2019 [125]	Perspectives and vision of communication and collaboration in automation
Kazuyoshi Ishii, Kanazawa Institute of Technology, Japan	e-Learning and e-Training systems	Various	Ishii and Tamaki, 2009 [99]; 2023 [100] Takahashi et al., 2020 [126]	Systematic approaches to e-Learning and e-Training
Avital Bechar, Volcani Institute, Israel	Precision collaboration, agriculture, and robotics automation	Volcani Institute; BARD	Bechar et al., 2012 [127] Nof et al., 2013 [128] Bechar et al. 2015 [129] Wang et al., 2019 [67] Nair et al., 2019 [66] Dusadeerungsikul et al., 2020 [71]	1. Foundation of precision collaboration 2. Implementation and validation of CCT, collaborative control theory, in cyber-physical precision ag. Robotic systems
J. Reinaldo Silva, University of San Paolo, Brazil	Service-oriented manufacturing, and Manufacturing as a Service	Industry	Moghaddam et al., 2015 [130] Silva and Nof, 2015 [131] Nof and Silva, 2018 [132]	Best matching of services in cloud-based manufacturing

(continued)

Table 3. (*continued*)

Pioneers of PRISM Center & PGRN	PRISM Lab Project (Samples)	Company/Sponsor	Sample References	Pioneered Topic/Key Significant Impact
Sigal Berman, Ben-Gurion University, Israel	Collaborative control of robotic assembly	Various	Berman and Nof, 2011 [133] Zhong et al., 2015 [134]	Reconfigurable robot grippers; robotics application in agriculture
Other PRISM/PGRN key participants during this period/Topic (1991--): **Yohanan Arzi** (Braude College, Israel/Work methods and automation); **Parasuraman Balasubramanian** (Theme Work Analytics Ltd., India/Automation of e-Service systems); **Pat Banerjee** (University of Illinois-Chicago/Medical and surgical computer simulators); **Miryam Barad** (Tel Aviv University, Israel/Flexibility measures of quality in collaborative supply systems); **Jim W. Barany** (Purdue IE/Work methods engineering); **Ruth Bars** (Budapest University of Technology and Economics, Hungary/Control theory and control models); **Luis Basanez** (Technical University of Catalonia, Spain/Telerobots design and control); **Octavian Bologa** (Lucian Blaga University of Sibiu, Romania/Machine design and materials engineering); **Gary J. Cheng** (Purdue IE/Laser-based manufacturing and control); **Byoung Kyu Choi** (KAIST Korea/Robot cell design); **Manfred Dangelmaier** (Fraunhofer Institute, Germany/Business and engineering systems); **Dan DeLaurentis** (Purdue AAE/System of systems for security); **Alexandre Dolgui** (IMT Atlantique, France/Engineering design and O.R. of supply chains); **Vincent G. Duffy** (Purdue ABE & IE/Human-automation interaction); **Luminita Duta** (Valahia University, Romania/Remanufacturing and disassembly); **Heinz H. Erbe** (TU Berlin, Germany/Collaborative learning and engineering automation); **Opher Etzion** (IBM Research Labs, Haifa, Israel/Middleware for activity control); **Sonia A. Fahmy** (Purdue CS/Design and evaluation of network architectures and protocols); **Jose A.B. Fortes** (Purdue ECE/Parallel computing design models for tasks integration); **Fujii Susumu** (Osaka University, Japan/Machine tool automation); **Mitsuo Gen** (Waseda University, Japan/Evolutionary optimization); **Boaz Golany** (IEM, Technion Israel/R&D and operations research); **Martin Haegele** (Fraunhofer Institute, Germany/Commuter integrated manufacturing); **Brad C. Harriger** (Purdue Polytechnic/Computer integrated manufacturing and material flow); **Christian Hernandez** (Kimberly Clark/Enterprise decision support systems); **Steven W. Holland** (General Motors Research/Industrial robotics and automation); **Sirkka-Liisa Jämsä-Jounela** (Helsinki University of Technology, Finland/Mineral and mining automation); **Troy E. Kostek** (Purdue Polytechnic/Facility control and sensor networks); **Steven Landry** (Purdue IE/Aerospace automation); **Seokcheon Lee** (Purdue IE/Distributed control and collaborative logistics); **Tae-Eog Lee** (KAIST, Korea/Microelectronic manufacturing); **Joachim Lentes** (Fraunhofer Institute/Manufacturing automation); **Eric T. Matson** (Purdue Polytechnic/Automation and information systems); **Wing B. Lee** (Hong Kong Polytechnic University, Hong Kong/Digital manufacturing systems); **Arturo Molina** (ITESM, Mexico/Enterprise networks); **Gerard Morel** (CRAN, Center for automation research Nancy, France/Manufacturing control architectures); **Carlos Moreno** (Kimberly Clark/Production and logistics systems); **Christopher O'Brien** (University of Nottingham, U.K./Decision making, operations management, and performance assessment of supply chains); **Anibal Ollero** (University of Sevilla, Spain/Mobile robots); **Jinwoo Park** (SNU, Korea/Enterprise resource planning and integration); **Namkyu Park** (IntelligenceWare, S. Korea/Shared process workflow); **Karthik Ramani** (Purdue ME and ECE/Augmented reality for future work); **David Romero** (ITESM, Mexico/Intelligent manufacturing); **Manuel B. Scavarda** (Kimberly Clark/Supply chain design and operations); **Kumares C. Sinha** (Purdue CE/Intelligent transportation systems); **Dieter Spath** (Fraunhofer Institute, Germany/Production automation and integration); **John P. Sullivan** (Purdue AAE and Discovery Park CAM/Intelligent Nano- and micro-sensor networks); **Jose M.A. Tanchoco** (Purdue IE/e-Work and AGVs); **Yang Tao** (University of Maryland/Machine vision and machine learning); **Lefteri H. Tsoukalas** (Purdue NE/Neuro-fuzzy controllers for collaborative control protocols); **Geanie Umberger** (Purdue Polytechnic/Computer security and training systems); **H. Van Brussel** (KUL Leuven, Belgium/Modeling flexibility in mfg. Systems); **Francois B. Vernadat** (University of Metz, France/Systems interoperability); **Juan P. Wachs** (Purdue IE/Human-robot interactions and machine vision); **James E. Ward** (Purdue Business/Teamwork collaboration); **Steven T. Wereley** (Purdue ME/Intelligent Nano- and micro-sensor networks); **Wilbert E. Wilhelm** (Texas A&M University/Industrial assembly systems modeling); **David K.Y. Yau** (Purdue CS/Communication networks); **Yeuhwern Yih** (Purdue IE/Collaboratorium initiative); **Constantin B. Zamfirescu** (Lucian Blaga University of Sibiu, Romania/Decision support systems and models); **Roi Zivan** (Ben-Gurion University, Israel/Multi agent systems and processes); **Eyal Zussman** (Technion, Israel/Remanufacturing systems)				

(*continued*)

Table 3. (*continued*)

Pioneers of PRISM Center & PGRN	PRISM Lab Project (Samples)	Company/Sponsor	Sample References	Pioneered Topic/Key Significant Impact
Other key impacts of the PRISM Center and PGRN projects, participants, and affiliates are with thirteen books: ***Information and Collaboration Models of Integration***, NATO ASI Series E: Applied Sciences, Vol. 259, Kluwer Academic Publishers (1994) ***Industrial Assembly,*** Springer (1997; 2013) ***Handbook of Digital Human Modeling***, CRC Press (2008) ***Springer Handbook of Automation*** (2009); 2nd edition (2022) ***Cultural Factors in Systems Design: Decision Making and Action***, CRC press (2012) ***Revolutionizing Collaboration through e-Work, e-Business and e-Service***, Springer Series on Automation, Collaboration, and E-Services (ACES) Vol. 2 (2015) ***Best Matching Theory and Applications***, Springer Series on Automation, Collaboration, and E-Services (ACES), Vol. 3 (2017) ***Computer-Supported Collaborative Decision-Making***, Springer Series on Automation, Collaboration, and E-Services (ACES) Vol. 4 (2017) ***Resilience by Teaming in Supply Chains and Networks,*** Springer Series on Automation, Collaboration, and E-Services (ACES) Vol. 5 (2018) ***Dynamic Lines of Collaboration - Disruption Handling & Control,*** Springer Series on Automation, Collaboration, and E-Services (ACES) Vol. 6 (2020) ***Distributed Heterogeneous Multi Sensor Task Allocation Systems***, Springer Series on Automation, Collaboration, and E-Services (ACES) Vol. 7 (2021) ***Network Science Models for Data Analytics Automation***, Springer Series on Automation, Collaboration, and E-Services (ACES) Vol. 9 (2022)				

* Distinguished PRISM Scholar

Fig. 14. PRISM researchers at the PRISM Lab in Grissom Hall, 1994, L-R: Orlena Nwoka, Howard Kang, Nitin Khanna, Jim Witzerman, Keyvan Esfarjani, Virginia Serna

Fig. 15. PRISM/PGRN researchers, 1996–97, L-R: Marco Lara, Naraye Williams, Jose Ceroni*, Shimon Nof, Masayuki Matsui* (PRISM Visiting Scholar), Chin-Yin Huang*

Fig. 16. PRISM/PGRN pioneers meet in Tokyo, Japan August 1997. L-R: M. Matsui, S.Y. Nof, M. Yamamoto, K. Tamaki

Fig. 17. PRISM meeting, 1999, at the original Gilbreth Library at Grissom Hall. L-R Jose A. Ceroni, Marco Lara, Jianhao Chen, John Gadient (lab technician), Ardi Octorio, Chin-Yin Huang, visitor, Jorge Avila, E. El-Khatib, Shimon Y. Nof, John Auer, Pornthep Anussornnitisarn

3 Intelligent Collaborative Automation with Collaborative Control Theory, Protocols, and Algorithms

Collaborative Control Theory (CCT) is the theory of optimizing the actions and interactions between and among systems integrating distributed humans, machines, robots, and software agents, and systems of such systems.

- Its purpose is to improve and optimize the quality and effectiveness of distributed and collaborative production and service systems and networks.
- CCT comprises design principles, models, algorithms, and protocols to accomplish this optimization.
- CCT integrates computer intelligence and information technology, control and decision theory, management science, human factors, and data science to enable its successful implementation at scale.
- CCT agents, algorithms, and protocols are designed and implemented for two main objectives:

JOURNAL AND COURIER
LOCAL EDITOR PHILLIP A. FIORINI
Phone / 420-5231 Fax / 420-5246
E-mail / fiorinijc@yahoo.com
Journal and Courier online: www.jconline.com

BUSINESS C6

SATURDAY, AUGUST 11, 2001

Purdue lab marks 10th anniversary

Researchers, educators and manufacturers gather for 3-day session

By Max Showalter
Journal and Courier

Automation techniques that will improve manufacturing are already being developed at a Purdue University laboratory that is celebrating its 10th anniversary this week.

In its second decade of operation, the Production, Robotics and Integration Software for Manufacturing/Management lab may include human-robot interaction, microelectromechanical systems and e-work among its research efforts.

"We don't like the cubbyhole. We aspire for ... having people from different groups working together," industrial engineering professor Shimon Nof said Friday during a presentation that is part of the PRISM symposium and reunion. "We can produce something new, something inspiring. That direction is where we're moving to."

More than 50 researchers, educators and manufacturers from around the world attended the three-day session which wraps up this morning with a three-hour meeting.

"With different applications, you can take your work to critical infrastructures," Dennis Engi, the head of Purdue's School of Industrial Engineering, told symposium participants. "We're very optimistic about working with you."

The Indiana 21st Century Research and Technology Fund is a potential source of financing for the development of projects that can improve efficiency in manufacturing and other businesses.

Since it was created in late 1999, the fund has invested $49.2 million in 43 statewide projects that also have leveraged more than $100 million from non-state-financed sources.

"PRISM's activities are exactly what Indiana needs to go splashing into the new economy. Information technology is a key to future growth," said Karl Koehler, the fund's administrative director. "Our partnerships go outside the boundaries of Indiana. There's no reason to limit your thinking.

"If your colleagues live in Arizona, involve them. Several of our awards ... involve bringing technology from The Netherlands and other countries. The idea is that Indiana, through it's program, will decide what it wants to do."

Changing look of Purdue technology

Participants in the PRISM symposium and 10-year reunion may not recognize portions of the Purdue University campus when they return for the laboratory's 25-year reunion.

Richard Schwartz, dean of the School of Engineering, said the university is embarking on an expansion and improvement plan over the next eight to 15 years that includes $250 million for new buildings and $150 million for renovation.

The first project is a $60 million nanotechnology building planned at the southeast corner of State Street and Airport Road, at the west edge of the Purdue campus on what now is home to Purdue Village apartments. Construction is slated to begin next spring and take two years to complete.

By Michael Heinz/Journal and Courier

ADDRESSING A GLOBAL AUDIENCE: Dennis Engi, the head of Purdue's School of Industrial Engineering, speaks Friday during the Production, Robotics and Integration Software for Manufacturing/Management symposium and reunion at Grissom Hall. More than 50 researchers, educators and manufacturers from around the world are attending the three-day session.

Fig. 18. a An Article in the Journal & Courier about PRISM 10th Year Anniversary Symposium and Reunion. b. PRISM 10th Year Symposium and Reunion, August 9–11. 2001, Purdue University, West Lafayette, IN Cover of Proceedings CD, ISBN 0–931682-92–4.

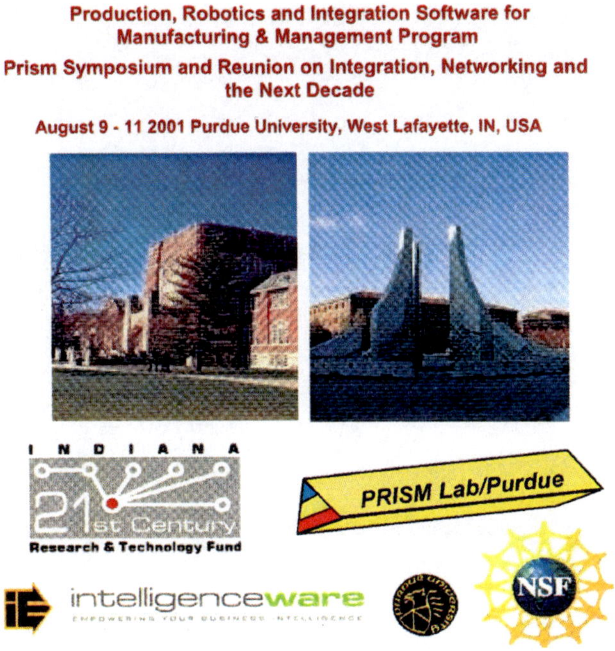

Fig. 18. (*continued*)

Fig. 19. Lab tour at 10th year anniversary PRISM Center Reunion, 2001, West Lafayette, IN, L-R Wayne Eubank (MGL labs manager), Juan D. Velasquez (former PRISM Center researcher); PRISM/PGRN affiliates Tetsuo Yamada (Japan), Stanislaw Raczynski (Mexico), Masayuki Matsui (Japan); former PRISM Center researchers Jose A. Ceroni (Chile), Robert G. Wilhelm (Virginia)

Fig. 20. PRISM Celebration, 2002, L-R: Yan Liu, Juan Velasquez, Diana Milner with daughter, Thomas Bellocci, visitor, Shimon Nof, Jianhao Chen, Pornthep Anussornnitisarn.

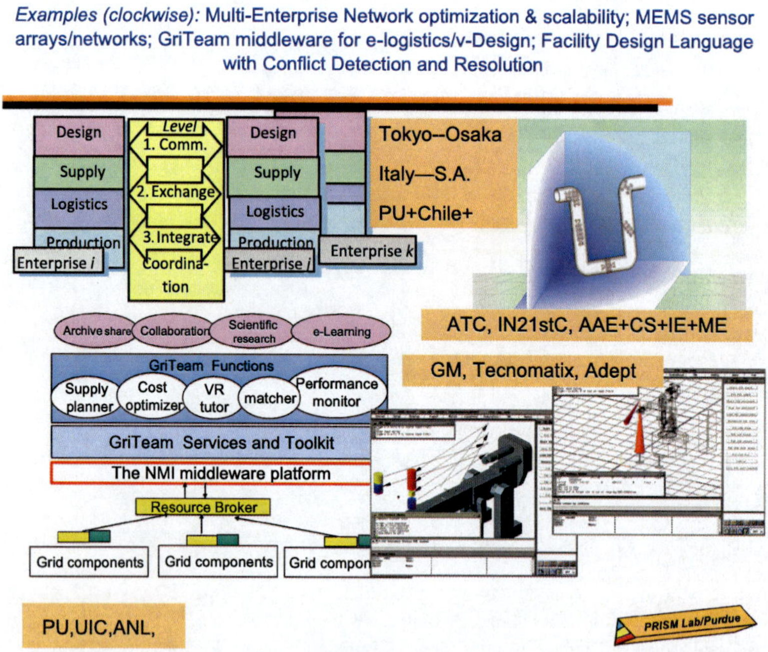

Fig. 21. Four PRISM projects in the first decade of the 21st century, dealing with e-Work, e-Mfg., and e-Logistics. Shown are also the participating labs and sponsors. Source: PRISM Center presentation, 2003. (VR: Virtual Reality; NMI: NSF Middleware Initiative for collaboration with computer grids.; GriTeam: Team collaboration over a fast communication grid.)

Fig. 22. PRISM project with Kimberly Clark Latin America, 2006 meeting in Costa Rica. L-R Mark R. Lehto, Sangwon Yoon, Ching-Yi Wu, Shimon Y. Nof

Fig. 23. PRISM meeting, 2007. L-R Hoosang Ko, Yu Yang (PRISM Visiting Scholar, from China), Wootae Jeong, Juan D. Velasquez, Shimon Y. Nof, Sangwon Yoon with Mrs. Yoon holding Grace, and Xin W. Chen

Fig. 24. PRISM end of semester party, May 2009, L-R front: Sangwon Yoon, Lina Uribe, Xin Chen, Meerant Chokshi; back: Hoosang Ko, Tao Hong, Shimon Nof, Ezhil Kanagaraju, Anurag Puranik, Cigdem Duru.

a. b.

Fig. 25. 26th year anniversary PRISM Center and PGRN Reunion, 2011, Stuttgart, Germany. Organizing Committee Chair: Juan D. Velasquez. a. CD-ROM of proceedings. b. Souvenir magnet.

Fig. 26. PRISM Meeting at the original PRISM Lab in Grissom Hall, Feb. 2014, L-R, front: Lu Zhang, Radhika Bhargava, Jiaxi Li, Glenn Candranegara; back: Hao Zhong, Shimon Nof, Rodrigo Reyes Levalle, Mohsen Moghaddam. Notice, we still used paper notebooks and slide projector.

Fig. 27. End of semester May 2015, PRISM celebration, L-R Zijian He, Vradharajan Mohan, Akhil Rankha, Rishab Vardhan Harikrishnan, Radhika Bhargava, Rohit Kshirsagar, Mohsen Moghaddam, Jiaxi Li, Shimon Nof, Lu Zhang, Hao Zhong, Arfinandi (Nandi) Ferialdy, Christopher Quinn, Vaneet Agarwal.

Fig. 28. Unforgettable, May 2015.

Fig. 29. PRISM Meeting April 2017 at the renovated Grissom Hall, to celebrate Jawahar's graduation. L-R front: Radhika Bhargava, Jawahar Krishna Gogineni, Ping Guo (PRISM Visiting Scholar); back: Shimon Nof, Ashwin Nair, Win PV Nguyen, Oak Puwadol Dusadeerungsikul, Meerant Chokshi (visiting after he graduated in 2009). The Rising Star Award in front was won by Hyesung Seok, mentored by Manuel Scavarda at the Kimberly Clark Summer 2011 Interns competition in Atlanta Georgia. The award was kept by the PRISM Center and given permanently to Dr. Seok when she came to present a PRISM Seminar later in April 2017.

(a) PRISM 30 Special Session Templete (b) PRISM 30 shot-glass

Fig. 30. 30th anniversary PRISM Center and PGRN Reunion, PRISM30, during ICPR-26, August 2021 Taichung, Taiwan Reference: This book. Shown above: The video meeting background for all PRISM30 Reunion participants.

Fig. 31. PRISM Meeting Feb. 2022 at Grissom Hall, L-R Churchill Sandana, Rashed Rabata, David Hongyi Chen, Shimon Nof, Divija Shweta (guest from CLAN Lab), Mahdi Moghaddam, Frederik Weber, Praditya Ajidarma.

a. Augment people and systems of individuals, teams, and organizations, by cyber support for collaborative intelligence;
b. Enable better results with physical tools, products, services, and infrastructure by applying cyber intelligence.

The following Table 4 summarizes the pioneering and development of CCT and its tools by the PRISM and PGRN researchers over the years.

Table 4. Collaborative Control Theory principles, models, algorithms and protocols for augmentation of computer-supported collaborative work (Source: Nof 2007, Nof et al., 2015)

CCT Principle	Augmentation by Cyber Tools	Rational
1. **CRP** Collaboration Requirement Planning	Advanced planning and on-going re-planning enable effective e-collaboration	"Think before you act"
2. **EWP** e-Work Parallelism & **KISS** Parallelize to "Keep it simple, system!"	Optimal parallelism of autonomous agents with simple interfaces and smooth interactions allow efficient delivery of results	"Divide and conquer"
3. **CEDP** Conflict & Error Detection and Prevention	Increase security, safety, deliverable gains by preventing or resolving errors and conflicts	"Learn from mistakes"
4. **CFT** Collaborative Fault-Tolerance by Teaming	Fault-tolerant collaboration of weak agents' team can outperform a single strong agent	"Team for synergy"
5. **ADP** Associate/Dissociate Protocols, a.k.a. **JLR** Join/Leave/Remain in a network	Ensure resilience by monitoring, reorganizing & reconfiguring a collaborative network	"Be selective"
6. **ELOCC** Emergent (Evolutionary) Lines of Collaboration & Command	Overcome disruptions over time by adaptive reorganization, interactions and ad-hoc best matching	"Trust the backup"

(*continued*)

Table 4. (*continued*)

CCT Principle	Augmentation by Cyber Tools	Rational
7. **BMP** Best Matching Protocols	Optimize pairing of best-fit members from two local or distributed sets of people, components, suppliers, robots, and agents	"Avoid mismatch"
8. **CVI** Collaborative Visualization & Insight	Integrate interactive visualization, analytics, and augmented reality to overcome complexity in humans-Humans, humans-machines, and machines-machines collaborations	"What you can see from there we cannot see from here"

Table 5. PRISM projects on Intelligent Collaborative Automation with Collaborative Control Theory, Protocols, and Algorithms over the years (sample)

CCT Design Principle, its Protocols and Algorithms (sample)	PRISM Participants; Where They Worked Later (sample)	Companies/Sponsors (sample, alphabetic)	Sample References
CRP, Collaboration Requirement Planning	**Venkat N. Rajan**, i2 Technologies; **Hao Zhong**, Facebook; **Sigal Berman**, Ben-Gurion University	• Alcoa • BARD • DOD • DOT • Factrol • General Motors • IBM • Indiana 21st Century Fund for Science & Technology • INDOT • Kimberly Clark • Kirby Risk • NSF • Purdue University • TAP • Volcani Institute	Rajan and Nof, 1992 [135], 1996b [54] Nof and Chen, 2003 [136] Zhong et al., 2015

(*continued*)

Table 5. (*continued*)

CCT Design Principle, its Protocols and Algorithms (sample)	PRISM Participants; Where They Worked Later (sample)	Companies/Sponsors (sample, alphabetic)	Sample References
Shared resource(s) Timeout Protocols, e.g., Client-Server model; TestLAN; Multi-robot cells	**Keyvan Esfarjani**, Intel; **Pornthep Anussornnitisarn**, Kasetsart University; **Jose A. Peralta,** Schlumberger; **Naraye P. Williams**, i2 Technologies		Esfarjani and Nof, 1998 Anussornnitisarn et al., 2002 Williams et al., 2002
• **EWP, E-Work Parallelism in tasks and actions, and Keep It Simple, System! TIE, Teamwork Integration Evaluation parallel computing simulators family**	**Nitin Khanna**, Oracle; **Jose A.B. Fortes,** Purdue University; **Jose A. Ceroni,** Pontifica Universidad de Valparaiso; **Chin-Yin Huang**, Tunghai University **Manuel B. Scavarda,** Kimberly Clark		Khanna et al., 1998 Ceroni and Nof, 1999, 2001, 2002, 2005 Huang et al., 2000 Huang and Nof, 2001, 2002 Scavarda et al., 2017
• **Conflict & Error Prevention, Detection, Recovery**	**James P. Witzerman**, General Motors; **Jianhao Chen**, Northern State University; **Marco A. Lara Garcia**, Tompkins and Assoc.; **Xin W. Chen**, University of Southern Illinois-Edwardsville **Hoosang Ko**, University of Southern Illinois-Edwardsville **Juan D. Velasquez**, Purdue University		Lara Garcia et al., 2000 Nof and Chen, 2003 Lara Garcia and Nof, 2003, 2009 Chen and Nof, 2007, 2010, 2011, 2012a, b Ko et al., 2011 Chen, 2022 Velasquez et al., 2008

(*continued*)

Table 5. (*continued*)

CCT Design Principle, its Protocols and Algorithms (sample)	PRISM Participants; Where They Worked Later (sample)	Companies/Sponsors (sample, alphabetic)	Sample References
Collaborative Fault Tolerance by Teaming, e.g., FTTP (fault-tolerant timeout protocol) for sensor arrays and networks; RBT (resilience by teaming) for supply chains and supply networks	**Yan Liu**, University of Dayton; **Wootae Jeong**, Korea Railroad Research Institute; **Rodrigo Reyes Levalle**, American Airlines		Liu et al., 2001, Liu and Nof, 2004, 2008 Jeong and Nof, 2009a, b Nof et al., 2009 Reyes Levalle, 2018 Reyes Levalle and Nof, 2015a, b, 2017
Associate/Dissociate a.k.a. Join/Leave/Remain on a team, or network of enterprises, e.g., Multi-robot teams; sensor networks; supply chains	**Claudia N. Chituc,** University of Porto; **Sangwon Yoon,** Binghamton University; **Manuel B. Scavarda,** Kimberly Clark **Hyesung Seok,** Hongik University		Chituc and Nof, 2007 Yoon and Nof, 2011a Scavarda et al., 2012 Moghaddam and Nof, 2015a Seok and Nof, 2014, 2015 Seok et al., 2016
ELOCC and DLOC, Dynamic Lines of Collaboration & Command	**Hao Zhong**, Facebook; **Win P.V. Nguyen,** Virginia Tech University		Zhong and Nof, 2015, 2020 Nguyen and Nof, 2018, 2019, 2020
Best Matching Protocols	**Howard Kang,** Anderson Consulting; **Juan D. Velasquez,** Purdue University; **Mohsen Moghaddam,** Northeastern University		Kang, 1994 Velasquez and Nof, 2008, 2009 Moghaddam and Nof, 2014, 2015a, 2017

(*continued*)

Table 5. (*continued*)

CCT Design Principle, its Protocols and Algorithms (sample)	PRISM Participants; Where They Worked Later (sample)	Companies/Sponsors (sample, alphabetic)	Sample References
Demand & Capacity Sharing Protocols	**Sangwon Yoon,** Binghamton University; **Mohsen Moghaddam,** Northeastern University; **Hyesung Seok,** Hongik University		Yoon and Nof, 2009, 2010, 2011b Moghaddam and Nof, 2013, 2014 Seok and Nof, 2014
HUB-CI, HUB for Collaborative Intelligence (CI)	**Hyesung Seok,** Hongik University; **Hao Zhong,** Climate; **Oak P. Dusadeerungsikul,** Chula University; **Ashwin S. Nair,** John Deere		Seok and Nof, 2011 Zhong et al., 2013, 2014 Dusadeerungsikul et al., 2019b Nair et al., 2019
CLAP Collaborative Location-Allocation Protocols	**Mohsen Moghaddam,** Northeastern University; **Itshak Tkach,** IAI		Moghaddam et al., 2016, Moghaddam and Nof, 2018 Tkach et al., 2017, 2018 Tkach and Edan, 2020
ARS, Cyber-physical agricultural robotic system with HUB-CI optimization and learning protocols	See in Table 2, last (most recent) entry		See in Table 2, last (most recent) entry
C2W, Cyber-Collaborative Warehouse of the Future, and cyber-collaborative task administration protocols for its competitive design and control	**Oak P. Dusadeerungsikul,** Chula University; **Maitreya Sreeram,** Decision Analytics; **Billy Xiang He,** Bastian Solutions		Dusadeerungsikul, et al., 2019b, 2021b, c

4 Conclusions and Vision

During the past 40 years, PRISM Lab, PRISM Center, and PGRN researchers have worked through all 5 generations shown in Table 6.

Four typical examples, out of many, for each generation can be illustrated as follows to indicate the specialization and contribution.

Gen 1: Material Flow Control (Huang and Nof 1998).

Interactive software tools are norms today. However, during 1990s, the computing power is still weak. Computing and demonstration can only occur one at a time at the computer. The research pioneered the concept to integrate two tools in an integrated modeling environment: Facility Description Language (FDL) for 3-D emulation and Concurrent Flexible Specifications (CFS) for developing the physical layout and material flows. The research paved the way to further development on computer-supported collaborative tool with a learning mechanism.

Gen 2: Distributed CIM data activities (Kim and Nof, 2001).

Table 6. Digital & Cyber Augmentation (including AI) of Work System

Digital & Cyber Augmentation (including AI) of Work Systems	
"Generation"	Significant Contribution
1.0	Computerized
2.0	Computer Integrated
3.0	Internetworked + Mobile
4.0	Cloud-Based + Machine Learning
5.0	Cyber-Physical + Cybernetics
6.0	Quantum Computing, Communication, and Intelligence

This research proposed two automatic data activities models: DAF-Net and AIMIS. DAF-Net is for coordinating interdependent data activities while AIMIS is for integrating distributed and heterogeneous information systems on demand. Both models were pioneering for today's cloud computing with distributed active data sources of Internet of Things.

Gen 3: Sensor network (Ko et al. 2010).

To conquer the problem of low reliability in Wireless Sensor Networks (WANs) in facility monitoring, this research developed an automation facility-specific WSNs, called facility sensor networks (FSNs). First, multiple sensing nodes were analyzed to see if FSNs are vulnerable to interference. Second, interference sources were identified by applying statistical methods to find the appropriate FSN configuration. Finally, the optimal deployment strategy that minimized influence of interference was developed. This was an innovative research concerning the best number of sensors and the configuration of sensor networks. The research provided a great reference for studying 5G network reliability today.

Gen 4–5-6: Early detection with cyber-physical agricultural robotics (Wang et al., 2019).

Combining advanced machine learning, AI-based collaborative control theoretic tools and integration of local and remote humans - robot manipulator - mobile robot - network with vision and spectral sensors in a farming greenhouse combined by a multinational team, aimed and succeeded to investigate and implement early detection of stress in food crops. With the goal of food security by precision agriculture, this work has also contributed to cyber-physical systems in agriculture.

As stated in the beginning of this chapter: The vision and mission of the PRISM Center and PGRN has been to foster innovation and creativity in the Center's scope by all participants affiliated with our Center, and inspire both current and emerging leaders and pioneers of industry, scholarship, and service to flourish. In this chapter we described with some detail how these vision and mission have been accomplished so far, and will hopefully continue in the future. Let us add two personal perspectives.

"The paper 'Decision integration fundamentals in distributed manufacturing topologies' (IIE transactions, 24(3), 27–42) written by Professors Papastavrou and Nof in 1992 was the beginning of my journey in the PRISM Lab. After reading the paper, I was eager to seek the opportunity to be taught by Professor Nof. Thankfully, I could join Professor Nof's research team (PRISM Lab) in 1994 as a master student. The initial research topic was a continuation of the research on Teamwork Integration Evaluation with Parallel Computing. Even now, the research on parallel computing in production management is still important and critical. Later, the research subject was shifted to agent-oriented issues, which also became the theme of my master and doctoral theses. The agent-based production system was in line with the enterprise virtualization through global sourcing. The study described the members of the supply network as agents, and discusses the autonomy and viability for individual agents and the whole network. Thinking about it now, isn't it influencing the study for today's supply chain resilience? Digital twins?

I was greatly enlightened during the study in the PRISM Lab. There was a doctoral seminar class. Prof. Nof used his book "Information and Collaboration Models of Integration" as the textbook for the whole semester. Many advanced research issues were investigated and discussed in the class. I am still deeply motivated by the class today. In another seminar class, Prof. Nof used "An Introduction to Distributed and Parallel Computing" (by Joel M. Crichlow) as the textbook. This course gave the students a sound theoretical knowledge on how a supercomputer can be utilized in the study of production management. At the same time, I also had the opportunity to use Paragon Super Computer (128 computing nodes) to conduct my doctoral research.

In addition, the Industrial Robotics course of PRISM Lab delivered a complete theoretical and practical knowledge on the integration of robotic automation and peripheral facilities. The lecture tutorials in the MGL's Lab of Industrial Robots is still vivid in my memory. At the same time, RobCAD was utilized to investigate the real robotic operations in association with their emulations from robot motions to the whole automated manufacturing system. Thinking about it now, isn't it the pilot study of cyber physical systems?

The cultivation that PRISM Lab gave me definitely did not just happen at Purdue University. In the following 20 + years of my academic career, I was inspired by PRISM Lab to the research on information sharing of supply chain, distributed manufacturing execution system, collaborative design, and even the current research on ontological manufacturing. The vision of Prof. Nof in the PRISM Lab is so high and far. Many issues of distributed collaborative systems were so well investigated for the last four decades and can still be applied today with different kind of technology contents. I can foresee in the future, the growth of the dispersed cyber-physical robot-human collaboration over sensor/knowledge networks on the soil paved by the PRISM Lab." Chin-Yin Huang.

"I joined in the PRISM Center in the fall of 2004 and worked on several research projects under the guidance of Prof. Nof, sponsored by Indiana Department of Transportation and Kimberly-Clark Corp. During my time at Purdue University, I was able to develop my research themes under decentralized/distributed decision making and support systems, which have numerous potentials in current/future emerging research domains, such as advanced manufacturing, healthcare systems, and logistics. After being part of PRISM/PGRN for almost two decades, I now understand that the foundation of the PRISM Center is to develop a scientific exploration via an integration of multi-disciplinary knowledge from traditional Industrial Engineering tools to other advanced technologies. This foundation has helped me establish a research lab for the Complex System Design and Analysis at the State University at New York at Binghamton. I joined the faculty of the Watson College in the Department of Systems Science and Industrial Engineering in 2010 and I became a professor in 2020. Also, I received the SUNY

Chancellor's Award for Excellence in Scholarship and Creative Activities in 2019 and have secured over $9 million from more than 80 industrial projects as of 2022. Recently, my research agenda has been studying how to extract useful insights from expanding data sets to support intelligent decision-making processes, which highlight not only better understanding large-scale data set by statistical learning methodologies, but also leveraging optimization, soft computing, simulation, and complex theories with various algorithms. As a final remark, future generation 6.0 aims to improve industry competitiveness with cutting-edge information and communication techniques, which would require massive amount of computations. Data have become valuable but cheaper and more accessible, but we still need to discover useful information and generate insightful knowledge from them. I believe that this follows the inspiring phase of the PRISM Center developed by Prof. Nof many years ago; i.e., Knowledge through Information; Wisdom through Collaboration. I believe that future research could be a little more complicate than past but the theme of the PRISM Center remains the same. We just need to be more collaborative to go one step further." Sang Won Yoon.

5 Distinguished Prism Scholar Award

As determined by the members of the PRISM Center Advisory Council after the second PRISM symposium and reunion in 2011, a Distinguished PRISM Scholar award will be created. The purpose of this award is to recognize prominent leaders and scholars in the research field of PRISM who have provided PRISM/PGRN researchers with significant advice on research directions and ideas for collaborative projects, over many years. So far, eight such awards have been bestowed.

The recipients are:

Dr. Juan Ernesto de Bedout, for his exceptional contributions to innovations in Supply Network Decisions (In 2011).

Dean Dr. Jose Arturo Ceroni Diaz for his exceptional contributions to innovations in Collaborative Automation and Control (in 2021).

Professor Yael Edan for her exceptional contributions to innovations in Agricultural, Biological, and Cognitive Robotics (in 2021).

Academician Florin G. Filip for his exceptional contributions to innovations in Collaborative Decision Support (In 2021).

Professor & Chairman Chin-Yin Huang for his exceptional contributions to innovations in Collaborative Systems and Robotics (In 2021).

President Douglass A. Mansfield for his exceptional contributions to innovations in Complex Assembly Services (In 2021).

Professor Masayuki Matsui for his exceptional contributions to innovations in Artifacts Science and Models (In 2021).

Professor Agostino Villa for his exceptional contributions to innovations in Collaborative SME Supply Networks (In 2021).

References

1. Bullers, W.I., Nof, S.Y., Whinston, A.B.: Artificial intelligence in manufacturing planning and control. AIIE Trans. **12**(4), 351–363 (1980). https://doi.org/10.1080/05695558008974527
2. Nof, S.Y., Whinston, A.B., Bullers, W.I.: Control and decision support in automatic manufacturing systems. AIIE Trans. (Am. Inst. Ind. Eng.) **12**(2), 156–169 (1980). https://doi.org/10.1080/05695558008974503
3. De, S., Nof, S.Y., Whinston, A.B.: Decision support in computer-integrated manufacturing. Decis. Supp. Syst. **1**(1), 37–55 (1985). https://doi.org/10.1016/0167-9236(85)90196-4
4. Balakrishnan, A., Kalakota, R., Whinston, A.B., Ow, P.S.: Designing collaborative systems to support reactive problem-solving in manufacturing. In: Nof, S.Y. (ed.) Information and Collaboration Models of Integration. NATO ASI Series, Vol 259, pp. 105–133. Kluwer Academic Publishers (1994)
5. Paul, R.P., Nof, S.Y.: Work methods measurement—a comparison between robot and human task performance. Int. J. Prod. Res. **17**(3), 277–303 (1979). https://doi.org/10.1080/00207547908919615
6. Paul, R., Nof, S.: Human and robot task performance. In: Dodd, G.G., Lothar, R. (eds.) Computer Vision and Sensor-Based Robots, pp. 23–50 (1979)
7. Paul, R.P.: Robot manipulators: mathematics, programming, and control: the computer control of robot manipulators. Richard Paul, (1981)
8. Paul, R.P., Luh, J.Y.S., Hayward, V., Nof, S.Y.: Advanced industrial robot control systems. In: Proceedings - Society of Automotive Engineers, pp. 167–176 (1983)
9. Moodie, C.L., Nof, S., ElGomayel, J.: Manufacturing productivity education data control by a general database system. Comput. Ind. Eng. **5**(4), 245–256 (1981). https://doi.org/10.1016/0360-8352(81)90023-1
10. Nof, S.Y., Moodie, C.L.: Flow control effects on facility planning. Mater. Flow, 109–120 (1983)
11. Ben-Arieh, D., Moodie, C.L., Nof, S.Y.: Knowledge based control system for automated production and assembly. In: International Conference on Production Research, Stuttgart, West Germany, pp. 285–293 (1985)
12. Weber, D.M., Moodie, C.L.: Distributed, intelligent information system for automated, integrated manufacturing systems. In: Nof, S.Y., Moodie, C.L. (eds.) Advanced Information Technologies for Industrial Material Flow Systems, pp. 57–79. Springer, Heidelberg (1989). https://doi.org/10.1007/978-3-642-74575-1_4
13. Hasegawa, Y., Yonemoto, K.: The Japanese scene. In: Hartley, J. (ed.) Flexible Automation in Japan, pp. 1–22. Springer, Heidelberg (1984). https://doi.org/10.1007/978-3-662-07249-3_1

14. Hasegawa, Y.: Industrial robot standardization. In: Nof, S.Y. (ed.) Handbook of Industrial Robotics, pp. 518–524. John Wiley and Sons, Hoboken (1985)
15. Hasegawa, Y.: Evaluation and economic evaluation. In: Nof, S.Y. (ed.) Handbook of Industrial Robotics, pp. 665–687. John Wiley and Sons, Hoboken (1985)
16. Hasegawa, Y.: Robotization of reinforced concrete building construction in Japan. In: Bijl, A., et al. (eds.) CAD and Robotics in Architecture and Construction, pp. 113–121. Springer US, Boston, MA (1986). https://doi.org/10.1007/978-1-4684-7404-6_11
17. Nof, S.Y., Deisenroth, M.P., Meier, W.L.: Using physical simulators to study manufacturing systems design and control. In: The 30th Annual Conference of AIIE, San Francisco, CA, pp. 219–228 (1979)
18. Nof, S.Y., Deisenroth, M.P., Meier, W.L.: Physical simulators of production systems: an industrial engineering. Aid IE J. **12**(10), 70–75 (1980)
19. Nof, S.Y., Knight, J.L., Salvendy, G.S.: Effective utilization of industrial robots—a job and skills analysis approach. AIIE Trans. **12**(3), 216–225 (1980). https://doi.org/10.1080/05695558008974509
20. Salvendy, G., Nof, S.Y.: Integration OF human operators in computerized manufacturing: some considerations. In: IIE Conference, Chicago, IL, pp. 179–188 (1984)
21. Salvendy, G.: Human factors in planning robotic systems. In: Nof, S.Y. (ed.) Handbook of Industrial Robotics, pp. 639–664. John Wiley and Sons, Hoboken (1985)
22. Sweet, A.L.: A collection of programs for plotting control charts: volume one. Technical Report No. 87–11. In. School of Industrial Engineering, Purdue University, West Lafayette, IN (1987)
23. Sweet, A.L.: Forecasting the size of a collection in workflow models. In: Paper presented at the the PRISM Symposium & Reunion on Integration, Networking, and the Next Decade, West Lafayette, IN, 9–11 August 2001 (2001)
24. Sweet, A.L.: Forecasting the size of a collection in workflow models. J. Intell. Manuf. **13**, 477–484 (2002)
25. Sweet, A.L.: Partitioning the production processes when assembling a bore and shaft, prism research memorandum No. 2007-PI. In. School of Industrial Engineering, Purdue University (2007)
26. Sweet, A.L., Tu, J.F.: Tolerance design for the fit between bore and shaft for precision assemblies with significant error-scaling problems. Int. J. Prod. Res. **45**(22), 5223–5241 (2007)
27. Yamamoto, M.: A program package for solving general scheduling problems. Bull. Coll. Eng. **17**, 63–73 (1981)
28. Yamamoto, M., Nof, S.Y.: A Scheduling method in automatic manufacturing systems (in Japanese). J. Jpn. Ind. Manag. Assoc. **33**(3), 189–195 (1982)
29. Yamamoto, M., Nof, S.: Scheduling/rescheduling in the manufacturing operating system environment. Int. J. Prod. Res. **23**(4), 705–722 (1985)

30. Doumeingts, G.: How to decentralize decisions through GRAI model in production management. Comput. Ind. **6**(6), 501–514 (1985)
31. Doumeingts, G., Vallespir, B., Darricau, D., Roboam, M.: Design methodology for advanced manufacturing systems. Comput. Ind. **9**(4), 271–296 (1987)
32. Doumeingts, G.: GRAI approach to designing and controlling advanced manufacturing system in CIM environment. In: Nof, S.Y., Moodie, C.L. (eds.) Advanced Information Technologies for Industrial Material Flow Systems, pp. 461–529. Springer, Heidelberg (1989). https://doi.org/10.1007/978-3-642-74575-1_21
33. Bullinger, H.-J., Lentes, H.: The future of work: technological, economic and social changes. Int. J. Prod. Res. **20**(3), 259–296 (1982)
34. Bullinger, H.-J., Warnecke, H.-J., Lentes, H.-P.: Toward the factory of the future. Int. J. Prod. Res. **24**(4), 697–741 (1986)
35. Bullinger, H.J., Ziegler, J. (eds.): Human-Computer Interaction: Communication, Cooperation, and Application Design. Lawrence Erlbaum Association, Mahwah (1999)
36. Nof, S.Y., et al.: Research needs and challenges in application of computer and information sciences for industrial engineering. IIE Trans. (Inst. Ind. Eng.) **21**(1), 50–65 (1989). https://doi.org/10.1080/07408178908966206
37. Bullinger, H.-J., Wagner, F.: A model-based methodology for management of concurrent simultaneous engineering. In: Nof, S.Y. (ed.) Information and Collaboration Models of Integration. NATO ASI Series, vol. 259, pp. 61–67. Kluwer Academic Publishers (1994)
38. Grant, F.H.: Simulation technology for the design and scheduling of material handling and storage systems. In: Nof, S.Y., Moodie, C.L. (eds.) Advanced Information Technologies for Industrial Material Flow Systems, pp. 563–580. Springer, Heidelberg (1989). https://doi.org/10.1007/978-3-642-74575-1_23
39. Grant, F., Nof, S., MacFarland, D.: Adaptive/predictive scheduling in real-time. In: Advances in Manufacturing Systems Integration and Processes: 15th Conference on Production Research and Technology, pp. 277–280. Society of Manufacturing Engineers, Dearborn (1988)
40. Grant, F.H., Nof, S.Y.: Automatic adaptive scheduling of multiprocessor cells. In: the 10th International Conference on Production Research, Nottingham, U.K. (1989)
41. Nof, S.Y., Grant, H.F.: Adaptive/predictive scheduling: review and a general framework. Prod. Plan. Control **2**(4), 298–312 (1991). https://doi.org/10.1080/09537289108919359
42. Nof, S.Y.: Control and decision support in automatic manufacturing systems. AIIE Trans. **12**(2), 140–144 (1980). https://doi.org/10.1080/05695558008974503
43. Nof, S.Y.: A methodology for computer-aided facility planning. Int. J. Prod. Res. **18**(6), 699–722 (1980). https://doi.org/10.1080/00207548008919701
44. Nof, S.Y.: On the structure and logic of typical material flow systems. Int. J. Prod. Res. **20**(5), 575–590 (1982). https://doi.org/10.1080/00207548208947788
45. Lechtman, H., Nof, S.Y.: Performance time models for robot point operations. Int. J. Prod. Res. **21**(5), 659–673 (1983). https://doi.org/10.1080/00207548308942402

46. Nof, S.Y.: Robot ergonomics: optimizing robot work. In: Nof, S.Y. (ed.) Handbook of Industrial Robotics, pp. 549–604. Wiley, Hoboken (1985)
47. Fisher, E.L., Nof, S.Y.: Knowledge-based economic analysis of manufacturing systems. J. Manuf. Syst. **6**(2), 137–150 (1987). https://doi.org/10.1016/0278-6125(87)90037-9
48. Nof, S., Lechtman, H.: Now it's time for rate-fixing for robots. Ind. Robot Int. J. **9**(2), 106–110 (1982)
49. Maimon, O., Nof, S.: Analysis and control of flexible multi-robot operations. In: Fall Industrial Engineering Conference, American Institute of Industrial Engineers, pp. 17–28 (1984)
50. Maimon, O.Z., Nof, S.Y.: Coordination of robots sharing assembly tasks. J. Dyn. Syst. Meas. Control Trans. ASME **107**(4), 299–307 (1985). https://doi.org/10.1115/1.3140740
51. Maimon, O., Nof, S.Y.: Analysis of multi-robot systems. IIE Trans. (Inst. Ind. Eng.) **18**(3), 226–234 (1986). https://doi.org/10.1080/07408178608974699
52. Rajan, V.N., Nof, S.Y.: A game-theoretic approach for co-operation control in multi-machine workstations. Int. J. Comput. Integr. Manuf. **3**(1), 47–59 (1990). https://doi.org/10.1080/09511929008944432
53. Rajan, V.N., Nof, S.Y.: Minimal precedence constraints for integrated assembly and execution planning. IEEE Trans. Robot. Autom. **12**(2), 175–186 (1996). https://doi.org/10.1109/70.488939
54. Rajan, V.N., Nof, S.Y.: Cooperation requirements planning (CRP) for multiprocessors: optimal assignment and execution planning. J. Intell. Rob. Syst. Theory Appl. **15**(4), 419–435 (1996). https://doi.org/10.1007/BF00437605
55. Rajan, V.N., Nof, S.Y.: Computation, AI, and multiagent techniques for planning robotic operations. In: Nof, S.Y. (ed.) Handbook of Industrial Robotics, 2nd edn, pp. 579–602 (1999)
56. Nof, S.Y., Rajan, V.N.: A workstation for robotic manufacturing systems design experimentation. In: International Robot and Vision Conference, Detroit, MI (1993)
57. Esfarjani, K., Nof, S.Y.: Client-server model of integrated production facilities. Int. J. Prod. Res. **36**(12), 3295–3321 (1998). https://doi.org/10.1080/002075498192076
58. Widmer, N.S., Nof, S.Y., Karlan, G.R.: An interactive robotic device with progress monitoring. Robotica **10**(1), 11–18 (1992)
59. Widmer, N.S., Nof, S.Y.: Design of a knowledge-based performance progress monitor. Comput. Ind. Eng. **22**(2), 101–114 (1992). https://doi.org/10.1016/0360-8352(92)90037-K
60. Remski, R.M., Nof, S.Y.: Analytic and empirical assessment models of on-line inspection technologies. Comput. Ind. Eng. **25**(1–4), 439–443 (1993). https://doi.org/10.1016/0360-8352(93)90315-O
61. Williams, N.P., Liu, Y., Nof, S.Y.: TestLAN approach and protocols for the integration of distributed assembly and test networks. Int. J. Prod. Res. **40**(17), 4505–4522 (2002). https://doi.org/10.1080/00207540210155873

62. Williams, N.P., Liu, Y., Nof, S.Y.: Analysis of workflow protocol adaptability in TestLAN production systems. IIE Trans. (Inst. Ind. Eng.) **35**(10), 965–972 (2003). https://doi.org/10.1080/07408170309342348
63. Chen, X.W., Nof, S.Y.: Error detection and prediction algorithms: application in robotics. J. Intell. Rob. Syst. Theory Appl. **48**(2), 225–252 (2007). https://doi.org/10.1007/s10846-006-9094-9
64. Zhong, H., Wachs, J.P., Nof, S.Y.: Telerobot-enabled HUB-CI model for collaborative life-cycle management of design and prototyping. Comput. Ind. **65**(4), 550–562 (2014). https://doi.org/10.1016/j.compind.2013.12.011
65. Guo, P., Dusadeerungsikul, P.O., Nof, S.Y.: Agricultural cyber physical system collaboration for greenhouse stress management. Comput. Electron. Agric. **150**, 439–454 (2018). https://doi.org/10.1016/j.compag.2018.05.022
66. Nair, A.S., Bechar, A., Tao, Y., Nof, S.Y.: The hub-CI model for telerobotics in greenhouse monitoring. Procedia Manuf. **39**, 414–421 (2019)
67. Wang, D., et al.: Early detection of tomato spotted wilt virus by hyperspectral imaging and outlier removal auxiliary classifier generative adversarial nets (OR-AC-GAN). Sci. Rep. **9**(1) (2019). https://doi.org/10.1038/s41598-019-40066-y
68. Sreeram, M., Nof, S.Y.: Human-in-the-loop: role in cyber physical agricultural systems. Int. J. Comput. Commun. Control **16**(2), 1–19 (2021). https://doi.org/10.15837/ijccc.2021.2.4166
69. Ajidarma, P., Nof, S.Y.: Collaborative detection and prevention of errors and conflicts in an agricultural robotic system. Stud. Inf. Control **30**(1), 19–28 (2021). https://doi.org/10.24846/v30i1y202102
70. Dusadeerungsikul, P.O., Nof, S.Y.: A collaborative control protocol for agricultural robot routing with online adaptation. Comput. Ind. Eng. **135**, 456–466 (2019). https://doi.org/10.1016/j.cie.2019.06.037
71. Dusadeerungsikul, P.O., Liakos, V., Morari, F., Nof, S.Y., Bechar, A.: Smart action. In: Agricultural Internet of Things and Decision Support for Precision Smart Farming, pp. 225–277 (2020)
72. Dusadeerungsikul, P.O., Nof, S.Y.: A cyber collaborative protocol for real-time communication and control in human-robot-sensor work. Int. J. Comput. Commun. Control **16**(3), 1–11 (2021). https://doi.org/10.15837/ijccc.2021.3.4233
73. Bechar, Avital (ed.): Innovation in Agricultural Robotics for Precision Agriculture. PPA, Springer, Cham (2021). https://doi.org/10.1007/978-3-030-77036-5
74. Mathur, A.P.: Parallel models in software life cycle. In: Zunde, P., Hocking, D. (eds.) Empirical Foundations of Information and Software Science V, pp. 65–79. Springer US, Boston, MA (1990). https://doi.org/10.1007/978-1-4684-5862-6_7

75. Mathur, A.P.: Performance, effectiveness, and reliability issues in software testing. In: 1991 The Fifteenth Annual International Computer Software & Applications Conference, pp. 604,605–604,605. IEEE Computer Society (1991)
76. Wiegner, R., Nof, S.: The software product feedback flow model for development planning. Inf. Softw. Technol. **35**(8), 427–438 (1993). https://doi.org/10.1016/0950-5849(93)90040-A
77. Eberts, R.: Neural Network Based Agents for Coordination of Interaction. In: Nof, S.Y. (ed.) Information and Collaboration Models of Integration, pp. 321–346 (1994)
78. Eberts, R.E., Nof, S.Y.: Distributed planning of collaborative production. Int. J. Adv. Manuf. Technol. **8**(4), 258–268 (1993). https://doi.org/10.1007/BF01748636
79. Eberts, R., Nof, S.: Tools for collaborative work. In: Industrial Engineering Research - Conference Proceedings, Nashville, TN, pp. 438–441 (1995)
80. Leimkuhler, F.F., Nof, S.Y., Pearson, J.T.: Student extern projects in manufacturing design and integration. In: DMI-NSF Proceedings, Albuquerque, New Mexico, pp. 51–52 (1996)
81. Villa, A., Brandimarte, P., Calderini, M.: Meta-models for integrating production management functions in heterogeneous industrial systems. In: Information and Collaboration Models of Integration. NATO ASI Series, vol 259, pp. 135–145. Kluwer Academic Publishers (1994)
82. Brandimarte, P., Villa, A. (eds.): Modeling Manufacturing Systems. Springer, Heidelberg (1999). https://doi.org/10.1007/978-3-662-03853-6
83. Huang, C.-Y., Nof, S.Y.: Model of material handling and robotics. In: Brandimarte, P., Villa, A. (eds.) Modeling Manufacturing Systems: From Aggregate Planning to Real-Time Control, pp. 139–159. Springer, Heidelberg (1999). https://doi.org/10.1007/978-3-662-03853-6_7
84. Edan, Y., Nof, S.Y.: Motion economy analysis for robotic kitting tasks. Int. J. Prod. Res. **33**(5), 1227–1231 (1995). https://doi.org/10.1080/00207549508930205
85. Edan, Y., Nof, S.Y.: Graphic-based analysis of robot motion economy principles. Rob. Comput.-Integr. Manuf. **12**(2), 185–193 (1996). https://doi.org/10.1016/0736-5845(96)00006-3
86. Edan, Y., Nof, S.Y.: Sensor economy principles and selection procedures. IIE Trans. (Inst. Ind. Eng.) **32**(3), 195–203 (2000). https://doi.org/10.1080/07408170008963892
87. Tkach, I., Edan, Y.: Distributed Heterogeneous Multi Sensor Task Allocation Systems. Springer, Cham (2020). https://doi.org/10.1007/978-3-030-34735-2
88. Tkach, I., Edan, Y., Nof, S.Y.: Multi-sensor task allocation framework for supply networks security using task administration protocols. Int. J. Prod. Res. **55**(18), 5202–5224 (2017). https://doi.org/10.1080/00207543.2017.1286047
89. Tkach, I., Jevtić, A., Nof, S.Y., Edan, Y.: A modified distributed bees algorithm for multi-sensor task allocation†. Sensors (Switzerland) **18**(3) (2018). https://doi.org/10.3390/s18030759
90. Matsui, M., Ceroni, J.A., Nof, S.Y.: A coordination consideration of manufacturing systems: job shop case. In: The 14th International Conference on Production Research, Osaka, Japan (1997)

91. Ceroni, J.A., Matsui, M., Nof, S.Y.: Communication-based coordination modeling in distributed manufacturing systems. Int. J. Prod. Econ. **60**, 281–287 (1999). https://doi.org/10.1016/S0925-5273(98)00196-0
92. Yamada, T., Matsui, M.: A management design approach to assembly line systems. Int. J. Prod. Econ. **84**(2), 193–204 (2003). https://doi.org/10.1016/S0925-5273(02)00413-9
93. Fujii, Y., Chen, X., Yang, C.L., Yamada, T., Matsui, M., Nof, S.Y.: Enterprise decision support with evolutionary organizational knowledge. In: The 18th International Conference on Production Research, Salerno, Italy (2005)
94. Matsui, M., Aita, S., Nof, S.Y., Chen, J., Nishibori, Y.: Analysis of cooperation effects in two-center production models. Int. J. Prod. Econ. **84**(1), 101–112 (2003). https://doi.org/10.1016/S0925-5273(02)00401-2
95. Yoon, S.W., Matsui, M., Yamada, T., Nof, S.Y.: Analysis of effectiveness and benefits of collaboration modes with information- and knowledge-sharing. J. Intell. Manuf. **22**(1), 101–112 (2011). https://doi.org/10.1007/s10845-009-0282-x
96. Vastag, G., Yerushalmy, M.: Automating serious games. In: Nof, S.Y. (ed.) Springer Handbook of Automation, pp. 1299–1311. Springer, Heidelberg (2009). https://doi.org/10.1007/978-3-540-78831-7_73
97. Warnecke, H.-J., Schweizer, M., Tamaki, K., Nof, S.Y.: Assembly. In: Salvendy, G. (ed.) Handbook of Industrial Engineering (2nd Ed.). John Wiley and Sons, (1992)
98. Tamaki, K., Nof, S.Y.: Design method of robot kitting sytem for flexible assemble. Robot. Auton. Syst. **8**(4), 255–273 (1991). https://doi.org/10.1016/0921-8890(91)90048-P
99. Ishii, K., Tamaki, K.: Automation in education/learning systems. In: Nof, S.Y. (ed.) Springer Handbook of Automation, pp. 1503–1527. Springer, Heidelberg (2009). https://doi.org/10.1007/978-3-540-78831-7_85
100. Ishii, K., Tamaki, K.: Automation in education, training, and learning systems. In: Nof, S.Y. (ed.) Springer Handbook of Automation, 2nd edn. Springer, Heidelberg (2023). https://doi.org/10.1007/978-3-030-96729-1_63
101. Tamaki, K., GODA, Y.: Collaborations among e-learning professionals for quality assurance. In: ASEM Conference on Lifelong Learning: e-Learning and Workplace Learning 2009 (2009)
102. Clark, D.R., Lehto, M.R.: Reliability, maintenance, and safety of robots. In: Handbook of Industrial Robotics, pp. 717–753. Wiley (1999)
103. Lehto, M.R., Lesch, M.F., Horrey, W.J.: Safety warnings for automation. In: Nof, S.Y. (ed.) Springer handbook of automation, pp. 671–695. Springer, Heidelberg (2009). https://doi.org/10.1007/978-3-540-78831-7_39
104. Proctor, R.W., et al.: Understanding and improving cross-cultural decision making in design and use of digital media: a research agenda. Int. J. Hum.-Comput. Interact. **27**(2), 151–190 (2011). https://doi.org/10.1080/10447318.2011.537175
105. Nanda, G., Lehto, M.R., Nof, S.Y.: User requirement analysis for an online collaboration tool for senior industrial engineering design course. Hum. Fact. Ergon. Manuf. **24**(5), 557–573 (2014). https://doi.org/10.1002/hfm.20551
106. Nof, S.Y.: Collaboration principles for networked manufacturing and logistics: state of the art and challenges (Plenary). In: Lefranc, G. (ed.) Management and Control of Production and Logistics 2004 (MCPL 2004): A Proceedings Volume from the IFAC/IEEE/ACCA Conference, Santiago, Chile, pp. 17–28 (2004)

107. Nof, S.Y.: Collaborative control theory for e-Work, e-Production, and e-Service. Ann. Rev. Control. **31**(2), 281–292 (2007). https://doi.org/10.1016/j.arcontrol.2007.08.002
108. Nof, S.Y., Velasquez, J.D., Partridge, B.K., Poturalski, J.M.: Conducting a mock drill in Indiana: state and university partnership gauges transportation security training needs. TR News **238**, 26–27 (2005)
109. Velásquez, J.D., Nof, S.Y.: Best-matching protocols for assembly in e-work networks. Int. J. Prod. Econ. **122**(1), 508–516 (2009). https://doi.org/10.1016/j.ijpe.2009.06.018
110. Yoon, S.W., Nof, S.Y.: Demand and capacity sharing decisions and protocols in a collaborative network of enterprises. Decis. Support Syst. **49**(4), 442–450 (2010). https://doi.org/10.1016/j.dss.2010.05.005
111. Yoon, S.W., Nof, S.Y.: Affiliation/dissociation decision models in demand and capacity sharing collaborative network. Int. J. Prod. Econ. **130**(2), 135–143 (2011). https://doi.org/10.1016/j.ijpe.2010.10.002
112. Yoon, S.W., Nof, S.Y.: Cooperative production switchover coordination for the real-time order acceptance decision. Int. J. Prod. Res. **49**(6), 1813–1826 (2011). https://doi.org/10.1080/00207540903567325
113. Reyes Levalle, R., Scavarda, M., Nof, S.Y.: Collaborative production line control: minimisation of throughput variability and WIP. Int. J. Prod. Res. **51**(23–24), 7289–7307 (2013). https://doi.org/10.1080/00207543.2013.778435
114. Scavarda, M., Seok, H., Puranik, A.S., Nof, S.Y.: Adaptive direct/indirect delivery decision protocol by collaborative negotiation among manufacturers, distributors, and retailers. Int. J. Prod. Econ. **167**, 232–245 (2015). https://doi.org/10.1016/j.ijpe.2015.05.006
115. Scavarda, M., Reyes Levalle, R., Lee, S., Nof, S.Y.: Collaborative e-work parallelism in supply decisions networks: the chemical dimension. J. Intell. Manuf. **28**(6), 1337–1355 (2017). https://doi.org/10.1007/s10845-015-1054-4
116. Scavarda, M., Seok, H., Nof, S.Y.: The constrained-collaboration algorithm for intelligent resource distribution in supply networks. Comput. Ind. Eng. **113**, 803–818 (2017). https://doi.org/10.1016/j.cie.2017.05.015
117. Seok, H., Nof, S.Y.: Intelligent information sharing among manufacturers in supply networks: supplier selection case. J. Intell. Manuf. **29**(5), 1097–1113 (2018). https://doi.org/10.1007/s10845-015-1159-9
118. Nof, S.Y., Morel, G., Monostori, L., Molina, A., Filip, F.: From plant and logistics control to multi-enterprise collaboration. In: IFAC Proceedings Volumes (IFAC-PapersOnline), pp. 218–231 (2005)
119. Filip, F.-G., Leiviskä, K.: Large-scale complex systems. In: Nof, S.Y. (ed.) Springer Handbook of Automation, pp. 619–638. Springer, Heidelberg (2009). https://doi.org/10.1007/978-3-540-78831-7_36
120. Nof, S.Y., Filip, F.G.: Sustainability in production and logistics: progress and methodological challenges (Plenary). In: Management and Control of Production and Logistics 2010 (MCPL 2010): A Proceedings Volume from the IFAC/IEEE/ACCA Conference, Coimbra, Portugal (2010)

121. Seok, H., Nof, S.Y., Filip, F.G.: Sustainability decision support system based on collaborative control theory. Ann. Rev. Control. **36**(1), 85–100 (2012). https://doi.org/10.1016/j.arcontrol.2012.03.007
122. Zhong, H., Nof, S.Y., Filip, F.G.: Dynamic lines of collaboration in CPS disruption response. In: IFAC Proceedings Volumes (IFAC-PapersOnline), pp. 7855–7860 (2014)
123. Pereira, C.E., Neumann, P.: Industrial communication protocols. In: Nof, S.Y. (ed.) Springer Handbook of Automation, pp. 981–999. Springer, Heidelberg (2009). https://doi.org/10.1007/978-3-540-78831-7_56
124. Nof, S.Y., Filip, F.G., Molina, A., Monostori, L., Pereira, C.E.: Advances in e-manufacturing, e-logistics, and e-service systems. In: IFAC Proceedings Volumes (IFAC-PapersOnline), Seoul, Korea (2008)
125. Morel, G., Pereira, C.E., Nof, S.Y.: Historical survey and emerging challenges of manufacturing automation modeling and control: a systems architecting perspective. Ann. Rev. Control. **47**, 21–34 (2019). https://doi.org/10.1016/j.arcontrol.2019.01.002
126. Takahashi, K., et al.: Special issue on present and future of production in Asia Pacific countries. Int. J. Prod. Res. **58**(8), 2433–2435 (2020)
127. Bechar, A., Wachs, J., Lumkes, J., Nof, S.: Developing a human-robot collaborative system for precision agricultural tasks. In: Eleventh International Conference on Precision Agriculture, Indianapolis, Indiana (2012)
128. Nof, S.Y., et al.: Laser and photonic systems integration: emerging innovations and framework for research and education. Hum. Fact. Ergon. Manuf. **23**(6), 483–516 (2013). https://doi.org/10.1002/hfm.20555
129. Bechar, A., Nof, S.Y., Wachs, J.P.: A review and framework of laser-based collaboration support. Ann. Rev. Control. **39**, 30–45 (2015). https://doi.org/10.1016/j.arcontrol.2015.03.003
130. Moghaddam, M., Silva, J.R., Nof, S.Y.: Manufacturing-as-a-Service: from e-work and service-oriented architecture to the cloud manufacturing paradigm. In: the 15th IFAC Symposium on Information Control Problems in Manufacturing, Ottawa, Canada, vol. 48, no. 3, pp. 828–833 (2015)
131. Silva, J.R., Nof, S.Y.: Manufacturing service: from e-work and service-oriented approach towards a product-service architecture. In: the 15th IFAC Symposium on Information Control Problems in Manufacturing, Ottawa, Canada, vol. 48, no. 3, pp. 1628-1633 (2015)
132. Nof, S.Y., Silva, J.R.: Perspectives on manufacturing automation under the digital and cyber convergence (invited). Polytechnica **1**(1–2), 36–47 (2018)
133. Berman, S., Nof, S.Y.: Collaborative control theory for robotic systems with reconfigurable end effectors. In: 21st International Conference on Production Research: Innovation in Product and Production, ICPR 2011 - Conference Proceedings, Stuttgart, Germany (2011)
134. Zhong, H., Nof, S.Y., Berman, S.: Asynchronous cooperation requirement planning with reconfigurable end-effectors. Rob. Comput.-Integr. Manuf. **34**, 95–104 (2015). https://doi.org/10.1016/j.rcim.2014.11.004

135. Rajan, V.N., Nof, S.Y.: Logic and communication issues in cooperation planning for multi-machine workstations (invited). Int. J. Syst. Autom. Res. Appl. (SARA) **2**, 193–212 (1992)
136. Nof, S.Y., Chen, J.: Assembly and disassembly: an overview and framework for cooperation requirement planning with conflict resolution. J. Intell. Rob. Syst. Theory Appl. **37**(3), 307–320 (2003). https://doi.org/10.1023/A:1025466401869

Forecasting the Size of a Collaborative Collection in Workflow Models

Arnold L. Sweet(✉)

School of Industrial Engineering, and Affiliate of PRISM Center, Purdue University, West Lafayette, IN 47907, USA
arnold.1@comcast.net

Abstract. This chapter is revised from the original paper, presented at our PRISM 10th anniversary symposium and reunion in 2001 (Sweet, 2001; a different version under the same title was published later, Sweet, 2002). Its perspective in this revised chapter is now influenced by the increasing role and value (since then) of automation and cyber-collaborative technologies in enabling and optimizing collaborative systems' and people's work.

A workflow management system (WFMS) is a software system that defines, manages and executes workflow through the use of software whose sequence of execution is driven by a computer representation of the workflow logic. Through software protocols, WFMS are also being applied today to manage the work of collaborating teams of workers, robots, software agents and systems, that collaborate with an objective to reduce overall time, cost, and errors. The correct sequence of tasks, which is given in a planning and control database, can be automatically discovered for performance tracking and improvement by applying data mining and machine learning technology to the audit-trail data recorded by the WFMS. During the collaborative collection process, an attempt is made to find all of the objects of interest. In this chapter, a model is presented which can be used to forecast: (1) The ultimate size of a complete collection; (2) The expected time necessary to complete the collection process, and (3) The probability of finding more items in a fixed interval of time. The data necessary to compute these three forecasts is the number of items which have already been found, and the time it took to collect them. A case study from the field of postcard collecting is presented. It can also apply as a framework for forecasting models in many other collection processes, common in current production, supply, assembly, storage & distribution, and service functions that rely on search, retrieval, and assembly tasks, and are often accomplished by teams.

Keywords: Assembly · Collaborative Teamwork · Collection Process · Retrieval · Search · Workflow Management System (WFMS)

1 Introduction

A workflow management system (WFMS) is a system used to define, manage, and operate work tasks and activities through the execution-guidance by software. The order by which the executed operation is driven is by a computer representation of the workflow

C.-Y. Huang and S. W. Yoon (Eds.): ICPR1 2021, ACES 14, pp. 51–60, 2023.
https://doi.org/10.1007/978-3-031-44373-2_2

logic (Lawrence, 1997). Workflow consists of a collection of human- and machine-based activities or tasks that must be organized and coordinated to perform the work of groups of collaborating human, machine, and software operators. The workflow of this collaborative team relies on multiple agents and sub-systems to complete the required work tasks (Huang, 2002).

WFMS is automated coordination, control and communication of work that is required to satisfy a workflow process (Georgakopoulos et al., 1995). It has become a common tool for business process modeling and business process re-engineering. Most research in WFPS focuses, however, on the execution of the system and its performance, e.g., transaction management, concurrency control, and scalability (Dogac *et al.*, 2000); supply chain management system integration (Huang et al., 2020; Zu and Liu, 2018). As a result, WFMS does not improve or accelerate a business process re-engineering (BPR) effort.

The first task in BPR is still a documentation of the current processes. Discovering existing business processes is a time-consuming and error-prone task. The effort associated with determining the correct sequence of the tasks in the database can be improved by adding the ability to determine the workflow sequence in WFMS (Leymann and Roller, 2000). The reason is that while the tasks and activities are carried out under the control of WFMS, those sequences can be automatically discovered, after an initial deployment phase, by recording them (by the WFMS) and applying data mining technology to the data of the audit trail recorded by WFMS. These concepts are represented in the diagram shown in Fig. 1, which is a modified figure from (Leymann and Roller, 2000).

During the process of data mining, an attempt may be made to find all of the objects of interest. The number of different objects that exist may not be known. This poses a problem for those who are carrying out the search, and who may eventually find less and less material as time passes, and cannot decide whether it is worth-while to continue a vigorous search.

In this chapter, a model will be presented which can be used to forecast the ultimate size of a collection and the expected time necessary to complete the collection. This model can also be used to estimate the probability of finding more items in a fixed interval of time. These forecasts are based on two data items:

1. The number of items collected so far and already available in the resulting collection; and
2. The time it took to collect those items.

Such a model can help the searchers evaluate whether observed times between finds are acceptable, and whether it is attractive to continue the search effort.

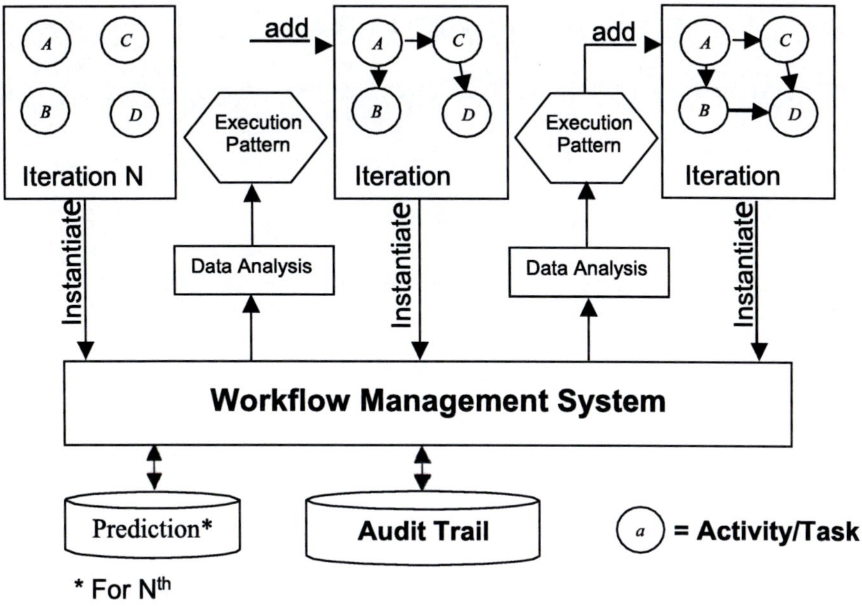

Fig. 1.

2 Development of the Model

In order to develop the model, let L be an integer valued variable that represents the size of a complete collection, and $N(t)$ an integer valued stochastic process which represents the number of items found in an interval of time of length t. It was assumed that the value of L is fixed at some time previous to the start of the search effort. Further, it was assumed that the search effort started at the time that the searcher obtained the first n_0 items. Thus, any time interval between the desire that searchers may have been harboring to begin the search and the time that the first items are actually found is ignored. It was assumed that in any small interval of time, the probability of the size of the collection increasing by one item is proportional to the product of the length of the time interval and the number of items not yet found. This concept can be made more precise by using the theory of a stochastic pure birth process, (e.g. Karlin and Taylor 1975).

When the size of the collection at time t is equal to $n_0 + n$, let the birth rate be given by

$$\lambda_n = \gamma(L - n_0 - n) \quad 0 \le n \le L - n_0, \gamma > 0. \tag{1}$$

Defining the conditional probability of the number of items found in an interval of length t as

$$Pn(t) = P[N(t) = n|N(0) = 0], \quad 0 \le n \le L - n_0 \tag{2}$$

the forward Kolmogorov differential equations can be solved to yield (Sweet, 1998)

$$P[N(t) = n|N(0) = 0] = \frac{L - n_0!}{n!(L - n_0 - n)!}(1 - p)^{L-n_0-n}p^n \quad 0 \leq n \leq L - n_0. \quad (3)$$

where

$$p = 1 - e^{-\gamma t}. \quad (4)$$

The expected number of items in the collection at the end of an interval of length t is given by

$$n_0 + E[N(t)|N(0) = 0] = n_0 + (L - n_0)(1 - e^{-\gamma t}) \quad (5)$$

and the variance of the number of items found is given by

$$o^2 = (L - n_0)e^{-\gamma t}(1 - e^{-\gamma t}). \quad (6)$$

The variance is a maximum at $t = (ln\ 2)/\gamma$ and the value of the maximum variance is equal to $(L - n_0)/4$. Note that as time increases, the variance approaches zero.

Given that the number of items in the collection is equal to m, if T_m is the time to find the next item, then it can be shown that the sequence of times $\{T_m\}$ for increasing values of m are mutually independent and exponentially distributed random variables with parameter λ_m (Karlin and Taylor 1975). Further, (3) may be interpreted as the probability that n successes" occur in $L - n_0$ independent trials, where each trial consists of drawing a sample from an exponential distribution with parameter γ. Thus, a success occurs if the sampled time is less than t, implying that an item has been found. The probability of success is p, as given in (4), and "failure" is the complementary event. Thus, the reciprocal of γ can be interpreted as the expected time for an item to be found.

Using the above interpretation, and the lack of memory property of the exponential distribution, if the size of the collection at any time is equal to m, then the expected number of items to be found in the next time interval of length t is given by (3), with n_0 replaced by m. Further, the expected time to find s more items is given by

$$\sum_{j-0}^{s-1} \frac{1}{\gamma(L - m - j)} = \frac{1}{\gamma} \sum_{k=L+1-m-s}^{L-m} \frac{1}{k} \qquad 1 \leq s \leq L - m;\ n_0 \leq m < L \quad (7)$$

Thus, the expected time to complete the collection, Y_s, can be obtained from (7) by setting $s = L - m$.

The cumulative distribution function for Y_s can be computed through the use of Laplace transforms and a partial fraction expansion, (e.g. Trivedi 1982). The result can be shown to be

$$P[Y_s \leq y] = \sum_{i=0}^{s} \frac{(-1)^i s! e^{i\gamma y}}{i!(s - 1)!} \quad (8)$$

3 Parameter Estimation and Forecasting

It is assumed that observations consist of noting the cumulative number of found items, denoted by n_i, at a sequence of increasing (non-random) times, denoted by t_i, $i = 0, 1, 2,..,r$, with $t_0 = 0$. Using the Markov property of $N(t)$ and (3), the likelihood function is given by

$$l(\gamma, L) = \prod_{i=1}^{r} P[N(t_i) - N(t_{i-1}) = x_i | N(t_{i-1} = n_{i-1}] = \prod_{i=1}^{r} \frac{(L - n_{i-1})!}{x_i!(L - n_{i-1} - x_i)!} e^{-\gamma u_i (L - n_{i-1} - x_i)} \left(1 - e^{-\gamma u_i}\right)^{x_i} \tag{9}$$

where x_i is the number of items collected in the time interval $u_i = t_i - t_{i-1}$, $i = 0, 1, 2,..,r$, and $t = 0$.

If the intervals between observations, u_i, are constant with value u, then taking the log of (8) and differentiating with respect to γ yields

$$\hat{\gamma} = \frac{1}{n} ln \frac{rL - S_r + n_r}{rL - S_r + n_0} \tag{10}$$

where

$$S_r = \sum_{i=0}^{r} n_i \tag{11}$$

Substitution of (10) into the log likelihood function yields

$$\ln l(L) = constant + ln \frac{(rL - S_r + n_0)^{rL - S_r + n_0}}{(rL - S_r + n_r)^{rL - S_r + n_r}} + \sum_{k=L+1-n_r}^{L-n_0} \ln k \tag{12}$$

A numerical search using (12) (*beginning with* $L = n_r$) will yield the maximum likelihood estimator $\hat{L}$.

To forecast the probability of finding n items during a forecast horizon of length h, given that the size of the collection at t is equal to m, use (3) and (4), replacing t by h, n_0 by m, L by $\hat{L}$ and γ by $\hat{\gamma}$.

4 Computing a Confidence Interval for L

Use can be made of the result that for large L, the binomial distribution can be approximated by a normal distribution, (e. g. Feller 1957). Thus, let

$$1 - \alpha = P\left[-z_{.5\alpha} \leq \frac{N(t) - E[N(t)]}{\sigma} \leq z_{.5\alpha}\right] \tag{13}$$

where z_α denotes the upper tail critical value in a standard normal distribution and σ can be computed using (6). Applying (5) and (6) to the inequalities in (13) yields two quadratic equations for $\sqrt{L-n_0}$, namely

$$(L-n_0)\hat{\underline{p}} + z_{.5\alpha}[(L-n_0)\hat{p}(1-\hat{p})]^{\frac{1}{2}} - (n_r - n_0) = 0 \qquad (14)$$

In (14), parameters have been replaced by their estimates. Using the positive root for the solution to each equation yields estimates for the upper and lower confidence limits for the true value of L.

5 Case Study

As a test of the model, use was made of data consisting of records that have been kept by the author of purchases of old picture postcards, starting in September 1991. One of the data sets used to illustrate the model consists of cards from Purdue University. Purchases were recorded on a monthly basis. The data is given in the Appendix. Since cards from this location are still being manufactured, the data shown is limited to cards whose dimensions are 3.5 inches by 5.5 inches. During the 1960's, manufacturers began to produce a larger card, and these are not included in the data set. It is not known how many different-looking cards of these types were produced during the time interval in question. The goal is to collect one of each of all of the different-looking cards that are under consideration.

Figure 2 shows a plot of the cumulative number of cards that were found (and purchased) versus time (in months), and also of the model fitted to the data. Estimates of L and γ were obtained by applying (10) and (12) to the first 106 months of data. Then the fitted curves were computed by sequentially applying the conditional mean given in (5), replacing n_0 by the number of items collected at the time of application, replacing t by the length of the next collection interval (one month), L by $\hat{L}$ and γ by $\hat{\gamma}$. The last 12 months of data were kept for purposes of comparison and updating of parameter estimates, as discussed later.

Table 1 shows estimates of L and γ, 90% confidence limits for L (computed by applying (14)), the expected time to complete the collection (computed by applying (7)), and the probability of not finding any items in the next month (computed by applying (3)). The purpose of this table is to show how the parameter estimates change as more data is accumulated. The first two rows show the estimates when approximately one-half and three-quarters of the total number of cards eventually accumulated (n_r were reached.) The next thirteen rows show the parameter estimates as they were updated, month-by-month.

Approximate confidence intervals for the time to complete the collection can be found by using (8) with replaced by $\hat{\gamma}$. Thus, at month 118, with $s = 1272 - 1235 = 37$, a 90% confidence interval is given by 219.4 months and 85.1 months.

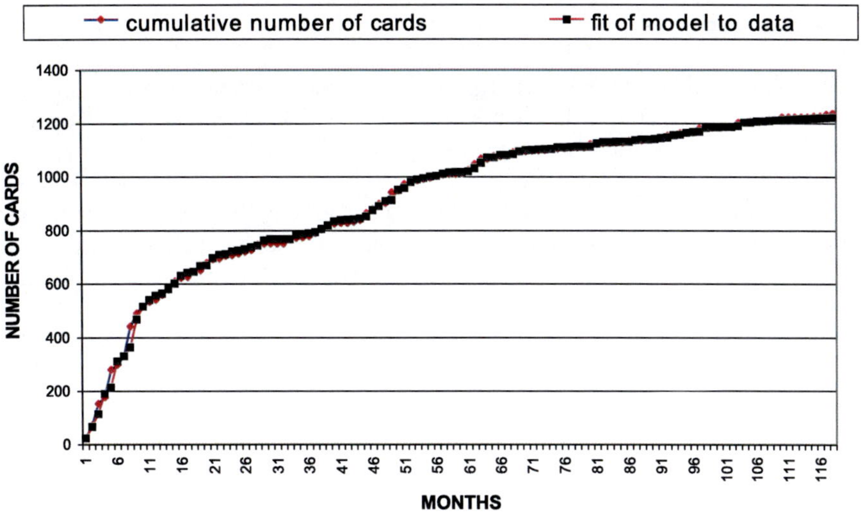

Fig. 2. Plot of data and fit of model. The forecast for the last 12 months is made at month 106.

Table 1. Updating the forecasts, making use of the Purdue data.

Month	Cumulative number	Est. of L	Est. of γ	90% UCL of L	90% LCL of L	Est. Months to completion	Prob. of zero cards in next month
16	620	788	.1005	814.3	765.8	56.7	.000
49	940	1032	.0655	1049.3	1016.1	102.5	.010
106	1205	1246	.0319	1259.2	1237.2	135.0	.271
107	1207	1247	.0318	1260.1	1238.4	134.6	.280
108	1210	1250	.0316	1262.8	1241.2	135.5	.283
109	1211	1252	.0314	1263.3	1241.8	137.1	.276
110	1220	1263	.0306	1275.3	1253.0	142.1	.268
111	1220	1262	.0307	1273.4	1251.6	141.1	.276
112	1220	1259	.0309	1270.7	1249.5	137.7	.300
113	1220	1254	.0313	1267.2	1246.9	131.5	.345
114	1220	1257	.0310	1267.3	1246.9	135.4	.317
115	1224	1262	.0306	1271.9	1251.4	138.0	.312
116	1224	1257	.0311	1268.6	1248.9	131.6	.359
117	1230	1267	.0303	1277.3	1256.9	138.7	.326
118	1235	1272	.0300	1282.9	1262.4	140.3	.330

6 Discussion

It can be seen from Fig. 2 that over the last 12 months the actual number of accumulated cards found was greater than that computed from the 12 monthly forecasts (all made at the end of month 106). However, the table shows considerable stability in the updated parameter estimates made for these 12 months. It might be thought that good estimates of L could be obtained be the use of curve fitting techniques. Examination of the figure shows that the greatest fluctuations occurred in the early months. For this reason, least square type-fitting procedures tended to produce curves that lay below the data over the time intervals studied.

While this case example is based on data from a single searcher / collector, it can represent the same for a process operated by a team of collectors.

7 Conclusions

Other similar data has been analyzed, and the model seemed to be a reasonable representation of the phenomenon of collecting postcards. It is hoped that it may prove useful in the process of mining data bases, although it has not been implemented so far in an actual WFMS. The model has only two parameters, and thus appears to be "robust". It is interesting to speculate on why this might be true, as independent exponential distributions play a central role in this model. In the application presented, it can be argued that the sequence in which early material is collected is not so important as the later search activity. As the search for new items becomes more difficult, it may depend on events such as sales due to the death of collectors, estate sales which uncover previously unknown material, and long searches through dealer stocks. Such events may be characterized as being "completely random" in nature, and hence described by exponential distributions. The parameter reflects the activity of the collector and the competition for the items to be collected. If these remain approximately constant, its value can be used for forecasting future finds. It remains to be seen in future research whether the searches carried out in WFMS also have, to some extent, this "completely random" nature.

From the perspective of multi-agent, collaborative collection process, there are several other future research opportunities:

a. What is the optimal number on collaborative, concurrent, collecting agents? Clearly, the larger the number, the faster the collection process can be completed. On the other hand, with too many collectors, the search may become less efficient and too costly.
b. What are the impacts of newer automation technologies for collection search and retrieval?

For example, Nof et al. (2015, pg. 31) consider three eras of library search for books: Manual only, before computers; supported by computer database catalogs; Internet era with systems' collaboration. A 4th era can be added with the emerging Industrial Internet (IoT/IoS), with the books themselves communicating (through their digital twin Internet of Service) to collectors "here I am, here are my Id and my location."

c. With the advent of cyber augmented work (Nair, 2019; and Dusadeerungsikul, 2020 in cyber-physical agriculture; Lu et al., 2019 in robotic manufacturing;

Dusadeerungsikul, et al., 2021 in cyber-augmented warehouses) other probability distributions may need to be considered by the model presented here, to better represent the search and retrieval tasks of collection processes by teams. Retrieval of physical items, in particular, is being transformed by AMRs, autonomous mobile robots, and by drones, integrated with IoT/IoS.

d. An interesting recent development of WFMS is by RPA, robotic process automation (Siderska, 2020). RPA so far has addressed workflow automation of routine work and procedures. It is anticipated that cyber-augmented RPA (with cyber including AI) would expand this approach to non-routine work and protocols, and enable further improvements in collaborative team collections.

About the process of postcards collection, it was my exciting hobby for many years as an active member of the Indianapolis Postcard Club. The comradery with other collectors and suppliers, and the frequent collectors' shows and meetings provided me a lot of happiness over many years. I have also enjoyed the transformation of the collection process by the Internet, e.g., by the convenient use of e-Bay. Beyond the pleasures of collecting, my card collection has proven productive in several historical publications related to Purdue and Lafayette, Indiana, e.g., Bill and Sweet, 2021, and others.

Acknowledgment. The author acknowledges helpful comments from Shimon Y. Nof on the value of stochastic modeling of collections in the workflow management systems, both in my original research on this topic 25 years ago, and now, with the updated version. I also want to share my happiness celebrating this book with the two co-editors, whom I remember fondly from their time as promising graduate researchers in our Purdue PRISM Center and School of IE.

Appendix

Purdue data: Accumulation by month: 23, 71, 151, 175, 277, 297, 331, 439, 488, 514, 531, 539, 557, 578, 608, 620, 623, 646, 648, 677, 691, 692, 703, 706, 711, 719, 725, 745, 749, 749, 749, 749, 765, 770, 772, 777, 789, 803, 818, 823, 825, 826, 830, 837, 862, 877, 897, 900, 940, 947, 970, 981, 985, 991, 994, 1002, 1007, 1008, 1008, 1012, 1023, 1045, 1065, 1065, 1075, 1075, 1078, 1089, 1093, 1094, 1096, 1097, 1098, 1104, 1104, 1106, 1107, 1107, 1107, 1119, 1124, 1124, 1125, 1126, 1126, 1132, 1135, 1135, 1137, 1140, 1143, 1152, 1154, 1161, 1164, 1165, 1179, 1183, 1183, 1184, 1184, 1187, 1201, 1202, 1205, 1206, 1208, 1211, 1212, 1221, 1221, 1221, 1221, 1221, 1225, 1225, 1231, 1236.

References

Bailey, N.T.J.: The Elements of Stochastic Processes, pp. 74–75. J. Wiley and Sons (1964)

Bill, P., Sweet, A.L.: Tippecanoe County and the 1913 Flood. History Press (2021)

Dogac, A., Kalinichenko, L., Ozsu, M.T., Sheth, A. (eds.): Workflow Management Systems and Interoperability. Springer, NATO Scientific Affairs Division (1998)

Dusadeerungsikul, P.: Operations Analytics and Optimization for Unstructured Systems: Cyber Collaborative Algorithms and Protocols for Agricultural Systems. (Doctoral dissertation, Purdue University Graduate School), (2020)

Dusadeerungsikul, P.O., He, X., Sreeram, M., Nof, S.Y.: Multi-agent system optimisation in factories of the future: cyber collaborative warehouse study. Int. J. Prod. Res. **60**, 1–15 (2021)

Feller, W.: An Introduction to Probability Theory and Its Applications, vol. 1, pp. 168–170. J. Wiley and Sons (1957)

Georgakopoulos, D., Hornick, M.F., Sheth, A.: An overview of workflow management: from process modeling to workflow automation infrastructure. Distrib. Parallel Databases **3**(2), 119–153 (1995)

Huang, B., Teng, Q., Tang, Y.: Collaborative workflow-based enterprise information system. In: International Conference on Application of Intelligent Systems in Multi- modal Information Analytics, pp. 22–27. Springer, Cham (2020)

Huang, C.Y.: Distributed manufacturing execution systems: a workflow perspective. J. Intell. Manuf. **13**(6), 485–497 (2002)

Karlin, S., Taylor, H.M. A First Course in Stochastic Processes, 2nd Edn., pp. 119–121. Academic Press (1975)

Lawrence, P. (Ed.): Workflow Handbook, Workflow Management Coalition, J. Wiley and Sons (1997)

Leymann, F., Roller, D.: Production Workflow: Concepts and Techniques. Prentice Hall, Upper Saddle River, New Jersey (2000)

Lu, H., Wang, H., Yoon, S.W., Won, D.: Real-Time stencil printing optimization using a hybrid multi-layer online sequential extreme learning and evolutionary search approach. IEEE Trans. Compon., Packag. Manuf. Technol. **9**(12), 2490–2498 (2019)

Nair, A.S.: A HUB-CI model for networked telerobotics. In: Collaborative Monitoring of Agricultural Greenhouses (M.S. Thesis, Purdue University Graduate School) (2019)

Siderska, J.: Robotic process automation—a driver of digital transformation? Eng. Manage. Prod. Serv. **12**(2), 21–31 (2020)

Sweet, A.L.: Forecasting the Size of a Collection, Research Memorandum 98–2. Purdue University, West Lafayette, Indiana, School of Industrial Engineering (1998)

Sweet, A.L.: Forecasting the size of a collection in workflow models. In: Proceedings of the PRISM Symposium & Reunion on Integration, Networking, and the Next Decade, August 9 – 11, 2001, p. 13. Purdue University, West Lafayette, Indiana (2001)

Sweet, A.L.: Forecasting the size of a collection in workflow models. J. Intell. Manuf. **13**(6), 477–484 (2002)

Zu, Q., Liu, Y.: Research on multi-agent distributed supply chain information collaboration based on cloud environment. In: International Conference on Human Centered Computing, pp. 156–168. Springer (2018)

The (not so) Little Robot that Could Foster Collaboration

José Ceroni(✉)

School of Industrial Engineering, Pontifical Catholic University of Valparaíso, Valparaíso, Chile
jose.ceroni@pucv.cl

Abstract. This chapter presents a series of facts that belong to my experience as a fellow PRISM member. From 1994 to 1998 I completed the dissertation for the degrees of master and PhD in industrial engineering at Purdue University. Those 5 years me and my family lived in West Lafayette, a lovely town in Indiana, USA. At the beginning of that period of five years, my daughter was only 1.5 years old and spent most of her time in the company of many children living in the student housing complex we lived in. That was the idyllic life for a professor from a Chilean university that wanted to get the most of the experience of the first world university in a safe environment. I was able to dedicate not only to taking courses but to get into the goal set to me by the head of my school of industrial engineering in Chile: to learn about robotics, and so I did. Little did I know how relevant robots will become in my life. A little robot (by standards of the manufacturing laboratory at Purdue University) followed me home once I had returned to Chile. An extremely caring and thoughtful donation by Professor Shimon Nof of the School of Industrial Engineering generated the most relevant impacts on my academic life as I will try to convey in the following pages.

1 Introduction: The Seed that Started All

During my doctoral studies at Purdue University in Indiana, United States, and once the funds of the scholarship the Chilean government had granted me were depleted, I had the chance for funding the remaining of the PhD program (about two years were still left for me to complete my dissertation) by performing the task of teaching assistant for graduate and undergraduate engineering students at Purdue. Among several courses I had the chance to serve as teaching assistant of the course IE-574 Industrial Robotics and Flexible Assembly. This course was key to the development of the story of the IBM 7547 SCARA robot. By the year 1997, the School of Industrial Engineering at Purdue University had available two robots, the IBM 7547 and a Cincinnati Milacron hydraulic articulated robot. Both robots had been donated to Purdue University by their respective companies/manufacturers. Originally the former was used mostly in electronics assembly by IBM and it was in service at the Lexington plant of IBM in Kentucky for the assembly of electronic typewriters, previous to its arrival at the Michael Golden Laboratory of Manufacturing at Purdue. The Cincinnati Milacron was new when donated to Purdue, and in service in the automotive industry as a painting or welding robot, common tasks for machines with that configuration. Few days before the semester started, I put on

C.-Y. Huang and S. W. Yoon (Eds.): ICPR1 2021, ACES 14, pp. 61–79, 2023.
https://doi.org/10.1007/978-3-031-44373-2_3

my oldest jeans and shirt to proceed to clean both robots and their working spaces. This duty I was undertaking surprised the workshop supervisor a great deal, the dearest and fondly remembered Wayne Eubanks, who could not believe that a PhD student would do such a thing. However, I have always considered normal for a teaching assistant or professor to service the students in the best conditions available at the moment and lots of dust on the machines to operate was not conceivable to me, of course I was a lot older than most of the PhD students at that time and had already five years of teaching experience at my university in Chile. However, this little detail that everybody could have not taken into consideration left a positive impression on Mr. Eubanks. I am certain of Mr. Eubanks' positive contribution to the great effort of Professor Shimon Nof for Purdue University to donate to our university the IBM robot in 1999, at a symbolic cost of US$99.00, as stated by Professor Nof communication of August 30th, 1999 (Appendix 1). In January 1999 I returned to my teaching position at the School of Industrial Engineering at Pontifical Catholic University of Valparaíso, Chile. The IBM robot came as a big surprise only 9 months since my return. Once the robot transportation arrangements from West Lafayette to Valparaíso were sorted out, the robot finally arrived in two big and extremely heavy boxes (Figs. 1, 2, 3 and 4). In Fig. 5 my friend and colleague John Clinton and myself appear evaluating the task of taking the robot to a small room the School was able to allocate for installing the robot and related equipment (mainly an air compressor for gripper actuation). The robot finally was available for teaching and research in the year 2000, both in undergraduate and graduate programs at our School.

Fig. 1. Arrival of the robot at PUCV

Fig. 2. The robot as arrived

Fig. 3. The robot specification

Fig. 4. Robot assembled and bolted to the laboratory floor

2 The Attractiveness of Hardware

During the year 1999, Professor Gastón Lefranc, a colleague at the Electrical Engineering School, proposed to me to join his Robotics, Artificial Intelligence and Advanced Automation Laboratory (RAIAAL). After careful consideration of the proposal with the Industrial Engineering School head at that time, Professor Dante Pesce, we came to the realization that the collaboration potential of participating in Professor Lefranc's laboratory will result in greater impact on research activities and results, which it did for a long time until Professor Lefranc's retirement in 2010 (Appendix 2 contains the communication letters to both school heads and the plan for the laboratory to operate as a multidisciplinary unit). So, there it was, the IBM robot, after traveling almost 5,200 miles from West Lafayette, USA to Valparaíso, Chile, it was now part of a university engineering multi school collaboration initiative that had many of us really excited about the future.

Fig. 5. Planning the installation of the robot

Later on, RAIAAL incorporated further academic participation from other Schools from the PUCV college of engineering. Professors Orlando Durán from the School of Mechanical Engineering and Professor Claudio Cubillos from the School of Informatics Engineering were now active participants of the laboratory research activities. Although the RAIAAL laboratory was part of the School of Electrical Engineering, it was clear that the decision regarding research and its management were now part of a collaborative initiative not only for professors and students from the Engineering College but also with national and international relationships of professors participating in the laboratory activities. Therefore, the scenario of collaborative research was coming into reality and started to show in the normal sequence of activities involved in this type of initiatives. Research grants proposals started to put together professors, students, national and international collaborators and resources available at the laboratory. Under the support from the schools of engineering participating in the laboratory, the group was able to slowly but steadily acquire new equipment to increase the range of research activities being covered. Important equipment added to the laboratory was a made in house automated storage and retrieval system that allowed the handling of parts to assist the operation of the SCARA robot as an assembly cell of variations of end or intermediate assembled products (Véliz and Lefranc, 2006; Leighton et al., 2011). Efforts for developing tools in the laboratory were also made with students and their capstone projects at the undergraduate and graduate levels. One of the interesting cases was the SCARA emulator developed by the industrial engineering undergraduate student Mr. Raúl Del Canto (Del Canto and Ceroni, 2005). The development by Mr. Del Canto allowed remote RAIAAL users to interact with the robot in a safe way by means of an emulated robot workspace and an internet- transmitted view of the lab to directly observe the motion of the robot.

This allowed to increase the area of influence of the lab to high school students or people from industry. Further areas of research were also explored for generating studies with industrial impact. The work in olive fruit recognition by Gatica et al., 2013 was an interesting development of vision systems in order to assist the decision of harvesting time of this product in particular. This had the potential to generate process improvements in the important industry of the olive oil in our country. Professor Claudio Cubillos was also an important contributor of the research performed at the RAIAAL lab. Claudio contributed from with his informatics point of view to computational applications (Urra et al., 2016, Schleyer et al., 2016) as well as to the robotics field research (Rojas et al., 2014, Dávila-Ríos et al., 2016). In the meantime, my colleague from the School of Industrial Engineering, professor Franco Guidi, also participated in a research project, increasing the participation of our school in the RAIAAL lab (Flen et al., 2011).

The previous review is a partial research activity being carried out at the RAIAAL lab, since it positioned itself as a meeting point for researchers from universities from all countries. This prolific activity was sustained until the departure of Gastón Lefranc from our university. Even though Gastón was able to continue doing top level research, he was not able to maintain his status of professor at the School of Electrical Engineering due to their faculty renewal policy applied fiercely to people at retiring age. Till today, Gastón continues publishing his research with his PUCV affiliation but not related to the School.

3 The Parallel Track in Collaboration Research

However, despite the frantic and ever-increasing level of activity and productivity at the RAIAAL lab, each of the participants also kept his own initiatives for performing research and contacts with industry, for capturing the eluding research funds. This two-level work also allowed us to manage the risk of having all our activity concentrated in one effort, regardless the promising future and seek to perform research in fields other than those convened at the RAIAAL lab. Therefore, I added to my activity at the lab two main focuses, on one hand, the search of funding opportunities for collaborative research projects involving industry and government agents and, on the second hand, the participation in the promotion of the production research activities through the participation and further on the organization of the International Conference of Production Research (ICPR). The ICPR conferences are managed and organized by the International Federation of Production Research (IFPR), both of them were made familiar to me by Professor Nof during my PhD studies at Purdue, I will be forever grateful to him for that. I must recognize that those were extenuating years, since the academic workload was kept in order to comply with the work share that I should satisfy as an academic of the School of Industrial Engineering. It should be noticed that these two main tasks also depicted a shift in my research interest in order to construct a wider scope of my scholar research activity and avoid focusing excessively in the robotics field. The main reason for this shift was motivated by a powerful lack of interest in the Chilean industry in robotics applications. However, in this writing the decision I made may look and sound as a very thoughtful one, nevertheless, uncertainties, doubts, the country's economic scenario, and other most relevant aspects including family, university, colleagues and friends made it look like a lifetime changing opportunity. Until nowadays (merely seven years until my retirement from academic work) I am not completely clear about the process that led me to the current state I find myself in today. A powerful feeling that some sort of gracious alignment of chances and opportunities has occurred in the past and I hope it will keep on happening.

The first task of developing collaboration based tools for industry and government agencies was the natural continuation of my PhD thesis, consequently I was prepared for tackling that objective. The distributed environment presents challenges for design, management, and operational functions in organizations. Integrated approaches for design and management of modern companies have become mandatory practices in the modern enterprise. Historically, management relied on a well-established hierarchy; however, the need for collaboration overshadows the hierarchy and imposes networks of interaction among tasks, departments, companies, etc. As a result of this interaction, three issues arise that make the integration problem critical: variability, culture, and conflicts. Variability will represent all possible results and procedures for performing the tasks in the distributed organizations. Variability is inherently present in the processes; however, distribution enhances its effects. Cultural aspects such as language, traditions, and working habits impose additional requirements for the integration process of distributed organizations. Lastly, conflicts may represent an important obstacle in the integration process. Conflicts here can be considered as the tendency to only organize locally for local optimizations in a dual local/global environment.

Collaborative relationships, such as user/supplier, are likely to present conflicts when considered within a distributed environment. Communication of essential data and decisions plays a crucial role in allowing organizations to operate in a cooperative manner. Communication must take place on a timely basis, in order to be an effective integration facilitator and allow organizations to minimize their coordination efforts and costs (Ceroni et al., 1999).

The organizational distributed environment has the following characteristics (Hirsch et al., 1995):

- Cooperation of different (independent) enterprises;
- Shifting of project responsibilities during the product life cycle;
- Different conditions, heterogeneity, autonomy, and independence of the participants' hardware and software environments.

With these characteristics, the following series of requirements for the integration of distributed organizations can be established for establishing the guidelines for the integration of the distributed organizations:

- Support of geographically distributed systems and applications in a multi-site production environment, and in special cases, the support of site-oriented temporal manufacturing;
- Consideration of heterogeneity of systems ontology, software, and hardware platforms and networks;
- Integration of autonomous systems within different enterprises (or enterprise domains) with unique responsibilities at different sites;
- Provision of mechanisms for business process management to coordinate the information flow within the entire integrated environment.

Among further efforts to construct a framework for collaborative manufacturing, there is Nof´s taxonomy of integration (Fig. 6). The integration taxonomy classifies collaboration in four types: mandatory, optional, concurrent, and resource sharing. Each of these collaboration types is found along an integration level between machines and humans, and interaction level (interface, Group Decision Support System, or Computer Supported Collaborative Work).

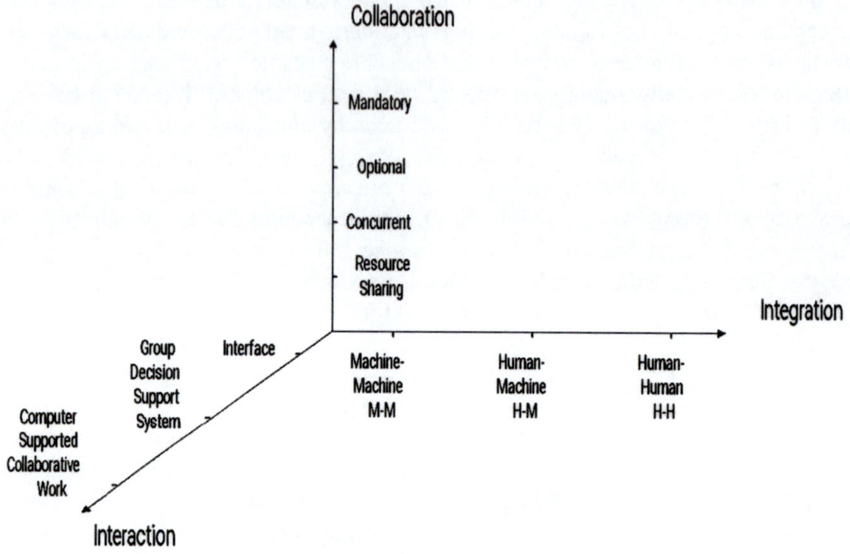

Fig. 6. Integration Taxonomy (Nof, 1994)

4 Facilitating and Implementing e-Collaboration

During the design stages of services and product, co-design (Eberts and Nof, 1995) refers to integrated systems implemented using both hardware and software components. Computer supported collaborative processes allow the integration and collaboration of specialists in an environment where work and co-designs are essential. The collaboration is accomplished by integrating CAD and database applications. Co-design protocols were established for concurrency control, error recovery, transaction management, and information exchange. The collaborative processes and tools support the following design steps:

- Conceptual discussion of the design project
- High-level conceptual design
- Test and evaluation of models
- Documentation

When deciding its operation, every enterprise is accounted for the following transaction costs (Busalacchi, 1999):

1. Searching for a supplier or customer
2. Finding out about the nature of the product or service
3. Negotiating the terms for the product or service
4. Making a decision on suppliers and vendors
5. Monitoring the design and supply to ensure quality, quantity, timeliness, etc.
6. Enforcing compliance with the agreement

Nowadays, the transaction costs have been highly affected by e-Work technologies and collaborative processes. These technologies are shifting transaction costs to: 1) coordination between potential suppliers or customers, 2) rapid access to information about products and services in progress, 3) means for rapid negotiation of terms between suppliers and consumers, 4) access to evaluation criteria for suppliers and customers, 5) mechanisms for ensuring the quality and quantity of products and services, and 6) mechanisms for enforcing compliance with contracts and agreements.

We define collaboration as: "A reciprocal and voluntary agreement between two or more distinct public sector agencies, or between public and private or nonprofit entities, to deliver government services." In general, these relationships involve a formal agreement about roles and responsibilities. The participating organizations share a common objective aimed at the delivery of a public service. They also share tangible and intangible risks, benefits, and resources.

Each collaboration rests on an understood (but often tacit) working philosophy. Collaboration has many meanings and different projects operate on different working assumptions. The underlying norms of each project shape how key roles and functions (such as leadership) are assigned and conducted. For many, the underlying normative structure reflects the historical evolution of the project. Some grew out of a grassroots community of interest while others started with a high-level mandate. As a consequence, the cases exhibit a wide range of work styles and working situations ranging from highly structured to quite informal. For some, equality is important, in others consensus among unequal partners drives decisions. Hierarchy remains a strong philosophy among others.

Collaborative relationships are evolving and dynamic. Each collaboration offers continuous opportunities for feedback and learning. They often employ trial-and-error experimentation, the outcomes of which strongly influence the growth of trust among the participants. In addition, existing and potential participants form and amend their perceptions of the initiative based on their experiences and observations. Roles and responsibilities shift in different stages of the life cycle of a project. In many instances, observations of early performance strongly affect later actions, perceptions, and results.

Data-intensive collaborations face issues of data ownership. In all of these collaborations, data is treated as a valuable asset. As a consequence, the collaborators are beginning to face issues about the data ownership rights of the private partners, the stewardship responsibilities of multiple public partners, and the basic question of whether anyone can actually "own" government information. Multi-organizational collaborations need an institutional framework. Because these initiatives stretch across the boundaries of distinct organizations, they need to establish a new kind of institutional legitimacy. Most often, legitimacy begins with a basis in law or regulation. This is commonly reinforced by the sponsorship of a recognized authority or by formal relationships with key external stakeholders. This formal institutional framework helps these dynamic initiatives weather political transitions and changes in key players. The formal structure also acts as the context for a rich array of complex, informal relationships. These informal relationships are the usual means for getting work done. They spur experimentation and creativity, and, for mature projects, are usually robust enough to resolve most problems. Technology choices affect participation and results. Technology tools and infrastructure are important to project performance and IT is generally well-managed by the collaborators. Technology choices also have consistently important effects on the participants and the results. The nature, cost, and cost distribution of the technologies strongly affect participation due to factors such as availability, affordability, and adaptability to different operating environments. Service performance and communication within the collaboration are strongly shaped by the capabilities of the chosen technical tools. Moreover, the ability of the collaboration to evolve to meet changing needs is significantly shaped by the flexibility of the tools (Dawes and Prefontaine, 2003).

Once developed the framework for collaboration, the research work continued with its application to specific industrial applications. The research works developed by Huang et al. (2000), Ceroni and Nof (2000), Ceroni and Nof (2001), Ascencio and Ceroni (2001), Ceroni and Nof (2002), Ceroni and Nof (2002), Velásquez and Ceroni (2003a, Velásquez and Ceroni, 2003.b), Ceroni and Nof (2005), Miranda et al. (2009), and Nof et al. (2004) contributed to deploy the framework into real life of production companies. A retrospective of decade of research reached when CONICYT, the Chilean government research agency granted to one of the research groups of the School of Industrial Engineering at PUCV the funds to undertake the project "Research Consortium on Foreign Trade Logistics Networks" a pioneering initiative for our country in order to collaborate among universities, research centers and companies all configuring a consortium for identify and solving problems of specific economic sectors of Chilean economy. The growing commercial trade between Asia and South America constitutes a great opportunity for Chile's central region transportation and logistics industry. On one hand, this industry operates supporting the various trade exporting sectors of the country: agricultural, forestry, fish farming, fishing, and mining. On the other hand, it concentrates most of the trade imports to the country. Additionally, it is the natural route for Argentina's foreign trade with Asia. This kind of operation has not yet reached its full possible development, and it will considerably increase the transportation and logistics activity.

This project seeks, as an expected impact, to build up a continuous improvement entity that will elaborate long term development strategies for the central region transportation and logistics industry. This should allow the system to handle higher volume traffic by attracting trade businesses between Argentina and Asia; this new type of business needs better and attractive conditions in price and quality of the logistics service than those currently available. Another of the expected impacts is to increase Chile's foreign trade by reducing the logistics service cost.

Coordination and collaboration among companies has become one of the most important logistics cost decreasing drivers. Coordination and collaboration allow the use of adequate and efficient capacities throughout the organizations, waiting time reductions, to share responsibilities on security and quality, among other key factors. Coordination and collaboration practices require enterprise directives' cooperation, but cooperation by itself is not enough. It is necessary to develop or to acquire technologies, to define standards, to develop management models, or to implement joint planning and communication methods.

Toward the purpose established in the project, a group of companies in the logistics and support service, together with the Pontifical Catholic University of Valparaiso, propose the implementation of a research consortium which will be composed of a group of researchers with fluent relationships with partner companies representatives.

The Consortium partner companies represent the different actors in the logistics business. Port activities are represented by the participation of Empresa Portuaria de Valparaíso, Empresa Portuaria de San Antonio, and Terminal Pacífico Sur. Transportation activities are represented by the railroad company FEPASA. Logistics technology providers are represented by SOLEM and TUXPAN, two of the most important solutions providers in the Fifth Region, which along with Telefónica-Empresas, all, independently and jointly, have an important experience in port logistics problems and solutions. The origin side of the logistics chain is represented by ASOEX, the association that gathers fresh fruit exporters. Also the Consortium has the promised sponsorship of ORACLE, Servicio Nacional de Aduanas, Regional Government, and research cooperation agreements have been signed with Purdue University (USA), Hong Kong University of Science and Technology (China) and the Torino Polytechnic Institute (Italy). The research cooperation agreements will involve participation of faculty at the identified institutions which have already collaborated with faculty of the PUCV Industrial Engineering School. The Consortium might add new partners and research fields if necessary.

The Research Consortium covered the following research topics:

1) Data and process Integration for Foreign Trade Logistics Networks (FORTRALN) (CILN: Computer Integrated Logistics Network).
2) Agent based optimization for Logistics Networks Operations Planning and Scheduling.
3) Radio frequency and PC technologies based identification and security for export trade logistics.
4) e-Work: the challenge for the next generation logistics ERP systems.
5) Operations optimization and integration in fruit export logistics chain.
6) Network Enterprise Economy Theory: applications to logistics networks.

All the previous research projects were aimed to generate technology and conceptual background to reinforce coordination and collaboration in logistics networks. By facilitating coordination and collaboration among the logistics partners, cost related to transportation and port operations, both of high participation in the seaways total logistics cost, will be decreased. It was expected to decrease in 4% the out of port costs (logistics before arrival to the port) and, reducing in the same proportion the port cost, to reduce in 2% the average of the seaways total logistics cost. A more competitive logistics system will foster the trade exporting initiatives and it will pull in trade between Argentina and Pacific Asia, setting the basis for a gateway between South America and the Pacific region of Asia.

The project was strongly related to the Master and PhD programs of the School of Industrial Engineering at PUCV, and by means of these, will establish contacts and cooperation agreements with other graduate programs. Among its academic objectives, the Consortium placed graduate and undergraduate students in research projects and partner companies for them to get a first hand experience on the business problems and situations. The Consortium also keep a continuous instruction program oriented toward the specific needs of professionals in the logistics companies and institutions.

At the core of the Consortium, a group of high level researchers operated. This research group had strong connections to research networks in the USA, Europe, and Asia. An important part of that research group was already participating in a FONDEF, companies and government funded projects, on port logistics operations optimization, a foundation research project for the ones proposed for the Consortium. Activity of the Consortium remained over time by means of service sales, generation of spin-off companies, and participation on international research projects. As the Head of the regional government stated, the Consortium had the potential to generate a great impact upon the regional economy and it contributed to the development of several other regions focused into trade exports and that ship their products through the Valparaiso or San Antonio ports. Consolidation of the Central Region Logistics System makes a clear contribution toward the national purpose of converting Chile into a service platform country. The Consortium had a duration of 2 years. This project constituted the first government funded initiative that strives to establish the collaboration of a foreign trade system as a key performance indicator for the betterment of the country's economic development.

Later on, in 2012 I decided to convey my collaboration efforts into the university management by presenting my candidacy to the university's College of Engineering Dean office. Once elected, again the possibility of contribution came from the government. The government funded project Engineering 2030 provided the most necessary funds to transform the teaching and processes related to the education of future generations of engineers at PUCV. The task was considerably challenging in every aspect from school heads behavior to faculty indifference toward the project. Fortunately, and perhaps because the government anticipated these difficulties, most of the universities participating in the first call of the overall initiative were forced to apply to the funds in an allied form of a consortium. Our university created a consortium with the schools of engineering from the universities of Santiago and Concepción. The three universities were similar in size, typical problems, programs and student bodies. The long and hard process of transformation of the engineering colleges at the three universities began in 2012, being Dean of Engineering at PUCV my friend, and main responsible for me to undertake his contributions, professor Edmundo López from the School of Electrical Engineering. There I was again using the concepts and means that collaboration offers to give a good try to the whole engineering college at PUCV. I must recognize the effort required both tolerance and patience for dealing with the obstacles for accomplishing the tasks set by the government in order to transform the engineering education in the country. After two periods of three years, I left the Dean position with a sensation of accomplishment for most of the duties carried out.

5 The Promotion of Production Research

Regarding the promotion of production research through the ICPR conferences, as the publishing reference profusely shows, the ICPR conferences, even before I had finished my PhD program, were a crucial part of my participation in first level research networks. The first time I contributed research work to the ICPR conference was in the version organized during August in Osaka, Japan in 1997 (later in December of 1998 I successfully defended my PhD thesis and returned to Chile at the beginning of 1999, after 5 years at Purdue University for a master and PhD degrees). My participation in ICPR conferences have extended from 1997 until 2011 due to my excursion into management life. Through all these opportunities, every two years it was a joyfully event to meet dear friends and colleagues from all over the world in beautiful places such as Limerick in Ireland, Prague in Czech Republic, Blacksburg, Virginia in USA, Salerno in Italy, Valparaíso in Chile, Shanghai in China, Stuttgart in Germany, Iguassu Falls in Brazil. During the conferences I had the chance not only to meet researchers from top universities but also had the opportunity to meet with Chilean friends from other universities in Chile as well. A funny story was in the plane to Ireland for the 14th ICPR was to find out that my very good friend Luis Quezada from University of Santiago was a former master and PhD student of Professor Christopher O'Brien from Nottingham University in UK, who happened to be a close friend of my PhD supervisor at Purdue University, professor Shimon Nof. Luis was traveling with some colleagues from the Department of Industrial Engineering at University of Santiago. Later on, the core of the group undertook the challenge of organizing the regional ICPR Americas in 2004 in Santiago, Chile and then the ICPR at international level in Valparaíso, Chile in 2007. In both instances, we had the opportunity to strengthening our cooperation and collaboration with friends such as professors Cecilia Montt, Pedro Palominos, Jorge Bravo and many others that will have my eternal gratitude for the most valuable contribution to the IFPR legacy. As organizer of a ICPR conference, IFPR grantees to the chair of the conference the incorporation to the board of the Federation. In such role Luis and I were part of the board of the IFPR. Luis later on became the President of the Federation. I am certain this shared experience helped in the process of constituting the consortium among University of Santiago, University of Concepción and Pontifical Catholic University of Valparaíso for the project Engineering 2030 in 2012. The attended ICPR conferences maintain the scope of my research throughout all these years. The details are show in the following references in Table 1.

Table 1. Participations in ICPR conferences

ICPR # and Conference venue	Reference
14th Osaka, Japan	Matsui et al., 1997
15th Limerick, Ireland	Ceroni and Nof, 2000a, 2000b
16th Prague, Czech Republic	Ceroni, 2001
17th Blacksburg, Virginia, USA	Velásquez and Ceroni, 2003a, 2003b
Americas Region, Santiago, Chile	Graboloza and Ceroni, 2004
18th Salerno, Italy	Ceroni et al., 2009
19th Valparaíso, Chile	Ceroni and Quezada, 2009
20th Shanghai, China	Piraino et al., 2009
21st Stuttgart, Germany	Ceroni et al., 2011

6 Conclusion

After the review of all the activities mentioned in this chapter, powerful sentiments of joy and thankfulness take over me. I am thankful for meeting the wonderful people I had the opportunity to meet in this journey of 23 years dedicated in body and soul to the academic life I precious the most. It is true that not everything has been only good moments, there were bad moments also but the experience forces you to not remember or get clinched to them. During this time I have had the chance to get to know my best friends and accomplish results I could not have had the chance to imagine when at the beginning of 1989 took the decision of staying at PUCV after finishing the bachelor degree in industrial engineering. Every class with my students, every project undertaken, and every challenge I had constituted enrichment experience for my and my family. Therefore, if something must be learned from the experience I have briefly narrated in these few lines is that discipline, willingness and perseverance are key to enjoy any professional life. Trusting people around you by being true in your behavior will make an important part of your personality, the rest will be given by life. Remember the power of little things like that humble IBM robot that still fondly remember from time to time.

Appendix 1. Letter of Donation of the SCARA Robot

Purdue University

SCHOOL OF INDUSTRIAL ENGINEERING

DR. SHIMON Y. NOF, PROFESSOR
EMAIL: nof@ecn.purdue.edu
TEL: +1 765 494 5427

Professor J. A. Ceroni
Department of Industrial Engineering
Catholic University of Valpareiso
Valpareiso, Chile

Dear Professor Ceroni,

It is my great pleasure to inform you that our school is most pleased to donate the used industrial robot IBM –SCARA for educational use in your laboratory. The value of this robot is USD 99.-

After about fifteen years of using this robot effectively in our robotics lab here at Purdue, I am sure it can now be useful for your students too. I hope that the robot and all its components will arrive safely in Chile.

Please let me know if you have any questions about the robot.

With best wishes

Sincerely,

Shimon Y. Nof

CC W. D. Compton, Head School of IE
W. Eubanks, Manufacturing labs

Appendix 2: Communication of the multidisciplinary operation of the RAIAAL laboratory (in Spanish)

Laboratorio de Robótica, Inteligencia Artificial y Automatización Avanzada

Escuela de Ingeniería Eléctrica – Escuela de Ingeniería Industrial

Establecimiento de Derechos y Responsabilidades

Abril 28, 2000

Antecedentes

El laboratorio fue creado en 1989 en la Escuela de Ingeniería Eléctrica de la Universidad Católica de Valparaíso como parte de las actividades del Grupo de Investigación en Robótica, Inteligencia Artificial y Automatización Avanzada. En 1999, el Grupo se constituyó en multidisciplinario con la incorporación de un investigador en la Escuela de Ingeniería Industrial de la Universidad. El requerimiento por los equipos del laboratorio, a ser satisfecho con aportes de la Universidad y el Proyecto Mineduc asignado en 1999, fue establecido según los siguientes objetivo general y objetivos específicos:

Objetivo General

Equipar un laboratorio que permita la experimentación e investigación en las áreas de interés establecidas por el Grupo de Robótica, Inteligencia Artificial y Automatización Avanzada, a través del ejercicio directo y simulación computarizada de sistemas flexibles de manufactura. La ejercitación directa se logrará por medio de la interacción de los alumnos con los equipos de una celda flexible de manufactura con el propósito que éstos aprendan su operación y coordinación. Adicionalmente, se utilizará tecnología de simulación y realidad virtual para ampliar el espectro de aplicaciones a las que estarán expuestos los alumnos en el laboratorio. El laboratorio podrá realizar actividades de asistencia técnica que se enmarquen dentro de las áreas de experimentación del Grupo de Investigación en Robótica, Inteligencia Artificial y Automatización Avanzada.

Objetivos Específicos

El laboratorio permitirá las siguientes actividades académicas:

1. Instrucción de alumnos de pre y posgrado, para su formación en el análisis, modelamiento y utilización de sistemas de manufactura flexible, incluyendo en éstos los conocimientos de las áreas de Inteligencia Artificial y Robótica. Esta instrucción estará orientada a alumnos de pregrado en las carreras de ingeniería, e ingeniería civil de las escuelas participantes (inicialmente Escuela de Ingeniería Eléctrica y Escuela de Ingeniería Industrial).
2. Investigación en las áreas prioritarias del laboratorio. La investigación se desarrollará en función de las tesis de grado de los alumnos de posgrado de las escuelas participantes, trabajo de los académicos integrantes del laboratorio, actividades efectuadas por y con académicos visitantes de la Universidad u otras universidades, y la realización de proyectos en conjunto con empresas.
3. Asistencia técnica. Los integrantes del laboratorio podrán realizar proyectos de asistencia técnica con empresas siempre y cuando dicha actividad no genere el desmedro de las actividades de instrucción e investigación.
4. Capacitación. Esta actividad considera servicios de asesoría a empresas que requieran de instrucción en la utilización de sus equipos de manufactura flexible y automatización. Esta actividad no debiese generar investigación y pudiese ser abordada por personal no académico.

Responsabilidades de las Escuelas

- Las Escuelas se comprometen a asumir el 50% del costo de equipamiento del laboratorio. Este costo según comunicación DIR.EIE 0156/00, del Sr. Director de la Escuela de Ingeniería Eléctrica, totaliza $ 15.008.681 y deberá ser solventado con aportes del Proyecto Mineduc para el Fondo de Desarrollo Institucional ($ 10.000.000) y aportes iguales de ambas escuelas por $ 2.500.000 cada una. Con fecha 2 de Mayo de 2000, la Escuela de Ingeniería Industrial transferirá los fondos correspondientes a la Escuela de Ingeniería Eléctrica ($ 2.500.000).

- La Escuela de Ingeniería Industrial aportará al equipamiento del Laboratorio el manipulador robótico IBM 7547 modelo SCARA, número de serie 40256 y su controlador IBM, número de serie 40256.
- La Escuela de Ingeniería Eléctrica aportará al equipamiento del Laboratorio dos cámaras de video y software de análisis de imágenes, un robot móvil, un centro mecanizado experimental, y diversos computadores. Adicionalmente, la Escuela de Ingeniería Eléctrica aportará el espacio físico en que funcionará el Laboratorio.

Derechos de las Escuelas

1. Las Escuelas, a través del o de los académicos integrantes del Grupo de Investigación en Robótica, Inteligencia Artificial y Automatización Avanzada, tendrán acceso permanente e indefinido a las instalaciones del Laboratorio y todo su equipamiento.
2. La composición de académicos del Grupo de Robótica, Inteligencia Artificial y Automatización Avanzada deberá reflejar un equilibrio razonable entre las escuelas participantes. Las escuelas se reservan el derecho a verificar y exigir que este equilibrio exista.
3. Las Escuelas no serán responsables de los costos operacionales ni de actualización de los equipos o infraestructura del Laboratorio, debiendo ser éstos de responsabilidad exclusiva de los académicos que compongan el Grupo de Investigación en Robótica, Inteligencia Artificial y Automatización Avanzada.
4. Las Escuelas se reservarán el derecho de participación, por concepto de dedicación de sus académicos, en los ingresos que genere la actividad de asistencia técnica del Grupo de Investigación en Robótica, Inteligencia Artificial y Automatización Avanzada. Las utilidades de los proyectos de asistencia técnica se entenderán como el remanente de los ingresos después de descontado el costo por la utilización, mantención y actualización de los equipos del Laboratorio y deberán repartirse entre los ejecutores del proyecto.

Prof. Gastón Lefranc   Prof. José Ceroni
Escuela de Ingeniería Eléctrica   Escuela de Ingeniería Industrial

References

Dávila-Ríos, I., López-Juarez, I., Méndez, G., Osorio, R., Lefranc, G., Cubillos, C.: A fuzzy approach for on-line error compensation during robotic welding. In: 6th International Conference on Computers Communications and Control (ICCCC) (2016)

Ascencio, L., Ceroni, J.: Modelo de Diagnóstico IV Congreso Chileno de Investigación Operativa para la Implementación de Sistemas de Manufactura Celular: Caso Textil Chileno, Conferencia Optima 2001. Curicó, Chile (2001)

Busalacchi, F.: The collaborative, high speed, adaptive, supply-chain model for lightweight procurement. In: Proceedings of the 15th International Conference on Production Research, Limerick, Ireland, pp. 585–588 (1999)

Ceroni, J.: Optimization methods of parallelism in distributed organizations. In: 16th International Conference on Production Research, Prague, Czech Republic (2001)

Ceroni, J., Alfaro, R., Cubillos, C.: Information retrieval in collaborative logistics decision making. In: 21 International Conference on Production Research, Stuttgart, Germany (2011)

Ceroni, J., Matsui, M., Nof, S.: Communication based coordination modeling in distributed manufacturing systems. Int. J. Prod. Econ. **60–61**, 29–34 (1999)

Ceroni, J., Nof, S.: Planning integrated enterprise production operations with parallelism. In: 15th International Conference on Production Research (ICPR), Limerick, Ireland (2000a)

Ceroni, J., Nof, S.: Planning integrated enterprise production operations with parallelism. In: Proceedings of the 15th International Conference on Production Research (Limerick, Ireland), pp. 457–460 (2000b)

Ceroni, J., Nof, S.: The parallel computing approach in production system integration modeling. In: IFAC/IFIP/IEEE 2nd Conference on Management and System Control of Production and Logistics (MCPL'2000) July 2000, Grenoble, France (2000)

Ceroni, J.A., Nof, S.Y.: Collaborative manufacturing. In: Salvendy, G. (ed.) Handbook of Industrial Engineering: Technology and Operations Management, pp. 601–619. Wiley (2001). https://doi.org/10.1002/9780470172339.ch20

Ceroni, José, Nof, Shimon, (2002). Models for integration with parallelism of distributed organizations, XV Congreso de la Asociación Chilena de Control Automático, Santiago, Chile

Ceroni, J., Nof, S.: Task parallelism in distributed supply organizations: a case study in the shoe industry. Product. Plann. Control **16**(5), 500–513 (2005)

Ceroni, J.A., Quezada, L.E.: Development of collaborative production systems in emerging economies. Int. J. Product. Econ. **122**(1), 255–256 (2009). https://doi.org/10.1016/j.ijpe.2009.06.033

Ceroni, J., Shimon, N.: A workflow model based on parallelism for distributed organizations. J. Intell. Manuf. **13**(6), 439–461 (2002)

Dawes, S., Préfontaine, L.: Understanding new models of collaboration for delivering government services. Commun. ACM **46**(1), 40–42 (2003)

Del Canto, R., Ceroni, J.: Creación de un sistema emulador de un manipulador robótico para procesos de instrucción. Industrial Engineering School Capstone Project, Pontifical Catholic University of Valparaíso, Chile (2005)

Eberts, R.E., Nof, S.Y.: Tools for collaborative work. In: Proceedings of IERC 4, pp. 438–441. Nashville, TN (1995)

Flen, T., Guidi-Polanco, F., Benedetti, R., Lefranc, G.: Modelization and identification of the Hot Blast Stove's heating cycle. 1267–1273. In: 9th IEEE International Conference on Control and Automation, ICCA 2011, Santiago, Chile, 19–21 Dec 2011

Gatica, G., Ceroni, J., Best, S., Lefranc, G.: Olive fruits recognition using neural networks. Procedia Comput. Sci. **17**, 412–419 (2013)

Graboloza, D., Ceroni, J.: SME Collaboration Model, ICPR Americas 2004. Santiago, Chile (2004)

Hirsch, B., Kuhlmann, T., Marciniak, Z.K., Maβow, C.: Information system concept for the management of distributed production. Comput. Ind. **26**, 229–241 (1995)

Huang, C.-Y., Ceroni, J., Nof, S.: Agility of networked enterprise-parallelism, error recovery and conflict resolution. Comput. Ind. **42**(2000), 257–287 (2000)

Leigthon, F., Osorio, R., Lefranc, G.: Modelling, implementation and application of a flexible manufacturing cell. Int. J. Comput. Commun. Control **6**(2), 278 (2011). https://doi.org/10.15837/ijccc.2011.2.2176

Lin, C., Prassana, V.: Analysis of cost of performing communications using various communication mechanisms. In: 5th Symposium Frontiers of Massively Parallel Computation, pp. 290–297. USA (1995)

Matsui, M., Ceroni, J., Nof, S.: A coordination consideration of manufacturing systems: Job-Shop case. In: 14th International Conference on Production Research (ICPR), Osaka, Japan (1997)

Miranda, P., Garrido, R., Ceroni, J.: A collaborative approach for a strategic logistic network design problem with fleet design and customer clustering decisions. Comput. Indust. Eng. Special issue on Collaborative e-Work Netw. Indust. Eng. **57**(1) (2009)

Nof, S.Y.: Integration and collaboration models. In: Nof, S.Y. (ed.) Information and Collaboration Models of Integration. Kluwer Academic Publishers, Dordrecht, Netherlands (1994)

Nof, S., Ceroni, J., Huang, C.-Y.: Collaborative e-Work: Three Design Principles. ICPR Americas 2004, Santiago, Chile, August (2004)

Piraino, E., King, T., Pascual, J., Ceroni, J.: Performance study of RFID feedback in MRP systems. In: 20th International Conference on Production Research, Shanghai, China (2009)

Rojas, D., Millán, G., Passold, F., Osorio, R., Cubillos, C., Lefranc, G.: Algorithms for maps construction and localization in a mobile robot. Stud. Inform. Control **23**(2), 189 (2014). https://doi.org/10.24846/v23i2y201407

Schleyer, G., Lefranc, G., Cubillos, C., Millán, G., Osorio-Comparán, R.: A new method for colour image segmentation. Int. J. Comput. Commun. Control **11**(6), 860 (2016). https://doi.org/10.15837/ijccc.2016.6.2747

Urra, E., Cubillos, C., Cabrera-Paniagua, D., Lefranc, G.: Automatic parameter configuration for an elite solution hyper-heuristic applied to the Multidimensional Knapsack Problem. In: 6th International Conference on Computers Communications and Control (ICCCC), pp. 213–219 (2016)

Velásquez, A., Ceroni, J.: Agent based collaborative processes. In: 17th International Conference on Production Research. Blacksburg, Virginia, USA (2003a)

Velásquez, A., Ceroni, J.: Conflict detection and resolution in distributed design. Product. Plann. Control **14**(8), 734–742 (2003b)

PRISM & PGRN Research, Discoveries, and Emerging Challenges [General]

Challenges and Contributions to Intelligent and Transformative Production

Before, During and Beyond Pandemic Times

Shimon Y. Nof(✉)

PRISM Center, PGRN, and School of Industrial Engineering, Purdue University, West Lafayette, IN, USA

nof@purdue.edu

Abstract. The theme of this ICPR-26 conference is “Intelligent and Transformative Production in Pandemic Times.” In this plenary, we explore the roles and challenges that production researchers have been facing before and during this pandemic, and how their contributions have enabled survival and continuity of operations, now and beyond the pandemic times. The following themes will be reviewed:

1. Survey of production research before and during the pandemic eruption.
2. Focus on supply chain and supply network resilience and security.
3. Focus on cyber collaborative production for disruption handling and control.
4. Lessons learned and agenda for the future – “beyond pandemic times?”.

We will particularly try to understand if our emerging themes of future work, labs, and factories have been on target: The cyber collaborative, augmented factories, suppliers, and services; and the human-in-the-loop cyber physical production and service. Have we been prepared to deliver on time and at scale what society and civilization need and expect? We will conclude with interesting lessons learned and suggest future research challenges.

C.-Y. Huang and S. W. Yoon (Eds.): ICPR1 2021, ACES 14, pp. 83–89, 2023.
https://doi.org/10.1007/978-3-031-44373-2_4

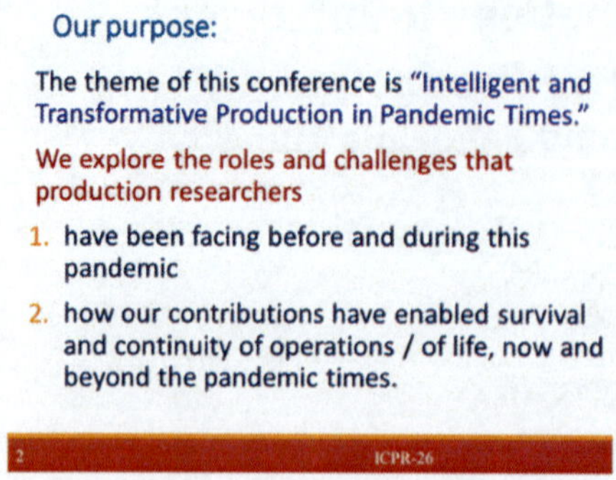
Our purpose:
The theme of this conference is "Intelligent and Transformative Production in Pandemic Times."
We explore the roles and challenges that production researchers
1. have been facing before and during this pandemic
2. how our contributions have enabled survival and continuity of operations / of life, now and beyond the pandemic times.
2
ICPR-26

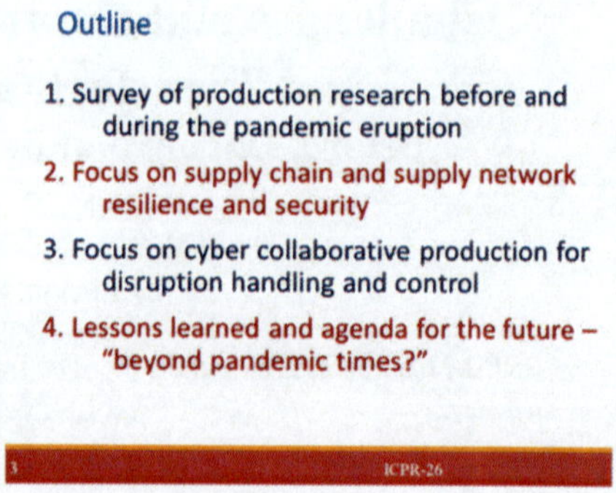
Outline
1. Survey of production research before and during the pandemic eruption
2. Focus on supply chain and supply network resilience and security
3. Focus on cyber collaborative production for disruption handling and control
4. Lessons learned and agenda for the future – "beyond pandemic times?"
3
ICPR-26

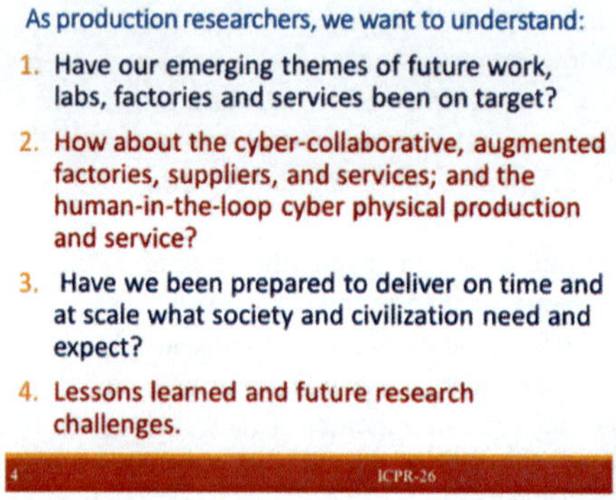
As production researchers, we want to understand:
1. Have our emerging themes of future work, labs, factories and services been on target?
2. How about the cyber-collaborative, augmented factories, suppliers, and services; and the human-in-the-loop cyber physical production and service?
3. Have we been prepared to deliver on time and at scale what society and civilization need and expect?
4. Lessons learned and future research challenges.
4
ICPR-26

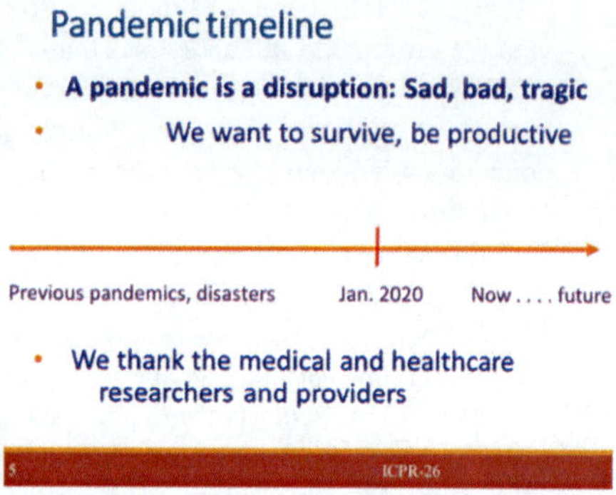
Pandemic timeline
A pandemic is a disruption: Sad, bad, tragic
We want to survive, be productive
Previous pandemics, disasters
Jan. 2020
Now . . . . future
We thank the medical and healthcare researchers and providers
5
ICPR-26

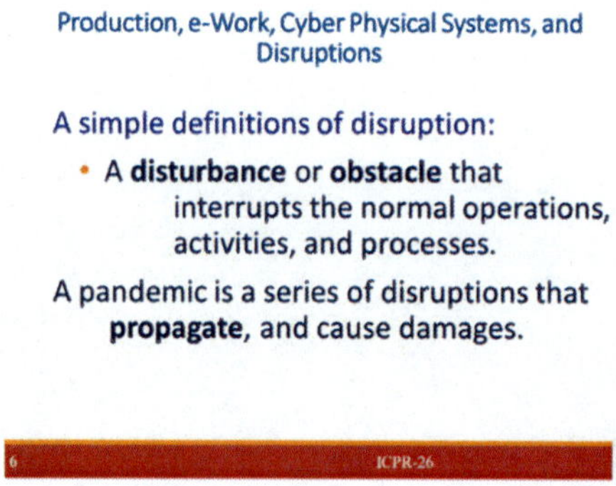
Production, e-Work, Cyber Physical Systems, and Disruptions
A simple definitions of disruption:
A disturbance or obstacle that interrupts the normal operations, activities, and processes.
A pandemic is a series of disruptions that propagate, and cause damages.
6
ICPR-26

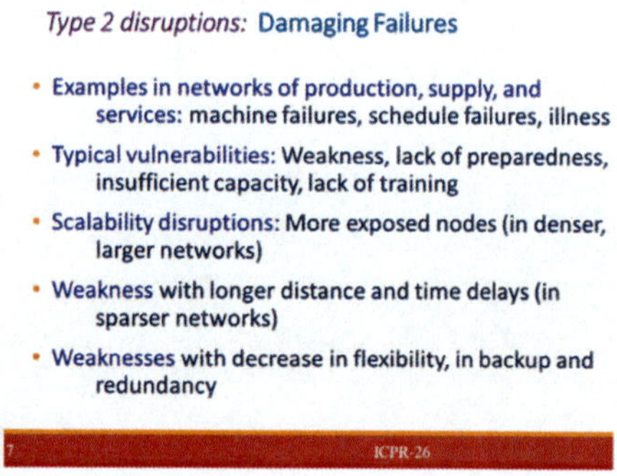
Type 2 disruptions: Damaging Failures
Examples in networks of production, supply, and services: machine failures, schedule failures, illness
Typical vulnerabilities: Weakness, lack of preparedness, insufficient capacity, lack of training
Scalability disruptions: More exposed nodes (in denser, larger networks)
Weakness with longer distance and time delays (in sparser networks)
Weaknesses with decrease in flexibility, in backup and redundancy
7
ICPR-26

Type 1 disruptions: Disruptive Innovations

Innovative disruption	Legacy industry	Disrupter	Disrupted vulnerability
Automobiles	Horses	Cars, trucks	Accessibility
Sharing economy	Cars and trucks	Electric cars & bikes	Congestion; parking
Cyber-collaborative production	Vaccine R&D + Mfg. + Logistics	PANDEMIC ⬇ Cyber collaborative meetings;	Delays; shortages
Cyber-collaborative v-meetings	F2F Meetings	Telemedicine	Accessibility

8 ICPR-26

Disruptive Innovations: Vaccines, Treatments:

Preventive, Predictive, Responsive Maintenance

e-Maintenance integrates cooperation, collaboration, and knowledge-sharing to evolve existing maintenance processes towards new enterprise concepts:
- Extended enterprise,
- supply-chain management,
- lean maintenance,
- distributed support and expertise centers.

Morel, Pétin, Johnson [2009]; Elsayed [2022]

9 ICPR-26

- Complexity and resilience in supply networks: physical, digital, service, combined
- Novel approach: Fault tolerance by teaming (CCT, 2007)
- Levels and dimensions of resilience in supply chains and in supply networks

Contribution:
- Resilience by Teaming framework, task protocols, multi-agent teams
- Application examples in real supply chains and networks
- Preparedness for supply problems.

2018

10 ICPR-26

Supply chain / Supply network Resilience and Security: Recent Research	
Multi-sensor task allocation framework for **supply networks security** using task administration protocols	*Tkach et al., 2017*
Resilience and agility: the crucial properties of humanitarian supply chain	*Dubey, 2019*
Viable supply chain model: integrating agility, **resilience and sustainability** perspectives—lessons from and thinking beyond the COVID-19 pandemic	*Ivanov, 2020*
Supply chain resilience for vaccines: review of modeling approaches in the context of the COVID-19 pandemic	*Golan et al., 2021*

11 ICPR-26

- Cyber physical systems and disruptions
- Types of disruptions and of mitigation
- Novel approach: Network-to-network (N2N) model of collaborative e-Work for prevention, repair, recovery, response.

Contribution:
- The DLOC model & protocols; network scheduling; critical performance metrics
- Applications with real disruptions
- Preparedness for disruptions.

Jan. 2020

12 ICPR-26

Collaborative / augmentative production for Disruption Handling and Control: Recent Research	
Handling disruptions in manufacturing systems: An immune perspective	*Darmoul et al., 2013*
Design of a reliable hierarchical **location-allocation model under disruptions** for health service networks: A two-stage robust approach	*Zarrinpoor, 2017*
Strategic lines of collaboration in **response to disruption propagation through cyber-physical systems**	*Nguyen et al., 2020*
Forecasting and planning during a pandemic: COVID-19 growth rates, **supply chain disruptions**, and governmental decisions	*Nikolopoulos, K., et al., 2021*

ICPR-26

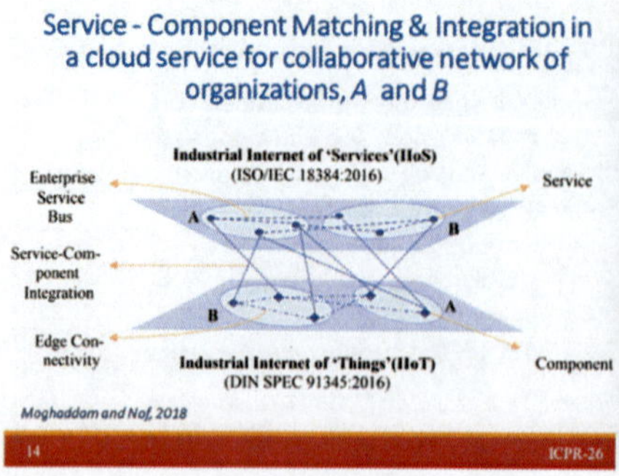
Service - Component Matching & Integration in a cloud service for collaborative network of organizations, A and B
Industrial Internet of 'Services'(IIoS)
(ISO/IEC 18384:2016)
Enterprise Service Bus
Service
Service-Component Integration
Edge Connectivity
Industrial Internet of 'Things'(IIoT)
(DIN SPEC 91345:2016)
Component
Moghaddam and Nof, 2018
14
ICPR-26

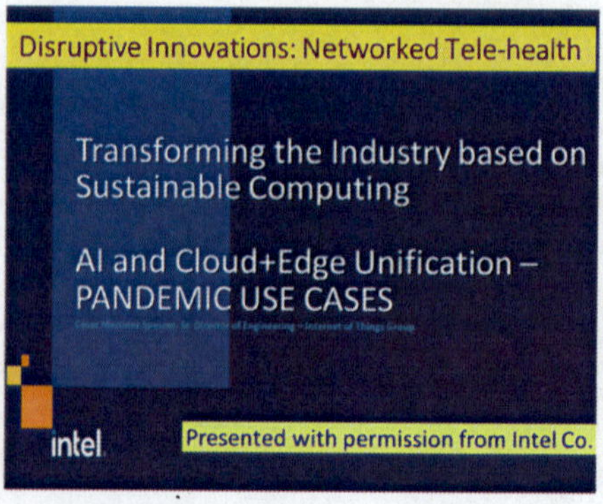
Disruptive Innovations: Networked Tele-health
Transforming the Industry based on Sustainable Computing
AI and Cloud+Edge Unification – PANDEMIC USE CASES
intel
Presented with permission from Intel Co.

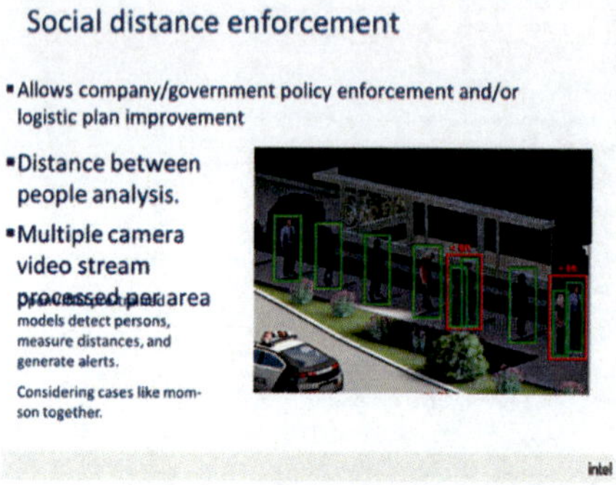
Social distance enforcement
▪Allows company/government policy enforcement and/or logistic plan improvement
▪Distance between people analysis.
▪Multiple camera video stream processed per area
models detect persons, measure distances, and generate alerts.
Considering cases like mom-son together.
intel

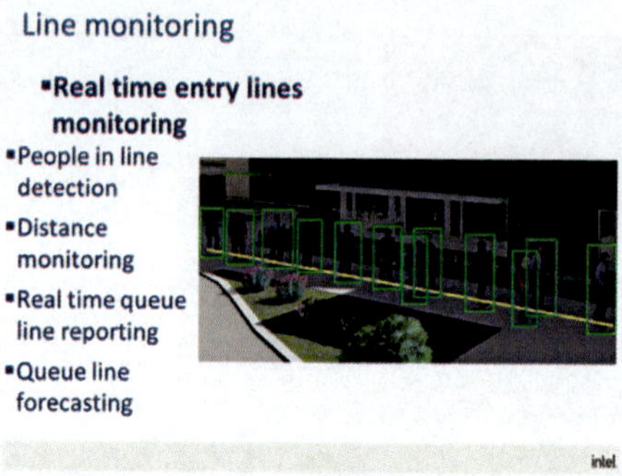
Line monitoring
▪Real time entry lines monitoring
▪People in line detection
▪Distance monitoring
▪Real time queue line reporting
▪Queue line forecasting
intel

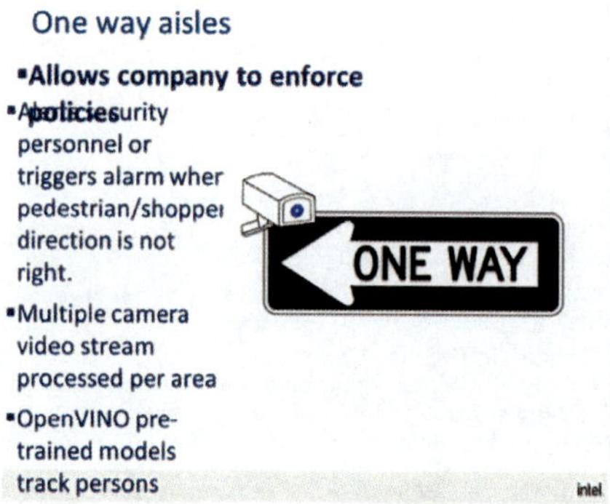
One way aisles
▪Allows company to enforce policies
personnel or triggers alarm wher pedestrian/shoppe direction is not right.
ONE WAY
▪Multiple camera video stream processed per area
▪OpenVINO pre-trained models track persons
intel

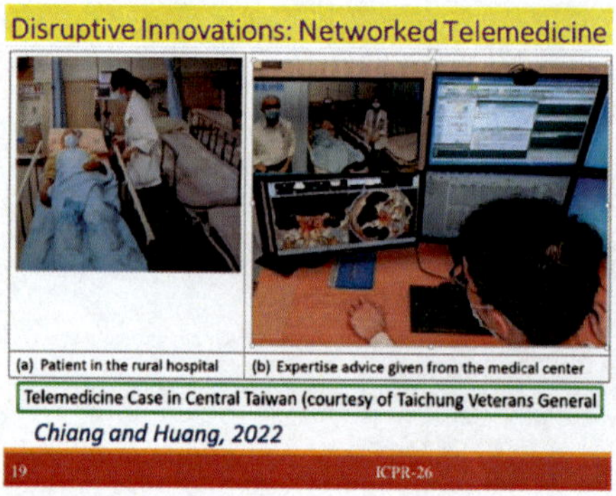
Disruptive Innovations: Networked Telemedicine
(a) Patient in the rural hospital
(b) Expertise advice given from the medical center
Telemedicine Case in Central Taiwan (courtesy of Taichung Veterans General
Chiang and Huang, 2022
19
ICPR-26

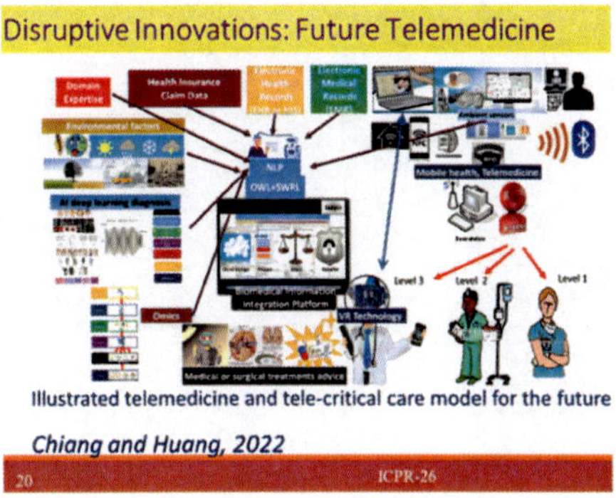
Disruptive Innovations: Future Telemedicine
Health Insurance Claim Data
Environmental factors
Mobile health, Telemedicine
Biomedical Information Integration Platform
VR Technology
Level 3
Level 2
Level 1
Illustrated telemedicine and tele-critical care model for the future
Chiang and Huang, 2022
20
ICPR-26

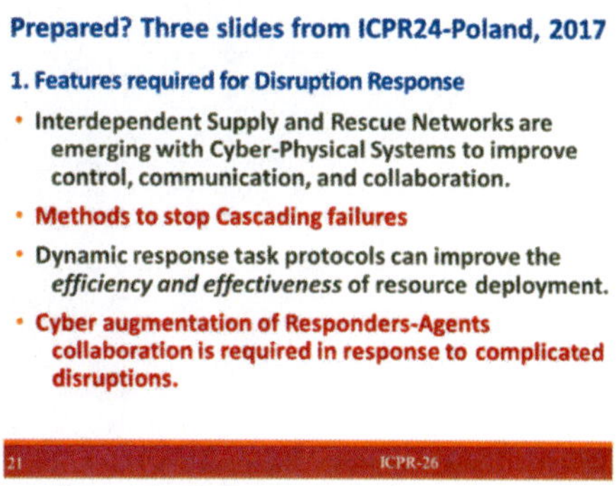
Prepared? Three slides from ICPR24-Poland, 2017
1. Features required for Disruption Response
• Interdependent Supply and Rescue Networks are emerging with Cyber-Physical Systems to improve control, communication, and collaboration.
• Methods to stop Cascading failures
• Dynamic response task protocols can improve the efficiency and effectiveness of resource deployment.
• Cyber augmentation of Responders-Agents collaboration is required in response to complicated disruptions.
21
ICPR-26

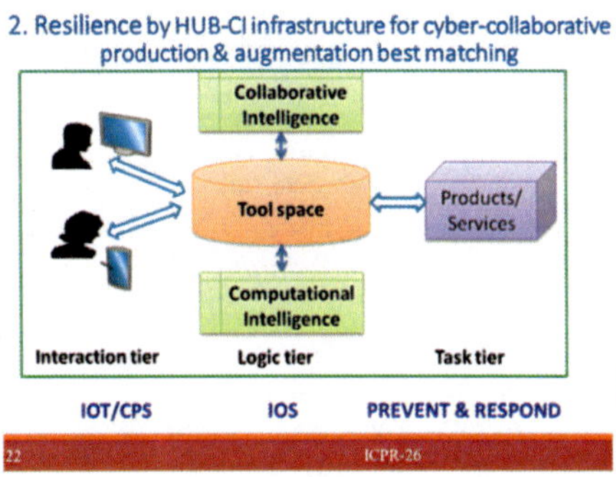
2. Resilience by HUB-CI infrastructure for cyber-collaborative production & augmentation best matching
Collaborative Intelligence
Tool space
Products/ Services
Computational Intelligence
Interaction tier
Logic tier
Task tier
IOT/CPS
IOS
PREVENT & RESPOND
22
ICPR-26

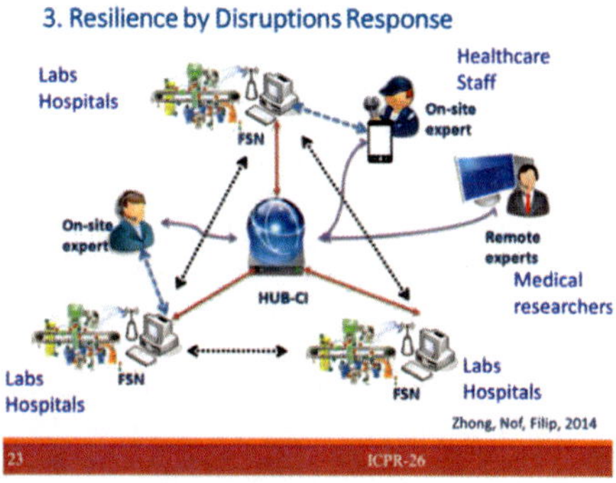
3. Resilience by Disruptions Response
Labs
Hospitals
Healthcare Staff
On-site expert
FSN
Remote experts
Medical researchers
On-site expert
HUB-CI
Labs
Hospitals
FSN
FSN
Labs
Hospitals
Zhong, Nof, Filip, 2014
23
ICPR-26

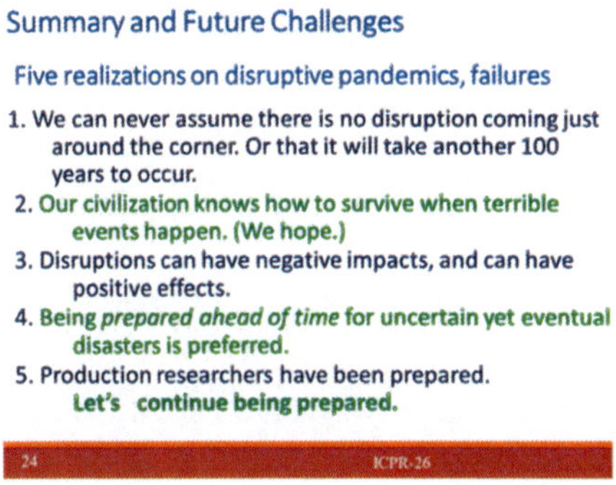
Summary and Future Challenges
Five realizations on disruptive pandemics, failures
1. We can never assume there is no disruption coming just around the corner. Or that it will take another 100 years to occur.
2. Our civilization knows how to survive when terrible events happen. (We hope.)
3. Disruptions can have negative impacts, and can have positive effects.
4. Being prepared ahead of time for uncertain yet eventual disasters is preferred.
5. Production researchers have been prepared.
Let's continue being prepared.
24
ICPR-26

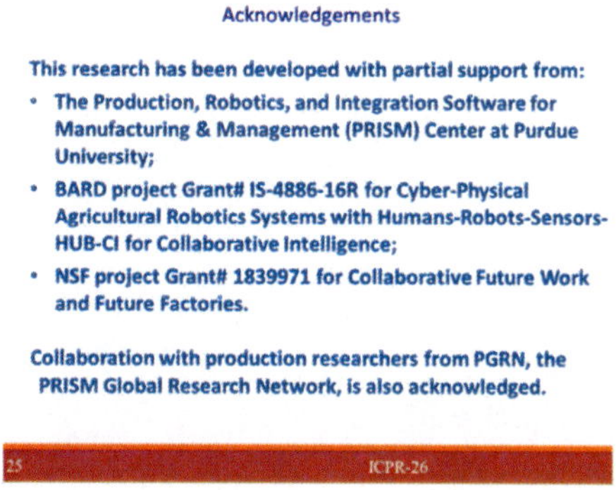
Acknowledgements
This research has been developed with partial support from:
• The Production, Robotics, and Integration Software for Manufacturing & Management (PRISM) Center at Purdue University;
• BARD project Grant# IS-4886-16R for Cyber-Physical Agricultural Robotics Systems with Humans-Robots-Sensors-HUB-CI for Collaborative Intelligence;
• NSF project Grant# 1839971 for Collaborative Future Work and Future Factories.
Collaboration with production researchers from PGRN, the PRISM Global Research Network, is also acknowledged.
25
ICPR-26

Invitation: "PRISM 30" special 30th Anniversary Celebration Sessions

26 ICPR-26

CCT, 2007: Nof, S.Y. Collaborative control theory for e-work, e-production, and e-service *Annual Reviews in Control*, 31(2), pp. 281-92.

Chiang, K.L., and Huang, C.Y. 2022: Precision medicine and telemedicine. In *Springer Handbook of Automation 2nd ed.*

Darmoul, S., Pierreval, H. and Hajri–Gabouj, S., 2013: Handling disruptions in manufacturing systems: An immune perspective. *Engineering Applications of Artificial Intelligence*, 26(1), pp.110-121.

Dubey R., 2019: Resilience and agility: the crucial properties of humanitarian supply chain. In *Handbook of Ripple Effects in the Supply Chain pp.* 287-308, Springer.

El Baz, J. and Ruel, S., 2021: Can supply chain risk management practices mitigate the disruption impacts on supply chains' resilience and robustness? Evidence from an empirical survey in a COVID-19 outbreak era. *International Journal of Production Economics*, *233*, p.107972.

Elsayed, E. 2022: Reliability, maintainability, safety, and sustainability. In *Springer Handbook of Automation 2nd ed.*

Golan, M.S., Trump, B.D., Cegan, J.C. and Linkov, I., 2021: Supply chain resilience for vaccines: review of modeling approaches in the context of the COVID-19 pandemic. *Industrial Management & Data Systems*.

Intel, 2021: https://software.intel.com/content/www/us/en/develop/articles/retail-pandemic-reference-implementations.html

Ivanov, D., 2020: Viable supply chain model: integrating agility, resilience and sustainability perspectives - lessons from and thinking beyond the COVID-19 pandemic. *Annals of Operations Research*, pp.1-21.

27 ICPR-26

Moghaddam, M., and Nof, S.Y. 2018: Collaborative service-component integration in cloud manufacturing *Int. J. of Prod. Res.*, 56 (1-2), pp. 677-691.

Morel, G., Pétin, J-F. and Johnson, T. L. 2009: Reliability, maintainability, and safety. In *Springer Handbook of Automation*, 735-747.

Nguyen, W.P. and Nof, S.Y., 2020: Strategic lines of collaboration in response to disruption propagation (CRDP) through cyber-physical systems. *International Journal of Production Economics*, 230, p.107865.

Nikolopoulos, K., Punia, S., Schäfers, A., Tsinopoulos, C. and Vasilakis, C., 2021. Forecasting and planning during a pandemic: COVID-19 growth rates, supply chain disruptions, and governmental decisions. *European j. of operational research*, *290*(1), pp.99-115.

Reyes Levalle, R. 2018: *Resilience by Teaming in Supply Chains and Networks*, Springer ACES Series.

Tkach, I., Edan, Y., and Nof, S.Y., 2017: Multi-sensor task allocation framework for supply networks security using task administration protocols, *Int. J. of Prod. Res.*, 55(18), pp. 5202-24.

Zarrinpoor, N., Fallahnezhad, M.S. and Pishvaee, M.S., 2017: Design of a reliable hierarchical location-allocation model under disruptions for health service networks: A two-stage robust approach. *Computers & Ind Eng.*, 109, pp.130-150.

Zhong, H., and Nof, S.Y., 2020: *Dynamic Lines of Collaboration: Disruption Handling & Control*. Springer ACES Series.

Zhong, H., Nof, S.Y., and Filip, F.G. 2014: Dynamic Lines of Collaboration in CPS Disruption Response. *Proc. 19th IFAC Worl Congress*, Cape Town, South Africa.

28 ICPR-26

Answers to FAQs

What is the Motivation for CPS production and supply?

Apply cybernetics/AI for significant value in three main dimensions:

1. **Reach** and **engage** workers, consumers, clients, and customers, including physical devices, and at remote locations, more effectively
2. Boost worker's / employee's ***productivity & safety***
3. Optimize operating ***efficiency & effectiveness***

Resilience & Sustainability

Motivation: Improve work / business / service processes by using computing & communication science & cyber technology → Economic supply of needed services

29 ICPR-26

Digital & Cyber in Production Automation:

Augments abilities of workers, robots, and organizations

CYBER (Cybernetics) = Computing + Communication + Real-time control + Brain models for collaboration protocols

Increasing levels of computers, communication, mobility, and computational intelligence lead to higher levels of automation intelligence, and of resilience to internal and external disruptions.

Digital & Cyber Augmentation	
1.0	Computerized
2.0	Computer Integrated
3.0	Internetworked + Mobile
4.0	Cloud-Based + Machine Learning
5.0	Cyber-Physical + Cybernetics

30 Digital & Cyber Convergence

Collaborative Decision-Making: Concepts, Methods, and Supporting Information and Communication Technologies

Florin Gheorghe Filip[1](✉), Constantin Bâlă Zamfirescu[2], and Cristian Ciurea[3]

[1] The Romanian Academy, Bucharest, Romania
ffilip@acad.ro

[2] "Lucian Blaga" University, Sibiu, Romania
zbc@acm.org

[3] University of Economic Studies, Bucharest, Romania

Abstract. Collaboration means that several entities such as humans, computers, robots, enterprises and so on jointly perform a certain task instead of working individually so that a good result could be obtained. Decision-making is a specific form of activity that is meant to eventually select a certain course of action which is expected to result in attaining a desired result by using the available. The chapter is meant to present a concise and balanced view of the basic concepts, methods, and main classes of supporting information and communication tools and systems regarding decision-making processes carried out by several collaborating agents called participants. The reasons for collaboration are briefly explained followed by an exposure of collaboration application in the multi-participant decision-making settings. Having presented the classification of decision problems and decision-making units, the main phases of a specific multi-participant form of Herbert Simon's decision process model are described followed by the presentation of two main forms of close and soft collaboration, namely consensus building and crowdsourcing, respectively. The need for technology support offered to collaborating participants is justified and two main classes of decision supporting systems, namely Decision Support Systems and collaboration platforms, are addressed. Three practical examples are briefly presented to illustrate the collaborative activities in public sector institutions, manufacturing, and culture economy, respectively. Open questions about the further role the information and communication tools in decision-making processes are eventually formulated from two perspectives, *digital humanism* and *dataism*, respectively.

Keywords: cyber-physical system · digital cognitive system · DSS · multi-criterion · virtual exhibition

1 Introduction

Collaboration is defined by the Merriam-Webster Dictionary [1] as performing a "work with another person or group in order to achieve or do something". The dictionary explains that it comes from the late Latin *collaboratus*, that represents the past participle

C.-Y. Huang and S. W. Yoon (Eds.): ICPR1 2021, ACES 14, pp. 90–106, 2023.
https://doi.org/10.1007/978-3-031-44373-2_5

of the verb *collaborare* (to labour together). In many languages, one can notice the saying „Two heads are better than one", which apparently comes from the Holly Bible (*Ecclesiastes 4;9*), together with the explanation "because they have a good return of their toil". [2] In his recommendation of reading old books, Clive Stapleton Lewis, an English scholar, gives an intuitive justification of the above saying. He stated that the above-mentioned saying was true, "not because either [person] is infallible, but because they are unlikely to go wrong in the same direction" [3]. Of course, the old saying and Lewis' explanation are true only if the persons involved in a collaborative work do not exhibit an irrational behaviour and act seriously and in good faith.

Collaboration is not restricted to the activities that are carried out by two or more persons who prefer to work together instead of individually acting and may involve other types of entities. For example, Herbert Simon, highlighted the advantages of making AI (*Artificial Intelligence*) tools and MS/OR (*Management Sciences/Operations Research*) models 'collaborate' to solve complex problems [4]. Nof et al. [5] explain in their book that, besides persons, other entities, such as computers, various automation devices, communication means, machines, and so on may be involved in collaboration. As Camarinha-Matos et al. [6] notice, there are several levels of collaboration, such as information exchange, harmonization of objectives, sharing resources and responsibilities, and advanced coordination in decision-making and acting. Consequently, a rather broad definition of collaboration is adopted in [7]:

> Collaboration is a specific class of interactions among several entities, such as organizations, humans, and machines that exchange information and knowledge for mutual benefits, harmonize their major goals and objectives and share resources, action plans, and responsibilities to attain the common goals.

According to the purpose of collaboration, in [5] and [8] three subclasses of the collaboration general class are identified: a) *mandatory* collaboration, when two or more entities must collaborate, because each one working independently cannot deliver the expected output, such as a product, a service, or a decision, and (b) *optional* (or progressive) collaboration in case the entities might start collaborating because all of them aim at improving the quality of their deliverables or/and to achieve higher values for all of them, and c) *concurrent collaboration* which is meant to improve the performance of the agents' work.

The chapter is about *collaborative decision-making,* which is a specific subclass of the more general class of collaborative activities and consists in a series of activities that are carried-out by more than one person that compose a *multi-participant decision unit*. The remaining part of the chapter is organized as follows: Next section contains a review of basic defining aspects of collaborative decision-making concept, such as decision problems, activities involved, collaborative group definition, and the process of adopting and releasing collaborative decisions. The particular cases of *consensus building* and *crowdsourcing* are described. The third section is about the specific information systems designed to support collaborative decision-making tasks. The need for such systems is explained and a formulation of main requirements and functions to be supported by multi-participant DSS (*Decision Support Systems*) are presented. The section also addresses the platforms which are ever more used nowadays. The fourth section briefly describes three projects in which the authors have participated deployed in public sector institutions, manufacturing, and culture economy, respectively. Two rhetoric questions about human-machine symbiosis are eventually formulated.

2 Collaborative Decision-Making: Basic Aspects and Special Cases

2.1 Decision Problems and Participants

A *decision problem* is associated with (a) a perceived or anticipated situation that requires action, and (b) several possible courses of action, called *alternatives*. The decision is defined in [9] as "the result of a series of human conscious activities that aim at choosing a course of action with a view to attaining a certain objective (or set of objectives)". It consists in processing information and knowledge by an empowered person or set of persons who have to make the choice and are accountable for the quality of the solution adopted to solve a particular problem or situation. *Making* (solving the decision problem) and *taking* (assuming the solution), releasing, and deploying the adopted solution normally imply allocating the necessary resources, such as people, time, information, and supporting technical means.

The decision problems and activities can be classified in accordance with several attributes as follows:

- The *pursuit objectives* that may be a) to obtain the result that is optimal or, at least satisfactory, with respect with a single criterion or set of criteria, b) to maintain the supremacy over competitors or to reduce the distance to the leader, c) to mitigate and recover from losses in case of disaster or crisis situations, and d) attaining a mutually accepted settlement in the case of negotiations of the parties who have conflicting objectives;
- The *number of decision units* that take part in decision-making and taking; one can distinguish *individual* and *multi-participant (collaborative)* units;
- The *number of people* that compose a decision unit; there may be a single person that makes and takes decisions, or there are several people that work together to make and take collaborative decision;

One can identify two subclasses of the more general class of multi-participant decision units as follows:

- *Collectivity*, which is characterized by an episodic composition of the unit and it is commonly met in a) *negotiations* when the parties typically pursuit different, sometimes conflicting, objectives [10], and b) *crisis management* situations [11];
- *Collaborative groups* that are characterized by several attributes, such as: a) *congruence* of goals and methods of group constituents with respect to the adopted objectives, activities and procedures of the group as a whole, b) *effectiveness* that measures the degree to which the objectives of group are attained, c) *efficiency* to measure the savings individual resources to attain the group goals, d) *cohesion* of group members that are willing and ready for further collaboration [12].

2.2 Multi-participant Decision-Making

The general model of decision-making activities was proposed by Herbert Simon [13]. It is a process composed of several phases such as: a) intelligence, b) design of alternatives and models, and c) choice of a solution to be released for implementation. A fourth phase, the evaluation of the decision implementation impact and possibly to re-start the process, was later added to model.

In collaborative decision-making, the Simon's process model can be adapted to multi-participant settings. A possible model consists in the following phases [14]:

- *Preparation* meant for a) defining the problem characteristic aspects, such as: purpose, the domain, current context, criteria, and possible constraints, and b) empowering the decision unit;
- *Collective understanding* of problem which can be viewed as a natural extension of the preparation phase and consists in activities such as: sharing a common vision of the problem with all the participants and agreeing on how to implement the designed process;
- *Solution generation* meant to identify or design alternatives and applicable models to solve the problem;
- *Negotiation and confrontation* of viewpoints to enable participants to elaborate their contributions and present them in order to win the support of the greatest number of other parties;
- *Decision* for selecting, according to the criteria previously agreed, the ideas which have been voted by most of participants, or which have received the consensus within the group;
- *Monitoring* phase which covers the entire decision-making process so that any problem can be solved in the allocated time period. It includes generating a report on the decision-making process and ensures the implementation of the adopted and assumed solution.

2.3 Collaboration Forms and Methods

In [15], the following forms of collaboration were identified:

- *close collaboration* established among the members of the decision group who exchange information to make and take decisions;
- *asymmetric* or *skew collaboration*, which is a particular form of the previous one and it is established among the *decision takers* (the persons accountable for the released decision impact) and their own human support team of assistants or/and hired consultants;
- *soft collaboration* of the decision makers and takers with commonly anonymous members of a crowd.

There are several approaches and methods that can be used in collaborative decision-making. Section 3 of the monograph dedicated to computer supported collaborate decisions [7] contains a review of the most commonly used methods. Among the methods for aggregating individual preferences, the chapter addresses social choice (including voting mechanisms), its axioms and paradoxes, and several extensions, such as: judgement aggregation, resource allocation, group argumentation, and collaboration engineering.

Voting is a very frequently encountered method used in adopting a decision. Various forms of voting are used almost in any phase when the decision group requires a collective view over the set of decision objectives, allocation of individual tasks, consensus building, etc.

Starting with Kenneth Arrow's influential book *Social Choice and Individual Values*, published first in 1951 by Yale University Press, the voting rules have been characterized by a set of axioms which highlight the desired conditions that they satisfy or not. Some axioms are fulfilled by almost any voting rule (e.g. anonymity, neutrality), while others are nearly impossible to be satisfied without additional constraints (i.e. independence of irrelevant alternatives, non-manipulability). Therefore, in collective decision-making the selection of the proper voting rule depends very much on the context, and should consider multiple variable such as: the group size, the number of alternatives, the way of ranking the alternatives (i.e., cardinal or ordinal), the availability of intermediate results during the voting procedure and possibility to change the individual preferences, the number of winning alternatives, tie-breaking procedures, etc. Moreover, they require dissimilar effort, both cognitive and physical, to elicit the preferences from the members of the decision group. Unsatisfying one of the desired axioms and losing the interest to express the individual preferences leads to voting paradoxes with unintuitive results in respect to collective view.

Judgement Aggregation and Resource Allocation. In collaborative decision-making, voting is not only used to aggregate the individual preferences, but is also used to aggregate judgments, to allocate resources or to aggregate opinions in argumentation frameworks. These extensions are already employed in several collaborative tools that support a rational decision-making process, such as reaching agreements in mixed human-agent teams, fusing different ontologies, collective annotation of data for mutual interpretation, etc.

Basically, the method tries to aggregate the logical assumptions that are behind an individual choice. These are expressed in a set of correlated logical propositions in respect to certain data. As may be expected, even for a simple reasoning process, judgment aggregation rules (e.g., majority, premise-based, conclusion-based, quota, distance-based) are not guaranteeing a consistent collective judgment. Therefore, prioritizing their axiomatic properties (e.g., anonymity, systematicity, collective rationality) is always needed depending on the context, such as competence levels of group members, completeness of the set of logical propositions and so on.

Preference aggregation is also used when different types of common resources (single divisible or multiple indivisible) are collaboratively allocated. This problem appears when the group needs to distribute natural or artificial resources in a way that maximize either the efficiency or the fairness of individual preferences. Due to its NP complexity, the available allocation methods are limited in satisfying certain criteria (i.e. envy-free, equitable) for a group size greater than three members. Therefore, in most cases fairness criteria are used as additional constraints to discriminate among multiple equally efficient solutions (i.e., Pareto-efficient, social welfare).

Consensus Mechanism. One of the approaches used in close collaborative decision-making, with or without moderator's support, that has received, from academia cercles, a lot of attention over the last decades is based on *consensus building* [16]. According to

Merriam-Webster Dictionary [1], the consensus term means “a general agreement about something: an idea or opinion that is shared by all the people in a group”. An example for *consensus* desideratum is a quotation from President Abraham Lincoln, who, before issuing the *Emancipation Proclamation*, wrote in his message to Congress:” We can succeed only by concert. It is not ‘can any of us imagine better?’ but, ‘can we all do better*?”* [17].

The process of aggregating participants’ individual preferences is composed of two main sub-processes: a) consensus building, and b) selecting a recommended solution [16, 18]. During consensus building, the participants might need to revise their opinions with a view to making them closer to one another in an interactive process, so that an acceptable level of consensus could eventually be reached. The process is viewed as composed, at each iteration, of several activities, such as: a) collecting from each participant his/her individual preference, b) aggregating individual preferences by using one of the available methods c) measuring the consensus level expressed as a distance of individual preferences either to the calculated collective one, or as the result of comparing pairs of preferences, d) implementing a revising scheme for the individual preferences with a view to improving the consensus level based either on identifying the participants whose further contribution to consensus reaching could be neglected or minimizing the number of preference revisions. Other methods based on the fuzzy approach have recently been proposed [19].

Crowdsourcing. There are many reported results concerning *large scale decision-making processes* [20] possibly using AI (*Artificial Intelligence*)-based methods and tools. However, an implicit assumption in many schemes for collaborative decision-making based on consensus building consists in limiting the number of participants involved, so that various methods proposed could be technically applicable. In addition, the expertise of the participants might not be appropriate or sufficient for the faced decision situation, or/and the problems could be too complex, or persistent. In such situations, a larger number of people could provide with the necessary information and knowledge for solving the problem. Such a soft collaboration form can be effective in many domains. A particular form of soft collaboration which has got traction over the last decades is *crowdsourcing*. Howe [21] coined the concept as “the act of taking a job traditionally performed by a designated agent (usually an employee) and outsourcing it to an undefined, generally large group of people in the form of an open call”. Crowdsourcing means, in substance, that an individual or collective initiating agent (called *crowdsourcer*) does not assign a certain known person or group of persons to work on a specific task, but he/she will hand over the task to the crowd composed of anonymous *crowdworkers* who will complete it.

Having carried-out an extensive study of systems that had claimed to support crowdsourcing solutions, Estelles Arolas and González-Ladrón [22] proposed a rather broadly accepted definition as follows:

> Crowdsourcing is a type of participative online activity in which an individual, organization, or company with enough means proposes to a group of individuals of varying knowledge, heterogeneity, and number, via a flexible open call, the voluntary undertaking of a task. The undertaking of the task, of variable complexity and modularity, and in which the crowd should participate bringing their work,

money, knowledge and/or experience, always entails mutual benefit. The user will receive the satisfaction of a given type of need, be it economic, social recognition, self-esteem, or the development of individual skills, while the *crowdsourcer* will obtain and utilize to their advantage that what the user has brought to the venture, whose form will depend on the type of activity undertaken.

Crowdsourcing is deployed in many various domains and its usage is nowadays facilitated by the *mobile computing* [23]. In the particular setting of collaborative decision-making through using crowdsourcing the following steps are carried-out [7, 24]:

- Identification of the decision problem to be solved by using the opinions collected from the crowd, and defining the corresponding task. The activity basically corresponds to the *Intelligence* phase of Simon's process model;
- Broadcasting the task to the crowd, commonly in the form of an open call;
- Idea generation by the crowd members in the form of various action alternatives or/and evaluation criteria. It basically corresponds to the *Design* phase of Simon's process model.
- Evaluation of collected ideas by the same members of the crowd that generated them or by another crowd or limited group of hired experts who may pursue the process of reaching the consensus, as described above;
- Choosing the solution through a voting mechanism or based on expert views [25].

3 Information and Communication Technologies for Collaborative Decision-Making

3.1 The Need for Technology Support

The multi-participant decision units, in case they are not supported by technology, may face a series of problems [7, 12] such as: a) *groupthink* caused by an authoritarian leader or a very vocal participant, time or/and external pressure, high homogeneity of the group, when the participants have similar interests, b) *cognitive overload* caused by excessive interactions, c) the fear for possible negative consequences, d) possible misunderstanding in case the participants possess different cultural or technical backgrounds or speak different native languages, e) high costs or sanitary restrictions to organize the meetings and so on.

At present, all collaboration forms can be enabled and supported by modern I&CT (*Information and Communication Technology*) tools, systems, and platforms. A review of several technologies and tools, such as AI (*Artificial Intelligence*), social networks, Data Science, web technology, mobile and cloud computing, the biometric tools, and serious games, and their relevance for supporting collaborative decision-making can be found in the third chapter of [7]. The MADM/MCDM (*Multi-attribute decision-making/multi-criterion decision-making*) methods [26, 27] and their computerised versions, possibly combined with *AI* [28] and *Big Data* methods and technologies [29], are very useful tools for solving the problems characterized by several aspects to be taken into.

3.2 Multi-participant Decision Support Systems

A *decision support system* (DSS) is defined in [9] as "an anthropocentric and evolving information system which is meant to implement the functions of a *human support system* [the team of assistants and possible external consultants] that would otherwise be necessary to help the decision-maker to overcome his/her limits and constraints that he/she may encounter when trying to solve complex and complicated decision problems that that matter". In the multi-participant setting, in order to overcome the problems that could be encountered by the decision unit as enumerated earlier, the specific subclass of *collaborative* (or *multi-participant*) DSS should possess the characteristic attributes of a collaborative [information] systems. The list of attributes includes: a) parallelism, in order to avoid the waiting time of participants who want to intervene by enabling all of them to simultaneously input into the system their ideas and views, b) *anonymity,* so that an idea could be accepted based on its value only no matter the proposer's professional reputation or social position, c) *memory of the group,* to accurately record the ideas and views expressed by individual participants and the solutions that were adopted, d) *unambiguous* and faithful presentation on participants' computer screen of the ideas and views of other attendants of the decision-making process [12]. Practical experience witnesses that the DSS are continuously evolving under the influence of several factors, such as the changes in business environment, available technology and methods, and developments in users' knowledge, skills, and willingness to use the system [28].

3.3 Platforms

The platforms have been traditionally representing a specific subclass of the more general class of I&CT means used to support various collaborative activities, including multi-participant decision-making, characterized by large numbers of people commonly working in different locations and organized in virtual teams [30]. The platforms are necessary enabling means for carrying out crowdsourcing and *crowdwork.* Nowadays, the pandemic and the associated sanitary restrictions have made the usage of platforms one of the most common styles of work and the *platform economy* is spreading and growing as a wild fire all over the world.

When one intends to use crowdsourcing, a decision problem is choosing the most appropriate platform. There are plenty of paid or free platforms available of the market, such as: *Idea Bounty, OpenIdeo, Innocentive, CrowdSpring, 99Designs, Cad Crowd, Design Crowd, Mikro Workers, Mechanical Turk* and so on. In the specific setting of collaborative decision-making based on crowdsourcing, a methodology inspired from [31] is proposed in [15] for choosing the platform. The following criteria (and derived subcriteria) are proposed and used to compare several available platforms: a) adequacy to the envisaged applications (informational transparency, accuracy of expected results, robustness to errors and low quality uncertain input data, response time), b) quality of implementation: (scalability, flexibility, functional transparency, documentation completeness), and c) delivery quality (price, service delivery time, provider's general reputation, easy adaptation, degree of independence on the technical assistance from the provider's specialists for implementation and usage).

4 Examples

4.1 iDS (Intelligent Decision Support)

iDS is a family of platforms developed by *Ropardo*, a Romanian company located in Sibiu. The family members have been designed over one and half decade with a view to supporting individual as well as multi-participant decision-making activities carried out in universities, local public administration, and digital factory milieux [7, 32]. It has been used to collaboratively define the performance indicators on various fields such as teaching, manufacturing, or investments analysis, and, based on such indicators, to identify and select the suitable action plans. The family comprises a series of successive versions developed under the influence of three main factors, such as: a) new I&CT tools, b) users' evolving needs and skills, and c) evaluation of results obtained in practical applications. These developments may also be noticed on similar platforms such as *XLeap* (https://www.xleap.net/), *DecisionRules* (https://www.decisionrules.io/), *FaciltatePro* (https://www.facilitate.com/), *Spilter* (https://spilter.nl/gdss/), etc.

The latest version of iDS is characterized by the following aspects: a) usage of web 3.0 technologies and social networks to support collaborative work, b) the possibility to integrate additional third-party modules via API (*Application Program Interface*) together with its own standard set of tools/functions such as a forum-like discussion list, a voting module, an electronic brain-storming, and c) facilitating asynchronous decisions through web 2.0 clients or dedicated mobile clients (Fig. 1). These technical updates increased the flexibility of using third-party tools in the tool chain needed to support a collaborative decision recorded as a business process model. At the same time, they introduce the challenge to exploit the intermediate results of some group decisions in subsequent or different decisions. Therefore, besides the registration of a third-party on the platform, iDS needed to record the tool's states on each session (i.e., configuration, working and report) and employ an ontology for tools description to facilitate the data transfer and communication with the tools (Fig. 2).

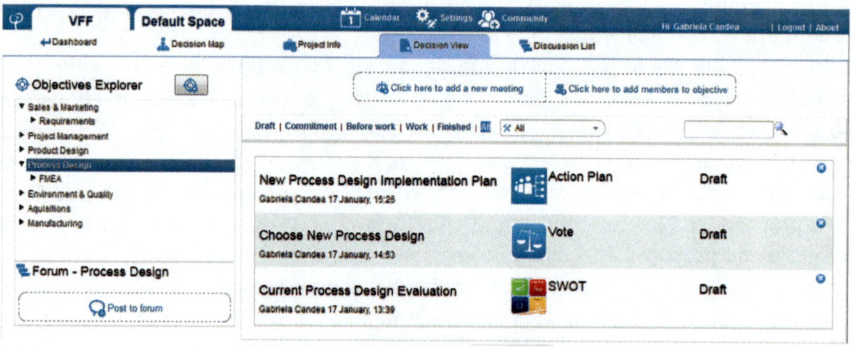

Fig. 1. The iDS Web Client [32]

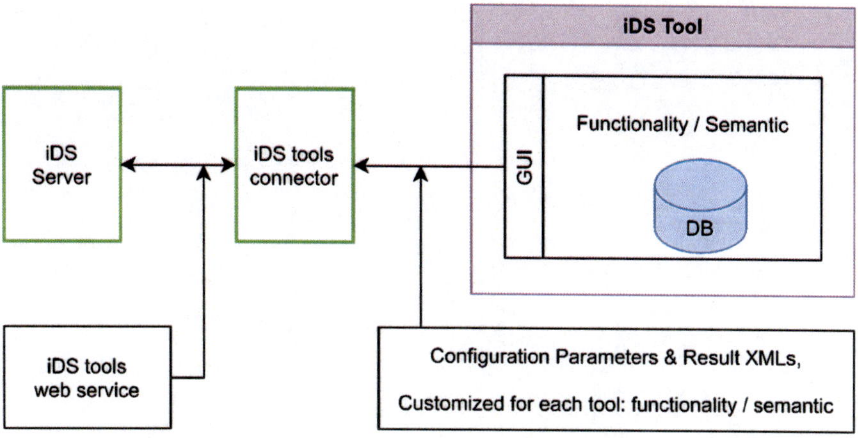

Fig. 2. The iDS connector with third-party tools [32]

4.2 FMI-A Functional Mock-up Interface

FMI is a standardized interface that allows to combine part models developed by specialized teams into a complex system that can be analyzed through co-simulation. *Co-simulation* is a common method to design heterogeneous systems, such as *Cyber-Physical Systems*, made up of different components, such as software, electronics, and mechanics. Firstly, it suppresses the lack of an integrated simulation tool able to model and simulate all the composite parts of a Cyber-Physical Systems, and secondly it supports the collaboration of engineers that use their familiar modelling tools to design the specific parts. The main challenge in this case is to identify the right interaction protocols (signals) among the models.

In [33], it is detailed the collaborative processes for defining the signals among the composite parts of a *Cyber-Physical Production System* (Fig. 3) and includes several steps meant to: a) identify the required messages and data from other models to properly exhibit one model behavior; b) reduce redundant information and combine simple messages into more complex ones; c) investigate aggregation alternatives for the messages in terms of names, generic data structure or codifications; d) evaluate the interaction protocols between the models by running co-simulations to identify and rectify unintended behaviors.

This collaborative process may be repeated until the desired behavior of the system is reached, either in the beginning when the low-fidelity models are produced using a single formalism [34, 35], or afterward when high-fidelity models are created using appropriate tools and formalisms for each part of the Cyber-Physical Production System [36]. The co-simulations can be executed on individual software such as INTO-CPS (https://into-cps.org/) or platform (https://hubcap-portal.eng.it/welcome/).

4.3 Virtual Exhibition Projects

Virtual exhibitions (VE) projects constitute one particular subclass of a more general class of collaborative projects and activities in the *culture economy* [37] In the recent

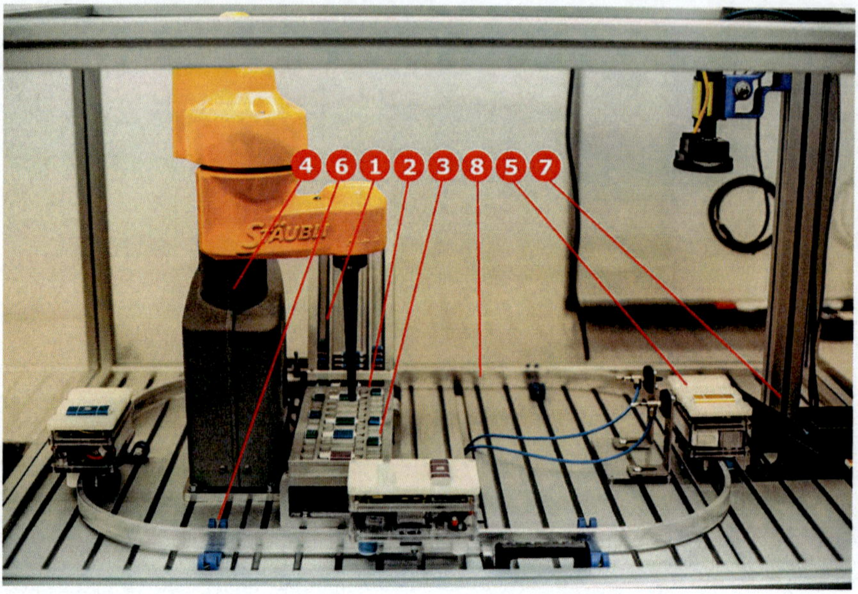

Fig. 3. The Cyber-Physical Production System containing: 1) the Warehouse stacks; 2) the Warehouse assembly box; 3) the Warehouse memory boxes; 4) the Robotic Arm; 5) the Wagons on the track; 6) the Loading Station; 7) the Test Station; 8) the circular track for the Wagons [34].

Covid pandemic time, the widespread implementation of social restrictions caused art exhibitions not be able to take place physically in museum buildings or halls as in the old calm "blue waters" times. The same phenomenon has been noticed in libraries. Consequently, the cultural actants, institutions and people, have been forced to look for alternative solutions that could be used as an exhibition space, or virtual reading rooms, where it is possible to bring together people and works of art or rare book collections to an accessible place. As practice has witnessed that the VE movement has got traction and made possible the culture ecosystem to continue to run with all the limits imposed by this pandemic and can help art workers in the mission they have during this period [38].

The VE project can be viewed as a collaborative system where many actors work together in order to attain beside their particular objectives, a common goal, namely to increase the number of visitors and improve the income of the parties involved. There are many actants and roles involved in the design, creation deployment and running a virtual exhibition that *communicate*, *coordinate* and *cooperate* (3C) with each other as a virtual decision-making unit team for the success of the project: *Data Manager*, *Curator, Vocabularies Manager*, *End-user / Visitor*, *Software Developer*, *Layout Designer*, *Media Expert* and so on (Fig. 4). The decision problem to be solved when a VE project is started is a multi-criterion one. It consists in deciding the VE content and technical deployment form so that several both qualitative and measurable criteria should be observed, such as: estimated interest of potential virtual visitors, the state and preservation needs of physical collections, special events envisaged and so on.

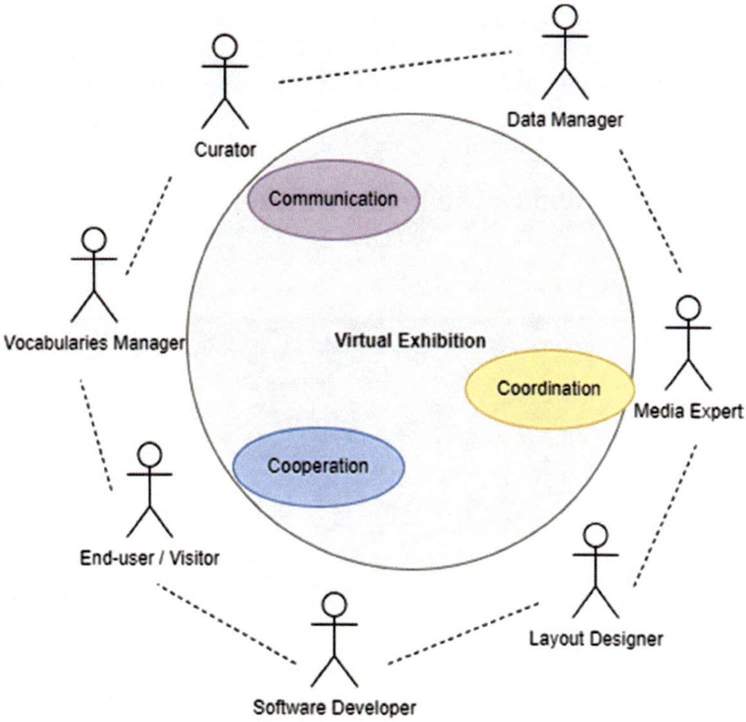

Fig. 4. Collaborating actants in VE projects

Beside the achievement of the 3C (communication, coordination and cooperation) associated with a collaborative system, there is a strong collaboration between the fields involved in the development of a virtual exhibition: the cultural field and the I&CT field. The collaboration inside a virtual exhibition is achieved at many levels, including the cooperation between many cultural institutions (GLAM – galleries, libraries, archives and museums) that interchange and put together their cultural collections inside the virtual exhibition. It can be realized a multicultural project, where different cultural institutions from different countries collaborate to achieve a common objective – an international virtual exhibition. Another level of collaboration related to a virtual exhibition is between the software and hardware components necessary to deploy the application, namely the server where the database and the application are running, the computers and mobile devices used to access the virtual exhibitions. In [39], the methodologies used to set up a VE, both as the web application and a native mobile one, are presented and in [40] several platforms are meant to support the creation of VE, such as *Movio, Prezi Digital Storytelling,* and *Omeka* are reviewed and compared by using the criteria presented in [31].

A selection of virtual exhibitions, created at the Library of the Romanian Academy by several professionals (curator, lay-out designer, IT engineer, platform provider), can be found at the address https://biblacad.ro/expoVirtuale.html. An example is presented in Fig. 5.

Fig. 5. A screenshot of the homepage of the virtual exhibition dedicated to historical seals (http://movio.biblacad.ro/SEALS/)

5 Conclusions

Collaborative networks that make humans collaborate with other humans, organizations, and various other artefacts, such as computers, machine tools and so on, constitute a class of essential enablers of the digital transformation [5, 41]. In the chapter, the presentation of collaborative decision-making was limited to only the processes consisting in the interactions among several human agents possibly supported be I&CT tools and systems.

At present, the human's work performance ever more depends on the I&CT tools and systems he /she uses to carry on his/her tasks. Many years ago, Licklider forecast a *man-machine symbiosis* meant to lead to high performances in solving hard decision problems [42]. In general, as noticed by Gerber et al. [43], in HMS (*human -machine systems*), where the nature of the various interacting agents is not the same, one might speak also about *mutualism,* a specific form of symbiosis, first introduced in the botany domain [44, 45], in which all parties involved benefit as partners.

At present, one may already notice the availability of *digital cognitive systems* [46, 47] which have evolved from simple information tools to *digital clones* of human advisers, consultants, and even mediators meant to augment human of intelligence, so that better capabilities and performances could be attained due to the collaboration between the humans and their digital decision-making 'partners' [47]. There are several companies that are already providing software products which deploy cognitive computing, such as: *Vantage Point AI* (in investment domain), *Welltok* (in healthcare), *Spark Cognition* (to support optimizing operations, preventing disasters and mitigating losses) [48].

At the same time, there are initiatives, research efforts, and reported results in AI domain aiming at making the artifacts ever more intelligent and even able to autonomously make decisions and act. In [49], it is noticed that "Some systems are self-teaching. They are able, partially in some cases and fully in others, to make decisions themselves without humans". Consequently, the result of the future developments in I&CT and their impact on human well-being and resilience are not too easy to predict [50]. Will the *Digital humanism* [51, 52] trends prevail and the future, possibly augmented, humans will take on board, in a recommended *service-dominant architecture*-SDA [53], the AI based artifacts as collaborators and additional participants in the decision-making activities and boards? The case of VITAL (*Validating Investment Tool for Advancing Life Sciences*) of *Aging Analytics*, a machine learning software that has been named member of the directory board of the Hong Kong's DKV (*Deep Knowledge Ventures*) company [49], has been, for several years, a practical example supporting a positive answer to the above question in spite of all debates and controversies. Or, will the people become mere *data feeders* of algorithms as in Harari's *Dataism* anticipations [54]? A rather realistic forecast for the nearest future has been articulated in [49] by D. Kaminskiy, managing partner of DKV, who does not think that "AI will fully replace people on boards of directors. Instead, it will probably be limited to augmenting human intelligence", and states that"the corporate winners will be so-called intelligent companies that combine'smart machines with smart people' using the latest AI technology to support management, but not to replace it".

Notes. This chapter is dedicated to the 30th anniversary of Purdue PRISM (*Production, Robotics and Integration Software for Manufacturing & Management*), a centre led by prof. S. Y. Nof. A preliminary and partial version of the chapter was published in April 2022 in the *International Journal of Computers Communications & Control*, vol 1, No2 [55].

References

1. Merriam-Webster Dictionary. Merriam-Webster.com. https://www.merriam-webster.com/dictionary/collaborate. Accessed 1 Jan 2022
2. Holy Bible English Standard Version-ESV® Text Edition: 2016. Copyright © 2001 by Crossway Bibles, a publishing ministry of Good News Publishers. https://www.biblegateway.com/passage/?search=ecclesiastes+4%3A9&version=ESV. Accessed 1 Jan 2022
3. Lewis, C.S.: On the reading of old books. First published as the introduction to R. P. Lawson's translation of The Incarnation of the Word of God, London: Bles, 1944). Reprinted in God in the Dock. 200–207 (1970)
4. Simon, H.A.: Two heads are better than one: the collaboration between AI and OR. Informs J. Appl. Anal. **17**(4), 8–15 (1987). https://doi.org/10.1287/inte.17.4.8

5. Nof, S.Y., Ceroni, J., Jeong, W., Moghaddam, M.: Revolutionizing Collaboration Through E-Work, E-Business, and E-Service. Springer (2015). https://doi.org/10.1007/978-3-662-45777-1
6. Camarinha-Matos, L.M., Afsarmanesh, H., Galeano, N., Molina, A.: Collaborative networked organizations – concepts and practice in manufacturing enterprises. Comput. Ind. Eng. **57**(1), 46–60 (2009). https://doi.org/10.1016/j.cie.2008.11.024
7. Filip, F.G., Zamfirescu, C.B., Ciurea, C.: Computer Supported Collaborative Decision-Making. Springer Cham (2017). https://doi.org/10.1007/978-3-319-47221-8
8. Nof, S.Y.: Collaborative control theory and decision support systems. Comput. Sci. J. Moldova **25**(2), 115–144 (2017)
9. Filip, F.G.: Control decision support and control for large-scale complex systems. Annu. Rev. Control. **32**(1), 62–70 (2008). https://doi.org/10.1016/j.arcontrol.2008.03.002
10. Schoop, M., Marc Kilgour, D. (eds.): Group Decision and Negotiation. A Socio-Technical Perspective: 17th International Conference, GDN 2017, Stuttgart, Germany, 14–18 Aug 2017, Proceedings. Springer International Publishing, Cham (2017)
11. Kapucu, N., Garayev, V.: Collaborative decision-making in emergency and disaster management. Int. J. Public Adm. **34**(6), 366–375 (2011). https://doi.org/10.1080/01900692.2011.561477
12. Nunamaker Jr. J.F., Romero Jr. N.C., Briggs R.O. (eds.): Collaboration Systems: Concept, Value and Use. Routledge, Taylor and Francis Group, London, (2015)
13. Simon, H.: The New Science of Management Decisions. Harper & Brothers, New York (1960). https://doi.org/10.1037/13978-000
14. Konaté, J., Gueye, A., Zaraté, P., Camillieri, G.: Collaborative decision-making: a proposal of an semi-automatic facilitation based on an ontology. Int. J. Inf. Technol. Decis. Mak. **22**, 447–470 (2022). https://doi.org/10.1142/S0219622022500420
15. Ciurea, C., Filip, F.G.: Collaborative platforms for crowdsourcing and consensus-based decisions in multi participant environments. Informatica Economică **23**(2), 5–10 (2019). https://doi.org/10.12948/issn14531305/23.2.2019.01
16. Herrera-Viedma, E., Herrera, F., Chiclana, F.: A consensus model for multiperson decision making with different preference structures. IEEE Trans. Syst., Man, Cybern.-Part A **32**(3), 394–402 (2002). https://doi.org/10.1109/TSMCA.2002.802821
17. Lincoln, A.: Annual Message to Congress. Concluding Remarks. In: Abraham Lincoln online Writings and Speeches. http://www.abrahamlincolnonline.org/lincoln/speeches/congress.htm. Accessed 1 Jan 2022
18. Dong, Y., et al.: Consensus reaching in social network group decision making: research paradigms and challenges. Knowl.-Based Syst. **162**, 3–13 (2018). https://doi.org/10.1016/j.knosys.2018.06.036
19. Kacprzyk, J., Zadrozny, S., H. Nurmi, H., Bozhenyuk, A.: Towards innovation focused fuzzy decision making by consensus. In: 2021 IEEE International Conference on Fuzzy Systems (FUZZ-IEEE), pp.1–6 (2021). https://doi.org/10.1109/FUZZ45933.2021.9494531
20. Ding, R.-X., Palomares, I., et al.: Large-scale decision-making: characterization, taxonomy, challenges and future directions from an artificial intelligence and applications perspective. Inform. Fusion **59**, 84–102 (2020)
21. Howe, J.: The rise of crowdsourcing. Wired Mag. **14**(6), 1–4 (2006). https://doi.org/10.1016/j.inffus.2020.01.006
22. Estelles Arolas, E., González-Ladrón De-Guevara. F.: Towards an integrated crowdsourcing definition. J. Inform. Sci. **32**(2), 189-200 (2012).https://doi.org/10.1177/0165551512437638
23. Yu, Z., Ma, H., Guo, B., Yang, Z.: Crowdsensing 2.0. Commun. ACM **64**(11), 76–80 (2021). https://doi.org/10.1145/3481621
24. Chiu, C.M., Liang, T.P., Turban, E.: What can crowdsourcing do for decision support? Decis. Support Syst. **65**, 40–49 (2014). https://doi.org/10.1016/j.dss.2014.05.010

25. Dodevska, Z.A., Kovacevic, A., Vukicevic, M., Delibašić, B.: Two sides of collective decision making - Votes from crowd and knowledge from experts. In: María Moreno-Jiménez, J., Linden, I., Dargam, F., Jayawickrama, U. (eds.) Decision Support Systems X: Cognitive Decision Support Systems and Technologies: 6th International Conference on Decision Support System Technology, ICDSST 2020, Zaragoza, Spain, May 27–29, 2020, Proceedings, pp. 3–14. Springer International Publishing, Cham (2020). https://doi.org/10.1007/978-3-030-46224-6_1
26. Zavadskas, E.K., Turskis, Z., Kildienė, S.: State of art surveys of overviews on MCDM/MADM methods. Technol. Econ. Dev. Econ. **20**(1), 165–179 (2014). https://doi.org/10.3846/20294913.2014.8920
27. Kou, G., Chao, X., Peng, Y., Xu, L., Chen, Y.: Intelligent collaborative support system for AHP-group decision making. Stud. Inform. Control **26**(2), 131–142 (2017). https://doi.org/10.24846/v26i2y201701
28. Filip, F.G.: DSS—a class of evolving information systems. In: Dzemyda, G., Bernatavičienė, J., Kacprzyk, J. (eds.) Data Science: New Issues, Challenges and Applications, pp. 253–277. Springer International Publishing, Cham (2020). https://doi.org/10.1007/978-3-030-39250-5_14
29. Shi, Y.: Advances in Big Data Analytics: Theory, Algorithms and Practices. Springer Nature, Singapore (2022). https://doi.org/10.1007/978-981-16-3607-3
30. Suduc, A.M., Bizoi, M., Filip, F.G.: Exploring multimedia web conferencing. Informatica Economică **13**(3), 5–17 (2009). http://citeseerx.ist.psu.edu/viewdoc/download?doi=10.1.1.463.4492&rep=rep1&type=pdf
31. Filip, F.G.: A decision-making perspective for designing and building information systems. Int. J. Comput. Commun. Control **7**(2), 264–272 (2012). http://www.univagora.ro/jour/index.php/ijccc/article/view/1408
32. Candea, C., Filip, F.G.: Towards intelligent collaborative decision support platforms. Stud. Inf. Control **25**(2), 143–152 (2016). https://sic.ici.ro/wp-content/uploads/2016/06/SIC_2016-2-Art1-1.pdf
33. Zamfirescu, C.B., Neghină, M.: Collaborative development of a CPS-based production system. Procedia Comput. Sci. **162**, 579–586 (2019). https://doi.org/10.1016/j.procs.2019.12.026
34. Neghina, M., Zamfirescu, C.B., Larsen, P.G., Lausdahl, K., Pierce, K.: Multi-paradigm discrete-event modelling and co-simulation of cyber-physical systems. Stud. Inform. Control **27**(1), 33–42 (2018). https://doi.org/10.24846/v27i1y201804
35. Neghina, M., Zamfirescu, C.B., Pierce, K.: Early-stage analysis of cyber-physical production systems through collaborative modelling. Softw. Syst. Model. **19**(3), 581–600 (2020). https://doi.org/10.1007/s10270-019-00753-w
36. Neghina, M., Zamfirescu, C.B., Larsen, P.G., Pierce, K.: Multi-paradigm modelling and co-simulation in prototyping a cyber-physical production system. In Tekinerdogan B., Blouin D., Vangheluwe H., Goulão M., Carreira P., Amaral V. (eds.) Multi-Paradigm Modelling Approaches for Cyber-Physical Systems, pp. 169–194. Academic Press (2020)
37. Filip, F.G., Ciurea, C., Dragomirescu, H., Ivan, I.: Cultural heritage and modern information and communication technologies. Technol. Econ. Dev. Econ. **21**(3), 441–459 (2015). https://doi.org/10.3846/20294913.2015.1025452
38. Ciurea, C., Filip, F.G., Zamfiroiu, A., Pocatilu, L.: Digital humanism: virtual exhibitions in the time of pandemic and evolving collaboration of relevant actant. In: Ciurea, C., Boja, C., Pocatilu, P., Doinea, M. (eds.) Education, Research and Business Technologies: Proceedings of 20th International Conference on Informatics in Economy (IE 2021), pp. 123–130. Springer Singapore, Singapore (2022). https://doi.org/10.1007/978-981-16-8866-9_11

39. Ciurea, C., Filip, F.G.: Virtual exhibitions in cultural institutions: useful applications of informatics in a knowledge-based society. Stud. Inform. Control **28**(1), 55–64 (2019). https://doi.org/10.24846/v28i1y201906
40. Ciurea, C., Filip, F.G.: Multi-criteria analysis in choosing IT&C platforms for creative digital works. Uncommon Cu. **6**(2), 20–27 (2015). https://uncommonculture.org/ojs/index.php/UC/article/view/6200
41. Camarinha-Matos, L.M., Fornasiero, R., Ramezani, J., Ferrada, F.: Collaborative networks: a pillar of digital transformation. Appl. Sci. **9**(24), 543 (2019). https://doi.org/10.3390/app9245431
42. Licklider, J.C.R.: Man-computer symbiosis. IRE Trans. Human Factors Electr. **HFE-1**(1), 4–11 (1960). https://doi.org/10.1109/THFE2.1960.4503259
43. Gerber, A., Derckx, P., Döppner, D.A., Schoder, D.: Conceptualization of the human-machine symbiosis a literature review. In: Proceedings of the 53rd Hawaii International Conference on System Sciences, pp. 289–298 (2020). https://hdl.handle.net/10125/63775
44. De Bary, A.: Die Erscheinung der Symbiose: Vortrag, Gehalten auf der Versammlung Deutscher Naturforscher und Aerzte zu Cassel (1879). https://onlinebooks.library.upenn.edu/webbin/book/lookupid?key=olbp70169
45. Pound, R.: Symbiosis and mutualism. The American Naturalist, Vol. 27, No. 318 (Jun., 1893), pp. 509–520. Faculty Publications in the Biological Sciences. 20. (1893). https://digitalcommons.unl.edu/bioscifacpub/20
46. Tecuci, G., Marcu, D., Boicu, M., Schum, D.A.: Knowledge Engineering: Building Cognitive Assistants for Evidence-based Reasoning, Cambridge University Press (2016). https://doi.org/10.1017/CBO9781316388464
47. Siddike, M.A.K., Spohrer, J. et al.: People's interactions with cognitive assistants for enhanced performances. In: Proceedings of the 51st Hawaii International Conference on System Sciences, pp. 640–1648 (2018). http://hdl.handle.net/10125/50092
48. Hager, P.: What is cognitive technology and how is cognitive computing and AI transforming industries. Gordon Flesc Company (2021). https://www.gflesch.com/elevity-it-blog/cognitive-technology
49. Burridge, N.: Artificial intelligence gets a seat in the boardroom. Hong Kong venture capitalist sees AI running Asian companies within 5 years. Nikei Asia, 10 (2017). https://asia.nikkei.com/Business/Artificial-intelligence-gets-a-seat-in-the-boardroom
50. Filip, F.G.: Automation and computers and their contribution to human well-being and resilience. Stud. Inform. Control **30**(4), 5–18 (2021). https://doi.org/10.24846/v30i4y202101
51. Nida Rümelin, J., Weidenfeld, N.: Digital Humanism. For a Humane Transformation of Democracy, Economy and Culture in the Digital Age. Springer Nature (2022). https://doi.org/10.1007/978-3-031-12482-2
52. Werthner, H., Prem, E., Lee, E.A., Ghezzi, C. (eds.): Perspectives on Digital Humanism. Springer, Cham (2022). https://doi.org/10.1007/978-3-030-86144-5
53. Spohrer, J., Maglio, P.P., Vargo, S.L., Varg, M. Service in the AI Era; Science, Logic, and Architecture Perspective, Business Experts Press LLS (2022). ISBN-13: 978-1-63742-304-2
54. Harari, Y.N.: Homo Deus: A Brief History of Tomorrow. Random House (2016)
55. Filip, F.G.: Collaborative decision-making: concepts and supporting information and communication technology tools and systems. Int. J. Comput. Commun. Control **17**(2), 4732 (2022). https://doi.org/10.15837/ijccc.2022.2.4732

Collaborative Requirement System Using Matrix and AI Approach

Tomohiro Nakada[1(✉)], Tetsuo Yamada[2], and Masayuki Matsui[2,3]

[1] Bunkyo Gakuin University, Tokyo, Japan
tnakada@bgu.ac.jp
[2] The University of Electro-Communications, Tokyo, Japan
[3] Kanagawa University, Kanagawa, Japan

Abstract. Diversification of consumer needs and shortening of product life cycles have led to the introduction of cell production methods using the Internet of Things (IoT) and industrial robots. Furthermore, the production system makes extensive use of humans, industrial robots, etc. to perform cooperative tasks. Therefore, future production planning requires a planning method that considers a collaborative requirement system that utilizes multiple industrial robots and humans. Collaborative requirement system output a number of results when various tools and components and their respective procedures are considered, making it difficult to compare the results. Therefore, in this research, we focused on methods using mathematical models such as matrix approach and AI approach. Mathematical modeling considers data from manufacturing sites, industrial robots and humans as artifacts and variables to the matrix approach and AI approach. The matrix approach is used in the field of various production systems, as it consists of rows and columns and shows the relationship between input and output as a matrix product. In addition, deep learning and neural networks are attracting attention in the field of artificial intelligence, and application techniques for production sites are being studied.

This paper presents a matrix approach and an AI-based approach for collaborative requirement system such as cell production systems, as well as case studies. As a result, the matrix approach and the AI approach expressed the changes in the elements numerically while considering the collaborative requirements system in their respective ways. Then, the matrix approach and the AI approach need to consider the case of adaptation where there is a collaborative system in social and economic systems.

Keywords: Matrix · AI · Collaborative Requirement System

1 Introductory CRS

1.1 Introduction

Companies are introducing cell production systems using Internet of Things (IoT), industrial robots, and other approaches owing to the diversification of consumer needs and a shortening of the product lifecycle [1]). The cell production system is one of the methods that installs industrial robots in a place called a "cell" and collects materials and

C.-Y. Huang and S. W. Yoon (Eds.): ICPR1 2021, ACES 14, pp. 107–121, 2023.
https://doi.org/10.1007/978-3-031-44373-2_6

parts using a conveyor to produce and process products. In addition, because multiple industrial robots and parts are procured at daily production sites, there is a risk of industrial robot failure. Therefore, in the production management of a company, a model is required to recalculate the cost and production period by adding the daily circumstances as a precondition for manufacturing.

In recent years, a matrix-based approach [2]) has been proposed as a modeling method that focuses on input and output data by regarding the data at the manufacturing site as an artificial body. The matrix-based approach is composed of rows and columns and is used in various fields [3–5]) because it shows the relationship between in-puts and outputs. Among them, Suzuki et al. [4] proposed a matrix-based approach that takes into consideration the operating rate of the manufacturing site and work in-process, and then examined the cost evaluation method for an assembly line. Nakada [5] proposed a matrix-based approach using cooperation requirement planning (CRP) and scheduling in industrial robots and conducted numerical comparisons.

In addition, deep learning and neural networks in the field of artificial intelligence are attracting attention, and applied technologies at production sites are being studied [6]. Okabe et al. [6] proposed a neural network model that uses the number of products of various shapes as input data and outputs the selection of multiple selectable packaging boxes in the product packaging.

This paper introduces the matrix approach and the AI approach method for collaborative requirement systems such as cell production systems. Section 2 outlines matrix approach to Nof's cooperation requirement planning. Section 3 outlines AI approach to Nof's cooperation requirement planning. Section 4 summarizes the current status and potential of the matrix and AI approach methods in collaborative requirements systems.

1.2 NOF's CRP

The concept of Previous paper is a cell production system using industrial robots [7–10]. This paper considers a planning method in which multiple industrial robots are placed around a central worktable. A task can use one industrial robot. Furthermore, the task can also use multiple industrial robots. The resource defines a combination of a single industrial robot and multiple industrial robots. And, previous research proposed a planning method based on industrial robots, resources, and tasks as a cooperation requirement planning (Hereafter referred to as CRM).

In addition, it is necessary to review the task plan for maintenance and breakdown of industrial robots. The multiple industrial robots can replace the role of other industrial robots. This paper examines a planning method based on industrial robot resources and tasks in the cell production process for multiple industrial robots. Nof proposed a cooperation requirement planning (CRP) based on multiple industrial robot tasks and available resources. This CRP consists of two parts (CRP-I and CRP-II) as shown in Fig. 1. The CRP-I defines single and multiple combinations using multiple industrial robots as resources. The CRP-II process selects a plan to meet each goal and resolve the interactions between them.

Nakada [5] applied the CRP [7, 8] in the production process proposed by Nof et al. using the matrix approach based on Matsui's equation, and expressed CRP-I and CRP-II in Fig. 1 as Fig. 2.

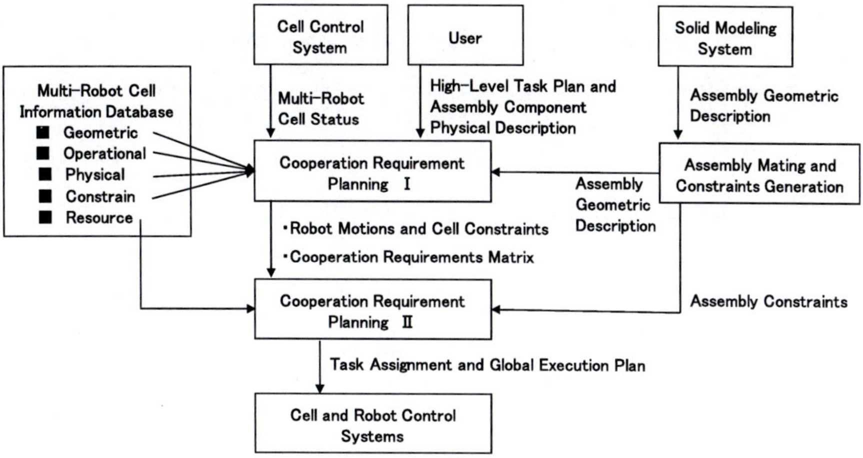

Fig. 1. Cooperation requirement planning system architecture developed by Nof et al. (1993).

r1={ROB1}
r2={ROB2}
r3={ROB3}
r4={ROB1, ROB2}
r5={ROB1, ROB3}
r6={ROB2, ROB3}
r7={ROB1, ROB2, ROB3}

	t1	t2	t3
r1	1	0	0
r2	1	0	0
r3	1	0	0
r4	1	1	0
r5	1	1	0
r6	1	1	0
r7	1	1	1

Fig. 2. Resource–robot relationship diagram (CRP-I, left) and resource–task relationship diagram (CRP-II, right). [5]

1.3 Cooperation Requirement Planning Using Matrix

Matsui's equation (matrix equation) is derived from the product × company decision-making mechanism and shows the formulation of the input/output relationship of the object (artificial body) required by the company [11]. As shown in Fig. 3, there are two matrix approaches: the structural matrix method (series-parallel system) and Matsui's equation (series method). The input/output relationship of the artificial body consists of the logical structure of the Introduction (I), Development (D), Transformation (T), and Conclusion (C) plus two matrices of Balance (B) and Goal (G). Expressing these in a diagram, in the case of an n × n matrix, as shown in Fig. 3 (a), the standard form is a table (series-parallel system), whereas in Fig. 3 (b), the standard is a compact form (series system). Figure 3 shows the status of the pair series in the form of I → D → T → C → B → G. This inverse problem is in the form of G → B → C → T → D → I. Therefore, in Matsui's equation, it is expressed through two methods, i.e., an objective equation and a proper form, as follows [11].

$$\text{Objective form}: \quad g = y^T D^T T^T C^T B^T \quad \text{TypeI} \tag{1}$$

$$\text{Unique equation}: y^T D^T T^T C^{T g} = \lambda g \quad \text{TypeII} \tag{2}$$

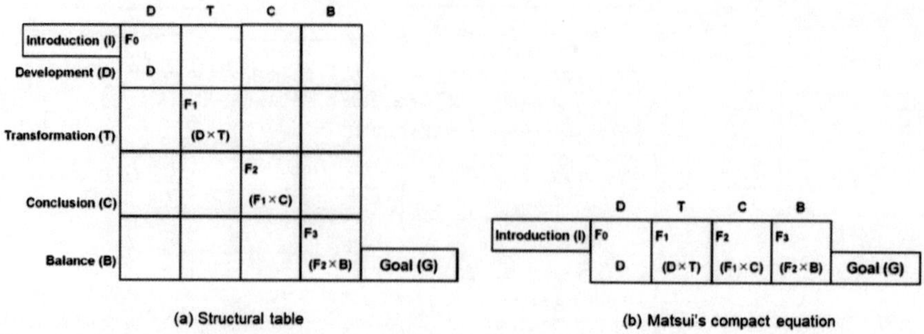

Fig. 3. Two types of forms: (a) structural table and (b) Matsui's compact matrix equation. [11]

The method for determining the mean flow time < FI (W) > and makespan < FII (L) > is as follows [12]. The mean flow time (make span) < FII (L) > of the OE type is related to Matsui's equation (W = ZL).

$$\mathrm{FI} = \Sigma\, \mathrm{Zi\, Li} = \Sigma(\mathrm{n} - \mathrm{i} + 1)\, \mathrm{xi} \tag{3}$$

$$\mathrm{FII} = (\Sigma \mathrm{Li})/\lambda = (\Sigma \mathrm{ci})/\lambda \tag{4}$$

$$\mathrm{xi} = \Sigma \mathrm{xij} \quad \mathrm{i} = 1, 2, 2, \ldots, \mathrm{n}. \tag{5}$$

2 Formulation Using Matrix Approach

2.1 Collaborative Scheduling Formulation

Scheduling in production control is classified into single-machine scheduling for a single machine, flow shop scheduling for multiple machines, job shop scheduling and so on, and methods of ordering using priority rules are used to improve the efficiency and reduce the burden of the manufacturing process. In order to improve the efficiency of the manufacturing process and to reduce the burden of the manufacturing process, methods of ordering using priority rules are studied [13]. The method of sequencing using priority rules is classified into SPT rule (Shortest Processing Time) which is implemented from the work with the shortest required time, LPT rule (Longest Processing Time) which is implemented from the work with the longest required time, and EDD rule (Earliest Due Date) which is implemented from the work with the earliest due date. There are such rules as EDD rules (Earliest Due Date), and we are considering the effective rules depending on the subject of scheduling [14, 15].

Formulation of Push-Type

In this paper, Eq. (6) based on Matsui's equation of push-type is given in order to examine the scheduling of a production plant [16, 17], which is one of the artificial bodies, as an example of AI matrix method. The push-type in a production control system is a method that examines the production schedule based on the production plan formulated by the demand forecast and sales forecast [18, 19]. The method to examine the production schedule based on the market demand such as the shipment request from customers and sales. This paper considers push-type as forward order and pull-type as reverse order and performs numerical calculations.

In this example, two machines (M_1 and M_2) are placed at the "introduction." The processing time (day) in Table 1 is converted to match the order of Eq. (6)'s "D × T" and converted to Eq. (6) at the matrix product's "Development" and "Transformation." Furthermore, the "conclusion" is a conversion from time to price, the left side of "B" represents the average flow time (L), the right side of "B" represents the cycle time (Z), and the "goal" is the demand price [17]. The sorting of the machines (M_1) in descending order is based on the scheduling theory's longest processing time (LPT).

$$\overset{\text{I}}{(M_1\ M_2)}\ \overset{\text{D}}{\begin{pmatrix} \overset{1}{3} & \overset{2}{1} & \overset{3}{8} & \overset{4}{4} & \overset{5}{11} \\ 2 & 6 & 7 & 6 & 4 \end{pmatrix}}\ \overset{\text{T}}{\begin{pmatrix} \overset{5}{} & \overset{3}{} & \overset{4}{} & \overset{1}{1} & \overset{2}{} \\ & & & & 1 \\ & 1 & & & \\ & & 1 & & \\ 1 & & & & \end{pmatrix}}\ \overset{\text{C}}{\begin{pmatrix} 1 & 1 & 1 & 1 & 1 \\ 1 & 1 & 1 & 1 & 1 \end{pmatrix}}\ \overset{\text{B}}{\begin{pmatrix} \overset{F_{I}}{1/5} & \overset{F_{II}}{5} \\ 1/5 & 4 \\ 1/5 & 3 \\ 1/5 & 2 \\ 1/5 & 1 \end{pmatrix}} = \overset{\text{G}}{\underset{(Z=5)}{(182/5\quad 182)}} \tag{6}$$

$$\text{"D} \times \text{T"} = \begin{pmatrix} \overset{M_1}{11} & \overset{M_2}{4} \\ 8 & 7 \\ 4 & 6 \\ 3 & 2 \\ 1 & 6 \end{pmatrix}.$$

Table 1. A scheduling problem of two machines. [16]

Job number	Work time (day)	
	Machine M_1	Machine M_2
1	3	2
2	1	6
3	8	7
4	4	6
5	11	4

Formulation of Pull-Type

In Eq. (6), Fig. 1 was assumed to be a push-type (forward order), but if it is assumed to be a pull-type (reverse order), Eq. (7) can be obtained [17]. In this Eq. (7), two machines (M_1 and M_2) are placed at the "introduction," and the processing time (day) in Table 1 is converted into Eq. (7) by the matrix product of "Development" and "Transformation" in the same order as "D × T" in Eq. (7) [17]. Furthermore, the "conclusion" can be converted from time to price, the left side of "B" represents the average flow time (L), the right side of "B" represents the cycle time (Z), and the "goal" is the demand price. Table 1 shows the processing time (day) of each machine, and the data is used to sort in descending order of machine (M_1) is Eq. (6)'s "D × T," and the data is sorted in ascending order of machine (M_1) is Eq. (7)'s "D × T" [17].

$$\overset{I}{(M_1\ M_2)}\ \overset{D}{\overset{\begin{matrix}1&2&3&4&5\end{matrix}}{\begin{pmatrix}3&1&8&4&11\\2&6&7&6&4\end{pmatrix}}}\ \overset{T}{\overset{\begin{matrix}2&1&4&3&5\end{matrix}}{\begin{pmatrix}&1&&&\\1&&&&\\&&&1&\\&&1&&\\&&&&1\end{pmatrix}}}\ \overset{C}{\begin{pmatrix}1&1&1&1&1\\1&1&1&1&1\end{pmatrix}}\ \overset{B}{\overset{\begin{matrix}F_{I}&F_{II}\end{matrix}}{\begin{pmatrix}1/5&5\\1/5&4\\1/5&3\\1/5&2\\1/5&1\end{pmatrix}}} = \overset{G}{\underset{(Z=5)}{(130/5\quad 130)}} \tag{7}$$

$$\text{"D × T"} = \overset{\begin{matrix}M_1&M_2\end{matrix}}{\begin{pmatrix}1&6\\3&2\\4&6\\8&7\\11&4\end{pmatrix}}$$

Comparison of Numerical Results Between Push-Type and Pull-Type

In this paper, we substituted the data in Table 1 into the AI matrix method using Eq. (6) in the descending order of "D × T" as the push-type and Eq. (7) in the ascending order of "D × T" as the pull-type, and the calculation results of the push-type and pull-type are shown in Table 2. The push-type could be understood with the LPT rule (Longest Processing Time) of the schedule in production management and the pull-type with the SPT rule (Shortest Processing Time) [16].

Table 2. Numerical results for push-type and pull-type

	push-type	pull-type
mean flow time	182/5	130/5
Makespan	182	130

The push-type makespan with the LPT rule was 182. The makespan of the push-type with the SPT rule was 130. Sobazek et al. [15] also showed a maximum for the LPT rule and a minimum for the SPT rule in the job shop. Thus, the AI matrix method showed the maximum and minimum value of makespan by using descending order (LPT rule) and ascending order (SPT rule).

2.2 Collaborative Robot Scheduling

A mathematical model based on Matsui's matrix equation for CRP among industrial robots to plan the coordinated operation of multiple industrial robots is proposed herein, as shown in Fig. 4 [17]. The equation shown in this figure expresses the resources and scheduling of multiple industrial robots and yields the maximum residence time (make-span) and the mean flow time for a particular job.

The constraints on this equation are as follows:

- Multiple industrial robots are used.
- Each job is processed according to the processing procedure.
- Each resource is either a single task performed by an industrial robot or is a cooperative requirement task that uses multiple industrial robots.
- The time for each processing step is the same, and each step has the same start and end times.
- There is no idle time in any processing step.
- Industrial robots process all work without any interruption.

Introduction

$$\begin{matrix} ROB1 & ROB2 & ROB3 \\ \end{matrix}$$
$$\begin{pmatrix} 1 & 1 & 1 \end{pmatrix} \times$$

Development Transformation Conclusion Balancing Goal

$$\begin{matrix} & r_1 & r_2 & r_3 & r_4 & r_5 & r_6 & r_7 \\ 1 \\ 2 \\ 3 \end{matrix}
\begin{pmatrix} 1 & & & 1 & 1 & & 1 \\ & 1 & & 1 & & 1 & 1 \\ & & 1 & & 1 & 1 & 1 \end{pmatrix}
\overset{t_1 \; t_2 \; t_3}{\begin{pmatrix} 1 & & \\ 1 & & \\ 1 & & \\ 1 & 1 & \\ 1 & 1 & \\ 1 & 1 & \\ 1 & 1 & 1 \end{pmatrix}}
\overset{c_1 \; c_2 \; c_3 \; c_4 \; c_5 \; c_6 \; c_7}{\begin{pmatrix} 1 & 1 & 1 & 1 & 1 & 1 & 1 \\ 1 & 1 & 1 & 1 & 1 & 1 & 1 \\ 1 & 1 & 1 & 1 & 1 & 1 & 1 \end{pmatrix}}
\overset{\bar{L} \quad \bar{Z}}{\begin{pmatrix} 1/7 & 7 \\ 1/7 & 6 \\ 1/7 & 5 \\ 1/7 & 4 \\ 1/7 & 3 \\ 1/7 & 2 \\ 1/7 & 1 \end{pmatrix}}
= \begin{pmatrix} F_{II} & F_{I} \end{pmatrix}$$

Fig. 4. CRP using the matrix approach and scheduling. [17]

This section focuses on CRP [4, 5] and describes the detailed procedure leading up to the construction of the equation shown in Fig. 4 [17]. The operation plans for CRP-I and CRP-II are expressed using Matsui's equation, as shown in Fig. 2. The left side of Fig. 2 illustrates CRP-I, where ROB indicates an industrial robot and "r" indicates a resource. For example, "r1" represents a single operation because only ROB1 is defined. Conversely, "r4" is defined as a combination of ROB1 and ROB2 and is a cooperative request. The right side of Fig. 2 illustrates CRP-II, which represents the relationship between tasks ("t1" to "t3") and resources ("r1" to "r7") using a cooperative requirement matrix. Herein, "t" represents the processing time rather than a given task. This matrix represents the use of a resource with the entry 1 and no use of that resource with 0. Thus, Nof's cooperative request approach is represented as a planning method comprising the two stages CRP-I and CRP-II.

The Introduction in Fig. 4 [17], which is based on Matsui's equation, represents industrial robots. The requirement for cooperation among multiple industrial robots is determined by the product of Introduction (I) and Development (D). The relationship between resources and tasks is determined by the product of Development (D) and Transformation (T). This matrix approach is illustrated by the examples discussed below.

In addition, Conclusion (C) is a matrix that transforms resources into time units to introduce the scheduling theory. Finally, balancing (B) is a matrix that represents the mean flow time and priority scheduling (the LPT rule) for each process illustrated in Fig. 4 [17]. The equation represented in Fig. 4 [17] can be used to calculate the make-span (F_I) and mean flow time (F_{II}) for multiple industrial robots and tasks.

2.3 Comparison of Calculation Results

This section performs calculations using the proposed mathematical model and compares various numerical results. First, Eq. (8) is used to calculate the numerical value for an application example of CRP [5]. The mean flow time and make-span result obtained by matrix multiplication are 24 and 672, respectively, which are as "standard values." Furthermore, the production plans for two cases resulting for CRP from for industrial robots based on these two values are examined herein.

Case 1 assumes that goods are manufactured by changing the order of resources and tasks. In other words, it is a modification of the Development and Transformation components of the mathematical model. Equation (9) below shows the results calculated with the beginnings and ends of the task changed [5]. The combination of resources is changed based on the order of the tasks. The mean flow time and make-span result obtained by matrix multiplication are 24 and 672, respectively. As the results of calculations using Eqs. (8) and (9) both yield the same value, changing the tasks and resources does not cause any outcome [5].

Case 2 assumes that ROB2 is broken. In other words, it is a modification of the Introduction factor of the mathematical model (the second entry from the left in that factor). Equation (10) shows the results of a calculation that assumes ROB2 $= 0$ [5]. The mean flow time and make-span obtained by matrix multiplication are now 16 and 448, respectively, which are lower than the standard values.

$$\begin{bmatrix}1\,1\,1\end{bmatrix}\begin{bmatrix}1\,0\,0\,1\,1\,0\,1\\0\,1\,0\,1\,0\,1\,1\\0\,0\,1\,0\,1\,1\,1\end{bmatrix}\begin{bmatrix}1\,0\,0\\1\,0\,0\\1\,0\,0\\1\,1\,0\\1\,1\,0\\1\,1\,0\\1\,1\,1\end{bmatrix}\begin{bmatrix}1\,1\,1\,1\,1\,1\,1\\1\,1\,1\,1\,1\,1\,1\\1\,1\,1\,1\,1\,1\,1\end{bmatrix}\begin{bmatrix}\frac{1}{7}&7\\\frac{1}{7}&6\\\frac{1}{7}&5\\\frac{1}{7}&4\\\frac{1}{7}&3\\\frac{1}{7}&2\\\frac{1}{7}&1\end{bmatrix}=\begin{pmatrix}24\;672\end{pmatrix} \tag{8}$$

$$\begin{bmatrix}1\,1\,1\end{bmatrix}\begin{bmatrix}1\,0\,0\,1\,1\,0\,1\\1\,1\,0\,1\,0\,1\,0\\1\,0\,1\,0\,1\,1\,0\end{bmatrix}\begin{bmatrix}1\,1\,1\\1\,0\,0\\1\,0\,0\\1\,1\,0\\1\,1\,0\\1\,1\,0\\1\,0\,0\end{bmatrix}\begin{bmatrix}1\,1\,1\,1\,1\,1\,1\\1\,1\,1\,1\,1\,1\,1\\1\,1\,1\,1\,1\,1\,1\end{bmatrix}\begin{bmatrix}\frac{1}{7}&7\\\frac{1}{7}&6\\\frac{1}{7}&5\\\frac{1}{7}&4\\\frac{1}{7}&3\\\frac{1}{7}&2\\\frac{1}{7}&1\end{bmatrix}=\begin{pmatrix}24\;672\end{pmatrix} \tag{9}$$

$$
\begin{bmatrix} 1\,0\,1 \end{bmatrix}
\begin{bmatrix} 1\,0\,0\,1\,1\,0\,1 \\ 0\,1\,0\,1\,0\,1\,1 \\ 0\,0\,1\,0\,1\,1\,1 \end{bmatrix}
\begin{bmatrix} 1\,0\,0 \\ 1\,0\,0 \\ 1\,0\,0 \\ 1\,1\,0 \\ 1\,1\,0 \\ 1\,1\,0 \\ 1\,1\,1 \end{bmatrix}
\begin{bmatrix} 1\,1\,1\,1\,1\,1\,1 \\ 1\,1\,1\,1\,1\,1\,1 \\ 1\,1\,1\,1\,1\,1\,1 \end{bmatrix}
\begin{bmatrix} \frac{1}{7}\;7 \\ \frac{1}{7}\;6 \\ \frac{1}{7}\;5 \\ \frac{1}{7}\;4 \\ \frac{1}{7}\;3 \\ \frac{1}{7}\;2 \\ \frac{1}{7}\;1 \end{bmatrix}
= \left(16\;448\right) \tag{10}
$$

3 AI Approach to CRS

3.1 CSPSystem Model

In this paper, we take the conveyor-served production system (CSPSystem Model) as one of the knowledge societies to study the AI matrix method in the knowledge society. The conveyor-serviced production system (CSPSystem Model) consists of a line type (serial type) and an ordered-entry type (OE: ordered-entry type), as shown in Fig. 5. OE (ordered-entry) type is a method in which materials and parts are lined up in series along a conveyor and assembled at a production station [20]. Even in the case of multiple production stations, the materials and parts are assembled in the order in which they are lined up on the conveyor. Therefore, the line type and the OE type (CSPSystem Model II) can be regarded as the dual system of the assembly production system using a conveyor [21, 22]. The OE type of this assembly production system by the conveyor and the hierarchical neural network of the artificial intelligence system [23] have affinity in that they can be divided into the input, intermediate, and output layers. Moreover, multiple inputs and outputs can be set.

The hierarchical neural network of the artificial intelligence system [23, 24] is divided into several layers (e.g., input layer, middle layer, output layer) as shown in Fig. 6, and it is a mechanism that receives data from the previous layer, assigns the weight of combination to each data, combines them, and sends the data to the next layer.

Matsui's equation type IDTC-BG system is similar to the characteristics of the layered neural network. As materials and parts are being assembled on a conveyor at a production station, as shown in Fig. 5, information data is transferred from the input layer (machine or line) to the intermediate layer (scheduling) and from the intermediate layer (scheduling) to the output layer (back-propagation), and through balancing, the goal is reached. Therefore, it can be used to plan and analyze production control by transferring and calculating the information data of the conveyor assembly production system.

On the other hand, in the corporate management in the knowledge society, the environment keeps changing with time, and "time" has become important in addition to 3M&I (i.e., Human, Material/Machine, Money, and Information). In this paper, we assume that [25] the horizontal axis of Fig. 5 is the cycle time (Z), and derive the total flow time of the line type (FI(W)) and the average flow time of the OE type (makespan) (FII (L)). Relationship.

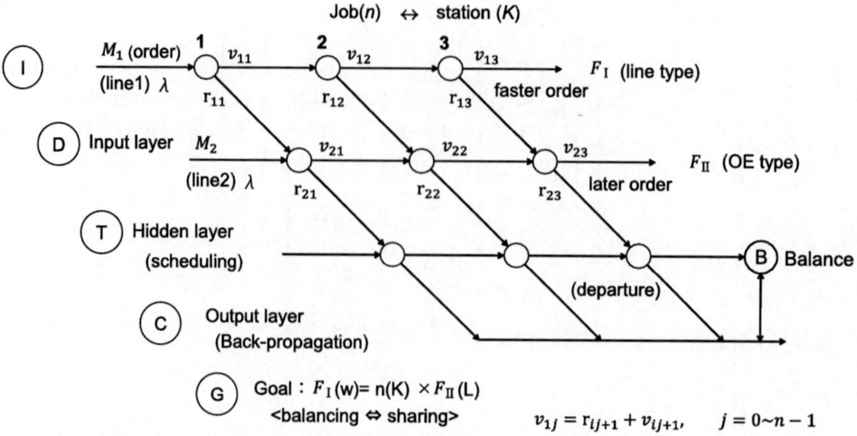

$$v_{1j} = r_{ij+1} + v_{ij+1}, \quad j = 0\sim n-1$$
$$r_{1j} = r_{i+1j} + v_{i+1j}, \quad j = 0\sim n-1$$

Fig. 5. Image overview of AI matrix method vs. conveyor system. [17]

3.2 Neural Network Type

In the matrix approach based on Matsui's equation proposed by Nakada [5], an industrial robot is expressed as a simple substance in the Introduction in Fig. 4, and the cooperative requirement of multiple industrial robots is expressed by the product of the Introduction and Development shown in the figure. Note that 1 in the matrix indicates a utilization, and 0 indicates a non-utilization. The relationship between resources and tasks (processing time) is expressed based on the product of the Development and Transaction shown in Fig. 4. The Development shows the cooperative requirement of an industrial robot as in the previous study [5] and can be expressed when a single industrial robot is used or when multiple industrial robots are used.

Furthermore, $\overline{L}$ and $\overline{Z}$ of Balancing applied Matsui's theory of shared balancing of equations [26], and in this case, there are seven resources, which express the processing time and temporal priority of each resource as a whole. Therefore, because $\overline{L}$ is expressed as 1/7 and resources 1 through 7 are used, it can be regarded as the reverse order of that shown in Fig. 4.

Goal uses F_{II} as the maximum residence time (makespan) and $\overline{F}$ as the average residence time. In addition, the maximum residence time (makespan) of F_{II} is consistent with $W(F_I) = \overline{L}(F_{II}) \times \overline{Z}$ of Matsui's theory of shared balancing. Nakada [5] express Fig. 4 using the matrix-based approach through Matsui's equation, with reference to the examples (Fig. 2) used in the CRP proposed by Nof et al.

Neural Network

Neural networks were devised with reference to nerve cells (neurons) in the brain and are information processing models based on an information transmission between neurons [27]. In the neuron model, as shown in Fig. 6, multiple inputs (x_1, x_2, ..., x_n) are multiplied by the weights (w_1, w_2, ..., w_n) and the sum is calculated. Furthermore, the sum is input into the sigmoid function and used as the output (y) of the neuron model.

These procedures can be expressed through the following Eqs. (11) and (12) [26].

$$\mu = \sum_i x_i w_i \tag{11}$$

$$y = \int (\mu) \tag{12}$$

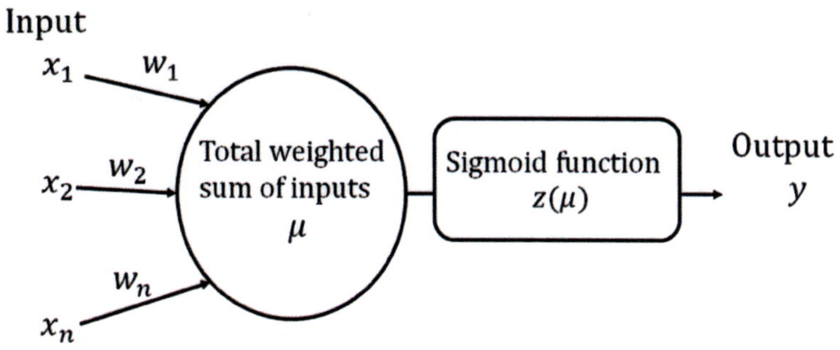

Fig. 6. Neuron model. [26]

A neural network involves the combination and networking of multiple neural models, as shown in Fig. 7 [28]. Neural networks are applied in various fields such as image processing and box selection in the packaging process [6]. Although there are various methods for constructing a neural network, the form in which multiple layers are layered is called a hierarchical neural network. Hierarchical neural networks are divided into multiple layers (e.g., input layer, hidden layer, and output layer) and is a mechanism that receives data from the previous layer, assigns a join weight to all data, joins the data, and sends them to the next layer.

Matrix Approach and Neural Networks

The matrix approach (IDTC-BG system of Matsui's equation) states that while the materials and parts are being assembled on the conveyor at the production station, information data are moved from the input layer (machine or line) to the intermediate layer (scheduling) and from the intermediate layer (scheduling) to the output layer (back propagation), and the goal is reached through the Balance [20]. Therefore, by decomposing Fig. 4, the matrix-based approach can be expressed through the following matrix product in Eq. (13).

$$|AB|_{i,j} = \sum_{m=0}^{M-1} a_{i,m} b_{m,j} \tag{13}$$

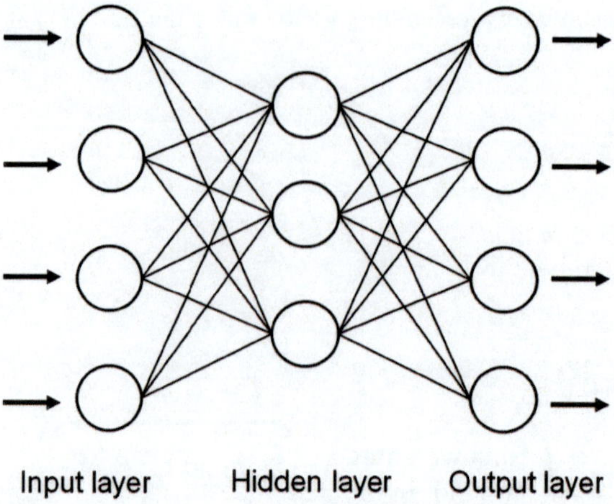

Fig. 7. Layered neural network [28].

The matrix-based approach and neural network are procedures similar to Eqs. (11) and (13) that multiply the input variables by each variable and then add them up. The matrix approach and the neural network also have similar expressions. For example, in a matrix-based approach[5] used in industrial robot cooperation requirement planning, each variable indicates a resource or task, but is defined as a "weight" in a neural network.

Therefore, in this paper, a model is proposed by which the matrix-based approach and a neural network are combined, as shown in Fig. 8. Because Fig. 8 corresponds to Introduction in Fig. 4, the first input data are from a single industrial robot. Next, Development, Transformation, Conclusion, and Balancing are assigned to the part that was set as "weight" in Fig. 6. Goal uses F_{II} as the maximum residence time (makespan) and $\overline{F}$ as the average residence time.

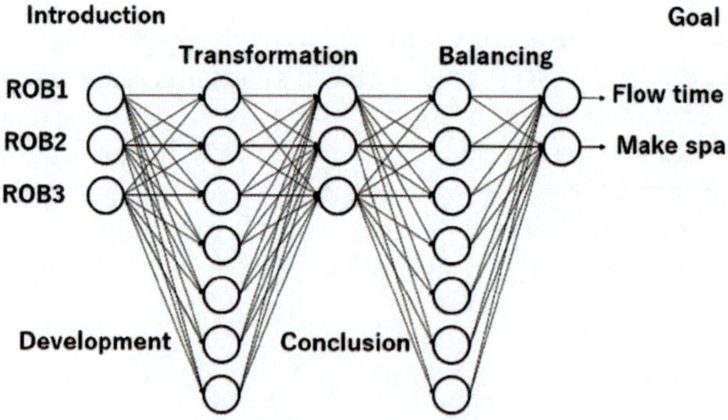

Fig. 8. Matrix and neuron model.

3.3 Case Study

In this paper, a numerical comparison of the matrix-based approach and a neural network were conducted based on CRP in an industrial robot. For the CRP in an industrial robot using the matrix-based approach, the average residence time was calculated as 24 and the maximum residence time (makespan) was calculated as 672 based on the calculation result of the matrix approach in Fig. 4 [5].

The CRP of the industrial robot using neural network (Fig. 9) was calculated by inputting the data ((Development, Transformation, Conclusion, and Balancing) shown in Fig. 4 into the neural network shown in Fig. 8. The neurons in the second layer of Fig. 9 showed values of 0.7, 0.9, and 1.0 in the results of the development of the Introduction in Fig. 4. The neurons in the third layer of Fig. 9 showed values of 1.0, 0.9, and 0.7 as a result of the transformation of Fig. 4. Similarly, when the neurons in the fourth and fifth layers were calculated for Conclusion and Balancing based on the calculation results thus far, the average residence time was 0.7 and the large residence time (makespan) was 1.0. When a neural network is used, because a sigmoid function is applied, it can be expressed up to a value of 1.

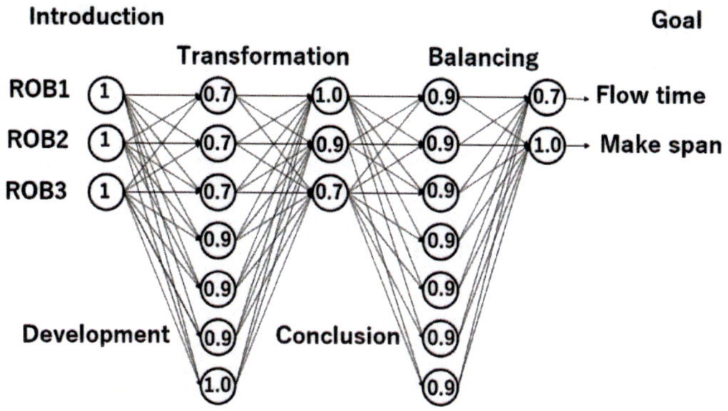

Fig. 9. Matrix and neural network case.

4 Concluding Remarks

This paper presents a formalization and examples of matrix and AI approaches in collaborative requirement system such as cell production systems. A collaborative requirement system is necessary to consider various combinations such as human and industrial robot, industrial robot and industrial robot, etc., in the context of globalization and changing production processes. The proposed matrix and AI approach is one way of cell production planning, considering a cooperative demand system with humans, industrial robots, etc.

The matrix and AI approaches presented in this paper use a variety of input data such as industrial robots and tasks. Moreover, in this paper, because the "weight" of the conventional neural network is changed into the resource and the processing time, all the matrices have meaning. Therefore, the result of each matrix product is meaningful, and the visualization from the input layer to the output layer. The difference between the matrix approach and the neural network is expressed from zero to 1 based on the influence of the sigmoid function after the matrix multiplication of the neural network.

In addition, various robots and artificial body systems will be introduced in future social systems, and a planning method that considers the collaborative requirement system is necessary. This method is expected to be used in social systems that use such cooperative demand systems. In the future, we would like to enhance the potential of the matrix and AI approach by examining examples of various social systems.

References

1. Shariatzadeh, N., Lundholm, T., Lindberg, L., Sivard, G.: Integration of digital factory with smart factory based on Internet of Things. Procedia CIRP **50**, 512–517 (2016)
2. Matsui, M.: Product × enterprise strategy: A matrix approach to enterprise systems for higher sustainability management. In: Proc. of APIEMS (2013)
3. Ishii, N., Ohba, M., Fujikawa H.: Evaluation of supply chain by matrix approach. In: Proceedings of the Conference of Transdisciplinary Federation of Science and Technology, pp. 1–4 (2017). (in Japanese)
4. Suzuki, Y., Yamada, T.: Evaluation of assembly line cost by matrix modeling considering busy rates and work-in-process. J. Soc. Plant Eng. Japan **33**(1), 1–13 (2021)
5. Nakada, T.: Matrix approach and scheduling for cooperation requirement planning in industrial robots. In: The International Conference on Production Research (26th ICPR 2021), pp. 1–6 (2021)
6. Okabe, D., Koide, Y., Onda, H., Kato, N., Akaho, S., Asoh, H.: Neural network with embedding layer for assistance of selecting boxes in packing task. In: The 34th Annual Conference of the Japanese Society for Artificial Intelligence, pp. 1–4 (2020). (in Japanese)
7. Nof, S.Y., Rajan, V.N., Lenz, E.: Automatic generation of assembly constraints and cooperation task planning. CIRP Ann. **42**(1), 13–16 (1993)
8. Rajan, V.N., Nof, S.Y.: Cooperation requirements planning (CRP) for multiprocessors: Optimal assignment and execution planning. J. Intell. Robo. Sys. **15**, 419–435 (1996)
9. Nof, S.Y., Ceroni, J., Jeong, W., Moghaddam, M.: Revolutionizing Collaboration throught e-Work, e-Business, and e-Service, in the Automation, Collaboration, and E-service. Springer, ACES) Book Series (2015)
10. Zhong, H., Nof, S.Y., Berman, S.: Asynchronous cooperation requirement planning with reconfigurable end-effectors. Robo. Comp.-Integr. Manuf. **34**, 95–104 (2015)
11. Matsui, M.: An enterprise-aided theory and logic for real-time management. IJPR Journal **51**(23–24), 7308–7312 (2013)
12. Matsui, M.: Fundamentals and Principles of Artifacts Science: 3M&I-Body Systems. Springer Briefs in Business, Springer (2016)
13. Hino, R.: Mathematical optimization models for job-shop scheduling problem. Systems Sci. **60**(1), 14–19 (2017). (in Japanese)
14. Tamaki, H., Ochi, M., Araki., M.: Introduction of a state feedback structure for adjusting priority rules in production scheduling. Trans. Soc. Instru. Cont. Eng. **35**(3), 428–434 (1999). (in Japanese)

15. Sobaszek, L., Gola, A., Kozlowski, E., U.: Effect of machine failure prediction on selected parameters of manufacturing schedule in a job-shop environment. Econ. Manage. Innov. **1**(1), 320–324 (2017)
16. Hitomi, K.: Introduction to Production Systems Engineering [6th Edition] -Pathway to Integrated Production Studies, pp. 133–134. Kyoritsu Publishing Co., Ltd. (2019). (in Japanese)
17. Nakada, T., Yamada, T., Matsui, M.: Proposal for an explainable AI matrix method in the knowledge society. J. Soc. Plant Eng. Japan **35**(1), 33–39 (2023). (in Japanese)
18. Takenouchi, T.: Outline of TOC (Theory of Constraint). Oper. Res. Soc. Japan **44**(6), 285–291 (1999). (in Japanese)
19. Serizawa, T., Karasawa, K., Muramatsu, K.: Push pull mixed type of optimal compound production planning method. Production Management **13**(1), 19–31 (2006). (in Japanese)
20. Feyzbakhsh, S.A., Kimura, A., Matsui, M.: Performance evaluation of a simple flexible assembly system and its 2-stage design. J. Japan Indus. Manage. Asso. **49**(6), 393–401 (1999). (in Japanese)
21. Matsui, M.: A Study on Optimum Work Strategies in Conveyor Production Systems, Doctoral Thesis. Tokyo Institute of Technology (1981). (in Japanese)
22. Matsui, M.: Conveyor-like networks and balancing. In: Savarese, A.B. (ed.) Manufacturing Engineering, pp. 65–87. Nova Science Publishers, Inc. (2011)
23. Aso, H.: Deep representation learning by multi-layer neural networks. J. Japanese Soc. Artif. Intell. **28**(4), 649–659 (2013). (in Japanese)
24. Akiyama, K., Kang, I., Kanno, T., Uchida, N.: Prediction of Residual Mg Contents in Ladle and Product after Graphite Spheroidizing Treatment by Using Artificial Neural Network. Mater. Trans. **62**(3), 461–467 (2021)
25. Matsui, M.: Development of factory science: Duality and balancing of job-shop (FS) vs. conveyor system (OE) by Matsui's flow approach. In: Proc. of 1 International Scheduling Symposium 2015 (2015)
26. Matsui, M.: A theory of modern economic growth toward sharing society. Theoretical Economics Letters **8**, 675–684 (2018)
27. Krogh, A.: What are artificil neural network? Nat. Biotechnol. **26**, 195–197 (2008)
28. LeCun, Y., Bengio, Y., Hinton, G.: Deep learning. Nature **521**, 436–444 (2015)

Collaborative Supply Chain Innovation Networks of Small-Mid Enterprises

Agostino Villa[1(✉)] and Gianni Piero Perrone[2]

[1] Politecnico di Torino, Corso Duca degli Abruzzi 24, 10129 Torino, Italy
villa.agostino@gmail.com

[2] Perrone Informatica, Corso Venezia 53, 14100 Asti, Italy

Abstract. The aim of this chapter is to illustrate some methods to support innovation and development of Small and Medium-sized Enterprises (SMEs) based on the experiences acquired on over 150 SMEs and SME networks operating in North-West Italy. The chapter will discuss, in a simple way such as to be understandable by SME owners or managers, graphic models of the SME network, rules for managing production flows and orders within the supply chain. Some new supply chain applications in the agro-food sector will also be presented.

1 Introduction

International and domestic markets are giving rise to problems for SMEs that affect their existence, such as: how to make their skills and efficiency known to potential customers; how to provide them the products at a competitive price. Their small size, in terms of design and production capacity, shows the motivation of their actual crisis: about 30% of Italian small enterprises have disappeared or have been greatly reduced, since 2017.

The growing weakness of Italian SMEs convinced the authors to promote the creation of the PMInnova Program, a strategic agreement between Politecnico di Torino and the banking group of Cassa di Risparmio di Asti, operating in Piedmont and part of Liguria and Lombardy regions, in North-West Italy, dedicated to support, through research and consultancy, the innovation and development of SMEs and SME networks. The PMInnova Program provides technical and organizational support to small mid companies with an average of 30 employees. An effective interaction with the SME managers – usually owners and often founders - requires the use, by the experts and consultants, of a simple language to make the SMEs managers able to understand the solutions of the problems proposed by themselves. "Simple language" means using graphic modeling of the SME network, easily usable production management rules, a secure and robust method of order control along the supply chain.

The approaches used to provide innovation and development support to SME networks through the three methods listed above are respectively described in the following three paragraphs.

However, we can anticipate the main problem encountered: egocentrism, especially in the owners and founders of small enterprises that, over time, have designed high value products (that is a typical characteristic of 95% of SMEs [1]); high consideration of the

C.-Y. Huang and S. W. Yoon (Eds.): ICPR1 2021, ACES 14, pp. 122–131, 2023.
https://doi.org/10.1007/978-3-031-44373-2_7

company and the quality of its products, associated with the fear that the qualifying aspects of their production techniques will be copied by competitors especially of larger dimensions [2].

Despite the desire to keep their design skills hidden, just because of the growing and growing globalization of markets of goods and the large disequilibrium between the labor markets around the world, a large part of Italian SMEs can no more be competitive in terms of labor cost and goods prices [3]. This so difficult and long crisis is gradually forcing a recent growth of "balanced SMEs networks", based on collaboration agreements that, in some cases, come to the mutual support of some SMEs in temporary difficulty through their inclusion in networks [4].

As soon as a network starts work, the real problem lies in the availability of network members to be cooperative, that means to act for the mutual benefit of the enterprises that compose the network itself [5].

Inspired by the problems of SME networks contacted by the PMInnova Program, this chapter first illustrates the graphical models used to evaluate the performance of an SME network (Sect. 2); then it presents the simple production management logics actually suggested and implemented (Sect. 3); finally it shows the applied innovation for the management of orders in supply chains of the agro-food sector, transforming traditional methods with the use of blockchains (Sect. 4). The conclusions (Sect. 5) summarize the main aspects of the SME network innovation supports, which have been adopted.

2 Graph-Based Models for SME Network Performance Evaluation

Utilization of a graph formulation appears to be the first tool both to provide a graphical illustration of a SME network, and to offer a starting point for modeling and analyzing flows and connections between the companies belonging to the network, as widely discussed in [6].

The most common configuration, denoted "marshallian-Italianate network" [7], is composed by a set of SMEs, each one providing and receiving material/information from the others. Examples are *the Valenza Gold District* in North-Italy and the *Shannon Soft*, network of software activity in the Shannon Region of Ireland. (Fig. 1.a).

A second type of SME aggregation is a "multi-stage supply chain" [8], an example of which is the *Fermo Footwear District* located in the Center of Italy, specialized in production of shoes (Figure 1.b composed by a sequence of stages, each one with a number of parallel SMEs).

Another type of SME aggregations can be modelled in terms of "hub-and-spoke" configuration [9], owing to the presence of a leader in the network that will affect the decisions of all other partners (Fig. 2.a). The *Belluno Eyewear District* in North Italy has a similar configuration: there are five leading firms corresponding to important brands, and around them about 1.500 small and medium enterprises specialized in the production of components.

A different kind of aggregation, mainly organized for high-tech production and/or service supply, is the so called "scientific park" or "pole of competitiveness" (Fig. 2.b), where enterprises are "nodes" of a pre-existing network (light gray in the figure) of

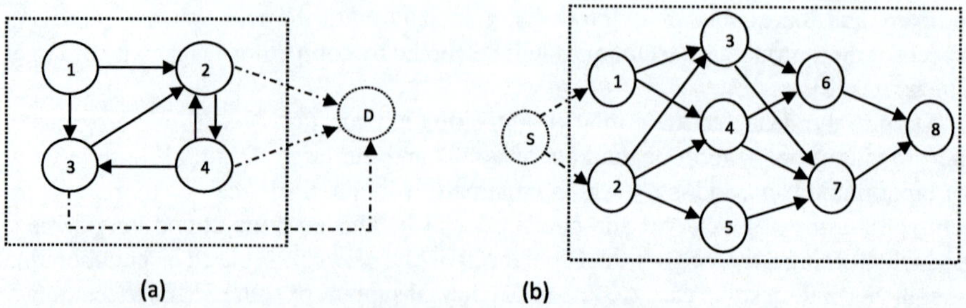

Fig. 1. Graphs of a Marshallian-Italianate network (a) and Multi-stage Supply chain network (b)

service links that can create contacts between the enterprises. In this configuration, connections are very flexible and more informal than in the others [6].

As illustrated in the above Figs. 1 and 2, the SME network structure can be represented in terms of a graph G = (V, E), where V is the set of vertices (nodes) and E is the set of edges or arcs. A vertex is referred to a component SME, while an edge represents a SME-to-SME logistic and information connection. From a theoretical point of view, the graph G modeling of a considered real SME network is described by four matrices [6]:

- the *incidence matrix M* [nodes vs edges] that identifies the links outgoing from each node, i.e. the existence of output flows from a given SME;
- the *adjacency matrix R* [nodes vs nodes] that specifies the existence of all the connections among the nodes, i.e. the existence of flows from a SMEs towards another SME;
- the path matrix P [paths vs edges], that specifies the input-output flows of parts for a pair of SMEs operating as suppliers and customers;
- the *distance matrix L* [nodes vs nodes], where each element is a certain "magnitude" associated to each edge, e.g. geographic distance, economic cost or time.

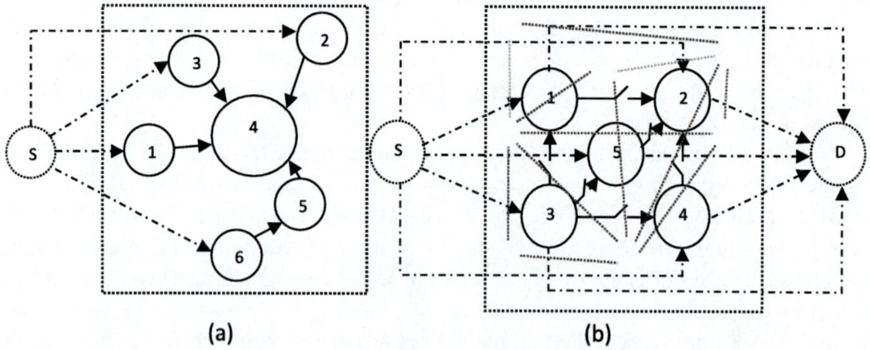

Fig. 2. Graphs of a Hub and Spoke network (a) and of a Scientific Park (b)

These matrices allow to recognize some conditions of either strong or weak collaboration of SMEs together, according to Key Performance Indicators (KPIs) as the following ones:

- *network connectivity index (NCO),* i.e. the number of non-null elements in matrix R, corresponding to the number of connections among SMEs;
- *network utilization balance (NUB),* in terms of the percentage number of SMEs for which the difference between the computed production capacity and the actual capacity value is greater than a given "sufficient utilization" lower bound;
- *network separation into chains (NSC),* i.e. percentage number of recognized independent supply chains, if any, referred to the number of component SMEs;
- *network chains independence (NCH),* in terms of the percentage number of links (i.e., cut-sets dimensions) connecting the recognized supply chains, if any;
- *number of network bottlenecks.*

These KPIs can support a SME manager in selecting an existing SME networks in which he could ask for the inclusion of his own SME. In case the "marshallian-italianate network" appears to be the most convenient, a measure of strong collaboration among SMEs is the high number of connections, then high value of NCO. As many the non-null elements in the path matrix P are, as large is the number of links, showing the possibility of good collaborations.

In case of a "multi-stage network" composed by a set of parallel supply chains, a low value of the NCH indicator and a high value of the NSC one can be found. In any type of SME network, existence of independent supply chains can cause the network subdivision into potentially competing and conflicting parts.

Opposite is the case of a "hub-and-spoke" network, where partial chains could exist, but all converging on a same hub (network-leading) SME. Then, the NSC indicator will be low.

Specific considerations should be done for analyzing a "scientific-park", whose network is modelled by two graphs: one composed by the SMEs already in operation, and the other defining the set of all links that the park management committee can make at disposal of other new SMEs (i.e. an underlying network whose links can be activated in the future). The former network can have small NCO and almost null NSC. The underlying network, on the contrary, must be characterized by high NCO.

3 Network Management Organizations for Maximizing Efficiency and Innovation

The possibility of maximizing the efficiency and innovation of an SME network is based on a clear vision of three elements: network structure, its functionality and management of the collaboration of the autonomous component SMEs.

In terms of structural aspects, the collaboration functions are implemented through processes of exchange of parts (components, products, etc.) and of information that occur in the connections between companies. These exchanges are generally managed by a collaboration management center, whose characteristics and management methods depend on the type of network interconnections. Hence, this is the structural element that differentiates the networks.

With reference to the functional aspects, the network of collaborative enterprises operates by organizing the interactions between the enterprises in terms of capital, goods and work. Therefore the transitions between the network and the markets must be functional to the same network, and also to all the companies included in it. This is perhaps the most critical aspect of managing a collaborative SME network.

The management aspects, referring also to the previous point, play a very important role in the organization of a network of companies, that may appear as a single operator on the markets but wants (must) guarantee a reasonable autonomy to each company. Therefore, the network management must certainly take place through a center that organizes collaboration between the companies (therefore, at the network level), but also through the managers of the various companies, who interact each other and with the aforementioned center. Hence it is often necessary to have a management organization at two levels, the central one for addressing market policies and network innovation, and the individual ones, for the operational management of each company [10].

These three aspects are the constitutive elements of the "collaboration of enterprises belonging to a same network" concept, as specified below:

- Collaboration is a way to interact together such to imply a very positive form of working in association with others for some form of mutual benefit, e.g. by applying strategic joint decision making about partnership and network design [11–13].
- In other words, one can say that collaboration is a way of doing so that organizations exchange information, share resources and enhance each other's capacity for mutual benefit, as well as for a common purpose, by sharing risks, responsibilities and rewards [14].

In order to clarify how to manage an effective collaboration between the SMEs in a network, two typical formulations of the network management problem – among the several ones presented in literature – can be stated by specifying, for each one of them, a proper collaboration goal [15]:

1st type: Maximize the average utilization of the network SMEs

- with respect to average workload assignments to each SME;
- In the case of constant demand;
- In the presence of graph constraints.

This problem is addressed in terms of constrained Linear Programming problem, the solution of which does not induces SMEs to collaborate but favor the most efficient SME in the network, thus requiring a balancing action by the network management center.

2nd type: Maximize the SMEs Average utilization on a given Mid-Term time horizon

- with respect to workload assignments to each SME, varying over time;
- with graphic constraints in terms of production flows and storage capacity of SMEs;
- with the hypothesis of variable demand to be satisfied.

This second problem, frequently applied in case of a linear supply chain, aims at avoiding the emergence of bottlenecks within the future time horizon. In addition, since the solution of this problem is a production plan for the medium-term, again a collaborative situation should be forced by the cluster management center.

By balancing workloads to SMEs, in the presence of a threshold of minimum utilization of each of them, the effect of the network management center is a mutual support that, by preventing individualisms, promotes mutual trust.

From research developed by the authors and from the data of ASSORETIPMI [16], an association of enterprise network (http://www.retipmi.it/pmi/), this behavior is often found in clusters of micro-enterprises in the Italian manufacturing sector.

4 SME Networks Innovation Suggestions from the PMInnova Program

As above mentioned, the PMInnova Program is developing consulting, research and training activities dedicated to develop technology and organization of SMEs and SME networks, in order pursuing their main innovation goals, by:

- studying the feasibility of innovation projects and alternative options;
- identifying funding opportunities for research and development calls in European, national or regional context;
- defining the necessary training programs for improving personnel competence.

About the main points to which the PMInnova program is dedicated, the following ones provide some suggestions:

1) need for "supply chain organizations" to allow real-time control of quality, costs and waste;
2) need for innovative logistics management, to maximize the level of customer service;
3) growing need to apply efficient automation, information and communication technologies, even in small production and/or service systems, with attention to the needs of effective communication and company promotion.
4) need for industrial security and cyber security;
5) requirements of energy saving, reduction of consumption, use of renewable energy sources;

Therefore this Sect. 4 is dedicated to showing how the owners of small businesses enrolled in the PMInnova Program must face the difficulties of managing production through balancing their need such as to respond quickly to customers, with the need to satisfy the constraints of belonging to the SME network which, in many cases analyzed, is a supply chain. In fact, as noted in the meetings with companies within the aforementioned Program, many supply chains have asked the experts of Politecnico di Torino to receive technical and organizational support to satisfy the orders of downstream stage customers as soon as possible, by mainly tacking into account the "importance" of each customer itself. In practice, they want to be able to assure prompt product delivery by an effective event-driven scheduling of production, but without any information on the occurrence of the next event, i.e. the acceptance of customer orders for any type of products that the SME is able to manufacture (as in points 1 and 2 of the above list).

Indeed, in the analyzed SME supply chains, since the production of any order from downstream in the supply chain requires the execution of a sequence of operations by the network SMEs, at the arrival time of a new order only some operations of the order under processing have already been completed. Therefore, at that time, the "state" of any order has to be updated, where the term "order state" indicates the number of operations to be still executed to complete the order itself.

At each event time, the production reorganization at every SME in the supply chain has to be based on the new state of each order, then characterizing an event-driven production scheduling problem.

A simplified version of the even-driven scheduling problem, that has been discussed with SME supply chain managers during the PMInnova Program, has been formulated as follows [17]:

Given any job under processing (i.e., an order already received by a SME in the supply chain):

a. an operation time equal to zero will be assigned to the operations already started and completed for this same order;
b. a new schedule will be assigned to the set of all the other operations, both the ones previously scheduled but not yet completed and the ones required to complete the new order.

The difficulty to apply an event-driven scheduling procedure in a real SME included into a supply chain pushes any manager to sequence their orders by using a very simplified logic, whose application seems him to be apparently useful and clear, even if frequently inefficient.

From the analysis of about 150 SMEs belonging to 32 different supply chains, it has been recognized the following event-driven scheduling rule at any arrival of a new order at a SME.

At the arrival of a new order at his own SME, the manager can:

i. decide to not modify the schedule under use, and insert the new order as the last in the queue to be processed; this rule is usually mentioned by SME managers as "weighted FIFO" or simply "FIFO;
ii. insert the new order as soon as one of those under processing will be completed, if the customer has a good level of importance; this is denoted as "weighted clients";
iii. allocate only the biggest orders, also considering some weights depending on the client importance, and include smaller orders in a random way (denoted "random entry of new orders").

Once the manager has realized that these criteria correspond to formulating a transaction in a blockchain, the blockchain procedure became the most interesting for managing order-driven interactions between two stages of the supply chain [18].

Blockchain applications have undoubtedly the potential to improve the supply chain operations because they provide an infrastructure that records, certifies and maps an asset that is transferred between often distant parts, connected between them through a chain of distribution between parties that are not necessarily bound by a bond of trust [19].

This has been verified by the blockchain application for two small supply chains of SMEs operating in the agri-food sector, registered in the PMInnova Program. For the two companies, named Agrocompany (www.agrocompany.it) and the Consorzio dei Produttori di Piccoli Frutti (Consortium of Small Fruit Producers, www.ciacuneo.org/frutta_verdura_piccoli_frutti.htm), in both cases, the main problem is to manage a "short" supply chain able to implement a retail sale via e-commerce, but with a large number of small transactions.

Indeed, one of the sectors that will benefit the most from the point of view of the consumer, is the food industry, where there are examples of cases of contamination of food chains due to poor control on suppliers or other reasons related to production, such as use of herbicides, fertilizers. And products storage by incorrect freezing.

With reference to the mentioned Italian supply chains producing valuable agri-food products, those with a "controlled designation of origin - DOC", the applied blockchain could help to counter frauds in the sale of controlled-source Italian goods. An accurate product record could also make the managed supply chain more efficient and send food to stores faster, thus reducing waste and waste.

5 Conclusion

This paper has discussed the main types of SME networks characterizing the Italian industry, where very small enterprises account for almost 93% of all enterprises in the non-financial business sector. The small dimension forces these enterprises to aggregate together.

The real strength of a SME network depends on an effective collaboration among SMEs: to apply the best conditions to promote collaboration is the crucial problem. To this aim, the paper presents a conceptual model of a SME network, by which a formal model in terms of mid-term constrained scheduling of orders arriving to each partner SMEs is formulated, such as to maximize the SME utilization and the SME loads balancing.

An acknowledgment of the importance of the "collaboration factor" in a SME network has been detected by the authors by analyzing network contracts between SME supply chains, stipulated in Italy from 2017 to the end of 2021. Neglecting the typical goal of expanding markets, present in the 38% of signed contracts, the aim is to collaborate in order to increase their innovation strength (17%) such to increase production capacity (20%) and their ability to compete (15%). On the other hand, some critical aspects should be underlined: only 7% of contracts are devoted to improving quality and certifications and a small 3% to share know-how and skills, the typical aspect of modern sharing economy.

Even if these data are referred to the Italian situation, some other European countries are approaching the autonomous creation of SME networks in similar way, so making more innovative and stimulating the industrial system [8].

The proposed model and its approximations, discussed in Sect. 4, can be used as an analysis tool. In industrial reality, analysis is just an initial step in the path of innovation.

In this perspective, from this model, support conditions to the SME network design should be derived, as well as be recognized the need for design criteria to integrate a structural and a dynamic vision of the production and organizational process in the SME networks.

Acknowledgments. This paper has been developed within the official agreement between Politecnico di Torino and Gruppo Banca di Asti, under the initiative Programma PMInnova - Pro-mote innovation and development in SMEs, signed in 2017, A. Villa Program Chair and G. P. Perrone Contract Manager.

References

1. Muller, P., Devnani, S., Ladher, R., et al.: Annual report on European SMEs 2020/2021: digitalization of SMEs: background document. In: Hope, K. (ed.) Publications Office (2021). https://doi.org/10.2826/120209
2. Collins-Dodd, C., Gordon, I.M., Smart, C.: Success without up-ward mobility: evidence from small accounting practice. J. Small Bus. Entrep. **18**(3), 327–342 (2005)
3. ISTAT - Istituto Italiano di Statistica, Enterprises Annual Report 2021, https://www.istat.it/en/enterprises
4. Herreros, S., Inoue, K., Mulder, N. (eds.): Innovation and SME internationalization in Korea and Latin America and the Caribbean - Policy experiences and areas for cooperation, LC/TS.2018/67. United Nations publication (2018)
5. Antonelli, D., Taurino, T.: Identifying and exploiting the collaboration factors inside SMEs networks. 1International Journal of Networking and Virtual Organizations **9**(4), 382–402 (2011)
6. Antonelli, D., Bruno, G., Taurino, T., Villa, A.: Graph-based models to classify effective collaboration in SME networks. Int. J. Prod. Res. **53**(20), 6198–6209 (2015)
7. Markusen, A.: Sticky places in slippery space: a typology of industrial districts. Econ. Geogr. **72**(3), 293–313 (1996)
8. Villa, A., Taurino, T., Ukovich, W.: Supporting collaboration in european industrial districts: the CODESNET approach. J. Intell. Manuf. **10**(4), 1–10 (2011)
9. Taurino, T., Antonelli, D.: An insight into innovation patterns of industrial districts. 6th CIRP International Conference on Intelligent Computation in Manufacturing Innovative and Cognitive Production Technology and Systems, pp. 23–25. Naples, Italy (July 2008)
10. Durugbo, C.: Collaborative networks: a systematic review and multi-level framework. Int. J. Prod. Res. **54**(12), 3749–3776 (2016)
11. Villa, A., Brandimarte, P., Calderini, M.: Meta-models for integrating production management functions in heterogeneous industrial systems. In: Nof, S.Y. (ed.) Information and Collaboration Models of Integration. NATO ASI Series. Kluwer, 1994. Based on an international workshop in Il Ciocco, Italy, pp. 135–145 (June 1993)
12. Nof, S.Y.: Integration and collaboration models: an overview. Studies in Informatics and Control **3**(4), 387–392 (1994)
13. Camarinha-Matos, L.M., Afsarmanesh, H.: On reference models for collaborative networked organizations. Int. J. Prod. Res. **46**(9), 2453–2469 (2008)
14. Villa, A.: (Chair), CO-DESNET – Collaborative Demand and Supply NETworks, Coordination Action, Commission of the European Communities, Directorate General Information Society, Contract Number 506673 (May 2004)

15. Villa, A., Bruno, G.: Promoting SME cooperative aggregations: main criteria and contractual models. Int. J. Prod. Res. **51**(23–24), 7439–7447 (2013)
16. ASSORETIPMI: Italian Association of SME Networks, Report of December 2016 (in Italian), www.retipmi.it
17. Villa, A., Taurino, T.: Event-driven production scheduling in SME. Production Planning and Control **29**(4), 271–279 (2017)
18. Genta, G., Villa, A., Perrone, G.P.: Supply chain management by blockchain, Int. Conference in Production Management Systems – APMS 2021, Nantes, France (Sept. 2021). https://www.apms-conference.org/wp-ontent/uploads/2021/08/APMS2021-Conference_Booklet-V2.pdf
19. Saberi, S., Kouhizadeh, M., Sarkis, J., Shen, L.: Blockchain technology and its relationships to sustainable supply chain management. Int. J. Prod. Res. **57**(7), 2117–2135 (2019)

CCT Principle of Error and Conflict Detection and Prevention

Xin W. Chen(✉)

Industrial Engineering, Southern Illinois University Edwardsville, Edwardsville, USA
xchen@siue.edu

Abstract. Errors and conflicts exist in many systems. It is important to detect and prevent errors and conflicts in systems. Collaborative Control Theory (CCT) provides principles and framework for the design and control of complex systems including error and conflict detection and prevention (ECDP). The purpose of this chapter is to illustrate advanced methods and state-of-the-art applications of ECDP. Eight key functions to detect and prevent errors and conflicts are identified and their theoretical background and applications in both production and service are explained with examples. As systems and networks become larger and more complex, ECDP becomes more challenging. There is a paradigm shift from detection to proactive prognostics and prevention of errors and conflicts.

Keywords: Collaborative Control Theory · Collaborative Control Protocol · Decentralized Error Prevention · Error and Conflict Prognostics · Prognostics

1 CCT Principle for ECDP

Collaborative Control Theory (CCT) includes principles and framework for domain experts to design and control a complex system with multiple agents [1–3]. Errors occur when the input, output, or intermediate result of a system does not meet specifications or expectations. A conflict refers to the difference between the information, goals, plans, tasks, operations, or activities of the collaborating agents [1]. The error and conflict detection and prevention (ECDP) principle enables effective and efficient detection and prevention of errors and conflicts. In addition, the error recovery and conflict resolution principle help to resolve conflicts and errors as early as possible.

A system usually has multiple units, some of which collaborate, cooperate, and/or coordinate to complete tasks. The most important difference between an error and a conflict is that an error involves only one unit, whereas a conflict involves two or more units in a system. An error at a unit may cause other errors or conflicts, for instance, a workstation that cannot provide the required number of products to an assembly line (a conflict) because one machine at the workstation breaks down (an error). Similarly, a conflict may cause other errors and conflicts, for instance, a machine that does not receive required products (an error) because the automated guided vehicles that carry the products collide when they move toward each other on the same path (a conflict). These phenomena, errors leading to other errors or conflicts, and conflicts leading to

C.-Y. Huang and S. W. Yoon (Eds.): ICPR1 2021, ACES 14, pp. 132–144, 2023.
https://doi.org/10.1007/978-3-031-44373-2_8

other errors or conflicts, are called error and conflict propagation. The CCT principle for ECDP aims at disrupting the propagation of errors and conflicts through detection and prevention.

2 Advanced Methods for ECDP

Methods for interactively preventing and detecting errors and conflicts through prognostics and diagnostics [4–7] include centralized and decentralized prevention and detection logic for real-world constraint networks: random networks (RN), scale-free networks (SFN), and Bose-Einstein condensation networks (BECN). These methods select an appropriate detection and prevention algorithm from a plurality of algorithms having either centralized or decentralized CEPD logic, based on analysis of the characteristics of the algorithms and the characteristics of the constraint network [4–7].

Two or more cooperative units in a system may periodically exhibit conflicts and errors. A detection and prevention algorithm has at least one control unit configured to receive a list of parameters associated with the cooperative units and interactions between the cooperative units. Control units perform detection and prevention, including: (a) providing a list of at least two constraints, each constraint defining a task to be accomplished or a requirement to be satisfied by one or more cooperative units; (b) identifying one or more constraints from the list, which need to be satisfied by a defined time; (c) identifying for each identified constraint whether any conflict or error exists; where a conflict occurs whenever an inconsistency between two or more cooperative units occurs, and an error is associated with any condition that is inconsistent with the list of parameters; (d) marking the constraints for which an error or conflict has been identified; (e) incrementing a mark count for each cooperative unit associated with each marked constraint; performing at least one of diagnosis and prognosis based at least in part on the marked constraints; (f) modeling dependencies between constraints to form at least one relationship table, wherein each node in the constraint network is represented by a constraint and links between nodes represent relationships between constraints; (g) diagnosing and marking constraints that have conflicts or errors through the analysis of the constraint network; and (h) predicting and marking constraints that have or will have conflicts or errors through the analysis of the constraint network.

A system for ECDP through prognostics and diagnostics comprises multiple autonomous agents or control units. These advanced ECDP methods model a system with a plurality of constraints that must be satisfied by cooperative units, wherein the constraints are indicative of potential conflicts and errors in the system and have relationships indicative of how conflicts and errors propagate between units. These methods apply ECDP logic configured to detect conflicts and errors that have occurred, and to identify conflicts and errors before they occur, based on whether the constraints are satisfied or unsatisfied.

The ECDP methods have been applied to vehicle traffic in air traffic and ground control scenarios [8]. The ECDP logic is applied to a constraint network generated to model the constraints pertinent to the vehicles (such as aircraft and ground-based vehicles) and resources (such as runways and taxiways) in the traffic control spaces. Time-based trajectory data for all vehicles in the traffic control space are continuously and interactively evaluated to detect conflicts that have occurred and/or identify potential conflicts before they occur among vehicles or between vehicles and resources. The ECDP methods also generate a conflict resolution using the ECDP logic and the resolution is communicated to the affected vehicles [8].

3 Definition and Examples

An error is any input, output or intermediate result that has occurred or will occur in a system and does not meet system specification, expectation, or comparison objective. A conflict is an inconsistency between different units' goals, plans, tasks, or other activities in a system. A system usually has multiple units, some of which collaborate, cooperate, and/or coordinate to complete tasks. The most important difference between an error and a conflict is that an error involves only one unit, whereas a conflict involves two or more units in a system. An error at a unit may cause other errors or conflicts, for instance, a workstation that cannot provide the required number of products to an assembly line (a conflict) because one machine at the workstation breaks down (an error). Similarly, a conflict may cause other errors and conflicts, for instance, a machine that does not receive required products (an error) because the automated guided vehicles that carry the products collide when they move toward each other on the same path (a conflict).

Tables 1 and 2 illustrate errors and conflicts in production and service systems with some typical examples. There are also human errors and conflicts that exist in systems. Figure 1 describes the difference between errors and conflicts in pin insertion.

Table 1. Examples of errors and conflicts in production systems (Source: [9])

Error	Conflict
• A robot drops a circuit board while moving it between two locations • A machine punches two holes on a metal sheet while only one is needed, because the size of the metal sheet is recognized incorrectly by the vision system • A lathe stops processing a shaft due to power outage • The server of a computer-integrated manufacturing system crashes due to high temperature • A facility layout generated by a software program cannot be implemented due to irregular shapes	• Two numerically controlled machines request help from the same operator at the same time • Three different software packages are used to generate optimal schedule of jobs for a production facility; the schedules generated are different • Two automated guided vehicles collide • A DWG (drawing) file prepared by an engineer with AutoCAD cannot be opened by another engineer with the same software • Overlapping workspace defined by two cooperating robots

This chapter provides a theoretical background and illustrates applications of how to prevent errors and conflicts in production and service. Different terms have been used to describe the concept of errors and conflicts, for instance, failure (e.g., [10–13]), fault (e.g., [12, 14]), exception (e.g., [15]), and flaw (e.g., [16]). Error and conflict are the most

Table 2. Examples of errors and conflicts in service systems (Source: [9])

Error	Conflict
• The engine of an airplane shuts down unexpectedly during the flight • A patient's electronic medical records are accidently deleted during system recovery • A pacemaker stops working • Traffic lights go off due to lightening • A vending machine does not deliver drinks or snacks after the payment • Automatic doors do not open • An elevator stops between two floors • A cellphone automatically initiates phone calls due to a software glitch	• The time between two flights in an itinerary generated by an online booking system is too short for transition from one flight to the other • A ticket machine sells more tickets than the number of available seats • An ATM machine dispenses $ 250 when a customer withdraws $ 260 • A translation software incorrectly interprets text • Two surgeries are scheduled in the same room due to a glitch in a sensor that determines if the room is empty

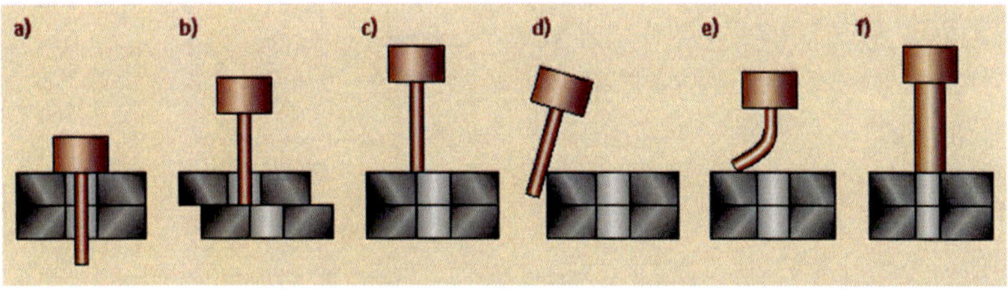

Fig. 1. Errors and conflicts in a pin insertion task: (a) successful insertion; (b–f) are unsuccessful insertion with (1) errors if the pin and the two other components are considered as one unit in a system, or (2) conflicts if the pin is a unit and the two other components are considered as another unit in a system (Source: [9])

popular terms appearing in literature (e.g., [11, 12, 14, 17–23]). The related terms listed here are also useful descriptions of errors and conflicts. Depending on the context, some of these terms are interchangeable with error; some are interchangeable with conflict; and the rest refer to both error and conflict.

Eight key functions have been identified as useful to prevent errors and conflicts automatically as described below [24–27]. Functions 5–8 prevent errors and conflicts with the support of functions 1–4. Functions 6–8 prevent errors and conflicts by managing those that have already occurred. Function 5, prognostics, is the only function that actively determines which errors and conflicts will occur and prevents them. All other seven functions are designed to manage errors and conflicts that have already occurred, although as a result they can prevent future errors and conflicts directly or indirectly. Figure 2 describes error and conflict propagation and their relationship with the eight functions:

1. Detection is a procedure to determine if an error or conflict has occurred.

2. Identification is a procedure to identify the observation variables most relevant to diagnosing an error or conflict; it answers the question: Which of them has already occurred?
3. Isolation is a procedure to determine the exact location of an error or conflict. Isolation provides more information than identification function, in which only the observation variables associated with the error or conflict are determined. Isolation does not provide as much information as the diagnostics function, however, in which the type, magnitude, and time of the error or conflict are determined. Isolation answers the question: Where has an error or conflict occurred?
4. Diagnostics is a procedure to determine which error or conflict has occurred, what their specific characteristics are, or the cause of the observed out-of-control status.
5. Prognostics is a procedure to prevent errors and conflicts through analysis and prediction of error and conflict propagation.
6. Error recovery is a procedure to remove or mitigate the effect of an error.
7. Conflict resolution is a procedure to resolve a conflict.
8. Exception handling is a procedure to manage exceptions. Exceptions are deviations from an ideal process that uses the available resources to achieve the task requirement (goal) in an optimal way.

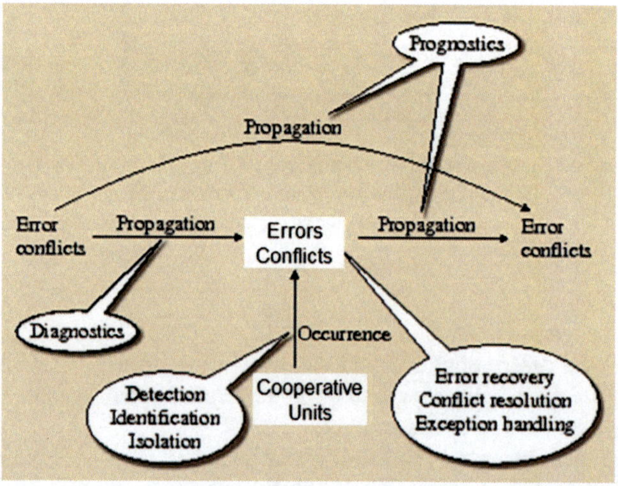

Fig. 2. Error and conflict propagation and eight functions to prevent errors and conflicts (Source: [9])

4 Error Detection in Assembly and Inspection

As the first step to prevent errors, error detection has attracted much attention, especially in assembly and inspection; for instance, researchers [11] have studied an integrated sensor-based control system for a flexible assembly cell which includes error detection function. An error knowledge base has been developed to store information about previous errors that had occurred in assembly operations, and corresponding recovery programs which had been used to correct them. The knowledge base provides support for both error detection and recovery. In addition, a similar machine-learning approach to error detection and recovery in assembly has been discussed. To enable error recovery, failure diagnostics has been emphasized as a necessary step after the detection and before the recovery. It is noted that, in assembly, error detection and recovery are often integrated.

Automatic inspection has been applied in various manufacturing processes to detect, identify, and isolate errors or defects with computer vision. It is mostly used to detect defects on printed circuit board [28–30] and dirt in paper pulps [31, 32]. The use of robots has enabled automatic inspection of hazardous materials (e.g., [33]) and in environments that human operators cannot access, e.g., pipelines [34]. Automatic inspection has also been adopted to detect errors in many other products such as fuel pellets [35], printing the contents of soft drink cans [36], oranges [37], aircraft components [38], and microdrills [39]. The key technologies involved in automatic inspection include but are not limited to computer or machine vision, feature extraction, and pattern recognition [40–42].

5 Error Detection in Software Design

The most prevalent method to detect errors in software is model checking. As Clarke et al. [43] state, model checking is a method to verify algorithmically if the model of software or hardware design satisfies given requirements and specifications through exhaustive enumeration of all the states reachable by the system and the behaviors that traverse them. Model checking has been successfully applied to identify incorrect hardware and protocol designs, and recently there has been a surge in work on applying it to reason about a wide variety of software artifacts; for example, model checking frameworks have been applied to reason about software process models, (e.g., [44]), different families of software requirements models (e.g., [45]), architectural frameworks (e.g., [46]), design models (e.g., [47]), and system implementations (e.g., [48–51]). The potential of model checking technology for (1) detecting coding errors that are hard to detect using existing quality assurance methods, e.g., bugs that arise from unanticipated interleavings in concurrent programs, and (2) verifying that system models and implementations satisfy crucial temporal properties and other lightweight specifications has led a number of international corporations and government research laboratories such as Microsoft, International Business Machines Corporation (IBM), Lucent, Nippon Electric Company (NEC), National Aeronautics and Space Administration (NASA), and Jet Propulsion Laboratory (JPL) to fund their own software model checking projects.

A drawback of model checking is the state-explosion problem. Software tends to be less structured than hardware and is considered as a concurrent but asynchronous system. In other words, two independent processes in software executing concurrently in either order result in the same global state [43]. Failing to execute checking because of too many states is a particularly serious problem for software. Several methods, including symbolic representation, partial order reduction, compositional reasoning, abstraction, symmetry, and induction, have been developed either to decrease the number of states in the model or to accommodate more states, although none of them has been able to solve the problem by allowing a general number of states in the system.

Based on the observation that software model checking has been particularly successful when it can be optimized by considering properties of a specific application domain, Hatcliff and colleagues have developed Bogor [52], which is a highly modular model-checking framework that can be tailored to specific domains. Bogor's extensible modeling language allows new modeling primitives that correspond to domain properties to be incorporated into the modeling language as first-class citizens. Bogor's modular architecture enables its core model-checking algorithms to be replaced by optimized domain-specific algorithms. Bogor has been incorporated into Cadena and tailored to checking avionics designs in the common object request broker architecture (CORBA) component model (CCM), yielding orders of magnitude reduction in verification costs. Specifically, Bogor's modeling language has been extended with primitives to capture CCM interfaces and a real-time CORBA (RT-CORBA) event channel interface, and Bogor's scheduling and state-space exploration algorithms were replaced with a scheduling algorithm that captures the particular scheduling strategy of the RT-CORBA event channel and a customized state-space storage strategy that takes advantage of the periodic computation of avionics software.

Despite this successful customizable strategy, there are additional issues that need to be addressed when incorporating model checking into an overall design/development methodology. A basic problem concerns incorrect or incomplete specifications: before verification, specifications in some logical formalism (usually temporal logic) need to be extracted from design requirements (properties). Model checking can verify if a model of the design satisfies a given specification. It is impossible, however, to determine if the derived specifications are consistent with or cover all design properties that the system should satisfy. That is, it is unknown if the design satisfies any unspecified properties, which are often assumed by designers. Even if all necessary properties are verified through model checking, code generated to implement the design is not guaranteed to meet design specifications, or more importantly, design properties. Model-based software testing is being studied to connect the two ends in software design: requirements and code.

The detection of design errors in software engineering has received much attention. In addition to model checking and software testing, for instance, Miceli et al. [16] has proposed a metric-based technique for design flaw detection and correction. In parallel computing, synchronization errors are major problems and a nonintrusive detection method for synchronization errors using execution replay has been developed [22]. Besides, concurrent error detection (CED) is well known for detecting errors in distributed computing systems and its use of duplications [17, 53], which is sometimes considered a drawback.

6 Conflict Prognostics and Prevention

Conflicts can be categorized into three classes [54]: goal conflicts, plan conflicts, and belief conflicts. Goals of an agent are modeled with an intended goal structure (IGS; e.g., Fig. 3), which is extended from a goal structure tree [55]. Plans of an agent are modeled with the extended project estimation and review technique (E-PERT) diagram (e.g., Fig. 4). An agent has (1) a set of goals which are represented by circles (Fig. 3), or circles containing a number (Fig. 4), (2) activities such as Act 1 and Act 2 to achieve the goals, (3) the time needed to complete an activity, e.g., T1, and (4) resources, e.g., R1 and R2 (Fig. 4). Goal conflicts are detected by comparing goals by agents. Each agent has a PERT diagram and plan conflicts are detected if agents fail to merge PERT diagrams or the merged PERT diagrams violate certain rules [54].

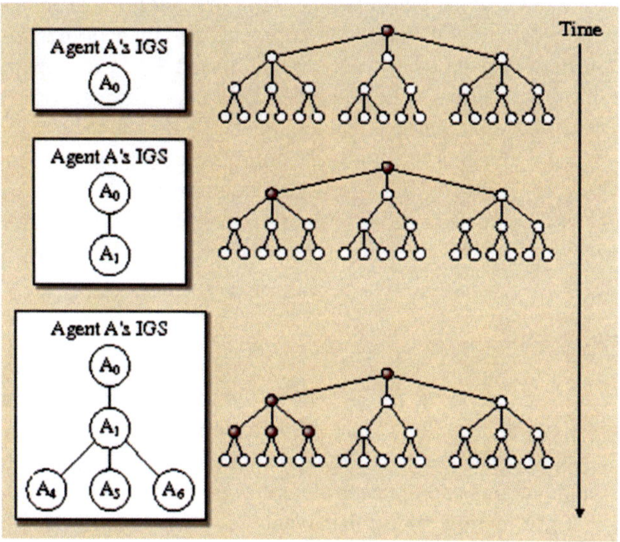

Fig. 3. Development of agent A's intended goal structure (IGS) over time (Source: [9])

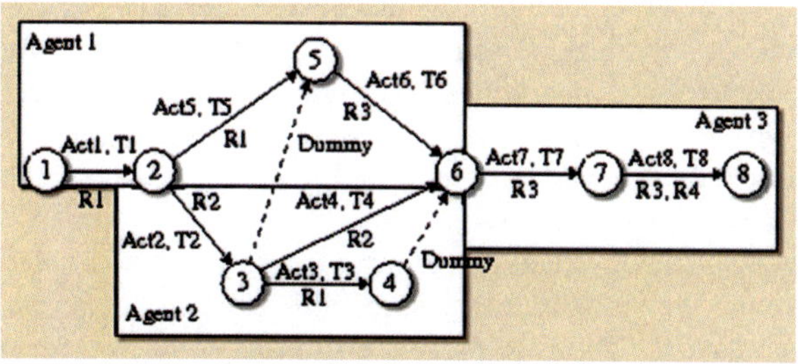

Fig. 4. Merged project estimation and review technique (PERT) diagram (Source: [9])

The three classes of conflicts can also be modeled by Petri nets with the help of four basic modules [56]: sequence, parallel, decision, and decision-free, to detect conflicts in a multiagent system. Each agent's goal and plan are modeled by separate Petri nets [57], and many Petri nets are integrated using a bottom-up approach [56] with three types of operations [57]: AND, OR, and precedence. The synthesized Petri net is analyzed to detect conflicts. Only normal transitions and places are modeled in Petri nets for conflict detection. The Petri-net-based approach for conflict detection developed so far has been rather limited. It has emphasized more the modeling of a system and its agents than the analysis process through which conflicts are detected.

The three common characteristics of available conflict detection approaches are: (a) they use the agent concept because a conflict involves at least two units in a system; (b) an agent is modeled for multiple times because each agent has at least two distinct attributes: goal and plan; and (c) they not only detect, but mainly prevent conflicts because goals and plans are determined before agents start any activities to achieve them. The main difference between the IGS and PERT approach, and the Petri net approach is that agents communicate with each other to detect conflicts in the former approach whereas a centralized control unit analyzes the integrated Petri net to detect conflicts in the latter approach [57]. The Petri net approach does not detect conflicts using agents, although systems are modeled with agent technology. Conflict detection has been mostly applied in collaborative design [58–60]. The ability to detect conflicts in distributed design activities is vital to their success because multiple designers tend to pursue individual (local) goals prior to considering common (global) goals.

7 Emerging Trends

Most ECDP methods developed so far are centralized approaches in which a central control unit controls data and information and executes some or all eight functions to detect and prevent errors and conflicts. The centralized approach often requires substantial time to execute various functions and the central control unit often possesses incomplete or incorrect data and information. These disadvantages become apparent when a system has many units that need to be examined for errors and conflicts.

To overcome the disadvantages of the centralized approach, the decentralized approach that takes advantage of the parallel activities of multiple agents has been developed [24, 61, 76]. In the decentralized approach, distributed agents detect, identify or isolate errors and conflicts at individual units of a system, and communicate with each other to diagnose and prevent errors and conflicts. The main challenge of the decentralized approach is to develop robust protocols that can ensure effective communications between agents. Further research is needed to develop and improve decentralized approaches for implementation in various applications.

Compared with humans, systems perform better when they are used to prevent errors and conflicts through the violation of specifications or violation in comparisons [21]. Humans, however, have the ability to prevent errors and conflicts through the violation of expectations, i.e., with tacit knowledge and high-level decision making. To increase the effectiveness degree of automation of error and conflict prognostics and prevention, it is necessary to equip systems with human intelligence through appropriate modeling techniques such as fuzzy logic, pattern recognition, and artificial neural networks. There has been some preliminary work to incorporate high-level human intelligence in error detection and recovery (e.g., [11, 62]) and conflict resolution [63]. Additional work is needed to develop self-learning, self-improving artificial intelligence systems for ECDP.

The performance of an error and conflict prognostics and prevention method is significantly influenced by the number of units in a system and their relationship. A system can be viewed as a graph or network with many nodes, each of which represents a unit in the system. The relationship between units is represented by the link between nodes. The study of network topologies has a long history stretching back at least to the 1730s. The classic model of a network, the random network, was first discussed in the early 1950s [64] and was rediscovered and analyzed in a series of papers published in the late 1950s and early 1960s [65–67]. Most recently, several network models have been discovered and extensively studied, for instance, the small-world network (e.g., [68, 75]), the scale-free network (e.g., [69–72]), and the Bose–Einstein condensation network [73]. Bioinspired network models for collaborative control have recently been studied by Nof [74, 75].

Because the same prognostics and prevention method may perform quite differently on networks with different topologies and attributes, or with the same network topology and attributes but with different parameters, it is imperative to study the performance of prognostics and prevention methods with respect to different networks for the best match between methods and networks. There is ample room for research, development, and implementation of ECDP methods supported by graph and network theories.

8 Conclusion

In this chapter we have discussed advanced ECDP methods and the eight functions that automate error and conflict prognostics and prevention and their applications in various production and service areas. ECDP methods for errors and conflicts are developed based on extensive theoretical advancements in many science and engineering domains, and have been successfully applied to various real-world problems.

As systems and networks become larger and more complex, such as global enterprises and the Internet, error and conflict prognostics and prevention become more necessary and important. The focus is shifting from passive response to active prognostics and prevention, and to intelligent predictive models and techniques.

References

1. Dusadeerungsikul, P.O., Nof, S.Y.: A collaborative control protocol for agricultural robot routing with online adaptation. Comput. Ind. Eng. **135**, 456–466 (2019)
2. Nof, S.Y.: Collaborative control theory and decision support systems. Comput. Sci. J. Moldova **25**(2), 115–144 (2017)
3. Zhong, H., Levalle, R.R., Moghaddam, M., Nof, S.Y.: Collaborative intelligence - definition and measured impacts on internetworked e-Work. Manag. Prod. Eng. Rev. **6**(1), 67–78 (2015)
4. Chen, X.W., Nof, S.Y.: Interactive, constraint-network prognostics and diagnostics to control errors and conflicts (IPDN) (U.S. Patent 9,009,530 (2015)
5. Chen, X.W., Nof, S.Y.: Interactive, constraint-network prognostics and diagnostics to control errors and conflicts (IPDN) (U.S. Patent 9,760,411 (2017)
6. Chen, X.W., Nof, S.Y.: Interactive, constraint-network prognostics and diagnostics to control errors and conflicts (IPDN) (U.S. Patent 10,496,463 (2019)
7. Nof, S.Y., Chen, X.: Failure repair sequence generation for nodal network (U.S. Patent 9,166,907 (2015)
8. Chen, X.W., Nof, S.Y.: Interactive conflict detection and resolution for air and air-ground traffic control (U.S. Patent 8,831,864 (2014)

9. Chen, X.W., Nof, S.Y.: Automating prognostics and prevention of errors, conflicts, and disruptions. In: Springer Handbook of Automation, 2nd Edition. Springer Publishers (2022)
10. Lopes, L.S., Camarinha-Matos, L.M.: A machine learning approach to error detection and recovery in assembly. In: Proceedings of IEEE/RSJ International Conference on Intelligent Robots Systems 95, Human Robot Interaction and Cooperative Robots, vol. 3, pp. 197–203 (1995)
11. Najjari, H., Steiner, S.J.: Integrated sensor-based control system for a flexible assembly. Mechatronics **7**(3), 231–262 (1997)
12. Steininger, A., Scherrer, C.: On finding an optimal combination of error detection mechanisms based on results of fault injection experiments. In: Proceedings of 27th Annual International Symposium on Fault-Tolerant Computing, FTCS-27, Digest of Papers, pp. 238–247 (1997)
13. Toguyeni, K.A., Craye, E., Gentina, J.C.: Framework to design a distributed diagnosis in FMS. Proc. IEEE Int. Conf. Syst. Man. Cybern. **4**, 2774–2779 (1996)
14. Kao, J.F.: Optimal recovery strategies for manufacturing systems. Eur. J. Oper. Res. **80**(2), 252–263 (1995)
15. Bruccoleri, M., Pasek, Z.J.: Operational issues in reconfigurable manufacturing systems: exception handling. In: Proceedings of 5th Biannual World Automation Congress (2002)
16. Miceli, T., Sahraoui, H.A., Godin, R.: A metric based technique for design flaws detection and correction. In: Proceedings of 14th IEEE International Conference on Automated Software Engineering, pp. 307–310 (1999)
17. Bolchini, C., Fornaciari, W., Salice, F., Sciuto, D.: Concurrent error detection at architectural level. In: Proceedings of 11th International Symposium on Systems Synthesis, pp. 72–75 (1998)
18. Bolchini, C., Pomante, L., Salice, F., Sciuto, D.: Reliability properties assessment at system level: a co-design framework. J. Electron. Test. **18**(3), 351–356 (2002)
19. Jeng, M.D.: Petri nets for modeling automated manufacturing systems with error recovery. IEEE Trans. Robot. Autom. **13**(5), 752–760 (1997)
20. Kanawati, G.A., Nair, V.S.S., Krishnamurthy, N., Abraham, J.A.: Evaluation of integrated system-level checks for on-line error detection. In: Proceedings of IEEE International Computer Performance and Dependability Symposium, pp. 292–301 (1996)
21. Klein, B.D.: How do actuaries use data containing errors: models of error detection and error correction. Inf. Resour. Manag. J. **10**(4), 27–36 (1997)
22. Ronsse, M., Bosschere, K.: Non-intrusive detection of synchronization errors using execution replay. Autom. Softw. Eng. **9**(1), 95–121 (2002)
23. Svenson, O., Salo, I.: Latency and mode of error detection in a process industry. Reliab. Eng. Syst. Saf. **73**(1), 83–90 (2001)
24. Chen, X.W., Nof, S.Y.: Conflict and error prevention and detection in complex networks. Automatica **48**, 770–778 (2012)
25. Gertler, J.: Fault Detection and Diagnosis in Engineering Systems. Marcel Dekker, New York (1998)
26. Klein, M., Dellarocas, C.: A knowledge-based approach to handling exceptions in workflow systems. Comput. Support. Coop. Work **9**, 399–412 (2000)
27. Raich, A., Cinar, A.: Statistical process monitoring and disturbance diagnosis in multivariable continuous processes. AIChE J. **42**(4), 995–1009 (1996)
28. Chang, C.-Y., Chang, J.-W., Jeng, M.D.: An unsupervised self-organizing neural network for automatic semiconductor wafer defect inspection. In: IEEE International Conference on Robotics and Automation ICRA, pp. 3000–3005 (2005)
29. Moganti, M., Ercal, F.: Automatic PCB inspection systems. IEEE Potentials **14**(3), 6–10 (1995)
30. Rau, H., Wu, C.-H.: Automatic optical inspection for detecting defects on printed circuit board inner layers. Int. J. Adv. Manuf. Technol. **25**(9–10), 940–946 (2005)

31. Calderon-Martinez, J.A., Campoy-Cervera, P.: An application of convolutional neural networks for automatic inspection. In: IEEE Conference on Cybernetics and Intelligent Systems, pp. 1–6 (2006)
32. Duarte, F., Arauio, H., Dourado, A.: Automatic system for dirt in pulp inspection using hierarchical image segmentation. Comput. Ind. Eng. **37**(1–2), 343–346 (1999)
33. Wilson, J.C., Berardo, P.A.: Automatic inspection of hazardous materials by mobile robot. Proc. IEEE Int. Conf. Syst. Man. Cybern. **4**, 3280–3285 (1995)
34. Choi, J.Y., Lim, H., Yi, B.-J.: Semi-automatic pipeline inspection robot systems. In: SICE-ICASE International Joint Conference, pp. 1166–1169 (2006)
35. Finogenoy, L.V., et al.: An optoelectronic system for automatic inspection of the external view of fuel pellets. Russ. J. Nondestr. Test. **43**(10), 692–699 (2007)
36. Ni, C.W.: Automatic inspection of the printing contents of soft drink cans by image processing analysis. Proc. SPIE **3652**, 86–93 (2004)
37. Cai, J., Zhang, G., Zhou, Z.: The application of area-reconstruction operator in automatic visual inspection of quality control. Proc. World Congr. Intell. Control Autom. (WCICA) **2**, 10111–10115 (2006)
38. Erne, O., Walz, T., Ettemeyer, A.: Automatic shearography inspection systems for aircraft components in production. Proc. SPIE **3824**, 326–328 (1999)
39. Huang, C.K., Wang, L.G., Tang, H.C., Tarng, Y.S.: Automatic laser inspection of outer diameter, run-out taper of micro-drills. J. Mater. Process. Technol. **171**(2), 306–313 (2006)
40. Chen, L., Wang, X., Suzuki, M., Yoshimura, N.: Optimizing the lighting in automatic inspection system using Monte Carlo method. Jpn. J. Appl. Phys., Part 1 **38**(10), 6123–6129 (1999)
41. Godoi, W.C., da Silva, R.R., Swinka-Filho, V.: Pattern recognition in the automatic inspection of flaws in polymeric insulators. Insight Nondestr. Test. Cond. Monit. **47**(10), 608–614 (2005)
42. Khan, U.S., Igbal, J., Khan, M.A.: Automatic inspection system using machine vision. In: Proceedings of 34th Applied Imagery and Pattern Recognition Workshop, pp. 210–215 (2005)
43. Clarke, E.M., Grumberg, O., Peled, D.A.: Model Checking. MIT Press, Cambridge (2000)
44. Karamanolis, C., Giannakopolou, D., Magee, J., Wheather, S.: Model checking of workflow schemas. In: 4th International Enterprise Distributed Object Computing Conference, pp. 170–181 (2000)
45. Chan, W., Anderson, R.J., Beame, P., Notkin, D., Jones, D.H., Warner, W.E.: Optimizing symbolic model checking for state charts. IEEE Trans. Softw. Eng. **27**(2), 170–190 (2001)
46. Garlan, D., Khersonsky, S., Kim, J.S.: Model checking publish-subscribe systems. In: Ball, T., Rajamani, S.K. (eds.) SPIN 2003. LNCS, vol. 2648, pp. 166–180. Springer, Heidelberg (2003). https://doi.org/10.1007/3-540-44829-2_11
47. Hatcliff, J., Deng, W., Dwyer, M., Jung, G., Ranganath, V.P.: Cadena: an integrated development, analysis, and verification environment for component-based systems. In: Proceedings of 2003 International Conference on Software Engineering (ICSE 2003), Portland (2003)
48. Ball, T., Rajamani, S.K.: Bebop: a symbolic model checker for boolean programs. In: Havelund, K., Penix, J., Visser, W. (eds.) SPIN Model Checking and Software Verification. SPIN 2000. Lecture Notes in Computer Science, vol. 1885. Springer, Heidelberg (2000). https://doi.org/10.1007/10722468_7
49. Brat, G., Havelund, K., Park, S., Visser, W.: Java PathFinder – a second generation of a Java model-checker. In: Proceedings of Workshop Advanced Verification (2000)
50. Corbett, J.C., Dwyer, M.B., Hatcliff, J., Laubach, S., Pasareanu, C.S., Robby, H.Z: Bandera: extracting finite-state models from Java source code. In: Proceedings of 11th International Conference on Software Engineering (2000)
51. Godefroid, P.: Model-checking for programming languages using VeriSoft. In: Proceedings of 24th ACM Symposium on Principles of Programming Languages (POPL 1997), pp. 174–186 (1997)

52. Robby, M.B., Dwyer, J.H.: Bogor: an extensible and highly-modular model checking framework. In: Proceedings of 9th European Software Engineering Conference held jointly with the 11th ACM SIGSOFT Symposium on the Foundations of Software Engineering (2003)
53. Mitra, S., McCluskey, E.J.: Diversity techniques for concurrent error detection. In: Proceedings of IEEE 2nd International Symposium on Quality Electronic Design IEEE Computer Society, pp. 249–250 (2001)
54. Barber, K.S., Liu, T.H., Ramaswamy, S.: Conflict detection during plan integration for multi-agent systems. IEEE Trans. Syst. Man. Cybern. B **31**(4), 616–628 (2001)
55. O'Hare, G.M.P., Jennings, N.: Foundations of Distributed Artificial Intelligence. Wiley, New York 1996)
56. Zhou, M., DiCesare, F., Desrochers, A.A.: A hybrid methodology for synthesis of Petri net models for manufacturing systems. IEEE Trans. Robot. Autom. **8**(3), 350–361 (1992)
57. Shiau, J.-Y.: A formalism for conflict detection and resolution in a multi-agent system. Ph.D. Thesis, Arizona State University, Arizona (2002)
58. Ceroni, J.A., Velásquez, A.A.: Conflict detection and resolution in distributed design. Prod. Plan. Control **14**(8), 734–742 (2003)
59. Jiang, T., Nevill, G.E., Jr.: Conflict cause identification in web-based concurrent engineering design system. Concurr. Eng. Res. Appl. **10**(1), 15–26 (2002)
60. Lara, M.A., Nof, S.Y.: Computer-supported conflict resolution for collaborative facility designers. Int. J. Prod. Res. **41**(2), 207–233 (2003)
61. Chen, X.W., Nof, S.Y.: A decentralised conflict and error detection and prediction model. Int. J. Prod. Res. **48**(16), 4829–4843 (2010)
62. Avila-Soria, J.: Interactive error recovery for robotic assembly using a neural-fuzzy approach. Master Thesis (School of Industrial Engineering, Purdue University, West Lafayette), (1999)
63. Velásquez, J.D., Lara, M.A., Nof, S.Y.: Systematic resolution of conflict situation in collaborative facility design. Int. J. Prod. Econ. **116**(1), 139–153 (2008)
64. Solomonoff, R., Rapoport, A.: Connectivity of random nets. Bull. Mater. Biophys. **13**, 107–117 (1951)
65. Erdos, P., Renyi, A.: On random graphs. Publ. Math. Debr. **6**, 290–291 (1959)
66. Erdos, P., Renyi, A.: On the evolution of random graphs, Magy. Tud. Akad. Mat. Kutato Int. Kozl. **5**, 17–61 (1960)
67. Erdős, P., Rényi, A.: On the strength of connectedness of a random graph. Acta Mathematica Academiae Scientiarum Hungarica **12**(1–2), 261–267 (1961). https://doi.org/10.1007/BF02066689
68. Watts, D.J., Strogatz, S.H.: Collective dynamics of 'small-world' networks. Nature **393**(6684), 440–442 (1998)
69. Albert, R., Jeong, H., Barabasi, A.L.: Internet: diameter of the World-Wide Web. Nature **401**(6749), 130–131 (1999)
70. Barabasi, A.L., Albert, R.: Emergence of scaling in random networks. Science **286**(5439), 509–512 (1999)
71. Broder, A., et al.: Graph structure in the Web. Comput. Netw. **33**(1), 309–320 (2000)
72. D.J. de Solla Price. Science **149**, 510–515
73. Bianconi, G., Barabasi, A.L.: Bose-Einstein condensation in complex networks. Phys. Rev. Lett. **86**(24), 5632–5635 (2001)
74. Nof, S.Y.: Collaborative control theory for e-Work, e-Production, and e-Service. Annu. Rev. Control. **31**(2), 281–292 (2007)
75. Chen, X.W.: Network Science Models for Data Analytics Automation. Springer Nature, ACES Series, vol. 9 (2022)
76. Chen, X.W., Nof, S.Y.: Agent-based error prevention algorithms. Expert Syst. Appl. **39**, 280–287 (2012)

Directed Graphs for Task Analysis of Human-Machine Systems

Steven J. Landry(✉)

The Harold and Inge Marcus Department of Industrial and Manufacturing Engineering, The Pennsylvania State University, 310B Leonhard Building, University Park, PA 16802, USA
slandry@psu.edu

1 Introduction

Task analysis is widely used within human factors, both within its practice and within research in human factors. Task analysis is a highly useful process for understanding and documenting a task, and can provide substantial insight into skill, personnel, and information requirements for accomplishing a task.

However, task analysis is predominately an exercise in enumeration, with few constraints on how to conduct the analysis or document its results. The product of task analysis is therefore highly idiosyncratic, making it impossible to validate, since no two enumerations will necessarily be similar, either in content or form. In addition, it is difficult, if not impossible, to compare two task analyses to determine if different ways of completing the task are similar or different.

There are also no methods that can reliably determine if the resulting enumeration is complete, exhaustive, or even correct. This problem is true of all existing task analysis methods, which limits their utility, particularly because such analyses are labor intensive.

One possible resolution is to formulate tasks as weighted directed graphs, derived from analysis, empirical data, and system data. In such a graph, each node is an elemental motion as identified in one of any of the established methods that decompose tasks into elements, such as THERBLIGS, the goals-operators-methods-selection method (GOMS), or the motion time measurement system (MTM). The (directed) edges between the nodes indicate the progression of these elemental motions toward accomplishing the goal of the task, and the weights of those edges are the probability of those two elemental motions being implemented in succession given repeated observations of the task being accomplished.

Formulating tasks as weighted directed graphs in this way would seem to address some of the limitations of existing task analysis methods. Using this method one can validate the task analysis graph, both its nodes and edges, by empirical and/or system data, and one can utilize the numerous methods and measures already developed for analyzing graphs.

In addition, a task analysis developed in this way is substantially more repeatable than other methods. While still challenging from a workload perspective, automated methods of data collection and graph development are possible with this new method.

C.-Y. Huang and S. W. Yoon (Eds.): ICPR1 2021, ACES 14, pp. 145–156, 2023.
https://doi.org/10.1007/978-3-031-44373-2_9

Lastly, this method is intended to identify a graph that shows **all** the ways the task can be accomplished, including failures. Typically, task analysis is intended to depict the ideal way the task can be accomplished. This distinction is important in that it means that different tasks can be compared as to their proneness to error, including the likelihood of those errors.

After a brief discussion of relevant work, it is shown how a task can be decomposed into a weighted directed graph. A manual example for a simple automated coffee-making task is provided, followed by conclusions.

2 Background

To support reader comprehension and the need for this work, a very brief review of task analysis and decomposition methods is provided, followed by a short discussion on weighted directed graphs. This section is not intended to be comprehensive on either of these topics, but only to provide sufficient information for the reader to understand and critically evaluate the work that follows.

2.1 Task Analysis and Decomposition Methods

Roughly speaking, task analysis consists of decomposing the work a person or persons must do to accomplish a goal. There are several purposes for such analysis, including as "an aid in modifying job operations so as to make their performance more simple and less liable to error" (Miller 1953, p. 2), for training and evaluation purposes, to determine the number of operators needed to accomplish a task, to determine the cost of labor for a product, and to identify the information requirements for decision making in the task.

There is a voluminous amount of work within the domain of human factors and ergonomics on various forms of "task analysis," including work that stretches back over many decades. This includes entire books (e.g., Jonassen, Hannum, and Tessmer 1989; Kirwan and Ainsworth 1992; Schraagen, Chipman and Shalin 2014), chapters in books (e.g., Adams, 1989; Annett, 2005; Hollnagel 2012; Stanton and Baber 2005), technical reports (Miller 1953), journal papers (e.g., Annett and Duncan 1967; Pinelle, Gutwin, and Greenberg 2003; Ramos et al. 2020), and conference papers (e.g., Crystal and Ellington 2004; Keller, Leiden, and Small 2003; Pirolli and Card 2005).

Moreover, task analysis is derived from much older work, at least as far back as the Gilbreths in the early part of the 20th century. The Gilbreths conducted work on motion studies, where the concept of "therbligs" was introduced to break down a complex task into a series of simpler motions. This was done to aid in analyzing and eliminating waste, mostly from manual, repetitive tasks.

There are many different "types" of task analysis, including (standard) task analysis, of which there are many documented methods; goal-directed task analysis; and cognitive task analysis. Standard task analysis is focused on the actions one must take and, sometimes, the decisions an operator must make to accomplish a task, along with the information needed to make those decisions. Goal-directed and cognitive task analyses are, roughly, focused on identifying the information and cognitive resources required to perform a task successfully.

Task analysis requires an analyst to carefully study a task, and then document the steps/actions an operator must take to accomplish the task. Usually, the documentation takes on a graphical or outline form, but there are also a substantial number of specific forms that are used for certain methods. Task analysis and its documentation can take a considerable amount of time, and usually involves repeated rounds of review and revision with subject matter experts.

What should be obvious from the above is that the resulting product of any task analysis is an enumeration. The enumeration may be good or bad, complete or incomplete, and any single task analysis is the product of the particular analyst and is unlikely to be repeatable. Moreover, there exists no method to validate the task analysis, although experience may invalidate it.

It would of course be preferable for a task analysis to be capable of being developed relatively quickly, validated with easy-to-obtain data, and for the resulting analysis to not be dependent upon the particular analyst. Using directed graphs as the basis for a task analysis seems capable of delivering these desirable features, while also improving the insight one gains from the task analysis.

2.2 Graph Theory

Graph theory is a set of mathematical methods to study structures that model entities that have relationships with one another. Graphs are closely related to networks, where networks are often depicted/recorded as graphs. Graph theory subsumes a very large body of work (see e.g., Bondy, Murty 2008; Deo 2016).

The particular subset of graph theory that is of relevance to this work is weighted directed graphs, which simply identifies "nodes," which are the entities, "links," which are the connections between the entities, if any connection exists, and the weights of those links. The links are "directed," in the sense that movement along the links from node to node represents a transition, as opposed to just a connection. The weight can be any meaningful information that represents a distinguishing characteristic of the edge to which it is attributed. As indicated above, there is a voluminous amount of work on directed graphs, with scores of attributes of graphs identified.

Without repeating that work, it seems useful to identify what advantages a directed graph depiction of a task would have. Specifically, below is a list of graph characteristics that seem helpful for evaluating a task, and that can be specifically computed when the task is put into the weighted directed graph format.

- Number of triangles
- Size of triangles
- Shortest path
- Parallelism
- Low/high probability paths
- Subsets of paths as procedures
- Number of steps in paths
- Number of choices/information quantity in paths

This is surely a very partial list. As a very rich area of research, it is likely that future researchers will find many useful additional tools, characteristics, and methods to apply to tasks once the task is instantiated as a weighted directed graph.

2.3 Task Analysis as Weighted Directed Graphs

The method for conducting task analysis using weighted directed graphs consists of the following steps:

1. Observe the task, if necessary, to obtain an initial set of steps for accomplishing the task.
2. Break down the steps into elemental motions.
3. Document each elemental motion as a node in a graph, connected by a directed edge indicating the sequential relationship between the nodes.
4. Repeat #1 as many times as possible, updating steps 2 and 3 as necessary. In doing so, the edges can be given a weight reflecting the probability that each edge emanating from a node is followed. (The sum of the probabilities of the edges emanating from any given node should add to 1).
5. Nodes that have only one edge emanating from it, with probability 1, can be collapsed into a single node.

Additional detail for each step is provided below.

Step 1: Observe the Task
Similarly to existing methods, an analyst must observe the task being accomplished, and identify the steps in the task. This is an initial step, and the analyst need only enumerate a structure sufficient to be evaluated, corrected, and detailed in the subsequent steps.

So, while there is an enumeration step to this analysis, as with other methods, that enumeration need not be as work-intensive or detailed as other methods, since this step serves as a structure to begin the graph, where the graph will be edited/corrected throughout the remaining steps. An example of the process and output of this step is provided later in the paper.

Step 2: Break Down the Steps into Elemental Motions
After the enumeration is completed, each task should be broken down into elemental ("atomic") elements. This is a common human factors practice, and there exist several methods for accomplishing it.

The simplest (and oldest) method is the use of "THERBLIGs." For each sub-task or motion, that element is broken down into a series of atomic elements, chosen from the following list (REF):

- *Grasp* object
- *Reach* with an empty hand
- *Move hand* while holding an object
- *Release* held object
- *Use* a tool
- *Assembly,* by placing one object into or onto another object
- *Select* from among two or more objects
- *Plan*
- *Rest*
- *Search* by hunting for an object with eyes, hands, or both
- *Find* an object
- *Hold* an object
- *Inspect* an object
- *Disassemble*
- *Position* an object
- *Pre-position* an object
- *Delay* (unavoidable)
- *Delay* (avoidable)
- *Transport loaded*
- *Transport unloaded*

The resulting enumeration is then a series of these atomic elements. Notably, this list is very short, was developed in the early 20th century, and is most applicable to manual labor tasks not requiring much movement or decision-making. (There are also multiple different versions of these elements.)

However, this method is the simplest to explain in this paper and is applicable to the task in the example. Other, more recently developed, and comprehensive methods exist to accomplish this same purpose and would likely be of more use for more complex tasks, particularly involving operator movement and decision-making.

Step 3: Document as a Graph

The enumeration from step 1, which is then made more detailed by step 2, can be depicted as a directed graph, where each step in the task is connected to the previous one. Steps where a selection must take place would necessitate a branch in the graph, where the succeeding steps depend upon the choice that was made.

The resulting graph is therefore "directed," as there is a direction to each connection ("edge") between two succeeding task elements. Once laid out in this way, all the methods and measures that are applied to directed graphs can be applied to the task analysis graph.

Step 4: Repeat and Update Graph

What is generated at this point is an anecdotal recording of the task, in that it was generated from one viewing of the task by one operator. It is necessary to repeat the task numerous times, with the graph being updated/edited each time to capture all the ways the operator completes the task.

In order for the method to be amenable to statistical analysis and estimation, it is desirable for the repetitions of the task to be selected randomly from the population of times the task will be accomplished. It is recognized that this is, however, unlikely since the task environment is not stationary (statistically) or ergodic. If it is not possible to randomly select the observations, then having a large number of observations and carefully recording the conditions under which the observations are taken are critical, so that the analyst can identify the conditions under which the estimated can be trusted to be accurate.

As repetitions are obtained, tallies of the traversal of each path can be recorded, resulting in estimates of the probability a path will be traversed by an operator in the future. That probability is of course computed as x/n, where x is the number of observations of that edge and n is the total number of observations. Such estimates allow analysts to estimate the likelihood of errors and the distribution of such measures as completion time.

Importantly, some paths may go unenumerated if they are so unlikely as to not appear in any of the observations. Such paths are not impossible, of course, but can only be estimated to be of less likelihood than the inverse of the number of observations. Path traversals are Bernoulli trials, whose probabilities are estimated using a binomial distribution.

For example, if 100,000 observations are made and a path is possible but not observed, then its probability is estimated to be less than 1/100,000 ($p < 1 \text{ x } 10^{-5}$). Moreover, a confidence interval around that estimate can be generated using the assumption that the path traversal are individual outcomes from a binomial distribution.

The resulting graph edges should contain these estimates, including confidence intervals on the estimates. Those estimates can form the "weight" of the edges for use in graph theory measures and analysis.

Confidence intervals can be computed using the standard formula for the confidence interval on the parameter p, where p is the "true" number of traversals of that edge, should all executions of the task be recorded. That formula is:

$$\hat{p} - z_{\alpha/2}\sqrt{\frac{\hat{p}(1-\hat{p})}{n}} \leq p \leq \hat{p} + z_{\alpha/2}\sqrt{\frac{\hat{p}(1-\hat{p})}{n}} \tag{1}$$

Therefore, if 100,000 observations are made and a particular edge is observed 5,000 times, the estimate for the probability that edge is traversed during the task is 5,000/100,000 = 0.05, with a 95% confidence interval of (0.049, 0.052). (The asymmetry of that interval is a result of rounding.)

2.4 Automated Data Collection

As might be obvious, the effort involved in observing, recording, decomposing, and documenting a task can be considerable. Fortunately, many applications might be able to take advantage of automated data collection and, perhaps, automated decomposition and recording.

For example, human-computer interfaces can record selections and other input actions. That data can form the basis for the initial graph, and repetitive use of the interface can generate data for updating/editing the graph. If the interface is used frequently, a very large volume of data can be obtained, and the graph can be continuously updated as new data is obtained.

For simple, stationary tasks such as interactions with computer interfaces, it may also be possible to automatically generate the graph itself. The software contains the possible paths within the interface code, which can then be identified as the nodes. Use of the interface by operators can then result in traversal of the various edges, and the simple selections and mouse/finger movements can be broken down into elemental motions, such as THERBLIGs, by software.

The following example, while conducted manually, points to this capability. While the interface analyzed in the example does not have this capability, it could easily be added by the developers of the system.

2.5 Validation

Once the graph is created and sufficient observations have been taken to obtain path traversal estimates of sufficient precision for the analyst, the graph should be validated. To conduct a validation, a new set of observations, not used in the creation of the graph, should be taken.

Prior to taking data, estimates and confidence intervals for edge traversals have been computed. If the graph is correct, then future observations should be consistent with those estimates. A chi-square test can be used to test if observed proportions are equal to the expected proportions.

If the p-value of the chi-square test is small, typically less than 0.05, then the validation has failed, otherwise the graph can be considered valid. If the validation fails, the analyst should determine why, correct the graph, and return to taking data to populate the graph edge estimates.

2.6 Graph Measures

Once the graph is created and validated, analysts can compute measures related to the graph in addition to estimates of probabilities of edge traversals. Such measures can have many valuable purposes, including:

(1) identifying paths through the task that are very rarely taken, which can indicate options that may not be need support;
(2) computing the number and size of "triangles," which are places where the operator "cancels" the current operation and returns to a prior step, as these can indicate errors or wasted effort on the part of the operator;
(3) identifying the number of decisions the operator must make, including the number of options from which the operator must choose, where more decisions and more options may reflect a more difficult or complex task;
(4) identifying the number of "wasted" nodes, such as delays; and
(5) identifying the number of parallel paths, where the operator is effectively doing more than one thing at once.

For (1), edges with very small probabilities should be identified. If there are paths that are unobserved, such that their probabilities are very small, that path should be investigated. If substantial resources are expended to support that path, those resources might be wasted, in that operators will not traverse that path frequently if at all.

For (2), edges that go from a step "later" in the task back to a "former" step reflect triangles. For human-computer interfaces these are easily identified by "cancel" or "back" operations. Generally, triangles are undesirable in tasks if those triangles are often traversed, as operators traversing the triangle have to repeat actions they have already taken.

Moreover, such actions often reflect cases where operators have executed actions that were either incorrect, perceived to be incorrect, or where the operator has lost track of the next actions to be taken.

Larger triangles are likewise usually undesirable, as they reflect operators having to go back to much earlier steps and re-doing more of the steps they had previously completed. If frequently traversed, the paths in the triangle should be investigated to elucidate the reasons for the reversals.

For (3), decision points are reflected in the graph as nodes where multiple paths could be taken. Simpler tasks typically require fewer choices, and tasks with more choices commonly require more experience/knowledge. In addition, more choices imply that those decision points must be supported by providing the operator with information to make those choices. Lastly, because each edge leading from a node is assigned a probability estimate, the entropy of the decisions can be computed.

For (4), most decomposition methods, including THERBLIGs, have identified node types that are typically wasteful. The most obvious example of this is "avoidable delays," but even "unavoidable delays" are usually wasteful if the operator cannot be productive during that delay. Other types of nodes, such as holding, finding, and pre-positioning are often wasteful and should be reviewed for efficiency improvements.

For (5), parallelism is a way to increase efficiency in that some steps might be able to be accomplished simultaneously with other steps, instead of perhaps enduring delays without being productive. However, multi-tasking in this way could impose burdens on the operator including task switching (e.g., Hartanto and Yang 2022; Vandierendonck 2018) and interruption penalties (e.g., Wirzberger et al. 2020).

Example. The simple example involves the making of coffee using a Cafection® Innovation Total Lite coffee machine. The procedure for making coffee is (mostly) specified directly on a touch screen monitor integrated into the machine. The procedure involves selecting the drink type, coffee type, size, and strength from among options, and dispensing the coffee into a cup the user has placed in the machine's receptacle. The menu sequence is shown in Fig. 1.

This procedure can be broken down into a set of elemental interactions using one of several options, as discussed previously. As also discussed previously, due to its simplicity and applicability to this task, the example here will decompose the task using THERBLIGs, the application of which is described above.

Using the method described in the previous section, the procedure can be broken down into a set of THERBLIGs, as (partially) shown in Fig. 2, which is depicted as a directed graph. This data was obtained using recordings of a video camera that was positioned to view operator interactions with the system, which were then documented by researchers. However, since this is a human-computer interface task, ideally the structure of the graph could be set up based on the software design, and interaction data would be recorded by the system.

Of particular note is that the system allows for "back" or "cancel" operations, which results in loops ("triangles") in the graph as described earlier. (Those triangles are not shown in Fig. 2 as they would make the graph very cluttered.) That is, after some selections are made but before the "go" button is pressed to start making the coffee, operators can "cancel" and return to previous menu items.

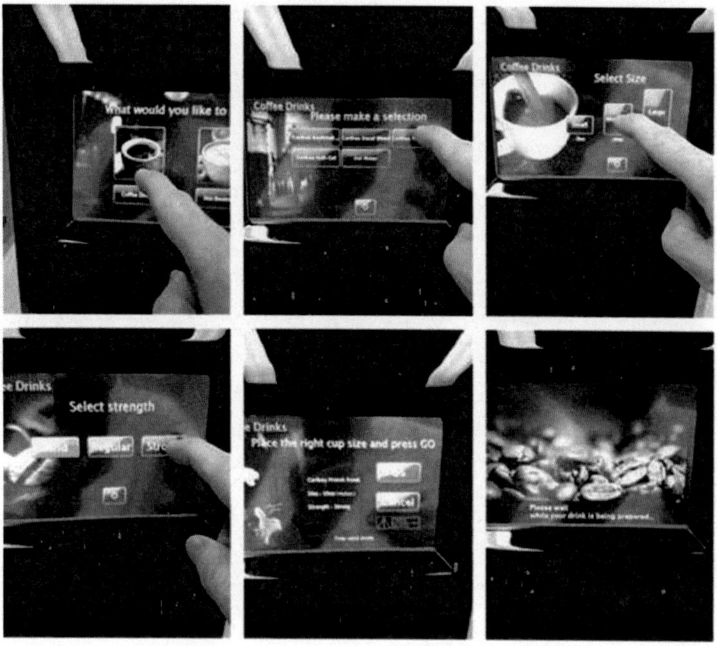

Fig. 1. Coffee making system interface.

As stated previously, such operations can be viewed as "errors" or at least undesired operations. Large triangles reflect having to go farther back into the menu hierarchy, where more selections would have to be re-done. Systems where triangles, especially large triangles, are utilized more frequently are likely less desirable.

As a manually recorded task using video, researchers could only exercise the method and not compute reliable statistics about user traversal of the graph. However, the following measures can be computed from the graph alone:

- Number of triangles = 40
- Size of triangles: 5 triangles of size 8, 27 triangles of size 10, and 8 triangles of size 14
- Minimum number of steps in paths: 5 step lengths of 19, 27 step lengths of 21, and 8 step lengths of 25
- Number of choices = 4 or 5 different decisions must be made along any given path, with the cardinality of those choice sets being between 2 and 9

These measures can be computed either manually, by viewing the graph, or automatically if the graph is stored in an electronically readable fashion. Alone, these measures do not carry much meaning, particularly as there is no base of task analyses using this method to which to compare. However, as more task analysis graphs are developed, comparisons can be made among these graphs to gain insight into what different values of these measures mean, if anything. In addition, as was mentioned, there are several

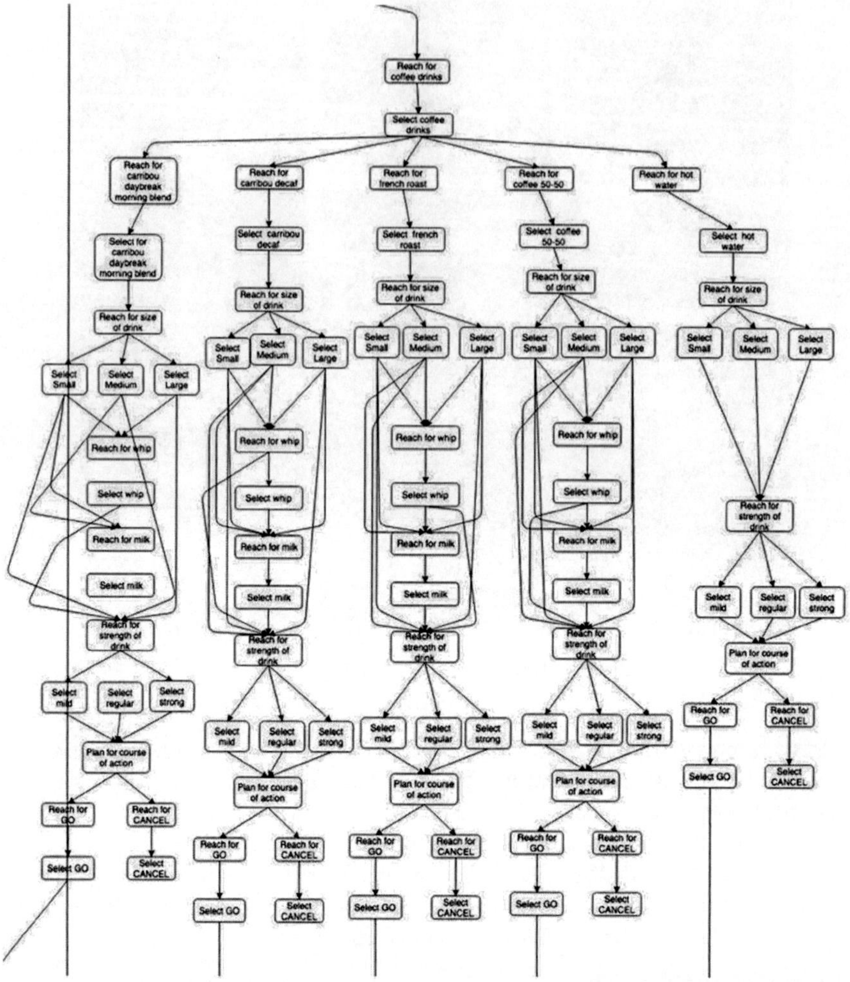

Fig. 2. Coffee making task directed graph (partial – actual graph is too large to depict).

aspects of the graph that can be examined for insight into how the task can be made more efficient, where operator load may be high, or where resources may be wasted.

User recordings would enable accurate calculation of the probability of traversal of any given edge, allowing for additional computations of such measures as:

- entropy, which if high would indicate few choices, if low would indicate that few paths were used regularly, and approximately 0.5 if all paths were traversed with approximately equal likelihood;
- statistics on path length traversed, which if high would indicate many steps needed statistics on to complete the task, which would be affected by cancellations/errors;

- estimates of the probability of particular paths being traversed, which would identify infrequently (or unused) paths where resources used to support completion of the task along those paths might be wasteful;
- statistics on time to complete task; and
- statistics on use of particular triangles, which would indicate potential sources of problems in the interface.

3 Discussion

Once sufficient data were obtained and the graph were validated, such computations would be reliable, in the sense that any analyst following the method would obtain the same results. In addition, the graph could be validated, as user operation data would indicate if there were paths being followed that were not found in the graph.

As reliable, verifiable data, there would be several important capabilities of such a method:

- A reliable figure for the time it takes an operator to complete the task could be computed, which can be translated into labor cost and/or productivity measures.
- A reliable set of decisions could be identified, which could be used to better analyze the cognitive demands of those decisions including the information needs of the operator to make those decisions.
- Given a second system for accomplishing the same task, a comparison of the graphs and, more importantly, of the measures, would enable a comparison of which system improved performance. For example, one could determine from such a comparison which system produced more traversed triangles, which could be indicative of a poorer-performing system since that would result from operators having to cancel and go back to previous menu items.

4 Conclusions

A method is introduced to conduct task analysis using directed graphs. The method appears to have significant advantages over existing methods, which are labor-intensive, rely on subject matter expert elicitation, and which are impossible to validate.

The existing method utilizes observed operator data to populate a directed graph, where the nodes are elemental tasks and the edges are the probabilities operators will traverse that node. Empirical data can be used to validate the graph.

The graph can produce numerous reliable estimates of important quantities, including "error rates," completion times, and number of choices to be made. The graph can also be used for comparison with other tasks or other methods for completing the same task.

Acknowledgements. The author thanks Dr. Harshwardhan Aggarwal, Michael Bailey, and Steven Ogbonna, who collected the data and compiled the results.

References

Adams, J.A.: Task Analysis Human Factors Engineering, pp. 161–182. MacMillan Publishing, New York (1989)

Annett, J.: Hierarchical Task Analysis (HTA). In: Stanton, N.A., Hedge, A., Brookhuis, K., Salas, E., Hendrick, H. (Eds.) Handbook of Human Factors and Ergonomics Methods, pp. 33–31–33–37. CRC Press, Boca Raton (2005)

Annett, J., Duncan, K.D.: Task analysis and training design. Occup. Psychol. **12**, 211–221 (1967)

Bondy, J.A., Murty, U.S.R.: Graph Theory. Springer (2008). ISBN 978-1-84628-969-9

Crystal, A., Ellington, B.: Task analysis and human-computer interaction: approaches, techniques, and levels of analysis. Paper presented at the Tenth Americas Conference on Information Systems, New York, August 2004

Hartanto, A., Yang, H.: Testing theoretical assumptions underlying the relation between anxiety, mind wandering, and task-switching: a diffusion model analysis. Emotion **22**, 493–510 (2022). https://doi.org/10.1037/emo0000935

Hollnagel, E.: Task Analysis: Why, What, and How. In: Salvendy, G. (ed.) Handbook of Human Factors and Ergonomics, 4th edn., pp. 385–396. John Wiley & Sons, Hoboken, NJ (2012)

Jonassen, D.H., Hannum, W.H., Tessmer, M.: Handbook of Task Analysis Procedures. Praeger Publishers, Westport (1989)

Keller, J., Leiden, K., Small, R.: Cognitive task analysis of commercial jet aircraft pilots during instrument approaches for baseline and synthetic vision displays. Paper presented at the 2003 Conference on Human Performance Modeling of Approach and Landing with Augmented Displays, Moffett Field, CA (2003)

Kirwan, B., Ainsworth, L.K.: A Guide to Task Analysis. Taylor & Francis, New York (1992)

Miller, R.A.: A method for man-machine task analysis. Retrieved from Wright-Patterson AFB, OH (1953)

Deo, N.: Graph Theory with Applications to Engineering and Computer Science. Dover Publications Inc, Mineola (2016)

Pinelle, D., Gutwin, C., Greenberg, S.: Task analysis for groupware usability evaluation: Modeling shared-workspace tasks with the mechanics of collaboration. ACM Trans. Comput.-Hum. Interact. **10**, 281–311 (2003)

Pirolli, P., Card, S.: The sensemaking process and leverage points for analyst technology as identified through cognitive task analysis. Paper presented at the International Conference on Intelligence Analysis McLean, Va, May 2005

Ramos, M.A., Thieme, C.A., Utne, I.B., Mosleh, A.: Human-system concurrent task analysis for maritime autonomous surface ship operation and safety. Reliab. Eng. Syst. Safety **195**, article 106697 (2020)

Schraagen, J.M., Chipman, S.F., Shalin, V.L.: Cognitive Task Analysis. Psychology Press, New York (2014)

Stanton, N.A., Baber, C.: Task analysis for error identification. In: Stanton, N.A., Hedge, A., Brookhuis, K., Salas, E., Hendrick, H. (Eds.) Handook of Human Factors and Ergonomics Methods, pp. 38–31–38–39. CRC Press, Boca Raton (2005)

Wirzberger, M., Borst, J.P., Krems, J.F., Rey, G.D.: Memory-related cognitive load effects in an interrupted learning task: a model-based explanation. Trends Neurosci. Educ. **20**, 100139 (2020). https://doi.org/10.1016/j.tine.2020.100139

Vandierendonck, A.: Further tests of the utility of integrated speed-accuracy measures in task switching. J. Cogn. **12**, 8 (2018). https://doi.org/10.5334/joc.6.

Human Factors and Sociotechnical Systems Integration

Barrett S. Caldwell[1](✉) and P. U. Grouper[2]

[1] School of Industrial Engineering and School of Aeronautics and Astronautics, Purdue University, West Lafayette, IN, USA
bcaldwel@purdue.edu
[2] Purdue University, West Lafayette, IN, USA

Abstract. This chapter describes areas of conceptual affiliation and shared interest between PRISM research and that of the Group Performance Environments Research (GROUPER) laboratory regarding the integration of humans, technological systems, and coordinated task performance. The history of GROUPER research builds on a sociotechnical systems tradition originally developed in the United Kingdom, as well as distributed expertise coordination and human-systems integration paradigms characteristic of cybernetic and human supervisory control models from the United States. This chapter describes important mathematical and engineering concepts describing system dynamics and performance measurement criteria that permit a quantitative study of teams, task, and time in complex settings. In addition, commonalities across application domains are utilized to capture and describe more generalizable principles with modeling value across a range of human-systems integration domains. This combination of applications, approaches, and criteria demonstrate this multidisciplinary approach to the design, evaluation and improvement of sociotechnical systems engineering analysis.

Keywords: Expertise · Feedback · Healthcare · Network/Security Operations Centers · Spaceflight · Systems dynamics · Task coordination · Team performance · Time delay

1 Introduction

In honor of the theme of this volume, the present chapter focuses the concepts of systems collaboration and integration on the many-to-many relationships associated with human performance in skilled groups and teams. The emphasis of the author's research laboratory, known as the Group Performance Environments Research (GROUPER) Laboratory, has long focused on the interplay of human expertise coordination, the challenges of human-systems-integration in complex engineering contexts, and the critical roles of emerging information technology systems to supplement other aspects of the engineering as well as organizational design and management process.

The complex interplay described above is alternatively presented in the literature as an *interdisciplinary*, *multidisciplinary*, or *transdisciplinary* approach to engineering. No attempt will be made to try to resolve or end this academic debate; the author

C.-Y. Huang and S. W. Yoon (Eds.): ICPR1 2021, ACES 14, pp. 157–180, 2023.
https://doi.org/10.1007/978-3-031-44373-2_10

only wishes to highlight how the integration of undergraduate degree discipline training experiences (in aeronautics and astronautics, as well as an eclectic "humanities" education emphasizing social psychology and sociology) and graduate degree specialization (in group dynamics and social psychology), focused in a faculty environment of industrial engineering, represents an approach combing individual, social, societal, and technological factors affecting information technology design and user experience in high-consequence task performance settings.

2 Conceptual Foundations: Sociotechnical Systems

A fundamental conceptual orientation that integrates these considerations is known as *sociotechnical systems* [1]. This orientation was developed at the Tavistock Institute in the United Kingdom during and after World War II, and addresses both psychological and social factors addressing the constraints affecting the development, use, and organizational integration of technology use. At the individual and small group level of analysis, this sociotechnical approach could be used to address work group norms and team-based differences in technology effectiveness (such as the original studies of differential adoption of mining technologies with different communities of miners). At broader social and societal levels of analysis, a sociotechnical approach can examine how different organizations, operating in a complex and dynamic environment, can utilize different information gathering and sharing strategies to determine how to adapt to a changing technological and economic landscape [2].

Similar concepts of individual, group, and organizational adaptation, response, and matching of skills and capabilities to environmental conditions may be seen across multiple research disciplines from the 1930s – 1980s. For instance, Murray's [3] theory of individual personality considered an array of over 20 distinct "manifest needs," with a similar number of "environmental press" conditions to which an individual's needs may be more or less well matched. Lewin [4, 5] considered dynamic and even "topological" processes to describe an individual's development and experience as a member of an affiliative reference group, which may influence how and with whom they may share particular types of information and norms affecting self- and group identity. Rogers' [6] conceptualization of "diffusion of innovations" emphasized dynamic processes of how a new process, product or concept (the innovation) is communicated and shared among members of a target group over time (concepts that are a continuation of Lewin's social communications and persuasion work during World War II). A collection of essays written in the 1980s were gathered into a broader discussion of technological innovation concepts ranging from Rogers' individual and social network considerations, to organizational-level staffing and work functions, to corporate- and governmental-level policy decisions, to even cybernetic discussions of corporate strategy [7, 8]. Additional cybernetic and system dynamics approaches to studying sociotechnical functions of systems from organs to multinational organizations are seen in writings of authors ranging from Ashby [9] to von Bertalanffy [10] to Smith [11, 12].

Despite this range of disciplinary backgrounds and application "grain sizes" (level of analysis ranging from body subsystems to individuals to groups to organizations and beyond), there was a growing appreciation (in some communities) to use mathematical descriptions and recognition of flow variables, energy balances, and equilibrium states as a common language for description, modeling and analysis of human-scale sociotechnical processes [12–14]. Innovators in computer hardware systems even envisioned the capability to conduct computer-based simulations and dynamic modeling of social, sociotechnical, and organizational systems over time [15, 16].

For a young systems engineer in the late 1980s and early 1990s, this seemed like a particularly fertile area of research. The explosion of personal-scale computers in office workplaces and homes, combined with advancing network communication technologies, allowed for an especially vivid description of information flow and knowledge coordination between both professional expert and social affiliation communities of interest or practice [17–19]. Although the lead author of this chapter (Caldwell) had been introduced to the mathematical techniques and underlying system dynamics considerations within the context of aerospace and other engineering analyses, these skillsets were not nearly as frequently represented in a graduate program in psychology. Authors such as Forrester recognized this potential tension between social contexts of non-mathematical discussions of systems dynamics, and technical contexts of engineering system behavior analysis [20].

Based on the lead author's educational emphases on space flight systems, there were particular highlights to help scope this burgeoning combination of disciplinary approaches and methodologies. When considering human spaceflight (in the United States) beyond the original Mercury astronauts, all space flight missions represented multiple astronauts coordinating with teams of ground-based mission control personnel. This "reference application" of highly trained experts (astronauts, flight controllers), working with highly complex engineering systems (spacecraft and ground control systems, with satellite-based communication networks transmitting both engineering status and science mission data), in a cooperative setting (rather than competitive settings ranging from political debate to sports contests to military operations) emphasizes a particular type of group task specialty application of relatively lower emphasis in the research literature [21]. This chapter will not focus on the evolution of spaceflight automation systems to conduct exploration and science; the reader is directed elsewhere for a brief history and sociotechnical emphasis of such automation approaches [22]. This chapter will instead highlight the sociotechnical considerations of information flow and knowledge sharing; human-systems integration; and team coordination and task performance in a variety of complex work applications, with the original human spaceflight use case as a primary exemplar of crucial sociotechnical systems factors.

3 Teams, Tasks, and Time

A team can be defined as a set of entities, both human and non-human (i.e., computers or animals), that work together to accomplish a goal. Analysis of team-level cognitive factors is significantly more complex than on the individual-level. Not only do successful teams need to engage in *taskwork* (i.e., the performance of tasks related to the overall goal of the team), but they must also perform activities related to *teamwork* (i.e., sharing information and managing the socio-emotional state of members) and *pathwork* (i.e., maintenance and support of communication channels to ensure that they meet the team's needs) [23]. Taskwork itself in complex systems contexts requires high levels of coordination and communication, and both teamwork and pathwork are meant to support critical information exchanges (though their associated tasks may themselves require coordination and communication).

Effective information exchange requires a shared language or vocabulary (to decode the information), shared expertise (to understand the information), and shared situational awareness (to understand how that information applies to the current system state). Breakdowns or mismatches in any of these dimensions decrease the quality of the information exchanged. In complex systems operating at the edge of human understanding or in highly time-critical contexts (i.e., spaceflight, healthcare, etc.), effective information exchange can be critical to preserving human life. One area of particular emphasis within GROUPER is the concept of how distributed (composed of both co-located and remote members) teams get, share, and use information.

While a number of authors have studied group behaviors and processes over the past 75 years, it is important to note, as McGrath [21] indicated, that relatively little of this published research had emphasized the types of task performances represented by highly skilled actors for whom deciding what to do, and then executing the decision, is a joint activity. Three considerations are especially important for these types of tasks:

- ***team*** members are skilled participants who engage in multiple cycles of activity and training (even if not performed with exactly the same teammates each time);
- ***tasks*** are performed in an uncertain environment where success is focused on effective achievement within physical and technological constraints, and not simply determined by the ability to "persuade," or "win," in a competitive setting against others;
- tasks are performed within time constraints where available ***time*** to complete required tasks may expire before gathering all required information, determining which actions to perform, and successfully executing those tasks.

These considerations might be seen as elementary for examination of coordinated behaviors in complex sociotechnical task settings. However, simply the issues of temporal factors affecting team performance were seen as a distinct and under-appreciated aspect of group processes, including both sociocultural and technical considerations [24–26].

Especially when considering the additional sociotechnical complexities engendered through the use of modern information technologies, the questions of how teams communicate and coordinate information, knowledge and task performance become increasingly complex [27]. The following sections address some additional details regarding the considerations of expertise distribution, task interdependence and coordination, and the effects of time as both a resource and constraint/cost when performing high criticality tasks.

4 Distributed Expertise

A primary source of the team performance literature regarding task performance (rather than decision making or persuasion) has been that of military team operations [21]. In this type of task setting, the concept of *command and control* (C2) has been a fundamental philosophy of management and manipulation of "own forces" at various levels of aggregation [28, 29]. Essential aspects of C2 include organizational level aggregations of situation awareness [30], including sensing, interpreting, and projecting states of the environment, and determining the needs for current and future actions. A historical view of C2 reflects *distribution* in this physical and sensorimotor sense, of the capabilities of the commander being augmented by the ability to gather information from, and direct actions to, a larger number of those commanded. Although one may claim that a military operation is distinct from other forms of organizational management because of the military objectives of actually destroying other resources or personnel [29], the C2 emphasis of coordinating the actions of the commanded by commanders reflects almost exactly the distinctions made in scientific management regarding the "management and the men" [31].

Originally, the C2 models of scientific management or traditional military command reflect an assumption of hierarchies of education and knowledge, and not just differences in experience and skill. (It is notable that Frederick Taylor's descriptions of scientific management had to contend with an organizational reality that many of the unskilled or semi-skilled workers in early 20th Century production organizations may not be literate in English; less than 10% of the workforce would be expected to have any education past high school [32]). Over the 20th Century, not only greater capabilities and resources for use of technology to improve sensing, perception and interpretation of conditions, but increasing education and skill development across levels of the sociotechnical organization, have served to create higher levels of available expertise for even lower organizational level participants in large sociotechnical entities.

The capabilities of C2 enabled by the increase in information and communications technologies (ICTs), then, enables greater development of expertise, and more effective sharing of expertise, across different levels of the organization; this, combined with increasing decentralization of command and coordination networks, enables a greater overall performance capability [28]. However, even military analysts such as Alberts note that non-military team performance (such as humanitarian assistance and disaster response, or HADR), with non-competitive outcome objectives, are significantly affected by the ability to decentralize and decouple sources of expertise to enable more adaptive and effective actions in response to time- and resource-constrained task settings.

As discussed above, effective teams require more than just accumulations of primary sensorimotor information and distributions of physical actions. Aspects of team effectiveness in complex settings (such as undersea, aviation, and spaceflight missions) integrating task performance and socioemotional or group maintenance functions have been studied since at least the 1960s [33–35]. It is worth noting that there are different types of expertise that may be present throughout a team as well [36]. The term is perhaps most frequently used to describe subject-matter expertise, which emphasizes domain knowledge. Other dimensions emphasize the recognition of environmental context (often described as situation awareness), awareness and identification of other team members' areas of expertise, awareness of available and appropriate communication channels, actual communication skills, and interface tool usage. This conceptualization, developed within GROUPER as "six dimensions of expertise," is broader than simply assigning task roles and training to criteria in specific areas of subject matter knowledge, physical capability, or accurate completion of operational rules; further, these distributions of expertise operate at individual, team, and multi-team levels of sociotechnical aggregation [37].

5 Coordinated Performances

Along with performance demands and capabilities enabled by increasing dimensions and distributions of taskwork and teamwork *expertise*, there are both challenges and gains associated with task performance *coordination*. Coordinated performance requires additional understanding between team members regarding issues of function allocation and handoffs, requiring both individual and team levels of situation awareness of who does what when, and how are task and knowledge elements synchronized [25, 37–40].

Traditional approaches to C2 and scientific management frequently design models of functional decomposition that intentionally limit the types and amounts of collaboration or handover coordination between actors. This is also why it has likely become the predominant method of functional decomposition in supervisory control operation with robotic or similarly foreign units, due to the simplicity of schemata and information flow pathways involved [41–43]. However, this type of allocation model lacks the sort of fidelity that can be found in true team-based collaboration, and thus to some extent limits the potential of the work and parties involved [44].

By contrast, effective teamwork requires a great deal of communication, and may have some distinct difficulties that must be assessed if all parties intend to functionally contribute [45]. For one, under most circumstances, there is not an individual supervisor of the project, but rather a supervisor designated by each entity [46]. This means that if an entity attempts to establish a single representative as the total administrator without previous agreements and much in the way of negotiations, this will go against the established structure of team-based collaboration and is thus highly likely to result in failure [47].

With regards to coordinating a number of disparate units, the style of collaboration and thus communication used is likely to impact the type and purpose of the resulting coordination. This is escalated if multiple sets of units are deployed at a single time, each with different initial conditions and update statuses or requirements. From a design and training perspective, robotic, foreign, or similarly auxilliary units are more likely to require high levels of task-based coordination and lower levels of team-based coordination, as these units are not confirmed (trusted through shared experience) to be able to act and behave as a separate entity [48]. As may be expected from this comparison, effective autonomous or commanding units are deemed more capable, thus requiring more team-based coordination to maintain their effectiveness and less task-based coordination to avoid micromanagement and conflict [49].

Under some circumstances, delivering all information to all parties may not be practical or even feasible, resulting in priorities and to some extent a disparity in distribution [50]. This in and of itself might not be an issue, so long as the task flow is represented in its best capacity, but severe problems can arise if this results in a misalignment in overall knowledge. If the fundamental objective of the collaborative efforts faces a certain level of divergence, the conclusion of some activities will likely follow one way or another

[51]. As team-based collaborators are much more susceptible to this type of error, this is one of the primary reasons why more communication is required as a minimal baseline for operation [52].

6 Time as Resource and Constraint

As described earlier, a fundamental constraint affecting task coordination and performance in complex sociotechnical systems settings is the time available for actors to complete their required performance in order to have a desirable outcome in the situation[25, 26]. Importantly, the consideration of such time available (as time to deadline) is an additional complicating factor that increases perceived stress, degrades actors' performance compared to known capability, and impairs decision making and judgment [53–57]. For high-criticality or extremely time-sensitive events such as HADR, or achieving a particular trajectory for a spaceflight launch operation, failure to meet an important deadline represents an overall mission failure mode that cannot be overcome with other resources of people, money, or equipment.

However, this concept of time as a performance constraint has considerable complexity not fully elaborated in traditional human factors or psychological laboratory studies of time pressure [37]. Despite the use of a particular research paradigm in laboratory studies addressing an objective measure of time pressure to a particular passage of clock time, there are multiple moderating and confounding factors addressed by situational, information, and expertise conditions. One clear impact of expertise in both physical and cognitive domains is that experts are clearly able to perceive, assess, and respond to task demands faster, with greater accuracy and efficiency, than novices [37, 54, 58–60]. As a result, a more accurate approach to considerations of time constraints and time pressure may be a dimensionless ratio, T_r/T_a, where T_r and T_a refer to time required and time available, respectively, *for those actors in that task situation* [61, 62]. Here, expertise can represent a time-saving or time-recovering resource; effective team coordination and expert performance integration are similar processes capable of enabling additional performance in a given period of time [63–65]. In addition to cognitive and physical speed and accuracy, high-performing team members in sociotechnical settings have several additional expertise-related skills and processes available to use time effectively.

When performing critical tasks in high-consequence settings, obtaining access to information and material resources is an important activity, which can be described as resource foraging [66]. Skilled task performers can use acquired expertise to recognize when and how to acquire resources "proactively," in advance of a specific critical event, or "latently," in the absence of a specific event demanding their use [66–68]. Increasing expertise allows task performers to anticipate task patterns and resource needs; increased communication among experts may create an additional capability for effective performance and "shared situation awareness" or "shared mental models" [39, 52, 69, 70].

7 Sociotechnical and System Dynamics: Analogs and Homologies

Models are particularly helpful when it comes to understanding the behaviors of complex systems under different conditions. It is rarely, if ever, possible to assess and monitor the effects of all relevant variables in the dynamic performance of a sociotechnical system. Environmental conditions and prior histories affect actors' beliefs, expectations, and choices of both decision and action. Individual and cultural variables (including non-task factors such as prejudice, tolerance, and valuing of personal characteristics as well as expertise contributions across multiple dimensions) cannot be experimentally controlled or reassigned. Therefore, all studies – both experimental laboratory efforts and real-world examinations – must build on models of varying levels of fidelity and representation of factors assumed to be of importance.

Both analogs and homologies are approaches to modeling systems; it is important to recognize that these terms have distinct meanings and levels of application in specific sociotechnical settings. Analogs, especially in a spaceflight context, are simplified models that emphasize a subset (though not all) of the attributes of the system of interest. For example, Mars analogs might be able to recreate the communication delays between Earth and Mars to allow for Earth-based research of the impact of delay [23, 71–73]. However, these analogs cannot and do not recreate or represent all conditions that would be present on a real Mars mission, such as differences in gravity, atmospheric conditions, or life-support equipment in use during a field exploration sortie. Thus, it is critical when developing any analog study of the performance of a complex sociotechnical system to understand construct and ecological validity of relevant system performance measures and behaviors of actors [74–77].

Homologies, in contrast, make use of system abstraction in a far more explicit manner when addressing and quantifying the dynamic behavior of particular system parameters. Meadows gets at this concept of a "systems zoo," where a number of different systems may be represented by the basic structural combination of stocks and positive (or self-reinforcing) and negative (or balancing) feedback loops [14]. Despite surface differences in structure, scale, or physical form, the mathematical similarity in function and performance described by the homology allows for both greater understanding of underlying conceptual similarities, and guidelines for applying similar mathematical analysis tools [11]. Thus, homologous are models that utilize comparisons between the system of interest and another system with similar underlying mathematical laws governing their behaviors [10], though the specific mechanisms at work may not resemble one another to the casual observer. For instance, physiological processes of an organism's homeostatic controls, ecological patterns of predator-prey dynamics, and business dynamics of supply chain instabilities share very few surface features, but can be shown to be described by very similar mathematical forms of second-order differential equation models [10, 16, 78]. In fact, Bertalanffy notes that while conceptual analogies might be "scientifically worthless", homologies by contrast "often present valuable models, and are therefore widely applied in physics" [pg. 84–85]. (One should note that Bertalanffy was considering "analogies" as describing casual metaphoric or superficial similarities, rather than the sense of "analogs" as emphasizing achievable and relevant selections of a subset of important model parameters for further study.) Within GROUPER, such homological approaches have been used to similarly describe human expectations or tolerance for transmission delays and acceptance of those delays and update cycles for coordinated information flow in various task settings and temporal processes [61, 79–81].

8 System Dynamics and Coupled Flow Models (Abstracted Concepts)

Although "systems engineering" is sometimes taken to refer to a single discipline, the qualitative and quantitative study of systems dynamics and coupled feedback flows has covered multiple disciplines addressing biological, ecological, technological, and sociotechnical systems over the 20th and 21st Century [14, 82]. Behaviors and stability of species populations have been shown to follow important differential models of coupling with predator, prey, food, and other environmental considerations [83, 84]. A critical primary stage of systems modeling includes the determination of how components are linked via energy, information, or material flows, and whether these links are balancing (representing stable, negative feedback flows) or reinforcing (representing potentially unstable, positive feedback flows) in their effects [14, 15].

As described above, it is important to understand the differences between metaphoric "analogies" of system dynamics models and actual mathematical "homologies" that provide insights as to how to improve the modeling and analysis of sociotechnical systems. An area of interest for a number of years is that of agent-based modeling, where the behaviors of individual actors in a complex sociotechnical system are represented by mathematical constructs with descriptors of quantitative states and behaviors over time [85–88]. When studying human system dynamics, the important criteria are to demonstrate that agents behave *plausibly* (representing the range of actual human behaviors – including boundedly rational and non-rational decisions and actions, with varying probabilities of accuracy or success) rather than *ideally* (always following a set of rational, optimal decision and behavior strategies). Some work in GROUPER has even addressed issues of social similarity and bias affecting communications effectiveness and cohesion in performance [89, 90].

Along these lines, it is also important to understand how seemingly simple deviations from ideal conditions can significantly affect agent and system performance. For instance, while time delay in information flow is a ubiquitous aspect of Human-Computer Interaction (HCI), the impacts of such delays may be incompletely addressed [61, 81, 91]. (Additional considerations in delays affecting sociotechnical performance will be presented in a later section.). One version of time delay affecting performance is an interruption, where progress on one task must be paused while another task (of higher urgency or immediacy) must be attended to and performed [92, 93]. Interruption studies are important in the study of human behavior and help make the world of HCI more seamless and realistic. In previous research, interruption studies show that inappropriate interruptions not only annoy users but also hinder their productivity, thereby affecting their emotional and social state [94, 95]. Interruptions can also lead to errors when a person is highly focused, leading to more stress and frustration [96, 97]. These types of studies suggest the potentially negative effects of previously unforeseen or unanticipated coupling between different levels of HCI interaction (such as tasks in Virtual Reality (VR) or eXtended Reality (XR) interaction environments). Future work is required to identify the role, impact, and mitigations of interruption in more immersive experiences, such as when users are faced with "real-world" tactile interruptions while carrying out a task in a 3D virtual environment.

9 Operations to Reference Cycles Over Time

Popular conceptualizations of expertise may imagine the development of expert skills as a monotonic improvement that occurs with repeated practice; this conceptualization is not borne out by research at individual or organizational levels [58–60, 98–100]. Patterns of rapid increase in capability or innovation may be separated by long periods of performance fixedness or relatively small incremental improvements. As a result, it is important to understand not just that expertise develops, but how and which particular processes may help support those increases in expertise.

Processes of team training and coordination in aviation, healthcare, and military operations contexts emphasize the importance of frequent performance cycles, where team members can obtain feedback on previous operational trials to help reference and distinguish effective from ineffective task activities [101–106]. An additional element of these processes is that of "scaffolding," where experience and feedback for learning are placed within an appropriate instructional context that allows the learner to effectively understand, integrate, and apply the learned material [107, 108].

Scaffolding and other instructional tools may be seen as a set of "reference supports" to enhance and structure the feedback associated with "operational experience". In most cases, the reference supports of textbook readings, historical documents and interviews, procedural manuals or other forms of received knowledge represent (to many) a set of fixed foundational materials. However, high-performing expert teams conducting performances at the edge of human experience do demonstrate a more explicit closed-loop cycle of utilizing and revising reference materials in the face of new operational experience [109]. These emergent capabilities, built on more explicit and ongoing updating of procedures and decision criteria, are described in some discussions of advanced military operations in the information age [28], as well as human spaceflight operations [64, 110].

An interesting, and unexpected, application of this multi-scale approach to operational experience and reference knowledge addresses considerations of enterprise- and individual-scale experiences of organization process cycles, with sociotechnical implications not unlike those suggested by Forrester or Wiener [15, 16, 111]. Consider the increasingly important challenge of supporting underrepresented minorities (URMs) in a higher education institution devoted to equity, diversity and inclusion. While a university (operating at the enterprise level) may keep longitudinal statistics and conduct "organizational feedback control" to improve URM retention or success over a number of years, this does not necessarily result in increased success outcomes *for individual students* experiencing that university in a single pass through the curriculum. At that level, each student is presented with decision and action policy choices that may expand, or restrict, their likelihood and range of success outcomes. Those outcomes are also dependent on the goals, contexts, and priorities of the student themselves. The concept of a *success pathway* represents an opportunity for an "operations to reference" guide for each student to apply to their own experiences in ways that expand their "operating range" of possible success outcomes [112]. It is also important to consider an appropriate cognitive framing approach [113] to consider those URM students as valuable assets to support, rather than just deficient raw materials to process at lower value using the same processing steps as other students.

10 Dynamics of and Tolerance for Delay

One of the most significant challenges facing distributed expert team coordination from a theoretical, operational, and modeling perspective is that of coordination in the face of information flow delays and lags. Each stage of information acquisition, sensemaking, performance execution, and communication/coordination is subject to production and transmission delays that include individual, social, and technological constraints [114]. When team members are physically distant, or data sources require significant time to process and synthesize to generate human-usable information, these time delays can become significant.

An axiom of complex system dynamics and cybernetics is that no system controller can operate instantaneously, either in terms of input or output processing [78, 111]. When forming a *decision policy*, the controller will always operate on a state of the world that has a lag of at least some τ_d, referring to the sensemaking and decision selection time after relevant data are acquired about the state of the world (which may have changed since the data acquisition process began). Similarly, an *action policy* can only be executed with a lag of at least τ_a, referring to the time required to perform intended actions, including physiological movements, command transmission, and completion of physical servomechanism or network activation actions in a cyber-physical context. Complex information processing or action coordination may require additional time delays to ensure that distributed actors are performing actions with the appropriate timing synchronization ("on my mark"). However, increasing lags and desynchronization affecting decision and action policies can also result in the potential for system control instability and performance degradation [115]. For example, τ_d and τ_a lags of only a few milliseconds when incorporating and synchronizing image processing data for the Mars helicopter *Ingenuity* resulted in highly unstable flight performance [116].

As a general rule, humans exhibit considerable performance instabilities when confronted with τ_d and τ_a lags across levels and stages of task performance, ranging from continuous control operations [117, 118] to incorporating the impact of feedback delay in complex decisions [20, 119, 120]. In fact, the famous "bullwhip effect" in organizational decision making can be described as an unstable controller (modeled with a second-order differential equation: see [121]) performing overcompensating responses to delayed input information, a performance mode that can be moderated by increased damping (in a mathematical sense) and moderating decision / action policies (in an operational sense) [61]. Depending on the task, expertise, and environmental conditions, there are dynamics in both the experienced lags and the damping ratio of supervisory controller processing of the costs and benefits of lagged information. This suggests that the energy cost (resistance or friction terms) as well as the energy benefit (voltage or spring terms) coefficients of the differential control model are functions of time, a formulation that is not closed-form solvable. Thus, numerical simulations of sociotechnical systems with time-varying values of the influence of system delays on performance are the primary forms of system dynamics modeling for description and analysis of such systems. This is an area of very limited research to date.

11 Sociotechnical Applications: GROUPER Emphasis

The multidimensional, transdisciplinary, sociotechnical study of distributed experts performing mission-critical tasks in context has been a longstanding characteristic of GROUPER research since the 1990s. However, rather than suggesting a hodgepodge of project activities, the GROUPER history represents an intentional consideration of how to develop and apply homological models to the processes of information access, sharing, and utilization by skilled experts. Concepts such as distributed supervisory coordination [23] specifically extend the paradigms of human supervisory control [115, 117] to consider human-human task coordination of multidisciplinary expertise through telecommunications channels, and not simply the traditional concept of human control and direction of mechanical servomechanisms.

It is especially important to understand that constraints of time delay and incomplete information are central to GROUPER examinations of sociotechnical systems, ranging over time scales from milliseconds to minutes to even months [68, 71, 122]. Increases in communications bandwidth or storage access may be imagined to overcome inefficiencies or coordination lags in variety of organizational contexts; by constrast, such lags are considered practically as well as theoretically insurmountable in many GROUPER applications. This allows a more explicit emphasis on how teams of experts respond to these constraints with social and operational adaptations, rather than assuming a reliance on technological tools to achieve complete solutions.

12 Spaceflight

As described above, the robustness of human spaceflight and exploration relies on the coordination of highly trained experts working with highly complex engineering systems both in space and on the ground. Event detection, isolation, and recovery (EDIR) in spaceflight mission control is a critical form of distributed expert troubleshooting of ever-present anomaly risks in on-board engineering systems, communications networks, or ground-based processing that may or may not result in catastrophic mission outcomes [71, 114, 123–126]. These troubleshooting processes occur using analog and digital telemetry signals, computer commands, and human-human voice communications between the on-board crew and the ground-based mission controllers [23, 71].

In space exploration, the crew and mission teams are spatially, functionally, and experientially dispersed, with unique ways of thinking, communicating, and operating in the high-pressure, unforgiving space environment. Workload for both the crew and ground teams can be lengthy, complex, and at times require creative problem solving. Additionally, as crews explore the Moon and Mars, communication delays will require increased crew independence and autonomy in day-to-day task coordination and critical time-dependent events. There are physical speed-of-light constraints for how quickly information can be transmitted between Earth ground stations and a crew in space; these lags are approximately 2–5 s for Earth-moon coordination, and may be as long as 20 min one-way for Mars message transmission, even without consideration of periods of opposition due to the Sun being directly in the path of Earth-Mars line of sight (and direct communications between the planets are impossible).

Agent-based modeling of spaceflight mission operations has attempted to understand processes of experts asking, sharing, learning, and problem solving potential EDIR challenges in this extreme sociotechnical setting [125, 126]. Mission controllers and spaceflight crews are trained in an extensive simulation environment on both taskwork and teamwork aspects of distributed expertise [64, 123, 127], with a common focus on respectful knowledge sharing and robust overall mission success, supported by multiple engineering and system dynamics displays in an extremely high-consequence environment. However, it is recognized that the complexity of the spaceflight environment and vehicle systems themselves can affect the processes of state determination, sensemaking, and event recovery. As such, they represent a unique opportunity for agent-based models of organizational performance [85].

Autonomy and function allocation in the spaceflight mission operations environment are sociotechnical considerations with dynamic flexibility. Rather than relying on fixed definitions or determinations of which technical components are declared to be independent, both human and automation systems perform according to *autonomy from whom*, and *autonomy to do what* [22, 126, 128, 129]. Such determinations can allow engines to automatically throttle at periods of maximum dynamic pressure with more accurate timing than humans; robots can shift directly to safe mode operations when vehicle health is compromised; and astronauts can pause exploration or repair activities while waiting for ground-based response to an anomaly. The capability for complex human-system integration operating under rules of dynamic function allocation and flexible autonomy remains a major systems development challenge [130, 131].

13 Healthcare Delivery Coordination

Healthcare team coordination to identify and reduce the risk of adverse events in patient care delivery is another example of distributed expertise and performance in a high consequence setting [132]. In some ways, healthcare delivery represents an antithesis of the type of sociotechnical setting seen in human spaceflight mission coordination. While a distinct, purpose-built engineering systems design capability and taskwork / teamwork training emphasis defines the "mission control" setting, healthcare operations (at least in the United States, where most GROUPER research has been conducted) operates with inconsistent and incompatible technologies, cultures, resources, and economics at different stages of aggregation—in essence, the "system of systems" of healthcare coordination [133, 134] is not a designed system at all.

GROUPER work in this area began in the 1990s with an examination of non-surgical healthcare delivery settings such as clinical laboratory information processes [135] and radiation therapy adverse events [136, 137]. In addition to the distributed expertise in care planning and delivery coordination integrating distinct skill sets of oncologists, nuclear physicists, and dosimetry technicians, these settings were also unique for required reporting of adverse events of excessive radiation doses to patients, unlike other medical specialty areas [137, 138]. This emphasis on care coordination and information exchange between various members of the care delivery team has become a major theme in GROUPER work in the healthcare sector. The different forms of coordination between nurses and physicians, physicians and pharmacists, and even technicians and clerical support help to identify distinctions of even when an "event" starts for different participants in clinical care delivery, and how resource foraging for materials, information and physical spaces (such as treatment rooms or surgical suites) combine proactive as well as reactive resource coordination strategies [67, 139–141].

A system-of-systems approach to considering a distributed care team can include not only primary and auxiliary healthcare professionals, but even the patient and formal or informal caregiver participants [142–144]. Effective coordination and communication between healthcare team members, including the patient, requires additional emphasis on improving patient literacy and their own ability to share lived expertise of their own condition [145] is especially critical in the case of chronic conditions (e.g., diabetes, TBI recovery, cancer), which require ongoing management and many care handovers across multiple time scales [143, 146–148]. Nonetheless, these sociotechnical approaches still describes the healthcare delivery settings in terms of inputs, outputs, time, people, tasks, and organization and environmental factors [143].

14 Cybersecurity and Network Operations

The rise of computer network security operations centers (frequently known as CSOCs) managing ICT systems in large organizations represents a new type of continuous process control setting. As in spaceflight operations, CSOCs represent a distributed expertise sociotechnical setting where the critical resource flow being managed is data rather than physical materials [71, 123, 149, 150]. Even with the rise in automated systems to support secure cyberphysical systems, responses to intrusions and system degradation are still frequently driven by human-intensive event detection, isolation and response (EDIR) activities [151, 152].

A particular challenge in CSOC operations is that it combines determination and analysis of a complex information technology system subject to degradation and unintended adverse couplings (such as software upgrade incompatibilities or accidental damage to systems due to electrical storms) with the adversarial game theoretic considerations of military "blue-team / red-team" operations. This combination represents a potential conflict of cognitive framing models where the limitations of information flow between experts may be seen as an additional security measure rather than an impediment to effective task coordination [151, 152]. Thus, it is important, when considering CSOC design and operations (which are further complicated by workforce shortages and increasing threats of damage or interruption of cyber-physical system performance), to further enhance the capabilities of transparent and viable training, information sharing, and coordinated task performance [149].

15 Aerospace and Critical Response

In the aerospace and critical response domain, it is essential that a decision maker receives timely, accurate, and complete information so that well-informed decisions can be made. It's been shown that "information delay limits an operator's ability to effectively respond to a dynamic and time-critical situation" { Houghton, 2022 #3946 \pg. 45} (see also [153]). In piloting and aerospace operations, inaccurate or delayed information flow can cause a decision maker (a pilot or air traffic controller) to make incorrect or risky decisions (for example, deciding to fly when weather conditions are hazardous). An incorrect decision here can begin a chain of events that can lead to flight into hazardous weather, a loss of aircraft control, and a severe or fatal aviation accident. In the critical response domain, including search and rescue operations, natural disaster relief operations, and emergency medical transport, it is more common to have incomplete situational information (e.g., the severity and scale of a hurricane can be largely unknown prior to the event occurring), greater information delay (e.g., reduced network coverage in operational areas or communication networks being damaged or destroyed entirely), and more rapidly changing response criteria (e.g., the hurricane causes flooding, which overwhelms dams and flood barriers, which causes power outages, leaving more individuals displaced and in need of rescue, which puts more strain on already scarce response resources, etc.). It would be unrealistic to attempt to remove all information uncertainty, delay, or incompleteness from aerospace or critical response operational decisions. Thus, the focus here is *not* to investigate how to increase the quantity of data received in piloting, search and rescue, or disaster response operations. Instead, we focus on how a decision maker can make timely, risk-adverse, and robust decisions *without* having all the necessary information.

Differences in task context and task performer expertise also have crucial impacts in these settings. Delays in obtaining weather information or minimal differences in temperature or wind speed may be of relatively little impact for a recreational pilot on a calm day. Winds might be observed at 5 knots at the nearest weather station, but perhaps they reach 8–10 knots at other points on the flight path. Similarly, cloud ceilings might be reported at 9,000', but they may be drop to 8,000' as the flight progresses. But do these changes in weather conditions matter? In other words, would the incompleteness or uncertainty in the weather information adversely affect the pilot's decision to fly? In this case, probably not. This is not the case for search and rescue or critical response operations pilots. For these pilots, successfully completing a flight could mean that a missing person is found (and their life saved), a natural disaster relief operation can be carried out, or a patient with declining health can be transferred to a medical facility where they can be properly treated. As well as increasing the amount of pressure put on pilots to carry out a flight, these critical response mission pilots must also coordinate with large, diverse, and complex teams to accomplish their mission objectives. It's been shown that high-risk team performance settings have "little availability for error, and there is a high penalty for failure to complete mission objectives" [37, 154].

16 Conclusion

The range of GROUPER performance domain areas of study is, admittedly, quite broad; without an appreciation of the sociotechnical systems homologies of distributed expertise coordination, technology-mediated information flow, or time-critical task performance, it would be difficult to make substantive conceptual or theoretical advancement. It would be incorrect, however, to simply view any of these applications as an accessibly convenient analog for some more "important" setting or "essential" theory. Each of these performance domains has its own challenges and limitations to access and effective study. More importantly, though, useful progress in any of these domains has undeniable value on its own to improve health, safety, performance, and scientific advancement.

References

1. Trist, E.: The evolution of socio-technical systems. Ontario Quality of Working Life Centre, Occasional paper 2 (1981)
2. Emery, F.E., Trist, E.L.: The causal texture of organizational environments. Human Relations **18**(1), 21–32 (1965)
3. Murray, H.A., (ed.): Explorations in Personality. New York: Oxford University Press, p. 742 (1938)
4. Lewin, K.: The Conceptual Representation and the Measurement of Psychological Forces. Duke University Press, Durham, NC (1938)
5. Lewin, K.: Field theory and experiment in social psychology: concepts and methods. Am. J. Sociol. **44**(6), 868–896 (1939)
6. Rogers, E.H.: Diffusion of Innovations, 4th edn. The Free Press, New York (1995)
7. Roy, R., Wield, D. (eds.): Product Design and Technological Innovation. Open University Press, Milton Keynes, UK/Philadelphia (1986)
8. Beer, S.: Brain of the Firm. Allen Lane the Penguin Press, London (1972)
9. Ashby, W.R.: An Introduction to Cybernetics. Chapman and Hall, London (1956)
10. von Bertalanffy, L.: General System Theory: Foundations, Development, Applications, p. 289. George Braziller, New York (1968)
11. Miller, J.G.: Living Systems. McGraw-Hill Inc, New York (1978)
12. Miller, J.G., Miller, J.L.: Introduction: the nature of living systems. Behav. Sci. **36**, 157–163 (1991)
13. Huismanvan, M., van Duijn, M.A.J.: Software for social network analysis. Mod. Meth. Soc. Network Anal. **28** (2005)
14. Meadows, D.H.: Thinking in Systems: A Primer, p. 240. Chelsea Green Publishing, White River Junction, VT (2008)
15. Forrester, J.W.: Principles of Systems, 2nd, preliminary MIT Press, Cambridge, MA (1968)
16. Forrester, J.W.: System dynamics-future opportunities. TIMS Stud. Manage. Sci. **14**, 7–21 (1980)
17. Caldwell, B.S., Uang, S.-T.: Macroergonomic design of information technology (IT): sociotechnical factors in adopting new IT systems. In: Proceedings of the 12th Triennial Congress of the International Ergonomics Association, Mississauga, ON, vol. 6, Toronto: Human Factors Association of Canada, p. 384 (1994)
18. Caldwell, B.S., Uang, S.-T.: Response surface analysis of effects of situation constraints on communications media choice in organizations. In: Proceedings of the Human Factors and Ergonomics Society 38th Annual Meeting, Santa Monica, CA, vol. 2, Nashville: Human Factors and Ergonomics Society, pp. 744–748 (1994)

19. Taha, L.H., Caldwell, B.S.: Social isolation and integration in electronic environments. Behav. Inf. Technol. **12**(5), 276–283 (1993)
20. Forrester, J.W.: System dynamics, systems thinking, and soft OR. Syst. Dyn. Rev. **10**(2–3) (Summer-Fall), 245–256 (1994)
21. McGrath, J.E.: Groups: Interaction and Performance. Prentice-Hall, Englewood Cliffs, NJ (1984)
22. Caldwell, B.S.: Space Exploration and Astronomy Automation. In: Handbook of Automation, S. Nof Ed.: Springer. in press, ch. 51 (2021)
23. Caldwell, B.S.: Analysis and modeling of information flow and distributed expertise in space-related operations. Acta Astronaut. **56**, 996–1004 (2005)
24. Jones, J.M.: Cultural differences in temporal perspectives: instrumental and expressive behaviors in time. In: McGrath, J.E. (ed.) The Social Psychology of Time: New Perspectives, pp. 21–38. Sage Publications, Newbury Park, CA (1988)
25. McGrath, J.E.: Time matters in groups. In: Galegher, J., Kraut, R.E., Egido, C. (eds.) Intellectual Teamwork: Social and Technological Foundations of Cooperative Work, pp. 23–61. Lawrence Erlbaum Associates, Hillsdale, NJ (1990)
26. McGrath, J.E., Kelly, J.R.: Time and Human Interaction: Toward a Social Psychology of Time. The Guilford Press, New York (1986)
27. McGrath, J.E., Hollingshead, A.B.: *Groups Interacting With Technology* (Sage Library of Social Research #194). Sage Publications, Newbury Park, CA (1994)
28. Alberts, D.S., Hayes, R.E.: Power to the edge: command... control... in the information age. Office of the Assistant Secretary of Defense Washington DC Command and ... (2003)
29. Lawson, J.: Command control as a process. IEEE Control Syst. Mag. **1**(1), 5–11 (1981)
30. Endsley, M.R.: Measurement of situation awareness in dynamic systems. Hum. Factors: J. Hum. Factors Ergonomics Soc. **37**(1), 65–84 (1995). https://doi.org/10.1518/001872095779049499
31. Taylor, F.W.: The Principles of Scientific Management. Harper & Brothers Publishers, New York (1911)
32. Caldwell, B.S.: Considering the future of land grant ergonomics education. Presented at the Proceedings of the Human Factors and Ergonomics Society 62th Annual Meeting, Philadelphia, PA, 2–5 October 2018, p. 359
33. Foushee, H.C., Helmreich, R.L.: Group interaction and flight crew performance. In: Weiner, E.L., Nagel, D.C. (eds.) Human Factors in Aviation, pp. 189–227. Academic Press, San Diego (1988)
34. Radloff, R., Helmreich, R.: Groups Under Stress: Psychological Research in SEALAB II. Appleton-Century-Crofts, New York (1968)
35. Helmreich, R.L., Merritt, A.C., Wilhelm, J.A.: The evolution of crew resource management training in commercial aviation. Int. J. Aviat. Psychol. **9**(1), 19–32 (1999)
36. Garrett, S.K., Caldwell, B.S., Harris, E.C., Gonzalez, M.C.: Six dimensions of expertise: a more comprehensive definition of cognitive expertise for team coordination. Theor. Issues Ergon. Sci. **10**(2), 93–105 (2009)
37. Caldwell, B.S.: Components of information flow to support coordinated task performance. Int. J. Cogn. Ergon. **1**(1), 25–41 (1997)
38. Bolstad, C.A., Endsley, M.R.: Shared mental models and shared displays: an empirical evaluation of team performance. In: Proceedings of the Human Factors and Ergonomics Society 43rd Annual Meeting, Houston, 1999: Human Factors and Ergonomics Society, pp. 213–217 (1999)
39. Gonzalez, C., Saner, L., Endsley, M., Bolstad, C.A.: Cuevas, H.M.: Modeling and measuring situation awareness in individuals and teams (2009)

40. Kelly, J.R.: Entrainment in individual and group behavior. In: McGrath, J.E. (ed.) The Social Psychology of Time: New Perspectives, pp. 89–110. Sage Publications, Newbury Park, CA (1988)
41. Parasuraman, R., Sheridan, T.B., Wickens, C.D.: A model for types and levels of human interaction with automation. IEEE Trans. Syst. Man and Cybern.-Part A: Syst. Hum. **30**(3), 286–297 (2000)
42. Sheridan, T.B.: Human and computer roles in supervisory control and telerobotics: Musings about function, language and hierarchy. In: Goodstein, L.P., Anderson, H.B., Olsen, S.E. (eds.) Tasks, Errors, and Mental Models, pp. 149–160. Taylor & Francis, London (1988)
43. Sheridan, T.B.: Human supervisory control. In: Handbook of Human Factors and Ergonomics, G. Salvendy Ed., 4th ed.: John Wiley & Sons, vol. 34, pp. 990–1015 (2012)
44. Killich, S., Luczak, H., Schlick, C., Weissenbach, M., Wiedenmaier, S., Ziegler, J.: Task modelling for cooperative work. Behav. Inf. Technol. **18**(5), 325–338 (1999). https://doi.org/10.1080/014492999118913
45. Salas, E., Burke, C., Cannon-Bowers, J.: Teamwork: emerging principles. Int. J. Manag. Rev. **2**(4), 339–356 (2000). https://doi.org/10.1111/1468-2370.00046
46. Cross, N., Clayborn Cross, A.: Observations of teamwork and social processes in design. Des. Stud. **16**(2), 143–170 (1995). https://doi.org/10.1016/0142-694x(94)00007-z
47. Locke, E.: Handbook of Principles of Organizational Behavior. John Wiley, Chichester, UK (2009)
48. Tsarouchi, P., Makris, S., Chryssolouris, G.: Human–robot interaction review and challenges on task planning and programming. Int. J. Comput. Integr. Manuf. **29**(8), 916–931 (2016)
49. Kluge, A.: The Acquisition of Knowledge and Skills for Taskwork and Teamwork to Control Complex Technical Systems. Dordrecht, Germany: Springer (2014)
50. Fang, F., Parameswaran, M., Zhao, X., Whinston, A.: An economic mechanism to manage operational security risks for inter-organizational information systems. Inf. Syst. Front. **16**(3), 399–416 (2012). https://doi.org/10.1007/s10796-012-9348-y
51. Somech, A., Desivilya, H., Lidogoster, H.: Team conflict management and team effectiveness: the effects of task interdependence and team identification. J. Organ. Behav. **30**(3), 359–378 (2009). https://doi.org/10.1002/job.537
52. Mullane, J.: The mission statement is a strategic tool: when used properly. Manag. Decis. **40**(5), 448–455 (2002). https://doi.org/10.1108/00251740210430461
53. Freedman, J.L., Edwards, D.R.: Time pressure, task performance, and enjoyment. In: McGrath, J.E. (ed.) The Social Psychology of Time: New Perspectives, pp. 113–133. Sage Publications, Newbury Park, CA (1988)
54. MacGregor, D.: Time pressure and adaptation: alternative perspectives on laboratory studies. In: Svenson, O., Maule, A.J. (eds.) Time Pressure and Stress in Human Judgment and Decision Making, pp. 73–82. Plenum Press, New York (1993)
55. Maule, A.J., Hockey, G.R.J.: State, stress, and time pressure. In: Svenson, O., Maule, A.J. (eds.) Time Pressure and Stress in Human Judgment and Decision Making, pp. 83–101. Plenum Press, New York (1993)
56. Svenson, O., Maule, A.J. (eds.): Time Pressure and Stress in Human Judgment and Decision Making. Plenum Press, New York (1993)
57. Zakay, D.: The impact of time perception processes on decision making under time stress. In: Svenson, O., Maule, A.J. (eds.) Time Pressure and Stress in Human Judgment and Decision Making, pp. 59–72. Plenum Press, New York (1993)
58. Charness, N.: Expertise in chess: the balance between knowledge and search. In: Ericsson, K.A., Smith, J. (eds.) Toward a General Theory of Expertise: Prospects and Limits, pp. 39–63. University Press, Cambridge, UK (1994)

59. Ericsson, K.A., Smith, J.: Prospects and limits of the empirical study of expertise: an introduction. In: Ericsson, K.A., Smith, J. (eds.) Toward a General Theory of Expertise: Prospects and Limits, pp. 1–38. Cambridge University Press, Cambridge, UK (1991)
60. Holyoak, K.J.: Symbolic connectionism: toward third-generation theories of expertise. In: Ericsson, K.A., Smith, J. (eds.) Toward a General Theory of Expertise: Prospects and Limits, pp. 301–335. Cambridge University Press, Cambridge, UK (1991)
61. Caldwell, B.S., Wang, E.: Delays and user performance in human-computer-network interaction tasks. Hum. Factors **51**(6), 813–830 (2009)
62. Wang, E., Caldwell, B.S.: Human information flow and communication pattern in NASA mission control system. In: Proceedings of the Human Factors and Ergonomics Society 47th Annual Meeting, 2016, no. 1, pp. 11–15 (2016). https://doi.org/10.1177/154193120304700103
63. Caldwell, B.S., Garrett, S.K.: Experience, task cycles and proactive resource allocation in organizational settings. In: Eighth Human Factors in Organizational Design and Management (ODAM) Symposium, Maui, HI, Carayon, P., Robertson, M.M., Kleiner, B., Hoonakker, P.L.T. (eds.), 22–25 June 2005: IEA Press, pp. 149–154
64. Garrett, S.K., Caldwell, B.S.: Mission control knowledge synchronization: operations to reference performance cycles. In: Proceedings of the Human Factors and Ergonomics Society 46th Annual Meeting, Santa Monica, CA, 2002, Baltimore: Human Factors and Ergonomics Society, pp. 849–853 (2002)
65. Handke, L., Schulte, E. M., Schneider, K., Kauffeld, S.: Teams, time, and technology: variations of media use over project phases. Small Group Res. **50** (2), 266–305 (2019). https://doi.org/10.1177/1046496418824151
66. Caldwell, B.S., Garrett, S.K.: Coordination of event detection and task management in time-critical settings. In: Informed by Knowledge: Expert Performance in Complex Situations, Mosier, K.L., Fischer, U.M. (eds.) 1 ed. New York: Psychology Press, vol. 22, pp. 339–351 (2010)
67. Caldwell, B.S., Garrett, S.K., Boustany, K.C.: Healthcare team performance in time critical environments: coordinating events, foraging, and system processes. J. Healthcare Eng. **1**(2), 255–276 (2010)
68. Garrett, S.K., Caldwell, B.S., Ebright, P.R.: Provider information and resource foraging in health care delivery. Int. J. Collaborative Enterp. **1**(3/4), 381–393 (2010)
69. Converse, S.A., Cannon-Bowers, J.A., Salas, E.: Team member shared mental models: a theory and some methodological issues. In: Proceedings of the Human Factors Society 35th Annual Meeting, Santa Monica, CA, 1991, San Francisco: Human Factors Society, pp. 1417–1421 (1991)
70. Cooke, N.J., Salas, E., Kiekel, P.A., Bell, B.: Advances in Measuring Team Cognition (2000). https://doi.org/10.1037/10690-005
71. Caldwell, B.S.: Multi-team dynamics and distributed expertise in mission operations. Aviat. Space Environ. Med. **76**(6), Sec II, B145–B153 (2005)
72. Hill, J.R., Caldwell, B.S.: Toward better understanding of function allocaiton requirements for planetary EVA and habitat tasks. In: Proceedings of the Human Factors and Ergonomics Society 2018 Annual Meeting, Philadelphia, 2–5 Octber 2018. Sage, pp. 29–33
73. Hill, J.R., Caldwell, B.S., Downs, M., Miller, M.J., Lim, D.S.S.: Remote physiological monitoring in a Mars Analog field setting. IISE Trans. Healthcare Syst. Eng. **8**(3), 227–236 (2019). https://doi.org/10.1080/24725579.2018.1501624
74. Lofland, J., Lofland, L.H.: Analyzing Social Settings: A Guide to Qualitative Observation and Analysis, 2nd edn. Wadsworth, Belmont, CA (1984)
75. Cooke, N.J., Shope, S.M.: Designing a synthetic task environment. In: Scaled Worlds: Development, Validation and Applications, Schiflett, S.G., Lelliott, L.R., Salas, E., Coovert, M.D. (eds.): Taylor & Francis (2004)

76. Rivlin, L., Wolfe, M., Kaplan, M.: Environmental psychology and action research: lewin's legacy. In: Stivers, E.H., Wheelan, S.A. (eds.) The Lewin Legacy: Field Theory in Current Practice, pp. 194–200. Springer-Verlag, Berlin (1986)
77. Stivers, E.: A Neo-Lewinian action research method. In: Stivers, E.H., Wheelan, S.A. (eds.) The Lewin Legacy: Field Theory in Current Practice, pp. 258–267. Springer-Verlag, Berlin (1986)
78. Forrester, J.W.: Principles of Systems, 2nd preliminary ed. Waltham, MA: Pegasus Communications (1990)
79. Caldwell, B.S.: Social implications of feedback and delay characteristics in electronic communication usage. In: Human-Computer Interaction: Software and Hardware Interfaces--Proceedings of the Fifth International Conference on Human-Computer Interaction, vol. 2, Salvendy, G., Smith, M. (eds.), (Advances in Human Factors / Ergonomics. Amsterdam: Elsevier, pp. 843–848 (1993)
80. Caldwell, B.S., Paradkar, P.V.: Factors affecting user tolerance for voice mail message transmission delays. Int. J. Hum.-Comput. Inter. **7**(3), 235–248 (1995)
81. Wang, E., Caldwell, B.S.: Modeling User Delay Perception and Tolerance Dynamics in Human-Computer Interaction (2005)
82. Caldwell, B.S.: Eleven years, five flavors: systems engineering education since IERC 2009. In: Proceedings of the 2020 IISE Annual Conference, Cromarty, L., Shirwaiker, R,. Wang, P. (eds.), IISE (2020)
83. Holling, C.S.: Resilience and stability of ecological systems. Ann. Rev. Ecol. Syst. **4**, 1–23 (1973). http://www.jstor.org/stable/2096802
84. von Bertalanffy, L.: The model of open system. In: General systems theory: Foundations, development, applications: George Brazillier, Inc., pp. 139–154 (1968)
85. Josslyn, C.: Semiotic Agent Models for Simulating Socio-Technical Organizations. Los Alamos National Laboratory, Los Alamos, NM, DS Project, PSL / NMSU September (1999)
86. Martin, C.E., Macfadzean, R.H., Barber, K.S.: Supporting dynamic adaptive autonomy for agent-based systems. In: Proceedings of the AI and Manufacturing Research Planning Workshop, AAAI, pp. 112–120 (1996)
87. Van Zandt, T.: Real-time decentralized information processing as a model of organizations with boundedly rational agents. Rev. Econ. Stud. **66**, 633–658 (1998)
88. Worm, A.: Evaluating tactical real-time interaction in multi-agent, dynamic, hazardous, high-stake operations. In: Proceedings of the Human Factors and Ergonomics Society 43rd Annual Meeting, Houston, pp. 199–203 (1999)
89. Nyre-Yu, M., Caldwell, B.S.: Using simulation modeling methods to study teams doing task work. In: Proceedings of the Human Factors and Ergonomics Society 2018 Annual Meeting, Philadelphia, PA, Sage, pp. 788–792 (2018)
90. Nyre-Yu, M., Caldwell, B.S.: Supporting advances in human-systems coordination through simulation of diverse, distributed expertise. Systems **6**(39) (2018). https://doi.org/10.3390/systems6040039
91. Wang, E., Caldwell, B.S.: Time delay tolerance in computer supported cooperative work. In: Proceedings of the 6th International Scientific Conference Work with Display Units: WWDU 2002--World Wide Work, Berchtesgaden, Germany, Luczak,H., Çakir, A.E., Çakir, G. (eds.), ERGONOMIC, pp. 199–201 (2002)
92. Altmann, E.M., Trafton, J.G.: Task interruption: resumption lag and the role of cues. In: Proceedings of the 26th Annual Conference of the Cognitive Science Society, Hillsdale, NJ: Erlbaum, pp. 42–47 (2004)
93. Monk, C.A., Boehm-Davis, D.A., Trafton, J.G.: The attentional costs of interrupting task performance at various stages, In: Proceedings of the Human Factors and Ergonomics Society 46th Annual Meeting, Baltimore, pp. 1824–1828 (2002)

94. Trafton, J.G., Altmann, E.M., Brock, D.P., Mintz, F.E.: Preparing to resume an interrupted task: effects of prospective goal encoding and retrospective rehearsal. Int. J. Hum. Comput. Stud. **58**, 583–603 (2003)
95. Yuan, F., Gao, X., Lindqvist, J.: How busy are you? Predicting the interruptibility intensity of mobile users. In: Proceedings of the 2017 CHI Conference on Human Factors in Computing Systems, pp. 5346–5360 (2017)
96. Züger, M., et al.: Reducing interruptions at work: a large-scale field study of flowlight. In: Proceedings of the 2017 CHI Conference on Human Factors in Computing Systems, pp. 61–72 (2017)
97. Züger, M., Müller, S.C., Meyer, A.N., Fritz, T.: Sensing interruptibility in the office: a field study on the use of biometric and computer interaction sensors. In: Proceedings of the 2018 CHI Conference on Human Factors in Computing Systems, pp. 1–14 (2018)
98. Kelly, P., Kranzberg, M., Rossini, F., Baker, N., Tarpley, F., Mitzner, M.: Introducing innovation. In: Roy, R., Wield, D. (eds.) Product Design and Technological Innovation, pp. 18–28. Open University Press, Milton Keynes, UK / Philadelphia (1986)
99. Roberts, E.B.: Generating effective corporate innovation. In: Roy, R., Wield, D. (eds.) Product Design and Technological Innovation, pp. 124–127. Open University Press, Milton Keynes, UK / Philadelphia (1986)
100. Abernathy, W.J., Utterback, J.M.: Patterns of industrial innovation. In: Roy, R., Wield, D. (eds.) Product Design and Technological Innovation, pp. 257–264. Open University Press, Milton Keynes, UK / Philadelphia (1986)
101. Rothwell, W.J., Whiteford, A.P.: Corporate employee training and development strategies. In: Oxford Handbook of Lifelong Learning, M. London Ed.: Oxford University Press, vol. 11, pp. 149–163 (2009)
102. Sundar, E., Sundar, S., Pawlowski, J., Blum, R., Feinstein, D., Pratt, S.: Crew resource management and team training. Anesthesiol. Clin. **25**(2), 283–300 (2007). https://doi.org/10.1016/j.anclin.2007.03.011
103. Swezey, R.W., Salas, E. (eds.) Teams: Their Training and Performance. Norwood, NJ: ABLEX (1992)
104. Wilson, G.F.: Air-to-ground training missions: a psychophysiological workload analysis. Ergonomics **36**, 373–389 (1993)
105. Rosen, M.E., et al.: Tools for evaluating team performance in simulation-based training. J. Emerg. Trauma Shock **3**(4), 353–359 (2010). https://doi.org/10.4103/0974-2700.70746
106. Salas, E., Dickinson, T.L., Converse, S.A., Tannenbaum, S.I.: Toward an understanding of team performance and training. In: Teams: Their Training and Performance, R. W. Swezey and E. Salas Eds. Norwood, NJ: ABLEX, pp. 3–29 (1992)
107. McCormack, C.M.: Information architecture and cognitive user experience in distributed, asynchronous learning: a case design of a modularized online systems engineering learning environment. Master's Thesis, School of Industrial Engineering, Purdue University, West Lafayette, IN (2021)
108. Tilley, D.S., Allen, P., Collins, C., Bridges, R.A., Francis, P., Green, A.: Promoting clinical competence: using scaffolded instruction for practice-based learning. J. Prof. Nurs. **23**(5), 285–289 (2007). https://doi.org/10.1016/j.profnurs.2007.01.013
109. Caldwell, B.S., Ghosh, S.K.: Supporting operations-reference knowledge development cycles for collaborative, distributed research (2003)
110. Caldwell, B.S., Wang, E.: Event cycle and knowledge development in NASA mission control center. In: Proceedings of the HCI International 2003, 10th International Conference on Human Computer Interaction, Crete, Greece, Mahwah, NJ, Jacko, J., Stephanidis, C. (eds.) 22–27 June 2003, vol. Part I: Lawrence Erlbaum, pp. 301–305
111. Wiener, N.: Cybernetics, 2nd edn. MIT Press, Cambridge, MA (1961)

112. Richards, M.A., Caldwell, B.S., Iannacone, A.C.: Social justice and equity as operating range: enriching success pathways for underrepresented minority students. In: IISE Annual Conference & Expo 2022, Seattle, Ellis, K., Ferrell, W., Knapp, J. (eds.), IISE (2022)
113. Caldwell, B.S.: Framing, information alignment, and resilience in distributed human coordination of critical infrastructure event response. Procedia Manuf. **3**, 5095–5101 (2015). https://doi.org/10.1016/j.promfg.2015.07.524
114. Caldwell, B.S.: Information and communication technology needs for distributed communication and coordination during expedition-class spaceflight. Aviat. Space Environ. Med. **71**(9, supp I), pp. A6–A10 ()2000
115. Sheridan, T.B.: Telerobotics, Automation, and Human Supervisory Control. MIT Press, Cambridge, MA (1992)
116. Messier, D.: Enigneers identify software solution for ingenuity mars helicopter anomaly. Parabolic Arc (2021). http://www.parabolicarc.com/2021/04/13/engineers-identify-software-solution-for-ingenuity-mars-helicopter-anomaly/
117. Sheridan, T.B.: Supervisory control. In: Handbook of Human Factors and Ergonomics, Salvendy, G. (ed.), 3rd ed. New York: John Wiley & Sons, pp. 1025–1052 (2006)
118. Tulga, M.K., Sheridan, T.B.: Dynamic decisions and workload in multi-task supervisory control. IEEE Trans. Syst. Man Cybern. **SMC-10**(5), 217–231 (1980)
119. Holweg, M., Bicheno, J.: Supply chain simulation - a tool for education, enhancement and endeavour. Int. J. Prod. Econ. **78**, 163–175 (2002)
120. Sterman, J.D. (n.d.).: Flight Simulators for Managment Education [web available manuscript document]
121. D'Azzo, J.J., Houpis, C.H.: Feedback Control System Analysis and Synthesis, 2nd edn. McGraw Hill Book Company, New York (1966)
122. Garrett, S.K., Caldwell, B.S.: Human factors aspects of planning and response to pandemic events. In: IIE Industrial Engineering Research Conference, Miami (2009)
123. Caldwell, B.S.: Distributed supervisory coordination with multiple operators and remote systems. In: IEEE International Conference on Systems, Man and Cybernetics, Washington, DC, IEEE, pp. 442–447 (2003)
124. Caldwell, B.S.: Issues of task and temporal coordination in distributed expert teams. In: Proceedings of the 16th World Congress of the International Ergonomics Association, Maastricht, Netherlands, Koningsveld, E.A.P. (ed.), 10–14 July 2006, pp. 2761–2766
125. Caldwell, B.S., Onken, J.D.: Simulation and human factors in modeling of spaceflight mission control teams. In: International Symposium on Resilient Control Systems, Salt Lake City, 14–16 August 2012
126. Onken, J.D., Caldwell, B.S.: Problem solving in expert teams: functional models and task processes. In: Proceedings of the Human Factors and Ergonomics Society 55th Annual Meeting -- 2011, Las Vegas, pp. 1150–1154 (2011). https://doi.org/10.1177/1071181311551240. http://pro.sagepub.com/content/55/1/1150
127. Garrett, S.K., Caldwell, B.S., Bruins, A.: Information sharing and knowledge management in MCC system evolution (2001)
128. Onken, J.D., Caldwell, B.S.: Towards information coordination and reduced team size in space flight mission operations. In: Proceedings of the Human Factors and Ergonomics Society Annual Meeting, vol. 53, no. 1, pp. 101–105 (2009).https://doi.org/10.1177/154193120905300122
129. Caldwell, B.S., Onken, J.D.: Modeling and analyzing distributed autonomy for spaceflight teams. In: 41st International Conference on Environmental Systems, Portland, OR (2011)
130. E. National Academies of Sciences and Medicine, Human-AI Teaming: State-of-the-Art and Research Needs (2021)

131. Sheridan, T.B.: Adaptive automation, level of automation, allocation authority, supervisory control, and adaptive control: distinctions and modes of adaptation. IEEE Trans. Syst. Man Cybern.-Part A: Syst. Hum. **41**(4), 662–667 (2011)
132. Garrett, S.K., Caldwell, B.S.: Healthcare systems engineering. In: Biomedical Engineering and Design Handbook, Kutz, M. (ed.), New York: McGraw-Hill, vol. 27 (2009)
133. Boustany, K.C., Caldwell, B.S.: Dimensions of information and resource flow in healthcare systems. In: Proceedings of the Human Factors and Ergonomics Society 51st Annual Meeting, Baltimore, MD, pp. 1268–1271 (2007)
134. Boustany, K.C., Caldwell, B.S: Information coordination in healthcare providers. In: Proceedings of the 2007 Industrial Engineering Research Conference, Nashville, TN, Bayraksan, G., Lin, W., Son, Y., Wysk, R. (eds.) (2007)
135. Caldwell, B.S., Anderson, T.J.: Design and active phases of medical process improvement systems. In: Examining Errors in Health Care: Developing a Prevention, Education and Research Agenda, Rancho Mirage, CA, Rancho Mirage, CA: Annenberg Center for Health Sciences (1996)
136. Thomadsen, B.R., et al.: Electronic intracavitary brachytherapy quality management based on risk analysis: the report of AAPM TG 182. Med. Phys. **47**(4), e65–e91 (2020). https://doi.org/10.1002/mp.13910
137. Thomadsen, B.R., Caldwell, B.S., Stitt, J.: Progress toward understanding human performance in medicine. Presented at the American Nuclear Society Winter Meeting, Washington, DC, 31 December 1998 (1998)
138. Thomadsen, B.R., et al.: Analysis of treatment delivery errors in brachytherapy using formal risk analysis techniques. Int. J. Radiat. Oncol. Biol. Phys. **57**(5), 1492–1508 (2003). https://doi.org/10.1016/s0360-3016(03)01622-5
139. Benedict, A.J., Caldwell, B.S.: Information flow paths between patients, providers, and pharmacists in the outpatient medication setting. In: Proceedings of the 2010 Industrial Engineering Research Conference, Johnson, A., Miller, J. (eds.) (2010)
140. Caldwell, B.S., Garrett, S.K.: Coordination of healthcare expertise and information flow in provider teams. In: Proceedings of the 16th World Congress of the International Ergonomics Association, Maastricht, Netherlands, Koningsveld, E.A.P. (ed.), 10–14 July 2006, pp. 3565–3570 (2006)
141. Caldwell, B.S., Garrett, S.K.: Experience, task cycles and proactive resource allocation in organizational settings. In: presented at the Eighth Human Factors in Organizational Design and Management (ODAM) Symposium, Maui, HI, June 2008 (2008)
142. Heiden, S.M., Caldwell, B.S.: Multilevel, multidiscipline, and temporally diverse handoffs in traumatic brain injury rehabilitation. In: Human Factors and Ergonomics in Healthcare Symposium, Boston, MA, 26–28 March 2018
143. Jahn, M.A., Caldwell, B.S.: Community health integration through pharmacy process and ergonomics redesign (CHIPPER). Ergonomics **61**(1), 69–81 (2018). https://doi.org/10.1080/00140139.2017.1353136
144. Vallette, M.A., Caldwell, B.S.: Activity cycles and infromation alignment in healthcare information flow. In: presented at the Symposium on Human Factors and Ergonomics in Health Care (2012)
145. Douglas, S.E., Caldwell, B.S.: Improving communication of health status. In: Proceedings of the Human Factors and Ergonomics Society 53rd Annual Meeting, 2009: Human Factors and Ergonomics Society, pp. 709–713 (2009). https://doi.org/10.1518/107118109X12524442636788
146. Heiden, S.M., Caldwell, B.S.: Systematic comparison of longitudinal brain injury care to cancer survivor and diabetes patient chronic care models. In: Human Factors in Healthcare 2016 Symposium, San Diego, CA, 13–16 April 2016

147. Heiden, S.M., Caldwell, B.S.: Considerations for developing chronic care system for traumatic brain injury based on comparisons of cancer survivorship and diabetes management care. Ergonomics **61**(1), 134–147 (2018). https://doi.org/10.1080/00140139.2017.1349932
148. Jahn Holbrook, M., Caldwell, B.S.: Development of a multi-layer systems engineering visualization for diabetes team coordination. In: Proceedings of the Human Factors and Ergonomics Society 2019 International Annual Meeting, Seattle, WA, 31 Oct–4 Nov 2019
149. Nyre-Yu, M., Sprehn, K., Caldwell, B.S.: Informing hybrid system design in cyber security incident response. In: presented at the Human Computer Interaction International 2019 Conference, Orlando, FL, 26–31 July 2019
150. Eldardiry, O.M., Caldwell, B.S.: Improving information and task coordination in cyber security operation centers. In: Industrial and Systems Engineering Research Conference 2015, Nashville, 30 May–2 June 2015
151. Nyre-Yu, M., Gutzwiller, R.S., Caldwell, B.S.: Observing cyber security incident response: qualitative themes from field research. In: Proceedings of the Human Factors and Ergonomics Society 2019 International Annual Meeting, Seattle, WA, Oct 31 October–4 November 2019
152. Nyre-Yu, M., Caldwell, B.S.: Observing cyber security incident response: qualitative themes from field research. In: Proceedings of the Human Factors and Ergonomics Society 2018 Annual Meeting, Seattle, WA, Sage, pp. 437–441 (2019)
153. Caldwell, B.S.: Cognitive challenges to resilience dynamics in managing large-scale event response. J. Cogn. Eng. Dec. Mak. **8**(4), 318–329 (2014). https://doi.org/10.1177/1555343414546220
154. Houghton, N.M.: Analyzing weather observation data to improve emergency services pilot risk assessment in marginal weather conditions. Master's Degree Thesis, School of Aeronautics and Astronautics, Purdue University, West Lafayette, IN (2022)

Design and Development of Collaborative Hub for Safety and Reliability Analysis

Gaurav Nanda[1] and Mark R. Lehto[2](✉)

[1] School of Engineering Technology, Purdue University, West Lafayette, USA
[2] School of Industrial Engineering, Purdue University, West Lafayette, USA
lehto@purdue.edu

Abstract. Effective utilization of product quality and reliability intellectual capital assists companies to avoid expensive errors, allows for streamlined product development, provides better customer satisfaction, creates better issue/process management and results in more robust and reliable products. For large multinational organizations, it is often challenging to record and reuse innovative problem-solving approaches developed by teams working in different divisions or geographical locations. To address this, we developed a knowledge management system based on the HUBzero platform for the Reliability Engineering (RE) division of a leading consumer goods company. HUBzero is a WEB 2.0 based scientific collaboration platform with various social networking and data management features like user groups, access-controlled file sharing and review mechanism, content tagging, online discussions, wikis and blogs that can be used by researchers and commercial organizations. It allows people in different teams to collaborate in a social networking manner and also saves time by reusing the work already done. The HUB developed by us created a searchable and reusable organizational memory of expert-verified good-quality reliability analyses performed using various RE tools such as Failure Mode Effects and Criticality Analysis, Reliability Growth Analysis and Shakedown Testing. We developed a mechanism of acquiring, publishing and sharing RE analyses in a semi-automated manner by interfacing HUBzero to other software being used in the organization. Post implementation, the number of RE files on HUBzero has been increasing at a steady pace. With growing number of files, effective organization and easy retrieval becomes a challenge for any knowledge management system. To address this, we developed automated tagging of files using linguistically extracted keywords and designed new navigational features based on the metadata information of RE tool files with sorting and filtering capabilities to improve the searchability and discoverability of past analyses.

Keywords: Collaborative HUB · Reliability Engineering · Knowledge Management · Organizational Memory · WEB 2.0

C.-Y. Huang and S. W. Yoon (Eds.): ICPR1 2021, ACES 14, pp. 181–198, 2023.
https://doi.org/10.1007/978-3-031-44373-2_11

1 Introduction

With the increasing adoption of Industry 4.0 technologies such as Internet-of-Things (IoT), cloud computing, and artificial intelligence (Fig. 1), by organizations, most modern manufacturing and supply chain operations functions integrate the physical, software, and network components as cyber-physical systems (CPS) [1]. Cloud-based information systems with Artificial Intelligence (AI) capabilities can help streamline the operations as well provide predictive insights for safety, reliability, and maintenance. These cloud-based information systems enable collaboration and organizational learning across geographical locations, which is particularly useful for large multinational organizations.

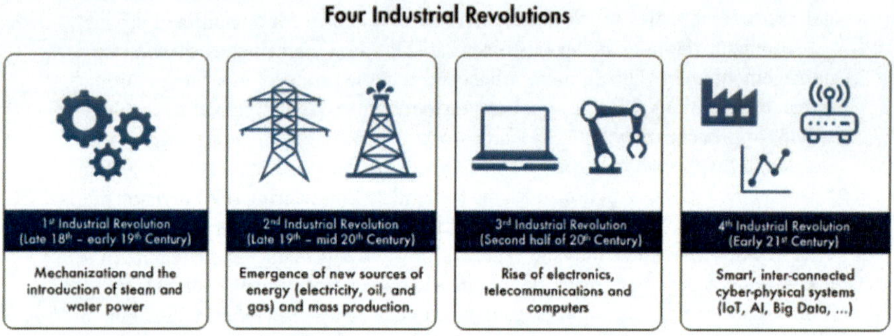

Fig. 1. Evolution of Industry 4.0 [2] (Courtsey: Mathworks)

Reliability of a system is typically defined as the measure of likelihood of performing its intended functions satisfactorily for a specified duration. For complex cyber-physical manufacturing systems, reliability engineering plays an essential role in decision making with uncertainties such as equipment health and production strategies. Low system reliability can lead to expensive delays, repairs, replacements, and sub-optimal operations [3, 4]. The overall system reliability of manufacturing CPS involves three major factors: hardware reliability, software reliability, and human reliability, and their interplays and interdependencies [5]. For interconnected manufacturing CPS spanning various geographical locations, it is important to assess the quality and relevance of data for predictive calculations and analyze the impact of local failures and disruptions on the overall manufacturing system reliability [5–7]. Along with hardware and software reliability, the human reliability plays a critical role as unlike software and hardware systems, there can be considerable variation in human behavior and interactions with the system, leading to improper or sub-optimal use of the system [5]. Therefore, it is important to carefully consider the human-machine interaction and human-human collaboration aspects for successful operations of a complex cyber-physical manufacturing system. To enable this, we developed and implemented a Collaborative Hub based on the online collaborative platform HUBzero in a leading consumer goods organization for their reliability engineering division [8, 9]. In this chapter, we discuss HUBzero's first industrial implementation, the challenges faced during implementation, and how they were addressed.

2 Enterprise Knowledge Management for Reliability Analysis

Enterprise knowledge management is often challenging for large multinational organizations as each step in the process involves considerable complexities starting from (a) acquisition of data from documents residing on internal servers, groupware, and email communications, (b) creating organizational memory by refinement of collected data by filtering the relevant data and storing it in an organized and structured manner and performing analytics on relevant data, and (c) disseminating the knowledge and insights collected in form training programs, structured repositories, and decision support systems. An effective organizational knowledge management setup powered by cloud-based collaborative information system can enable efficient use of expert knowledge, reuse of past and learnings resulting in improved organizational decision making, cost savings, and more streamlined business operations. Additionally, this system can also improve collaboration across teams resulting in seamless exchange of ideas and developing innovative approaches for problem solving. It is particularly important in context of reliability engineering as most reliability and safety analysis such as root cause analysis and fault-tree analysis typically involve a lot of use of expert knowledge, which should be recorded and stored with relevant details and be easily searchable, accessible, and reusable through the cloud-based collaborative systems.

Effective utilization of product quality and reliability intellectual capital assists companies to avoid expensive errors, allows for streamlined product development, provides better customer satisfaction, creates better issue/process management and results in more robust and reliable products. This product quality and reliability information can arrive from different sources within the organization, including product manufacturing and testing divisions, customer support centers, service centers, sales and marketing units, etc. Management and easy access of such information becomes essential, not only for the reliability engineers, but also for the design engineers, management, sales and marketing personnel and other entities within the organization. The collection, categorization, analysis and presentation of reliability data to the whole organization can be achieved using the principles of knowledge management with the help of current software technology and the high levels of IT infrastructure in most enterprises [10]. With the reliability data being generated and used by different groups within the organization, use of collaborative knowledge management approach will be appropriate and beneficial to the organization.

3 Design of Collaborative HUB

In a project with the reliability engineering (RE) division of a large consumer products company, HUBzero (also referred to as HUB) was successfully implemented as a knowledge base of different reliability tools being used in the company such as Failure Mode Effects and Criticality Analysis (FMECA), Reliability Growth Analysis, and Shakedown Testing. The FMECA tool analyzes failures of a system by identifying and classifying the failure modes and the causes and effects of each potential failure mode on system service and subsequently defining appropriate detection procedures and corrective actions. The Reliability Growth Analysis tool uses Logistics model to determine various developmental data such as time-to-failure, discrete (success/failure) and reliability values at different times or stages. The Shakedown Testing tool is used for recording the results of equipment tests made during installation work. These RE tools are used within the organization in form of Microsoft Excel files with a set of predefined columns where engineers have to fill their analysis depending on the tool being used, a sample of FMECA Excel template commonly used in industrial settings is shown in Fig. 2 [11].

POTENTIAL
FAILURE MODE AND EFFECTS ANALYSIS
Front Door L.H.

FMEA Type ______ | FMEA Number 1450
Page 1 of 1
Item 1.1.1 - Front Door L.H. | Process Responsibility Body Engineering | Prepared By J. Ford - X6521 - Assy Ops
Model Year(s)/Vehicle(s) 20XX/Lion 4dr/Wagon | Key Date 3/10/2015 | FMEA Date (Orig.) 3/10/2015 (Rev) 3/21/2015
Core Team A. Tate Body Engrg, J. Smith - OC, R. James - Production, J. Jones - Maintenance

Name / Function Requirements	Potential Failure Mode	Potential Effect(s) of Failure	SEVi	Classification	Potential Cause(s) of Failure	OCCi	Current Process Controls (Prevention)	Current Process Controls (Detection)	DETi	RPNi	Recommended Action(s)	Responsibility & Planned Completion Date	Action Results: Actions Taken & Actual Completion Date	SEVr	OCCr	DETr	RPNr
1.1.1 - Front Door L.H.																	
Op. 70 Manual application of wax inside door/ cover inner door, lower surfaces with wax to specification thickness.	Insufficient wax coverage over specified surface	Allows integrity breach of inner door panel. Corroded interior lower door panels. Deteriorated life of door leading to: - Unsatisfactory appearance due to rust through paint over time - Impaired function of interior door hardware	7		Manually inserted spray head not inserted far enough	8		Visual check each hour - 1/shift for film thickness (depth meter) and coverage.	5	280	Add positive depth stop to sprayer.	Mfg Engrg - 3/10/2003	Stop added, sprayer checked on line.	7	2	5	70
											Automate spraying.	Mfg Engrg - 3/10/2003	Rejected due to complexity of different doors on same line.				
					Spray head clogged- Viscosity too high- Temperature too low- Pressure too low.	5	Test spray pattern at start-up and after idle periods, and preventive maintenance program to clean heads.	Visual check each hour - 1/shift for film thickness (depth meter) and coverage.	5	175	Use Design of Experiments (DOE) on viscosity vs. temperature vs. pressure.	Mfg Engrg - 3/10/2003	Temp and press limits were determined and limit controls have been installed - control charts show process is in control Cpk = 1.85.		1	5	35
					Spray head deformed due to impact	2	Preventive maintenance program to maintain heads.	Visual check each hour - 1/shift for film thickness (depth meter) and coverage.	5	70					2	5	70

Fig. 2. Example of FMECA Analysis [11] (Courtesy: Weibull.com)

3.1 HUBZero Overview

HUBzero is a WEB 2.0 based scientific collaboration platform with various social networking and data management features such as user groups, access-controlled file sharing, rating and review mechanism, content tagging, online discussions, wikis and blogs, that can help researchers, educational institutions and commercial organizations to efficiently manage their work [12, 13]. HUBzero was created by the NSF-funded Network for Computational Nanotechnology starting in 2002 with the development of their HUB at nanoHUB.org. In 2007, HUBzero was spun out from nanoHUB.org as a separate project and software package to power new hubs. It has been adopted as a new portal development framework in several scientific and engineering domains including pharmaceutical product development, cancer research, and earthquake engineering [9]. A leading example is HUB-CI (HUB system with Collaborative Intelligence) developed by PRISM center at Purdue University that enhances HUB infrastructure capabilities with intelligent agents supporting collaboration activities. With its advanced features, HUB-CI enables cyber-augmented collaborative interactions over cyber-supported complex systems facilitating both physical and virtual collaboration between several groups of human users along with relevant cyber-physical agents [14]. HUB-CI has been used for developing solutions to various real-life complex industrial problems including (a) improving operational robustness by translating and aggregating hand gesture commands from multiple operators into a single control stream for collaborative telerobotics in manufacturing [15], (b) improving system productivity using collaborative intelligence of an agricultural telerobotic system for early detection of anomalies in pepper plants grown in greenhouses [14], (c) requirement planning, scheduling, and optimizing resource utilization in Industry4.0-enabled factories and warehouses using collaborative intelligence approach to process diverse local information and signals obtained from robots, humans, and warehouse components, and developing real-time resource-assignments and schedules [16, 17].

3.2 HUBZero Features

HUBzero provides a turn-key collaboration platform in which users can easily access, contribute and share content, data and tools on the Internet or on the intranet of the organization. It also provides out-of-box support for a number of popular collaboration tools including user groups, discussion forum, and wiki. Users are not only able to network and share information, but also create, publish and access interactive visualization tools powered by a rendering farm and other cluster computing resources, as well as engage in online collaboration fostering the development of a scientific or engineering community on the world wide web or within the organization [18].

HUBzero combines unique middleware with WEB 2.0 functionality, providing a platform that aims at decentralizing the creation /editing /organization of contents where any user can become an author of contents, can make these contents available to all or a certain group of users, and can create and share their own customized view of the content [19]. WEB 2.0 principles overlap considerably with the principles of knowledge management in content generation and collection, but they differ mainly in the centralization and controlled aspects of knowledge management versus the decentralized and uncontrolled nature of WEB 2.0. WEB 2.0 tools such as wikis, blogs, and tagging can be used to enrich the knowledge management systems [20] because they enhance collaboration within the organization, and also because people have gotten well-accustomed to using these tools outside the organizations on the internet with popular sites such as Wikipedia, Facebook, WordPress, and YouTube. HUBzero presents a unique mix of these features [12].

The various features of HUBzero which make it an exciting and unique collaboration tool are discussed below:

- ***Mechanism for Uploading New Resources:*** HUBzero is a forum for users to come together and share information. One important way to accomplish this is by encouraging all users to upload their own resources in form of files, tools, presentations, and other materials onto the HUB in a user-friendly manner.
- ***Ratings and Citations*:** The HUB philosophy is not to judge the quality of each resource before deciding to post, but rather, to post resources and let the community judge the quality. Registered users are allowed to post 5-star ratings and comments for each resource. The ratings and citations for each resource are combined with web

statistics (measuring the popularity of the resource) to produce a single number on a scale of 0 to 10, called the ranking, which defines the overall quality of the resource. Registered users can also post citations that reference the resource in the literature, so everyone can see other work that builds upon the resource.

- ***User Groups for Private Collaboration***: Any registered user can create a group and invite others to join it. The creator can accept or reject group members and can promote various members to help manage the group. Resources can be associated with a group and kept private, meaning that their access is limited to other members of the group.
- ***Content Tagging***: Each of the resources on a HUB is categorized by a series of tags, which are arbitrary strings defined by the user when uploading content. Each tag has an associated page on the HUB where its meaning is defined, and its resources are listed.
- ***Wikis and Blogs***: Each HUB supports the creation of "topic" pages, which is similar to the Google "knol" model for knowledge articles. Other users can be allowed to add comments to the page or even suggest changes. The original authors are notified of changes suggested by other users. The changes can be incorporated, and the users suggesting them can be added as co-authors for the page, so they can make further changes without approval.
- ***Interactive Simulation Tools***: The signature service of a HUB is its ability to deliver interactive, graphical simulation tools through an ordinary web browser. In the world of portals and cyber-environments, this capability is completely unique. Unlike a portal, the tools in a HUB are interactive in real time without waiting for a web page to refresh. Users can visualize results without having to reserve time on a supercomputer or wait for a batch job to engage. They can also deploy new tools without having to rewrite special code for the web.
- ***Online Presentations***: Along with the tools, each HUB features a series of online seminars, which are PowerPoint slides combined with voice and animation. HUBzero recommends Adobe Presenter® for producing online seminars and supports other video formats. Online seminars can also be distributed as podcasts, so users can access them on-the-go via their video or audio iPod.
- ***Usage Statistics:*** Each HUB reports statistics about how its resources are being used, including the total number of users in a given period, the number of web hits, simulation jobs launched, CPU hours used, etc.
- ***News and Events***: Each HUB includes a calendar and a mechanism for any registered user to post events. This helps the HUB become a focal point for the community.
- ***Feedback mechanisms***: Each HUB includes a feedback area where users can respond to a poll question, share a success story, or provide other comments and suggestions. [12, 13, 21–23]

4 Implementation of Collaborative HUB

For its implementation in the reliability engineering division of a large consumer products company, the HUB-in-a-box version of HUBzero [12] was successfully implemented to play the role of a knowledge base of different reliability tools being used in the company such as FMECA, Reliability Growth Analysis and Shakedown Testing. In this implementation, HUB was used for collecting, organizing and reusing analyses done through various reliability tools used in the organization. The schematic diagram of implementation of HUBzero is shown in Fig. 3.

Fig. 3. Overview of implementation of HUB

In this implementation, the HUB acted as a medium for building organizational memory by allowing people working in different areas to collaborate in a social networking manner not only to improve the quality of the work but also to save a lot of time by reusing the work done by colleagues to solve a similar problem. The various steps involved in organizational knowledge management are knowledge creation, validation, presentation, distribution, and application [24]. We discuss below how each of these steps were accomplished using HUBzero, some of the key challenges faced during the implementation of HUB, and how they were addressed:

- ***Collecting data from people:*** Based on the previous studies done in the area of knowledge management [21] and our own experience, one of the major challenges is to get data from people for building a knowledge base. We addressed this issue in the current implementation by creating a mechanism within the reliability tools to automatically transfer the data from the tool to a central server until the tool user specifically marks the data as highly restricted/confidential. The reliability tools we dealt with were based on MS Excel, hence a macro was written in the tool which sent a copy of the excel file along with the metadata information of the data in the file to a central server, when the file is closed by the user, and it meets certain criteria designed to ascertain the quality of data is good and the security of the project is not violated.
- ***Getting owner's consent before publishing:*** One of the important things to be taken care of was to get user's permission before publishing the data on a portal like HUBzero. Some users might not be comfortable publishing their data on the HUB until they feel their analysis is complete and accurate. A mechanism was built through which the users were contacted via email asking their permission to publish their data and if they were not willing to publish then, they were supposed to provide a valid reason for that, such as work in progress, or the project data is highly restricted and hence not appropriate to be published.

- ***Selecting good quality resources for publishing:*** To ensure the quality of the knowledge base on the HUB, the workflow was designed such that a team of subject matter experts did a round of review of all the incoming resources keeping only the good quality ones for further processing. A dashboard facility was developed for facilitating the review work to be done by the experts/administrators. Once the administrators found the resource appropriate to be published, it could be done with a click of button.
- ***Interfacing HUBzero with other Software/Groupware:*** One important aspect of implementing HUBzero in a large enterprise setting is its interaction with the existing software which are already being used in the organization. One of the main interfaces developed was between HUB and a popular Microsoft groupware through customization. Overall, it was observed that it takes a considerable amount of time and effort to build a seamless interface between the two/multiple systems.
- ***Access Control of the files:*** One of the main challenges was to keep the HUBzero application open for everybody in the organizations and letting everyone view the metadata of a resource created in HUBzero but restricting the access of the original file which contains the whole data of the reliability tool usage to only a limited group of users who work in that area. In order to accomplish this, we provided the metadata information of the resource on HUBzero in an unrestricted manner using the custom fields but provided a link to the original excel file located on an access restricted central server where employees with valid credentials can only access the file.
- ***Selection of server to host HUBzero:*** The full installation of HUBzero requires Debian Linux and depending on the anticipated usage of the HUB within the organization, the specifications of the server required to host the HUB can be obtained with the help of HUBzero support team. For our implementation, the HUB was required to be hosted within the company intranet and should not be accessed from outside world. Since we were using the HUB-in-a-box version (based on Debian Linux virtual machine) and did not use any simulation tools on HUBzero, a large computing capacity was not required. We used a Windows 2003 Server to host the HUB-in-a-box virtual machine. It should be noted while installing HUBzero that all the HUBs supported by Purdue University are hosted on the Purdue Grid Computing system but if the HUB is hosted outside Purdue and the scale of implementation is large, then a grid system within the organization along with a powerful sever is required to host the HUB.
- ***Maintaining security of the HUBzero server:*** Since the data residing on the HUB will be company confidential, hence the HUB should be hosted on a server behind appropriate firewall settings and not accessible from outside world.

4.1 Customizations on HUBzero During Implementation

Based on the specific requirements of the organization and for usability enhancements, several customizations were also done on the base version of HUBzero, as discussed below:

- Sophisticated search mechanisms were developed by listing the metadata information of the resources in a tabular format equipped with filtering and sorting capabilities in addition to keyword search functionality so that users can easily navigate and find the resource they are looking for.

- Multiple views of the information were made available to the end user based on different parameters like resource or reliability tool type, tags, project name or subgroup within organization etc.
- Different navigation layouts were made available to the users for them to efficiently browse through the HUB and find the information they are looking for.
- Automated tagging based on content was accomplished by developing a mechanism to generate the tags automatically for each resource during its creation based on the metadata of the resource. This development saved the time and effort of the original author of the content to create tags.
- Social networking features of reviews and comment on resource page were enhanced by changing the user interface in order to increase user participation and hence collaboration.
- User interface enhancements were made based on layout principles of some popular websites like Youtube, Amazon etc. and thus utilize user's comfort level in using these websites to increase the usability of HUBzero. A sample of interface is shown in Fig. 4.

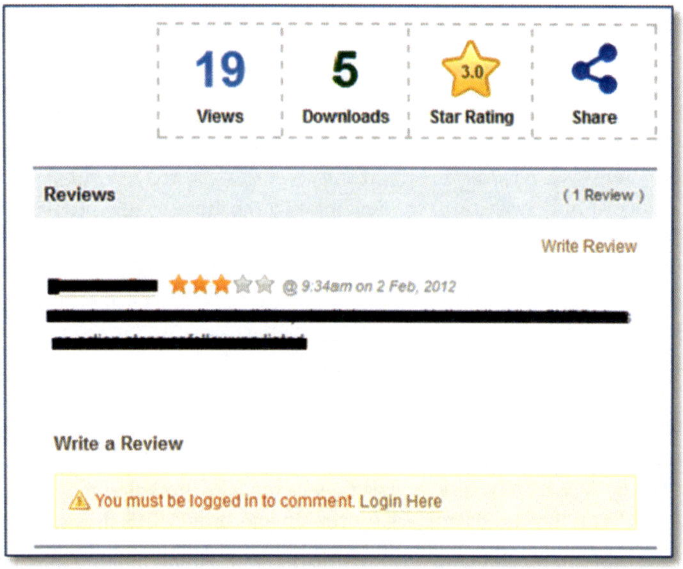

Fig. 4. Screenshot of user interface for RE tool file on the HUB

In summary, with the successful implementation of HUBzero for managing reliability data in a large enterprise, we were able to create a knowledge repository containing the results of past reliability analyses done in various RE areas. We also designed and developed efficient workflow for acquiring, publishing and sharing reliability data of the organization in an automated fashion. The users were able to use the portal for downloading and using different reliability tools and templates for doing analyses for their individual projects across different geographical locations. To improve HUB's utility and usability, easy-to-use interfaces were developed for users to look for completed reliability analyses, review and rate them in a social networking manner.

5 Data Management and Retrieval on Collaborative HUB

Post implementation, the number of RE files on HUBzero have been increasing at a steady pace. With growing number of files, effective organization and easy retrieval becomes a challenge for any knowledge management system. The screenshot of search tool developed on HUB for FMECA files including the key fields of FMECA analyses is shown in Fig. 5.

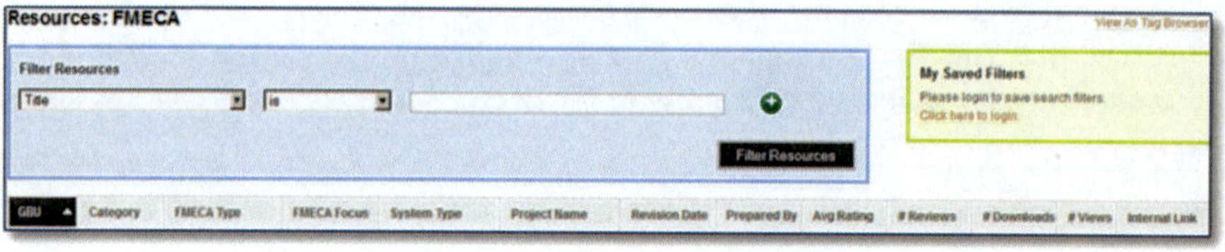

Fig. 5. Screenshot of search functionality on HUB for RE tool files (FMECA in this case)

The WEB 2.0 features of HUBzero such as tags, wikis etc. help overcome the problems of organizing content as well as of retrieval. Keywords or tags provide a fair idea about the content as well as provide opportunity to browse through related resources. In the initial phase of implementation, we developed a mechanism to automatically create tags for RE files published on the HUB using metadata information. But we observed that the metadata provided only high-level information about the RE file. So, while the metadata information was good for categorizing the RE files on the HUB, it did not provide a fair idea about the content of the RE analysis and limited the browsing capabilities. This provided us the motivation to develop new techniques based on statistical text mining for assigning keywords to RE files published on HUBzero. We developed an operational workflow to automatically generate a list of recommended keywords for a RE file, from which the subject matter expert can choose which ones to keep. The selected keywords will then be attached as tags to the RE files when they are published on HUBzero, helping the end user to easily search and browse RE files for reuse.

5.1 Keywords for Knowledge Management

Marking content with descriptive terms, also called as keywords, tags, or hashtags in WEB 2.0 and social media context, is a common way of organizing content for navigation, filtering, or searching [26]. Keywords summarize a document concisely and give a high-level description of the document's content. Keywords provide rich semantic information for many text mining applications, for example: document classification, document clustering, document retrieval, topic search, and document analysis [27]. Tags or Hashtags provide not only a method for organizing contents to facilitate the users who create it, but also a navigation mechanism for users to discover interesting resources. It has become a new interface of Web and has drawn much attention from both research and industrial communities [28]. Tags function as both resource organizers and discoverers. As resource organizers, tags allow tag creators to annotate and categorize a resource that would be easily retrieved later. As resource discoverers, tags can be used to make serendipitous discoveries of additional relevant resources [29]. Hence, tags are a useful tool for Knowledge Management systems where efficient organization and easy retrieval of content are most important objectives. Ontologies or controlled keywords for specific domains have been proven to be good additions to knowledge management systems. Domain ontologies have a good potential to improve information organization, management and understanding [30]. Ontologies provide a shared understanding of certain domains between people and application systems and support knowledge sharing and reuse [31]. Hence, a number of knowledge management applications have been developed with use of ontologies. For example, FRODO (a Framework for Distributed Organizational Memories) uses ontologies for knowledge description in organizational memories [32], CoMMA(Corporate Memory Management through Agents) investigates agent technologies for maintaining ontology-based knowledge management systems [33].

5.2 Automated Tag/Keyword Assignment

There has been ongoing research in effective automatic tag recommendation techniques that can assist users in identifying appropriate tags for their content which are representative of the meaning of the content. Tagging of a textual document involves identifying appropriate entities in the document that best summarize its content. The effective automation of this process requires from the system to be able to distinguish between the entities that play a central role to the document's meaning and those that are just complementary to it [34]. Tag recommendation can be addressed in two different aspects: user-centered approaches and document-centered approaches. User-centered approaches aim at modeling user interests based on their historical tagging behaviors and recommend tags to a user from similar users or user groups. On the other hand, document-centered approaches focus on the document-level analysis by grouping documents into different topics. The documents within the same topic are assumed to share more common tags than documents across different topics [35].

User-centered approaches are not effective if number of users tagging the same object is less [35], which is typically the case in organizations. Collaborative tagging, a common method of assigning tag to web content, can succeed when a lot of people are tagging, so that are enough similar tags for various users. While collaborative tagging works well on internet, organizational world is much smaller and therefore the rules would be different. "The world has already experienced this difference, while trying to copy internet forums to organizational internal discussion groups, which yielded much smaller success" [36]. Additionally, most of the existing collaborative tagging systems of manual tagging suffer from the vague meaning problem when users retrieve resources with keyword-based tags. This happens mainly because: (a) the tagging systems do not allow for identification of synonyms and are not able to distinguish between polysemies (words with multiple meaning) or between homonyms (similar pronunciation), (b) they do not filter grammatical and spelling errors, and (c) the relationships between tags cannot be structured [37]. So, in order to reduce the impact of these drawbacks and to aid tag-convergence, systems that assist the user in the task of tagging are required [38].

In an organizational setting, it would be extra work for engineers working on a reliability tool to assign keywords at the end of their analysis. The absence of multiple people tagging the same RE file would also lead to the tag related issues discussed above. Hence, it seemed useful to develop an automated tag recommendation system in our case. Simple probabilistic models have been tested of duplicating the performance of a human indexer in assigning subject index terms to documents [39]. We adopted a semiautomated process for keyword assignment to RE files. An automated tag recommendation system recommends a set of possible keywords for each RE file to a subject matter expert, who will then choose the best keywords for that file. The keywords discarded by the subject matter expert for that file will be recorded in database and will be used while making future recommendations. Manual selection of tags by subject matter experts will ensure high quality and excellent precision.

A combination of automatic and manual approaches for assigning keywords ensures consistency of tags and also reduces manual work [40]. Zhang [41] categorized the existing approaches for assigning tags/keywords in an automated manner broadly into two categories: keyword extraction and keyword assignment. In keyword extraction, words occurring in the document are analyzed to identify apparently significant ones, on the basis of properties such as frequency and length. In keyword assignment, keywords are chosen from a controlled vocabulary of terms, and documents are classified according to their content into classes that correspond to elements of the vocabulary. Further, the existing methods about automatic keyword extraction can be divided into different categories such as: simple statistics, linguistics, machine learning, or a combination of these methods. Simple statistics approaches do not require a set of training data; they use statistical information of the words to identify the keywords in the document. Some of the commonly used statistics methods include n-gram, word frequency, term frequency*inverse document frequency (TF-IDF), word co-occurrences etc. Linguistic approaches such as lexical analysis, syntactic analysis etc. use the linguistic features of the words, mainly sentences and document. Supervised machine learning approaches such as Naïve Bayes, Support Vector Machine, and Neural Network, use the extracted keywords from a set of training documents to learn a model and apply the model to find keywords from new documents [41].

In our case, the collection of RE tool files was not very large; hence using a set of training data would not be very effective. The amount of text in the each RE file is also small and in form of unstructured short sentences; hence the effectiveness of linguistic approaches would be limited. Thus, we used a customized statistical approach for keyword extraction based on the term frequency for identifying keywords from the RE files. "The most important knowledge source for finding important descriptors for a document is the document itself" [40]. Through the approach discussed in next section, a set of keywords are identified for each RE file and presented to a subject matter expert, who then makes the final selection of keywords for that file.

6 Approach for Keyword Assignment to RE Tool Files

The objectives the keywords assigned to a particular RE file were to help users find the RE files and give users an idea of the content of the RE file without opening it. To accomplish this, we assigned two sets of keywords to each RE tool file: a) global, that represented the broader area of RE tool file, and b) file-specific, that gave an idea about detailed analyses covered in that RE file. These keywords are determined using two types of relevance scores associated with each word in the RE file: file score and global score. The file score of a word indicates the association strength of the keyword with a particular file and would be displayed along with the file information to the user. The global score for a particular keyword indicates if the word has presence across a group of files and hence can be listed as a popular keyword. When the user would visit the RE knowledge base HUBzero website in the organization, a list of most popular keywords will be displayed to help browse or find different types of RE analysis. The schematic diagram of the interface is shown in the Fig. 6.

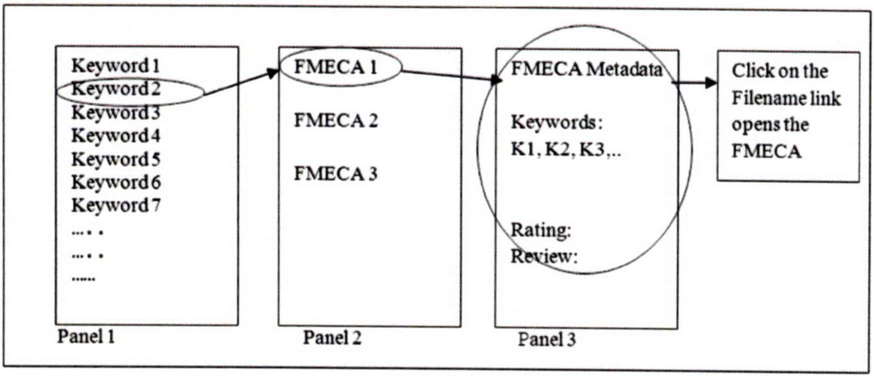

Fig. 6. Tag Browser for FMECA Resources on HUBzero website

In Fig. 6, Panel 1 contains a list of top keywords in form of tags sorted in the order of popularity. The popularity is judged using the global score of the keywords, discussed later in this section. When the user will click on a particular keyword, Panel 2 will be loaded with list of all FMECA files which are associated with it, rank-ordered according to their file scores for that keyword, reflective of the association strength of file with the keyword. When user clicks on a particular file in the Panel 2, Panel 3 will be loaded with a quick summary of the FMECA resource including file keywords, other metadata information, user rating, number of views, number of downloads, etc. to give an idea about the content of the file to the user.

6.1 File Keywords

In the workflow of acquiring, quality inspection, and publishing RE files on HUBzero, when the file is selected for publishing on HUBzero, keywords to be linked with the file would be assigned. The steps of keyword assignment process for the RE files are discussed below and schematically shown in Fig. 7.

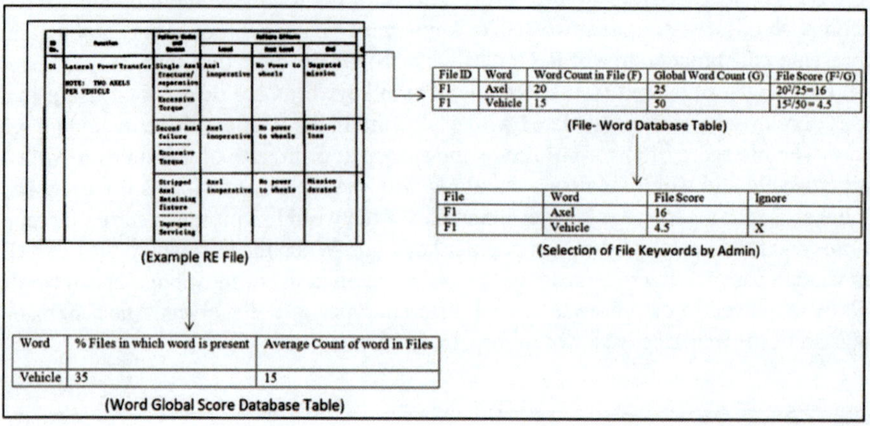

Fig. 7. Schematic Diagram of Keyword Assignment Process for sample FMECA [42]

As shown in Fig. 7, the steps involved in keyword assignment for RE tool files include:

1. Each RE file is read by a program and the words in those files are parsed, counted and stored in database tables.
2. The database table containing the basic data about the words has the following columns: Word, File ID, Word Count in File (F), Global Word Count (G), and File Score. F is the number of times the word occurs in that particular file, and G is the number of times the word occurs in all the files. An example RE file [20] as shown in Fig. 7, is parsed and the words are entered the File-Word database table.
3. The ratio F/G indicates the likelihood of a word being present in a particular file. We have used the formula F^2 /G for calculating the File score for each file-word combination, which is the product of the likelihood of the word being present in a file and the number of times the word occurs in that file. This formula gives a higher value of words which have strong association only with that file and lower values for words which are commonly used across all files. Thus, it eliminates the need of maintaining a list of stop-words for excluding the most common words we do not want to consider as keywords.
4. Based on the file scores, top keywords for a particular file will be recommended to a subject matter expert or administrator. The administrator can make the final selection by ignoring the keywords which do not seem appropriate for that RE file.

The purpose of 'Popular Global Keywords' shown in Tag browser of Fig. 6 is to direct the users from the top level to a collection of RE files with some common features. Hence, the global keywords are not associated with a particular file, but these words would represent a particular group of RE files. For example, if there are 10 FMECAs out of 100 related to vehicles, and each have the word "Vehicle" as a keyword, then it should qualify as a Global keyword. So, one of the possible ways of identifying such words would be to set a window with upper and lower limits for the percentage of files in which the word occurs and a minimum cutoff for the number of times the word occurs in the files it is present. This approach is explained in the steps below:

1. Regarding the window for percentage of files, having an upper limit for the percentage of files in which the word occurs would help eliminate the words which have a uniform presence in all the files. Similarly, having a lower limit for the percentage of files in which the word occurs would eliminate the words which occur in very few files and hence would not qualify to become global keywords.
2. A minimum cutoff for the number of times the word occurs in a particular file would ensure that the word has a minimum level of association with the file. For example, if the keyword "Vehicle" occurs in a FMECA file 8–10 times, then it means it has a strong association with the file and can qualify as a keyword for that file. But if the word occurs only 1 or 2 times in the file then the word does not qualify to become a good keyword to be associated with that file.
3. The cut-off values for the upper and lower limits of percentage window for number of files and minimum occurrence of word in a file have to be decided after assessment of the data and may need to be changed from time to time as the collection of FMECA file grows. So, they can be kept as system variables which the administrator can change with time.
4. The data for global keywords can be stored in a Database table with columns: Word, Percentage of Files in which word is present, and Average Count of word in Files as shown in Fig. 7. The percentage of files for each word can be calculated as:

   $$(number\ of\ files\ with\ word\ present)/(Total\ number\ of\ files)$$

   The Average Count of word in Files can be calculated by averaging the individual number of times the word occurs in files. The global score would be a weighted linear combination of the percentage of files, average count, and some other user-defined parameters.
5. We can extract the top-100 or most popular keywords from this table after applying the percentage window and cut-off criteria mentioned above. These selected global keywords would be then presented to a subject matter expert or administrator who can make the final decision on which keywords can be selected as 'Popular Keywords'. One of the main issues with any organizational database is managing the constant growth of data. As new files are added to the database, the global count for a particular word will keep increasing. Hence, the system recommendation of global keywords and even file keywords for the existing RE files might change with time. In order to deal with this, our current approach is to run the file and global keyword algorithms for all the RE files once in every few weeks. This will help the subject matter expert or administrator to choose the appropriate file and global keywords for the RE file database at different stages.

7 Summary and Future Work

Through HUBzero's implementation in the organization, we have been able to create a repository containing the results of past RE analyses. Efficient workflows were developed for acquiring, publishing, and sharing reliability data of the organization in a semi-automated fashion. Through additional customizations, easy-to-use interface were developed for users to look for completed reliability analyses, review, and rate them in a social networking manner. Overall, HUBzero has been effective to address the main issues of managing the reliability data in a large organization.

We also developed an efficient way of organizing and reusing reliability engineering data through an operational workflow for assigning keywords to RE tool files by extracting words from files based on the relative word frequency thresholds. The usage of complex data mining and linguistic approaches did not seem beneficial given the unique file structure of RE tools containing short text, absence of training dataset with pre-assigned keywords, and relatively smaller size of file collection. Therefore, we developed a semi-automated system which will help the subject matter experts to choose the keywords from recommendations made by the system. With growing database of RE files, the complexities of keyword assignment problem can be better handled by implementing more sophisticated algorithms. For example, there may be different words which have the same meaning or may be misspelled and have been used at different places in the FMECA files. There might be some word sequences or combinations as good keywords instead of the individual words. It would be useful to have capabilities to handle such situations.

While the HUB has been effective to address the main issues of managing the reliability data in a large organization, it can be made better by analyzing the results of usability testing of the new features developed on the HUB and overall beta testing by the subject matter experts. One of the limitations of the implementation is that the RE tools for which HUBzero has been implemented are MS excel based tools, hence the approach followed here would change if the tool is non-excel based and there may be some further complexity involved but the basic data flow from the author to the HUB will not change, as it has proven to address some of the main concerns of collecting data, separating good quality data and presenting data effectively. Enhancement of RE tool formats in form of cloud-based mobile applications can enable better collaboration, data collection, sharing, and analysis, and exchange of ideas sharing between users from different groups within the organization, vendors, suppliers and customers for that organization to design and deliver better products/services.

References

1. Liu, C., Vengayil, H., Zhong, R.Y., Xu, X.: A systematic development method for cyber-physical machine tools. J. Manuf. Syst. **48**, 13–24 (2018)
2. Industry 4.0? - MATLAB & Simulink. https://www.mathworks.com/discovery/industry-4-0.html. Accessed 03 Jul 2022
3. Zhang, L., Zhang, L., Shan, H.: Evaluation of equipment maintenance quality: a hybrid multi-criteria decision-making approach. Adv. Mech. Eng. **11**(3), 1687814019836013 (2019)

4. Vogl, G.W., Weiss, B.A., Helu, M.: A review of diagnostic and prognostic capabilities and best practices for manufacturing. J. Intell. Manuf. **30**(1), 79–95 (2016). https://doi.org/10.1007/s10845-016-1228-8
5. Lazarova-Molnar, S., Mohamed, N.: Reliability assessment in the context of industry 4.0: data as a game changer. Procedia Comput. Sci. **151**, 691–698 (2019)
6. Slon, C., Pandey, V., Kassoumeh, S.: Mixture distributions in autonomous decision-making for industry 4.0. SAE Int. J. Mater. Manuf. **12**(2), 135–148 (2019)
7. Souza, M.L.H., da Costa, C.A., de Oliveira Ramos, G., da Rosa Righi, R.: A survey on decision-making based on system reliability in the context of Industry 4.0. J. Manuf. Syst. **56**, 133–156 (2020)
8. Nanda, G., Tan, J., Auyeung, P., Lehto, M.: Evaluating HUBzero (TM) as a collaboration platform for reliability engineering. In: Proceedings of the IIE Annual Conference, p. 1. Institute of Industrial and Systems Engineers (IISE) (2012)
9. Nanda, G., Tan, J., Auyeung, P., Lehto, M.: Improving efficiency of organizational reliability engineering knowledge using keywords. In: Proceedings of the IIE Annual Conference, p. 3414. Institute of Industrial and Systems Engineers (IISE) (2013)
10. Mettas, A., Rock, D.: Intellectual capital: utilizing the Web for knowledge management and data utilization in reliability engineering. In: Proceedings of the Annual Reliability and Maintainability Symposium, pp. 379–385 (2002)
11. Figure. https://www.weibull.com/basics/fmea_fig1.htm. Accessed 03 Jul 2022
12. HUBzero - Platform for Scientific Collaboration. http://hubzero.org/. Accessed 03 Jul 2022
13. McLennan, M.: Managing data within the HUBzero™ platform. OMICS J. Integr. Biol. **15**(4), 247–249 (2011)
14. Nair, A.S.: A HUB-CI model for networked telerobotics. In: Collaborative Monitoring of Agricultural Greenhouses. M.S. Thesis, Purdue University Graduate School (2019)
15. Zhong, H., Wachs, J.P., Nof, S.Y.: HUB-CI model for collaborative telerobotics in manufacturing. IFAC Proc. Volumes **46**(7), 63–68 (2013)
16. Dusadeerungsikul, P.O., et al.: Collaboration requirement planning protocol for HUB-CI in factories of the future. Procedia Manuf. **39**, 218–225 (2019)
17. Dusadeerungsikul, P.O., He, X., Sreeram, M., Nof, S.Y.: Multi-agent system optimisation in factories of the future: cyber collaborative warehouse study. Int. J. Prod. Res., 1–15 (2021)
18. Zhao, L., et al.: Bring integrated GIS data and modeling capabilities into HUBzero platform. In: Proceedings of the ACM SIGSPATIAL Second International Workshop on High Performance and Distributed Geographic Information Systems, HPDGIS 2011, pp. 30–33. ACM, New York (2011)
19. Marlow, C., Naaman, M., Boyd, D., Davis, M.: HT06, tagging paper, taxonomy, Flickr, academic article, to read. In: Proceedings of the seventeenth conference on Hypertext and hypermedia, pp. 31–40 (2006)
20. Levy, M.: WEB 2.0 implications on knowledge management. J. Knowl. Manag. **13**(1), 120–134 (2009)
21. McLennan, M., Kennell, R.: HUBzero: a platform for dissemination and collaboration in computational science and engineering. Comput. Sci. Eng. **12**(2), 48–53 (2010)
22. Streveler, R.A., Magana, A.J., Smith, K.A., Douglas, T.C.: CLEERhub.org: Creating a digital habitat for engineering education researchers. In: Proceedings of the 2010 American Society for Engineering Education Annual Conference and Exposition (2010)
23. nanoHUB.org - Simulation, Education, and Community for Nanotechnology. https://nanohub.org/ Accessed 03 Jul 2022
24. Bhatt, G.D.: Knowledge management in organizations: examining the interaction between technologies, techniques, and people. J. Knowl. Manag. **5**(1), 68–75 (2001)
25. Abecker, A.: Corporate memories for knowledge management in industrial practice: prospects and challenges. J. Univ. Comput. Sci. **3**(8), 929–954 (1997)

26. Specia, L., Motta, E.: Integrating folksonomies with the semantic web. In: Franconi, E., Kifer, M., May, W. (eds.) ESWC 2007. LNCS, vol. 4519, pp. 624–639. Springer, Heidelberg (2007). https://doi.org/10.1007/978-3-540-72667-8_44
27. Zhang, K., Xu, H., Tang, J., Li, J.: Keyword extraction using support vector machine. In: Yu, J.X., Kitsuregawa, M., Leong, H.V. (eds.) WAIM 2006. LNCS, vol. 4016, pp. 85–96. Springer, Heidelberg (2006). https://doi.org/10.1007/11775300_8
28. Li, R., Bao, S., Yu, Y., Fei, B., Su, Z.: Towards effective browsing of large scale social annotations. In: Proceedings of the 16th International Conference on World Wide Web, WWW 2007, pp. 943–952 (2007)
29. Razikin, K., Goh, D.H., Chua, A.Y.K., Lee, C.S.: Social tags for resource discovery: a comparison between machine learning and user-centric approaches. J. Inf. Sci. **37**(4), 391–404 (2011)
30. Baruzzo, A., Dattolo, A., Pudota, N., Tasso, C.: Recommending new tags using domain ontologies. In: Proceedings of the 2009 IEEE/WIC/ACM International Joint Conference on Web Intelligence and Intelligent Agent Technology, vol. 03, pp. 409–412 (2009)
31. Fensel, D.: Ontologies: A Silver Bullet for Knowledge Management and Electronic Commerce. Springer, Heidelberg (2003). https://doi.org/10.1007/978-3-662-09083-1
32. Abecker, A., et al.: Toward a technology for organizational memories. IEEE Intell. Syst. **13**(3), 40–48 (1998)
33. Gandon, F., Dieng-Kuntz, R., Corby, O., Giboin, A.: Semantic web and multi-agents approach to corporate memory management. In: Musen, M.A., Neumann, B., Studer, R. (eds.) IIP 2002. ITIFIP, vol. 93, pp. 103–115. Springer, Boston, MA (2002). https://doi.org/10.1007/978-0-387-35602-0_10
34. Alexopoulos, P., Wallace, M.: Improving automatic semantic tag recommendation through fuzzy ontologies. In: 2012 Seventh International Workshop on Semantic and Social Media Adaptation and Personalization (SMAP), pp. 37–41 (2012)
35. Song, Y., Zhang, L., Giles, C.L.: Automatic tag recommendation algorithms for social recommender systems. ACM Trans. Web (TWEB) **5**(1), 4 (2008)
36. Levy, M.: WEB 2.0 implications on knowledge management. J. Knowl. Manage. **13**(1), 120–134 (2009)
37. Huang, S.-L., Lin, S.-C., Chan, Y.-C.: Investigating effectiveness and user acceptance of semantic social tagging for knowledge sharing. Inf. Proces. Manage. **48**, 599–617 (2012)
38. Musto, C., Narducci, F., De Gemmis, M., Lops, P., Semeraro, G.: STaR: a social tag recommender system. In: Proceeding of ECML/PKDD 2009 Discovery Challenge Workshop, pp. 215– 227 (2009)
39. Zhu, W., Lehto, M.R.: Decision support for indexing and retrieval of information in hypertext systems. Int. J. Hum. Comput. Interact. **11**(4), 349–371 (1999)
40. Hulth, A., Karlgren, J., Jonsson, A., Boström, H., Asker, L.: Automatic keyword extraction using domain knowledge. In: Gelbukh, A. (ed.) CICLing 2001. LNCS, vol. 2004, pp. 472–482. Springer, Heidelberg (2001). https://doi.org/10.1007/3-540-44686-9_47
41. Zhang, C., Wang, H., Liu, Y., Wu, D., Liao, Y., Wang, B.: Automatic keyword extraction from documents using conditional random fields. J. Comput. Inf. Syst. **4**(3), 1169–1180 (2008)
42. McKinney, B.T.: FMECA, the right way. In: Reliability and Maintainability Symposium, Annual Proceedings, pp. 253–259 (1991)

The Principle-Based EMS Logistics Policies

Seokcheon Lee(✉)

School of Industrial Engineering, Purdue University, 315 N Grant Street, West Lafayette, Indiana 47907, USA
lee46@purdue.edu

Abstract. This chapter introduces decision principles and policies for EMS (emergency medical services) logistics that essentially requires systems collaboration and integration for the effective support of patients' survivability. EMS logistics involves three major operational decisions which play an important role in reducing the response time: call-initiated ambulance dispatching, ambulance-initiated ambulance dispatching, and hospital selection. Simple logistics decision policies are usually adopted in practice in support of real-time, practical decision making, which are usually myopic in nature focusing only on immediate performance. Three EMS logistics policies enabling to secure long-term performance have recently been designed, namely the Preparedness policy, Centrality policy, and 3C policy. These policies are in a simplistic form yet showing a great potential in reducing response time. The three EMS logistics policies (called principle-based policies) establish an excellent foundation for intelligent and dependable EMS logistics and the objective of this chapter is twofold. First, behavioral properties of the EMS systems are analyzed with all the three policies in activation together. Note that this is a first attempt to study all the three different types of decisions in a single experimental framework. Second, the EMS policies are extended to the priority EMS systems (called principle-based priority policies) that take into consideration the heterogeneity of patients in terms of urgency, recognizing that many real life systems adopt priority classes of patients.

Keywords: EMS logistics · response time · ambulance dispatching · hospital selection · patient priorities

1 Introduction

EMS (emergency medical services) logistics requires systems collaboration and integration to support, with limited EMS resources, the survivability of emergency patients, and this chapter introduces such decision principles and policies that are practical yet effective in the dynamic and uncertain emergency situations. Response time in emergency medical services (EMS) is the time taken for an ambulance to arrive at the scene after receipt of an emergency call, and it is a critical performance metric influencing the survivability of patients (Andelius et al. 2020; Gonzalez et al. 2009; Sánchez-Mangas et al. 2010; Stiell et al. 2008; Vukmir 2006). EMS logistics involves three major operational decisions which play an important role in reducing the response time: call-initiated

C.-Y. Huang and S. W. Yoon (Eds.): ICPR1 2021, ACES 14, pp. 199–220, 2023.
https://doi.org/10.1007/978-3-031-44373-2_12

ambulance dispatching, ambulance-initiated ambulance dispatching, and hospital selection. Ambulance dispatching is a process of assigning an ambulance to an emergency call, and it can be either call-initiated or ambulance-initiated (Lee 2012). When several ambulances are available upon the arrival of a new call, a decision has to be made to select a unit (ambulance) among them (call-initiated). On the other hand, if a unit has just got freed and finds several calls waiting, a call among them has to be chosen (ambulance-initiated). The busyness of the system determines the relevance of the two types of dispatching decisions. The hospital selection decision determines a hospital among those having an eligible emergency department (ED), when a patient needs to be transferred to an ED.

The EMS logistics is relevant to a large amount of EMS resources and patients worldwide, e.g., in the United States alone, 18,200 EMS agencies, 73,500 ambulances and fire engines, and 1.03 million EMS personnel (NASEMSO 2020), and 4,577 EDs and 143.5M EMS patients annually (AHA 2020). The efficiency of EMS logistics is, therefore, a serious demanding issue, and the significance is even rising because 1) EMS demand is increasing but resources are getting scarce (63% increase in demand per ED between 1995 and 2018 in USA) (Fig. 1a), 2) healthcare expense is growing (247% increase in real dollars between 1980 and 2018 in USA) (Fig. 1b). In order to provide an acceptable level of EMS services in this trend, expanding EMS resources may be necessary, but this strategy requires securing and maintaining additional resources and a more viable option would be to enhance the EMS logistics efficiency to prevent the rise of medical expenses.

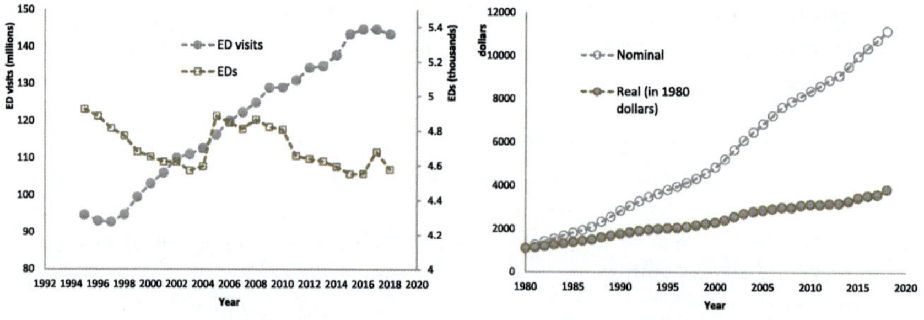

a. demand and supply of EDs (NASEMSO 2020)

b. per capita national health expenditures (AHA 2020)

Fig. 1. Trends supporting the need for EMS logistics efficiency

Simple logistics decision policies are usually adopted in practice in support of real-time, practical decision making, e.g., dispatch the closest unit available, dispatch to the closest call waiting, and select the nearest hospital. However, these policies are myopic in nature focusing only on immediate performance (e.g., in a simplified scenario as in Fig. 2 where an ambulance has to visit 11 patients, a myopic decision of Path B gives total response time of 144min while Path A results in 76 min). Recently, Lee developed the "principle-based design procedure" for producing decision policies that enable to take into account long term consequences. The design procedure consists of: 1) identify governing decision principles enabling to secure both short-term and long-term performance; 2) define performance measurements associated with the principles; and 3) synthesize a metric that aggregates and balances all the measurements. In each decision making, the metric is evaluated for each alternative decision and a decision is chosen such that the metric is optimized. A decision policy formed from this design procedure is mostly free from assumptions and projections, therefore fitting in the dynamic and uncertain emergency situations.

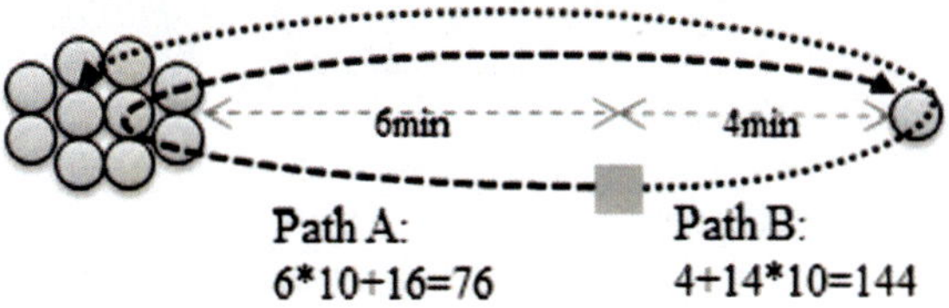

Fig. 2. Impact of different routes

Following this design procedure, three EMS logistics policies have been designed, namely the Preparedness policy for call-initiated dispatching (Lee 2011, 2015), Centrality policy for ambulance-initiated dispatching (Lee 2012, 2013), and 3C policy for hospital selection (Lee 2014a). These policies are in a simplistic form yet showing a great potential in reducing response time. The Preparedness policy reduces response time by up to 18.7%, the Centrality policy by up to 86.0%, and the 3C policy by up to 99.6%, over the myopic policies (discussed above) in simulation-based experiments.

The three EMS logistics policies (called principle-based policies) establish an excellent foundation for intelligent and dependable EMS logistics and the objective of this chapter is twofold. First, behavioral properties of the EMS systems are analyzed with all the three policies in activation together. Note that this is a first attempt to study all the three different types of decisions in a single experimental framework. Second, the EMS policies are extended to the priority EMS systems (called principle-based priority policies) that take into consideration the heterogeneity of patients in terms of urgency, recognizing that many real life systems adopt priority classes of patients. The rest of this chapter is organized as follows. Section 2 provides a background of prior developments along with an experimental result. Section 3 introduces the priority-based EMS logistics policies and the resulting behavior is analyzed. Section 4 conclude this work with discussions on challenges and issues in implementing the logistics policies in real life scenarios, and Sect. 5 provides research opportunities for further advancing the logistics policies.

2 Principle-Based EMS Logistics Policies

This section presents, in a chronological order of development, the EMS logistics policies developed by Lee following the principle-based design procedure. Before we proceed, the EMS logistics is overviewed (Fig. 3) as follows to provide the context of coming discussions. Only a portion of emergency patients actually call an ambulance. The probability of calling an ambulance (*amb_prob*) varies in space and time and it is reported that about 15.7% of patients arrive in ambulance in the United States (CDC 2011) and 23% in the United Kingdom (Lowthian et al. 2012). The patients who do not call are transported without assistance of ambulance to the nearest hospital (called walk-in patients). Ambulances are dispatched to emergency calls according to a certain dispatching policy and once arriving to a call site, the ambulance serves the patient for a certain onsite service time. The ambulance then, if necessary, transfers the patient to a hospital with an eligible ED following a hospital selection policy. The probability of transferring to hospital (*hospital_prob*) varies as well. The *hospital_prob* is reported to be 25% in the United States (Blackstone et al. 2007), while it varies in 47–72% in the United Kingdom (UKDH 2015). After arriving at the ED, if no patient is waiting and an ED bed is available, the ambulance crew transfers the patient to the bed and the ambulance gets discharged becoming available to other emergency patients. However, if there are patients waiting in the ED, the crew along with the patient waits in a queue that operates in first-come, first-served (FCFS) discipline until a bed is available for them.

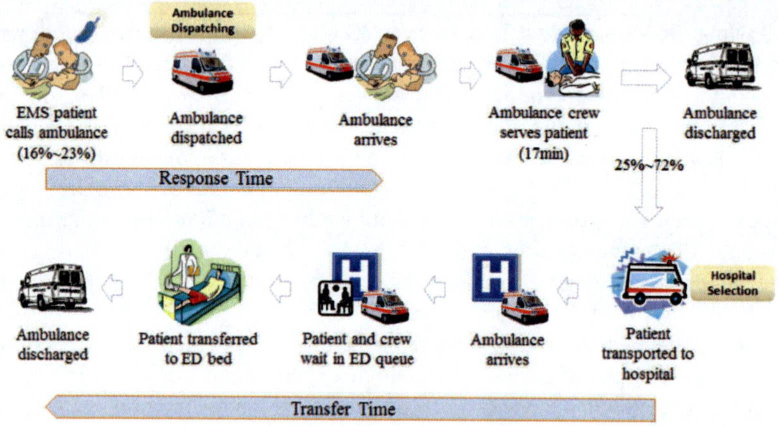

Fig. 3. EMS logistics

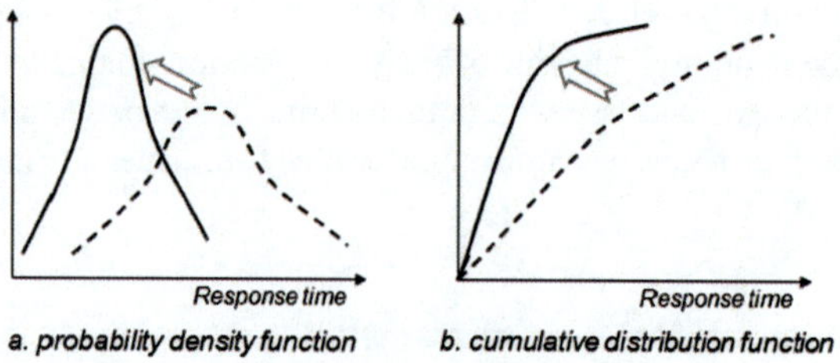

Fig. 4. Average and variation of response time

Coverage level, which is defined as the portion of emergency calls responded to within a certain response time threshold, is often used for evaluating the performance of EMS systems. However, no uniform definition exists for the response time threshold, varying from 7–30 min depending on country, urbanization, and urgency (Ball and Lin 1993; Black and Davies 2005; Gendreau et al. 2001; Holloway et al. 1999; McGrath 2002; Pons and Markovchick 2002; Woollard et al. 2003). Also, when using the coverage metric, the differences in response time distribution among scenarios within or beyond a threshold cannot be effectively counted, though significant differences actually exist (McLay and Mayorga 2011). Despite the diversity and incompleteness of the coverage metric, if a policy improves in both average and variation of response time, it implies that the policy is likely to have a better coverage level for any definition of response time threshold (Fig. 4). The principle-based EMS logistics policies we present next reduce both average and variation of response time, therefore possessing dominance over various possible definitions of response time threshold.

2.1 Preparedness Policy – "Choose a Unit That is Closer and Results in Higher Preparedness"

Sending the closest unit available in call-initiated dispatching (choosing among available units) is the most common policy in practice (Chaiken and Larson 1972; Dean 2008; Hayes et al. 2004; Repede and Bernardo 1994). However, if the closest unit is located in a dense area with a high rate of calls, the area will become ill-prepared for future calls and dispatching a unit located in a farther but sparse area with a relatively low rate of calls would be more desirable (Carter et al. 1972; Cunningham-Green and Harries 1988; Repede and Bernardo 1994; Weintraub et al. 1999). A dispatching policy based on a quantitative definition of preparedness is proposed by Andersson and Värbrand (2007), and the Preparedness policy has been subsequently reinforced by Lee (2011, 2015). The final version of the Preparedness policy (Lee 2015) adopts a metric combining closeness and preparedness so as to choose a unit that is closer to current call and at the same time results in higher preparedness for future calls, as follows:

1. A service area is divided into zones Z and when a new call arrives from a zone c, a set A of available (idle) ambulances are identified.
2. For each ambulance $i \in A$, preparedness level $p_{A\backslash i}$ is computed by setting ambulance i unavailable (resulting from dispatching it), where λ_j represents call rate in zone j and t_{kj} is the travel time of ambulance k to zone j.

$$p_{A\backslash i} = \frac{1}{\sum_{j\in Z} \lambda_j (1 + \min_{k\in A\backslash i} t_{kj})}$$

3. Dispatch to the call site c the ambulance i^* that maximizes fitness based on the preparedness $p_{A\backslash i}$ weighted by preparedness parameter α (≥ 0) and expected response time t_{ic} for a unit i to reach the call site c.

$$i^* = \arg\max_{i\in A} \frac{p_{A\backslash i}^{\alpha}}{1 + t_{ic}}$$

Consider the call rate of each zone as a population of potential patients in the zone, and let's represent the system in a network where patients and ambulances are nodes and each patient is connected to the ambulances with edge weight corresponding to the distance (in time). Upon this network representation, one can compute the operational efficiency (i.e., preparedness) of ambulances in reaching out the potential patients by the so called "node centrality" which is used in the study of complex networks that has emerged over the past decade as a theme of a wide range of disciplines (see Newman 2003 for detailed review of this area). Various centrality measures have been proposed, such as degree, transitivity, weighted degree, weighted transitivity, distance centrality, and betweenness centrality. The preparedness pi in Step ii is defined by adapting the distance centrality (Sabidussi 1996) among others due to its alignment with the notion and intent of preparedness. The distance centrality is originally defined for individual nodes and the preparedness metric used here is designed for a group of nodes by taking the minimal distance as the distance to the group, since the unit closest to the call is likely to serve the patient. Weight on preparedness α in step iii is a calibration parameter of which value has to be chosen appropriately for the EMS system at hand. The Preparedness policy with $\alpha = 0$ is exactly same as the myopic policy of sending the closest unit; however, when the weight is positive the policy incorporates preparedness into decision by the extent corresponding to the weight. Upon the right choice of the parameter, the Preparedness policy reduces average response time by up to 18.7% and variation of response time (measured in standard deviation) by up to 21.2%, over the myopic policy.

2.2 Centrality Policy – "Choose a Call That is Closer and More Centrally Located"

When call rate becomes high or the size of ambulance fleet gets small, a queue of calls would form and ambulance-initiated dispatching decisions (choosing among calls waiting) have to be made. A myopic policy for this decision problem is to choose the closest call and this policy is known well performing in various special cases of the problem (Bertsimas and van Ryzin 1991; Conway 1967). This policy, however, pursues only immediate performance without taking into account long-term consequences. The Centrality policy incorporates the principle of centrality to overcome this inefficiency, where the centrality of a call represents the efficiency of the call site in reaching out other calls with respect to the geographical call distribution over the service area. The centrality is computed upon a network where nodes represent waiting calls and an edge between every pair of calls has a value of distance (in time). The Centrality policy prioritizes calls based on the closeness as well as centrality, enabling to pursue short-term as well as long-term performance at the same time and thereby keeping the completion rate at maximum over time. The Centrality policy is presented in three steps as follows:

1. When an ambulance v gets freed, identify all unassigned calls U.
2. Compute centrality c_u of each call $u \in U$ upon the network of calls U with the edge between every pair of calls from u to i having a value of distance τ_{ui} (in time).

$$c_u = \sum_{i \in U, i \neq u} \frac{1}{(1 + \tau_{ui})}$$

3. Dispatch the freed unit v to the call u^* that maximizes fitness based on the centrality c_u weighted by centrality parameter β ($\geq$0) and expected response time t_{vu} for a unit v

to reach the call site u.

$$u^* = \arg\max_{u \in U} \frac{c_u^{\beta}}{(1 + t_{vu})}$$

The centrality c_u in Step ii is represented by the weighted degree (Barrat et al. 2004; Newman 2004) among other centrality measures due to its computational efficiency and ability of supporting robust performance. A calibration parameter in step iii, weight on centrality β, is associated with the Centrality policy. The Centrality policy with $\beta = 0$ is exactly same as the myopic policy; however, when $\beta > 0$ the policy incorporates centrality into decision by the extent corresponding to the weight. The weight value has to be chosen according to the characteristics of the EMS system at hand. The centrality policy, upon the right choice of the parameter value, outperforms the myopic policy across different conditions with up to 86.0% reduction in average response time and with up to 94.0% reduction in variation (measured in standard deviation). The reduction in both average and variation implies that the Centrality policy is likely to have a better coverage level for any definition of response time threshold and mitigate the risk of exposing patients to excessively tardy responses.

2.3 3C Policy – "Select a Hospital That is Closer, Less Congested, and More Centrally Located"

When transferring patient to an ED after onsite service, a decision has to be made in selecting an appropriate hospital among those having an eligible ED. One natural policy for hospital selection is the closest policy that transfers to the closest hospital and the majority of patients are in practice transferred in such a way (CDC 2011). However, ambulances are often redirected to an ED other than the closest ED when the closest one is extremely busy, and this practice is called "ambulance diversion" with over 33.4% of hospitals having gone on diversion during 2011 (CDC 2011). One diversion policy is that each ED goes on diversion if the ED has no bed available and the ambulance chooses the closest hospital among those not on diversion (Deo and Gurvich 2011). Two other hospital selection policies are also available in literature: Join the Shortest Queue (JSQ) policy (Adan et al. 1994; Enders 2010; Whitt 1986) and Shortest Transfer Time (STT) policy (Lee 2014a). The JSQ policy utilizes real-time queue length information of ED waiting rooms and the ambulance goes to the ED with the shortest queue. The SST policy is to choose a hospital that provides the shortest expected transfer time where transfer time consists of transport time (from the scene to ED) and turnaround time (interval between arrival at the ED and the time the ambulance becomes discharged).

All the four policies discussed above consider two factors (separately or jointly) in prioritizing hospitals: closeness (minimizing transport time) and congestion (minimizing turnaround time). The 3C policy, however, takes into account the "hospital centrality" additionally which represents the operational efficiency of a hospital in responding to patients after getting discharged from the hospital. In order for synthesizing the three factors of closeness, congestion, and centrality, the 3C policy uses two measures, transport time and queue length, in a three-step procedure as follows:

1. When transferring patient to an ED, identify all hospitals H having an eligible ED.

2. Estimate expected transport time t_h to reach each hospital $h \in H$, and acquire queue length q_h of ED waiting room from each hospital $h \in H$.
3. Transport patient to the hospital h^* that maximizes fitness based on expected transport time t_h and queue length q_h weighted by congestion/centrality parameter γ.

$$h^* = \arg\max_{h \in H} \frac{1}{(1 + t_h)(1 + q_h)^\gamma}$$

The queue length is an indicator of both congestion level and centrality of ED, and the weight on queue length γ determines tradeoff among the three conflicting factors. The 3C policy with $\gamma = 0$ is exactly same as the closest policy and the weight value has to be determined appropriately for the operational scenario. The 3C policy reduces response time by up to 99.6% over the closest policy, 90% over the diversion policy, 68.8% over the JSQ policy, and 67.7% over the STT policy. The 3C policy reduces variation as well by up to 99.2% over the closest policy, 87.7% over the diversion policy, 65.1% over the JSQ policy, and 66.4% over the STT policy.

2.4 Experimental Analysis

The three principle-based policies have been developed and evaluated independently, and this section analyzes them in an integrated experimental framework. The overall experimental scenario is similar to the EMS logistics scenario presented in the introductory part of this section and other details are as follows. The service area is represented in a 5*5 square grid with two hospitals as shown in Fig. 5. A total of 2,000 emergency calls are generated with an average inter-arrival time of 15min, and they are placed in one of the 25 vertices randomly and uniformly. Ten ambulances are deployed to serve the calls and they move from 25 vertex to vertex through edges each with average 1 min of travel time. The units are randomly and uniformly located in the beginning of each simulation run. Once arriving to a call site, the ambulance serves the patient with an onsite service time of 17 min on average (Budge et al. 2010; Carr et al. 2008). The ambulance then, with a probability of *hospital_prob*, transfers the patient to one of the two hospitals. EDs are modeled in a simple, single-queue system with 5 servers (beds), based on the fact that the capacity of ED beds is the primary source of ED crowding (Fatovich and Hirsch 2003; Fatovich et al. 2005; Moskop et al. 2009; Olshaker and Rathlev 2006; Schull et al. 2003). Patient care time (bed occupying time) is assumed to be 150 min on average (CDC 2006). All time distributions above are assumed to be exponential (Alanis et al. 2013; Erkut et al. 2008; Singer and Donoso 2008).

Different test conditions are composed factoring in the *hospital_prob* to expose to different load conditions. For each test condition, the three principle-based EMS logistics policies are applied altogether for different types of decisions. As discussed before, the Preparedness policy has parameter α, the Centrality policy β, and the 3C policy γ. In order to obtain optimal performance from parameter calibration, we utilize the OptQuest provided in the Arena simulation software upon which our simulator is built. The OptQuest is an improvement-type optimizer which searches for optimal solution starting from an initial solution provided by the user. A 3-dim discrete search space is set up with each dimension representing a weight parameter ranging from 0–2 with an

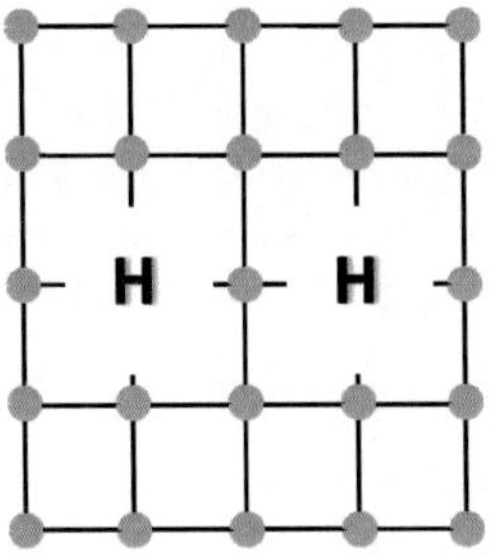

Fig. 5. Service area configuration

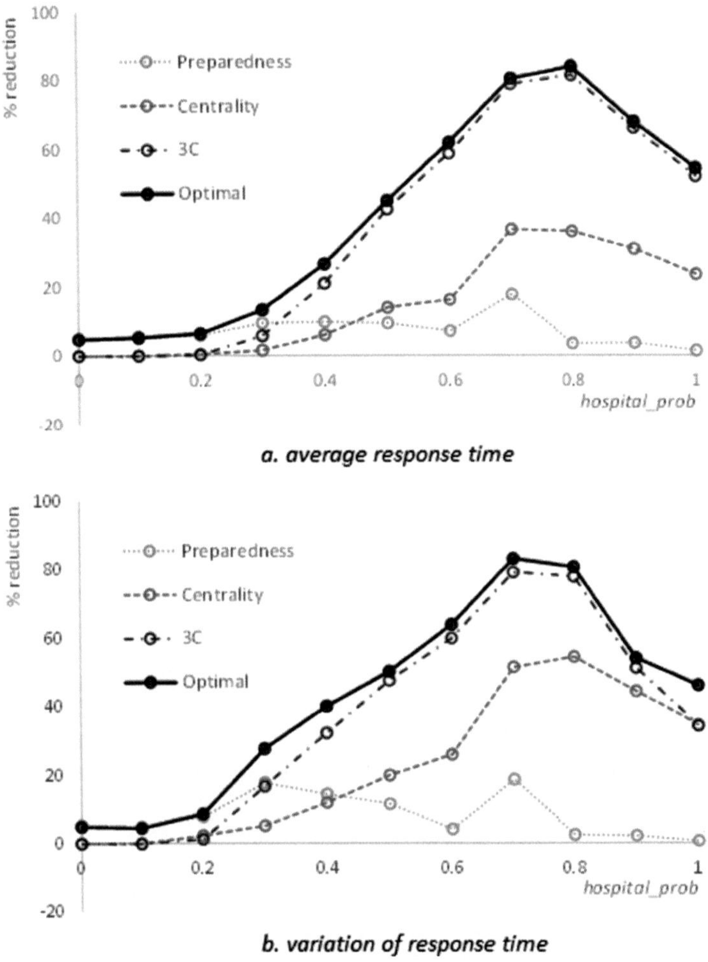

Fig. 6. Reduction in average and variation of response time by principle-based policies

increment 0.1. The initial solution is selected as the best solution obtained from applying the three policies individually, with each parameter varying from 0–2 with an increment 0.1 while fixing other parameter values to 0. The number of iterations allowed for the calibration is set to 100. A hundred simulation runs are replicated for each scenario and average response time is used as the performance metric of the scenario (Fig. 6).

This experimental result indicates that the principle-based policies synergistically enhance the logistics efficiency and reduce the response time as well as its variation, resulting from that they are individually effective and complement each other. Another interesting observation is that there is a strong linear relationship ($R^2 = 98.8\%$) between average and variation of response time as shown in Fig. 7, where all data points are obtained during the course of parameter calibration in a test condition with *hospital_prob* = 0.5. This high-correlation phenomenon supports an argument that endeavors to reduce response time naturally lead to the reduction in variation as well, therefore enabling to secure dominance over various possible definitions of response time threshold.

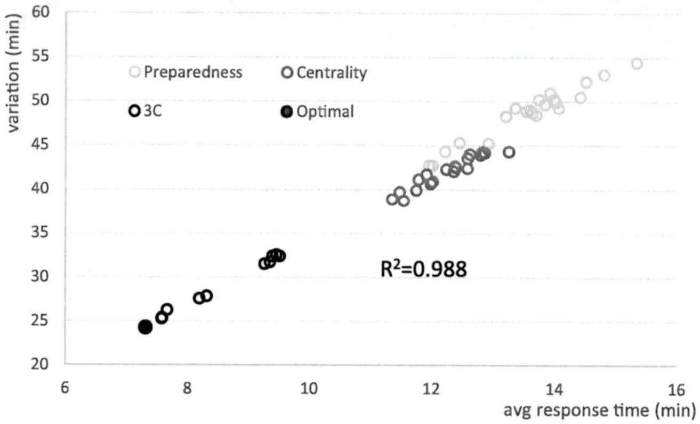

Fig. 7. Correlation between average and variation of response time, *hospital_prob* = 0.5

3 The Principle-Based Priority EMS Logistics Policies

The EMS logistics policies introduced in the previous section do not consider heterogeneity of patients in urgency, but many EMS systems in practice adopt and implement priority classes of patients. Therefore, the objective of this section is to extend the principle-based policies to the priority systems. Note that triaging and prioritizing of patients in emergency call centers (for dispatching purpose) and EDs (for ordering ED patients in queue) are not the same, and the number of priority levels vary in regions though there exist standardization efforts. For example, the number of priority classes in ambulance dispatching ranges from 2–4 (Black and Davies 2005; Henderson and Mason 2005; Kuisma et al. 2004; Nicholl et al. 1999) and, in the United States, there has been a trend toward standardization of triage acuity scales used in EDs since 2000 and it is reported in 2009 that 57% of EDs are using a five level triage system, the

Emergency Severity Index (ESI) (Gilboy et al. 2011). Therefore, the priority policies we develop have to be in a generic form allowing two separate sets of priority classes without restriction in size, one for call center (dispatch priority classes) and the other for EDs (ED priority classes).

3.1 Priority Policies

When an emergency call is received in a call center, its dispatch priority class (r) is identified and this information is used for dispatching decisions. When transferring patient to hospital, it is assumed that ambulance crew identifies the ED priority class (ω) of the patient so that this information can be used for hospital selection decision. The general directions for designing the priority policies are as follows:

The Preparedness policy can be extended to priority dispatching systems by assigning a higher weight on preparedness to a less urgent patient class, so that more immediate performance can be pursued for more urgent patients while preparedness for future calls is more taken into account for less urgent patients.

The Centrality policy can be adapted by applying the policy to the nonempty, highest-priority class of patients alone, and also by assigning different weights on centrality to different classes. There would be no specific relation between weights of different classes because different classes of patients are considered somewhat separately and independently.

The EDs will serve those patients first with higher priority when managing their queues. In addition, the 3C policy can be extended by assigning a less weight on queue length to a more urgent patient class, since more urgent patients need to be quickly taken care of while less urgent ones can be distributed to farther locations for load balancing purpose.

These guidelines lead to the principle-based priority EMS logistics policies as follows:

Priority Preparedness Policy:

1. A service area is divided into zones Z and when a new call arrives from a zone c, the priority r of the call (less r implies higher priority) is determined and a set A of available (idle) ambulances are identified.
2. For each ambulance $i \in A$, preparedness level $p_{A\setminus i}$ is computed by setting ambulance i unavailable (resulting from dispatching it), where λ_j represents call rate in zone j and t_{kj} is the travel time of ambulance k to zone j.

$$p_{A\setminus i} = \frac{1}{\sum_{j \in Z} \lambda_j (1 + \min\limits_{k \in A\setminus i} t_{kj})}$$

3. Dispatch to the call site c the ambulance i^* that maximizes fitness based on the preparedness $p_{A\setminus i}$ weighted by preparedness parameter α_r (≥ 0) of priority class r (α_r non-decreasing with r) and expected response time t_{ic} for a unit i to reach the call

site c.

$$i^* = \arg\max_{i \in A} \frac{p_{A\setminus i}^{\alpha_r}}{1 + t_{ic}}$$

Priority Centrality policy:

1. When an ambulance v gets freed, identify the nonempty, highest-priority, unassigned calls U and denote the priority of chosen calls as r (less r implies higher priority).
2. Compute centrality c_u of each call $u \in U$ upon the network of calls U with the edge between every pair of calls from u to i having a value of distance τ_{ui} (in time).

$$c_u = \sum_{i \in U, i \neq u} \frac{1}{(1 + \tau_{ui})}$$

3. Dispatch the freed unit v to the call u^* that maximizes fitness based on the centrality c_u weighted by centrality parameter β_r (≥ 0) of priority class r and expected response time t_{vu} for a unit v to reach the call site u.

$$u^* = \arg\max_{u \in U} \frac{c_u^{\beta_r}}{(1 + t_{vu})}$$

Priority 3C policy:

1. When transferring a patient of priority ω to an ED (less ω implies higher priority), identify all hospitals H having an eligible ED.
2. Estimate expected transport time t_h to reach each hospital $h \in H$, and acquire queue length q_h of ED waiting room from each hospital $h \in H$.
3. Transport patient to the hospital h^* that maximizes fitness based on expected transport time t_h and queue length q_h weighted by congestion/centrality parameter γ_ω (≥ 0) of priority class ω (γ_ω non-decreasing with ω).

$$h^* = \arg\max_{h \in H} \frac{1}{(1 + t_h)(1 + q_h)^{\gamma_\omega}}$$

When all parameters are set to zero, the Preparedness policy has no difference among classes, but the Centrality policy and 3C policy yet treat different classes differently in that the Centrality policy considers higher priority classes first and EDs serve patients according to the priority of their patients whatsoever. The policies with all parameters having zero form the "base priority policy" which is a myopic policy in priority systems. In the base priority policy, the closest unit is dispatched in call-initiated dispatching, the closest call among highest-priority calls is chosen in ambulance-initiated dispatching, and the nearest hospital is selected where patients are prioritized according to their priority classes.

3.2 Experimental Analysis

The principle-based priority policies are applied to the same test scenarios as presented in Sect. 2.4, but with two classes of patients for both dispatching classes and ED classes. Each arriving call is classified in either Class I or Class II each with 50% and the class designation persists for ED as well. The base priority policy by itself reduces response time of Class I patients by up to 97.7% over the optimal performance of non-priority policies (i.e., OptQuest solution obtained in Sect. 2.4), by its inherent prioritization mechanisms as mentioned before. Hereafter, our focus is on the performance improvement by the three priority policies over the base priority policy. To obtain optimal performance for Class I patients, we set up a 6-dim discrete search space for the OptQuest with each dimension representing a weight parameter ranging from 0–2 with an increment 0.1, and the initial solution is selected as the OptQuest solution found in the non-priority system (same parameter value applied to both classes). The number iterations allowed for the search is set to 200.

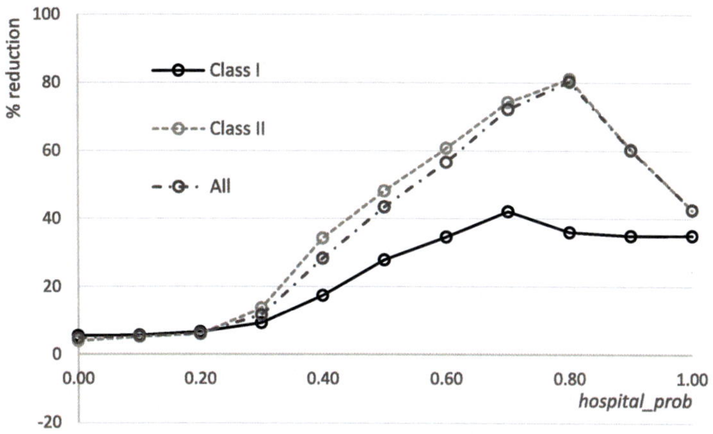

Fig. 8. Reduction in response time by principle-based priority policies

Figure 8 shows the reduction in response time over the base priority policy by the optimization. The response time of Class I calls is reduced by up to 42.1%, and it is interesting to observe that the response time of Class II calls is also reducing even larger than Class I calls (reduction for Class II is up to 81.1%). The reason is twofold. One is that the performance of Class I calls is hard to improve without logistics efficiency improvement for Class II calls because both classes are closely interconnected by sharing same resources. The other reason is that improving Class II is easier than improving Class I as Class I calls are already somewhat optimized by the inherent priority mechanisms. The overall improvement is closer to the one of Class II because the response time scale of Class II is much larger than the scale of Class I.

Figure 9 shows the response time of Class I patients for the base priority policy, initial solution, and optimal one. The performance of initial solution is significantly better than the base priority policy and it is even close to the optimal one. Therefore, the optimal solution of non-priority system can be considered to be a good candidate as the initial solution for calibrating parameters in priority systems. Though the performance improvement by optimization from the initial one looks small, in fact the improvement is up to 7.8% (0.73 min at *hospital_prob* = 0.6) which could be significant from the practical perspective.

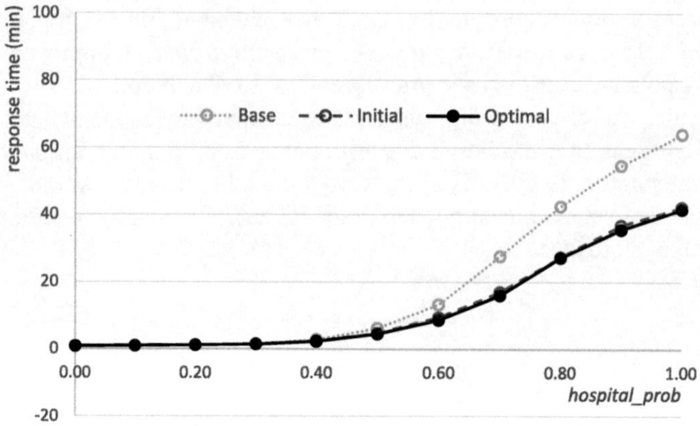

Fig. 9. Reduction in response time by principle-based priority policies – Class I

4 Discussion

This chapter introduces the principle-based EMS logistics polices and their generalizations to priority systems. The three principle-based priority policies have a potential to greatly enhance the efficiency of EMS logistics and thereby improve safety and welfare of EMS patients. Our logistics policies accomplish collaboration and integration among EMS resources and personnel, thus evidence the impact of collaboration and integration in the EMS logistics domain. This section discusses issues and challenges in implementing the policies in real life scenarios.

4.1 Performance Metrics

The research on EMS logistics usually takes performance metrics associated with response time when designing and evaluating EMS systems (e.g., coverage level, average response time, variation of response time (fairness)). However, other metrics might be also important to pursue in real life, such as:

- transfer time (from scene to hospital)
- energy cost (for ambulance traveling)
- crew workload and fairness
- financial fairness (among EMS providers and among hospitals)

At the same time, robustness properties to inaccurate information (e.g., travel time estimates) have to be incorporated. These various aspects of performance have to be synthesized into a weighted aggregate performance metric, where the weights reflect relative importance. The weight structure needs to be surveyed and estimated to guide, in a practical way, the design and evaluation processes of EMS logistics.

In order to reduce the complexity of considering various possible performance aspects, it is important to reduce the number of performance metrics if possible at all. This can be accomplished in two ways: correlation analysis and opinions of people. We observed the existence of a strong relationship between average and variation of response time in Fig. 7, supporting the possibility of avoiding a separate effort for variation reduction. This kind of correlations needs to be explored throughout all metrics. Another way is to rely on the opinions of people gathered from the survey on weight structure mentioned above. If some metrics are considered trivial consensually they can be excluded from the analysis.

4.2 Evaluation in Real Life Scenarios

The evaluation of the principle-based EMS logistics policies has been conducted in hypothetical scenarios, and it will be beneficial to validate them in real life scenarios also involving various performance metrics mentioned above. Opportunities exist to take full advantage of EMS data available in NEMSIS (National EMS Information System) database (http://www.nemsis.org). NEMSIS is the national repository that is used to store comprehensive EMS data from every state in the United States. The NEMSIS database structure is comprehensive enough to reconstruct EMS operational scenarios, including EMS agency information (e.g., service area, service area population, call volume, base station locations, ED locations, number of ambulances) and response records (e.g., time stamps, odometer readings, patient/vehicle locations, dispatch priority).

It is important, among others, to characterize empirical distribution of time variables (e.g., call inter-arrival time, ambulance travel time, onsite service time, patient care time) because the distributional assumptions on these variables would affect the performance of EMS logistics. Especially, the ambulance travel time is closely associated with not only logistics behavior but also all the logistics policies; therefore accurate modeling of the travel time with explanatory factors such as distance, congestion, speed limit, and weather condition, is essential in validity of decision making as well as performance evaluation. The research on the characterization so far, however, limited to specific regions, e.g., Calgary, Alberta, Canada (Budge et al. 2010) and Stockholm, Sweden (Jenelius and Koutsopoulos 2013). The characterization effort with the NEMSIS database will enable to identify various possible distributional models of time variables across different EMS systems in the United States, therefore leading to systematic and holistic research on EMS logistics at large.

4.3 Policy Calibration

The principle-based EMS logistics policies have algorithmic parameters that enable to achieve portable performance in various EMS scenarios. However, parameter calibration is laborious and computationally demanding, since the parameters strongly depend on each problem instance at hand and therefore the calibration process has to be repeated for each problem instance. For example, the search space of OptQuest in Sect. 2.3 is roughly $21^6 = 85{,}766{,}121$ even in a simplistic case with only two priority classes. Running the OptQuest with only 200 iterations in the small 5*5 grid takes about 6 h (Intel Core 2 Quad Processor Q6700). Therefore, the calibration process needs to be facilitated for practical use.

One way is to classify EMS systems by features through EMS databases such as NEMSIS and build a table that matches the best parameter values for each class of EMS systems. This is a straightforward, effective way for the calibration process, however, if there is an EMS system other than the classification scheme because of the lack of data or emergence of new EMS systems in future, a new endeavor has to be repeated to rebuild the table over and over again. Therefore, to cope with this potential problem, the second method can be to analyze and construct optimal parameter matching rules that relate characteristics of operational scenarios to the best parameter values. The discovery of the matching rules would require utilizing statistical and data mining techniques, and the techniques used in parameter calibration would be useful (Bartz-Beielstein 2006; Czarn et al. 2004; Eiben et al. 1999; Francois and Lavergne 2001; Freisleben 2002; Grefenstette 1986).

4.4 Enabling System Architecture

Implementing the EMS logistics policies requires an enabling system architecture. The architecture outlines information and communication capabilities, decision rights and methods of individual EMS entities, and coordination mechanisms for information and decisions among EMS entities. The system architecture once developed must be evaluated for feasibility in various practical aspects: economic, cultural, technological, and political.

5 Further Exploration Opportunities

This section provides research opportunities and directions to further improve or extend the principle-based EMS logistics policies.

5.1 Parallelism

The dispatching policies we propose (Preparedness policy and Centrality policy) involve only idle ambulances in making dispatching decisions, but it is possible that a busy unit can respond more quickly even after the completion of currently assigned service. The parallelism is a notion to consider both idle and busy units in parallel, and it leads to an assignment problem at each dispatching decision that matches between multiple (idle/busy) units and multiple unassigned calls. The consideration of parallelism will also lead to the capability of addressing incidents where several patients are introduced at the same time. A logistics policy, called the Parallelism policy, was proposed that incorporates the parallelism into dispatching decision (Lee 2014b) but this policy has to be extended to allow patient priorities along with infusing preparedness and centrality.

The parallelism would require computing expected response time for busy units as well, and then it becomes essential to estimate turnaround time at hospital. There are sophisticated models for ED available in literature (Laskowski et al. 2009; Wylie et al. 2015), however for the purpose of estimating turnaround time, the G/G/s queue with priority classes (where s is the number of ED beds) may work in some cases because the capacity of ED beds is known to be the primary source of ED crowding. In such a case the turnaround time can be estimated as follows: 0 if there is at least one bed available, otherwise (queue length of patients with higher or same priority + 1) * (average patient care time) / (number of beds).

5.2 Heterogeneous Ambulances

The logistics policies devised in this chapter assumes that all ambulances are homogeneous, but ambulances could be different in capability in terms of crew skills and equipment and some patients may be allowed only for some ambulances with a certain level of capability. In general, ambulances can be either a basic life support (BLS) unit or an advanced life support (ALS) unit. This heterogeneous situation may be deemed trivial if we simply consider only eligible ambulances or eligible patients in dispatching decisions. However the problem gets complicated, for example, when facing situations where an ALS unit is the best choice for a low-priority class patient (probably because of close distance). Dispatching the ALS unit could be beneficial to the patient on hand but we are losing the chance to serve high-priority class patients that might be arriving next. Therefore, it could be better to dispatch a low capability unit though it is located farther, and in case only that ambulance is available, to wait until a low capability unit becomes available.

5.3 Relocation

Relocation decisions enforce ambulances to move to different locations according to temporal and geographical demand patterns. However, there are several issues in implementing relocation models in real EMS systems. First, the relocation decision problem becomes complex with increase in size of EMS systems and thus efficient approaches have to be developed to obtain real-time relocation decisions (Andersson and Värbrand 2007). Second, confusions and mistakes that can be caused by frequent relocations are an important issue to be addressed (Haghani et al. 2004). Third, the preference of ambulance crews for spending time at a base station rather than on the road or relocation sites is also an issue in practice. However, despite such barriers, relocation strategies are employed in more EMS systems indicating that the EMS community is becoming more aware of the benefits of relocation. For example, the percentage of North American EMS operators using a relocation strategy increased from 23% in 2001 (Cady 2002) to 37% in 2008 (Williams 2009).

An implication of the research presented in this chapter is the use of the preparedness metric as a criterion for making relocation decisions. The system can relocate ambulances such that the preparedness is maximized, whenever resource configuration changes (i.e., an ambulance gets dispatched or freed). The goodness of the preparedness measure was indirectly shown through the performance of the Preparedness policy, but a more solid and direct evidence is required by evaluating in comparison with other strategies available in literature (Andersson and Värbrand 2007; Daskin 1983; Gendreau et al. 2001; Jagtenberg et al. 2015; Kolesar and Walker 1974).

5.4 Rerouting

The EMS logistics policies proposed in this chapter do not consider rerouting of ambulances from their current routes when the rerouting can reduce response time. Rerouting is a supplementary method that can be applied over an existing dispatching policy transforming it into a reroute-enabled dispatching policy (Lim et al. 2011). Similarly to ambulance dispatching, we can classify rerouting decisions in either call-initiated or ambulance-initiated. In call-initiated rerouting, an ambulance already assigned to a call is rerouted to a new call possibly because the new call is more urgent and the ambulance is closest to it (Gendreau et al. 2001; Andersson and Värbrand 2007). On the other hand, in ambulance-initiated rerouting, a just freed unit reroutes other ambulance taking over its service possibly because the freed unit is closer to the call (Lim et al. 2011). When and how either type of rerouting should be activated needs to be carefully designed in conjunction with the characteristics of the dispatching policy and the way the policy considers patient priorities.

5.5 Citizen Responders

In recent years, community-based programs are formed to recruit, train, and manage citizen responders (CRs) who are capable of providing basic EMS responses, such as hands-only cardiopulmonary resuscitation (CPR) or automated external defibrillator (AED) operation for out-of-hospital cardiac arrests (OHCA), naloxone spray administration for opioid overdoses, bleeding control for severe traumatic injuries, and epinephrine injection for allergic emergencies. CRs, as they are bystanders to patients in emergency situations, can provide a higher chance to survive (Andelius et al. 2020; Hansen et al. 2015; Khalemsky and Schwartz 2017; Lancaster and Herrmann 2020; Paz et al. 2021). The proposed EMS logistics policies need to be extended for the citizen-responders integrated EMS systems, to take or measure full advantage of such community-based programs.

Acknowledgements. Dr. Seokcheon Lee is the director of the Distributed Control (DC) laboratory, School of Industrial Engineering, Purdue University. DC lab's research focus is on scheduling and logistics, and this chapter contributes to the emergency logistics within the logistics thrust. DC lab and PRISM Center are affiliated with strong research overlaps and joint projects.

References

Adan, I., van Houtum, G.J., van der Wal, J.: Upper and lower bounds for the waiting time in the symmetric shortest queue system. Ann. Oper. Res. **48**(2), 197–217 (1994)

AHA (America Hospital Association): Trendwatch Chartbook 2020 (2020)

Alanis, R., Ingolfsson, A., Kolfal, B.: A Markov chain model for an EMS system with repositioning. Prod. Oper. Manag. **22**(1), 216–231 (2013)

Andelius, L., et al.: Smartphone activation of citizen responders to facilitate defibrillation in out-of-hospital cardiac arrest. J. Am. Coll. Cardiol. **76**(1), 43–53 (2020)

Andersson, T., Värbrand, P.: Decision support tools for ambulance dispatch and relocation. J. Oper. Res. Soc. **58**(2), 195–201 (2007)

Ball, M.O., Lin, L.F.: A reliability model applied to emergency service vehicle location. Oper. Res. **41**(1), 18–36 (1993)

Barrat, A., Barthelemy, M., Pastor-Satorras, R., Vespignani, A.: The architecture of complex weighted networks. Proc. Nat. Acad. Sci. U.S.A. **101**(11), 3747–3752 (2004)

Bartz-Beielstein, T.: Experimental Research in Evolutionary Computation: The New Experimentalism. Springer, Berlin (2006)

Bélanger, V., Kergosien, Y., Ruiz, A., Soriano, P.: An empirical comparison of relocation strategies in real-time ambulance fleet management. Comput. Ind. Eng. **94**, 216–229 (2016)

Bertsimas, D.J., van Ryzin, G.: A stochastic and dynamic vehicle routing problem in the Euclidean plane. Oper. Res. **39**(4), 601–615 (1991)

Black, J.J.M., Davies, G.D.: International EMS systems: United Kingdom. Resuscitation **64**(1), 21–29 (2005)

Blackstone, E.A., Buck, A.J., Hakim, S.: The economics of emergency response. Policy Sci. **40**(4), 313–334 (2007)

Budge, S., Ingolfsson, A., Zerom, D.: Empirical analysis of ambulance travel times: the case of Calgary emergency medical services. Manage. Sci. **56**(4), 716–723 (2010)

Cady, G.: 200 City survey. JEMS annual report on EMS operational & clinical trends in large, urban areas. J. Emerg. Med. Serv. **27**(2), 46–71 (2001)

Carr, B.G., Caplan, J.M., Pryor, J.P., Branas, C.C.: A meta-analysis of prehospital care times for trauma. Prehosp. Emerg. Care **10**(2), 198–206 (2008)

Carter, G., Chaiken, J., Ignall, E.: Response areas for two emergency units. Oper. Res. **20**(3), 571–594 (1972)

CDC (Centers for Disease Control and Prevention): National hospital ambulatory medical care survey: 2006 emergency department summary (2006)

CDC (Centers for Disease Control and Prevention): National hospital ambulatory medical care survey: 2011 emergency department summary tables (2011)

Chaiken, J.M., Larson, R.C.: Methods for allocating urban emergency units: a survey. Manage. Sci. **19**(3), 110–130 (1972)

Conway, R.W., Maxwell, W.L., Miller, L.W.: Theory of Scheduling. Addison-Wesley, Reading (1967)

Cunningham-Green, R., Harries, G.: Nearest-neighbour rules for emergency services. Zeitschrift Oper. Res. **32**, 299–306 (1988)

Czarn, A., MacNish, C., Vijayan, K., Turlach, B., Gupta, R.: Statistical exploratory analysis of genetic algorithms. IEEE Trans. Evol. Comput. **8**(4), 405–421 (2004)

Daskin, M.S.: The maximal expected covering location model: formulation, properties, and heuristic solution. Transp. Sci. **17**(1), 48–70 (1983)

Dean, S.F.: Why the closest ambulance cannot be dispatched in an urban emergency medical services system. Prehospital Disaster Med. **23**(2), 161–165 (2008)

Deo, S., Gurvich, I.: Centralized vs. decentralized ambulance diversion: a network perspective. Manage. Sci. **57**(7), 1300–1319 (2011)

Eiben, A.E., Hinterding, R., Michalewicz, Z.: Parameter control in evolutionary algorithms. IEEE Trans. Evol. Comput. **3**(2), 124–141 (1999)

Enayati, S., Özaltın, O.Y., Mayorga, M.E., Saydam, C.: Ambulance redeployment and dispatching under uncertainty with personnel workload limitations. IISE Trans. **50**(9), 777–788 (2018)

Enders, P.: Applications of stochastic and queueing models to operational decision making. Ph.D. thesis, Tepper School of Business, Carnegie Mellon University (2010)

Erkut, E., Ingolfsson, A., Erdoğan, G.: Ambulance location for maximum survival. Nav. Res. Logist. **55**(1), 42–58 (2008)

Fatovich, D.M., Hirsch, R.L.: Entry overload, emergency department overcrowding, and ambulance bypass. Emerg. Med. J. **20**(5), 406–409 (2003)

Fatovich, D.M., Nagree, Y., Sprivulis, P.: Access block causes emergency department overcrowding and ambulance diversion in Perth Western Australia. Emerg. Med. J. **22**(5), 351–354 (2005)

Francois, O., Lavergne, C.: Design of evolutionary algorithms – a statistical perspective. IEEE Trans. Evol. Comput. **5**(2), 129–148 (2001)

Freisleben, B.: Meta-evolutionary approaches. In: Bäck, T., Fogel, D. B., and Michalewicz, Z. (eds.) Evolutionary Computation 2: Advanced Algorithms and Operators, pp. 212–223. CRC Press (2002)

Gendreau, M., Laporte, G., Semet, F.: A dynamic model and parallel tabu search heuristic for real-time ambulance relocation. Parallel Comput. **27**(12), 1641–1653 (2001)

Gilboy, N., Tanabe, T., Travers, D., Rosenau, A.M.: Emergency Severity Index (ESI): A Triage Tool for Emergency Department Care, Version 4. Implementation Handbook 2012 Edition. AHRQ Publication No. 12–0014, Agency for Healthcare Research and Quality (AHRQ) (2011)

Gonzalez, R.P., Cummings, G.R., Phelan, H.A., Mulekar, M.S., Rodning, C.B.: Does increased emergency medical services prehospital time affect patient mortality in rural motor vehicle crashes? A statewide analysis. Am. J. Surg. **197**(1), 30–34 (2009)

Grefenstette, J.: Optimization of control parameters for genetic algorithms. IEEE Trans. Syst. Man Cybern. **16**(1), 122–128 (1986)

Haghani, A., Tian, Q., Huijun, H.: Simulation model for real-time emergency vehicle dispatching and routing. Transp. Res. Rec. J. Transp. Res. Board **1882**, 176–183 (2004)

Hansen, C.M., et al.: The role of bystanders, first responders, and emergency medical service providers in timely defibrillation and related outcomes after out-of-hospital cardiac arrest: results from a statewide registry. Resuscitation **96**, 303–309 (2015)

Hayes, J., Moore, A., Benwell, G., Wong, B.: Ambulance dispatch complexity and dispatcher decision strategies: implications for interface design. In: Lecture Notes in Computer Science, vol. 3101, pp. 589–593. https://doi.org/10.1007/978-3-540-27795-8_60

Henderson, S.G., Mason, A.J.: Ambulance service planning: simulation and data visualization. In: Brandeau, M.L., Sainfort, F., Pierskalla, W.P. (eds.) Operations Research and Healthcare: A Handbook of Methods and Applications, pp. 77–102. Kluwer Academic Publishers, Dordrecht (2005)

Holloway, J., Francis, G., Hinton, M.: A vehicle for change? A case study of performance improvement in the 'new' public sector. Int. J. Public Sect. Manag. **12**(4), 351–365 (1999)

Jagtenberg, C., Bhulai, S., van der Mei, R.: An efficient heuristic for real-time ambulance redeployment. Oper. Res. Health Care **4**, 27–35 (2015)

Jenelius, E., Koutsopoulos, H.N.: Travel time estimation for urban road networks using low frequency probe vehicle data. Transp. Res. Part B **53**, 64–81 (2013)

Khalemsky, M., Schwartz, D.G.: Emergency response community effectiveness: a simulation modeler for comparing emergency medical services with smartphone-based Samaritan response. Decis. Support Syst. **102**, 57–68 (2017)

Kolesar, P., Walker, W.: An algorithm for the dynamic relocation of fire companies. Oper. Res. **22**(2), 249–274 (1974)

Kuisma, M., Holmström, P., Repo, J., Määttä, T., Nousila-Wiik, M., Boyd, J.: Pre-hospital mortality in an EMS system using medical priority dispatching: a community based cohort study. Resuscitation **61**(3), 297–302 (2004)

Lancaster, G., Herrmann, J.: Simulating cardiac arrest events to evaluate novel emergency response systems. IISE Trans. Healthc. Syst. Eng. **11**(1), 38–50 (2020)
Laskowski, M., McLeod, R.D., Friesen, M.R., Podaima, B.W., Alfa, A.S.: Models of emergency departments for reducing patient waiting times. PLoS ONE **4**(7), e6127 (2009)
Lee, S.: The role of preparedness in ambulance dispatching. J. Oper. Res. Soc. **62**(10), 1888–1897 (2011)
Lee, S.: The role of centrality in ambulance dispatching. Decis. Support Syst. **54**(1), 282–291 (2012)
Lee, S.: Centrality-based ambulance dispatching for demanding emergency situations. J. Oper. Res. Soc. **64**(4), 611–618 (2013)
Lee, S.: The role of hospital selection in ambulance logistics. IIE Trans. Healthc. Syst. Eng. **4**(2), 105–117 (2014)
Lee, S.: The role of parallelism in ambulance dispatching. IEEE Trans. Syst. Man Cybern. Syst. **44**(8), 1113–1122 (2014)
Lee, S.: A new preparedness policy for EMS logistics. Health Care Manag. Sci. **20**(1), 105–114 (2015). https://doi.org/10.1007/s10729-015-9340-4
Lim, C.S., Mamat, R., Bräunl, T.: Impact of ambulance dispatch policies on performance of emergency medical services. IEEE Trans. Intell. Transp. Syst. **12**(2), 624–632 (2011)
Lowthian, J.A., Curtis, A.J., Jolley, D.J., Stoelwinder, J.U., McNeil, J.J., Cameron, P.A.: Demand at the emergency department front door: 10-year trends in presentations. Med. J. Aust. **196**, 128–132 (2012)
McGrath, K.: The Golden Circle: a way of arguing and acting about technology in the London ambulance service. Eur. J. Inf. Syst. **11**(4), 251–266 (2002)
McLay, L.A., Mayorga, M.E.: Evaluating the impact of performance goals on dispatching decisions in emergency medical service. IIE Trans. Healthc. Syst. Eng. **1**(3), 185–196 (2011)
Moskop, J.C., Sklar, D.P., Geiderman, J.M., Schears, R.M., Bookman, K.J.: Emergency department crowding, part 1-concept, causes, and moral consequences. Ann. Emerg. Med. **53**(5), 605–611 (2009)
NASEMSO (National Association of State EMS Officials): 2020 National Emergency Medical Services Assessment (2020)
Newman, M.E.J.: The structure and function of complex networks. SIAM Rev. **45**, 167–256 (2003)
Newman, M.E.J.: Analysis of weighted networks. Phys. Rev. E **70**(5), 056131 (2004)
Nicholl, J., Coleman, P., Parry, G., Turner, J., Dixon, S.: Emergency priority dispatch systems: a new era in the provision of ambulance services in the UK. Pre-hosp. Immediate Care **3**, 71–75 (1999)
Olshaker, J.S., Rathlev, N.K.: Emergency department overcrowding and ambulance diversion: the impact and potential solutions of extended boarding of admitted patients in the emergency department. J. Emerg. Med. **30**(3), 351–356 (2006)
Paz, J.C., Kong, N., Lee, S.: Evaluating a real-time ambulance redeployment framework under implementation of community-based citizen responder program. In: Proceedings of the 2021 IISE Annual Conference, pp. 351–357 (2021)
Pons, P.T., Markovchick, V.J.: Eight minutes or less: does the ambulance response time guideline impact trauma patient outcome? J. Emerg. Med. **23**(1), 43–48 (2002)
Repede, J., Bernardo, J.: Developing and validating a decision support system for locating emergency medical vehicles in Louisville Kentucky. Eur. J. Oper. Res. **75**(3), 567–581 (1994)
Sabidussi, G.: The centrality index of a graph. Psychometrika **31**, 581–606 (1996)
Sánchez-Mangas, R., García-Ferrer, A., de Juan, A., Arroyo, A.M.: The probability of death in road traffic accidents. How important is a quick medical response? Accid. Anal. Prev. **42**(4), 1048–1056 (2010)

Schull, M.J., Lazier, K., Vermeulen, M., Mawhinney, S., Morrison, L.J.: Emergency department contributors to ambulance diversion: a quantitative analysis. Ann. Emerg. Med. **41**(4), 467–476 (2003)

Singer, M., Donoso, P.: Assessing an ambulance service with queuing theory. Comput. Oper. Res. **35**(8), 2549–2560 (2008)

Stiell, I.G., Nesbitt, L.P., Pickett, W., Munkley, D., Daniel, W.: The OPALS major trauma study: impact of advanced life-support on survival and morbidity. Can. Med. Assoc. J. **178**(9), 1141–1152 (2008)

UKDH (United Kingdom Department of Health): Ambulance quality indicators data 2015–16 (2015)

Van Barneveld, T.C., Bhulai, S., van der Mei, R.: The effect of ambulance relocations on the performance of ambulance service providers. Eur. J. Oper. Res. **252**(1), 257–269 (2016)

Van Barneveld, T.C., Jagtenberg, C., Bhulai, S., van der Mei, R.: Real-time ambulance relocation: assessing real-time redeployment strategies for ambulance relocation. Socioecon. Plann. Sci. **62**, 129–142 (2018)

Vukmir, R.B.: Survival from prehospital cardiac arrest is critically dependent upon response time. Resuscitation **69**(2), 229–234 (2006)

Weintraub, A., Aboud, J., Fernandez, C., Laporte, G., Ramirez, E.: An emergency vehicle dispatching system for an electric utility in Chile. J. Oper. Res. Soc. **50**(7), 690–696 (1999)

Whitt, W.: Deciding which queue to join: some counterexamples. Oper. Res. **34**(1), 55–62 (1986)

Williams, D.M.: JEMS 2008 200 city survey: the future is your choice. J. Emerg. Med. Serv. **34**(2), 36–51 (2009)

Woollard, M., Lewis, D., Brooks, S.: Strategic change in the ambulance service: barriers and success strategies for the implementation of high-performance management systems. Strateg. Chang. **12**(3), 165–175 (2003)

Wylie, K., et al.: Emergency department models of care in the context of care quality and cost: a systematic review. Emerg. Med. Australas. **27**, 95–101 (2015)

Robotic Assembly with Deformable Objects

Ran Shneor(✉) and Sigal Berman

Department of Industrial Engineering and Management, Ben-Gurion University of the Negev, Beer-Sheva, Israel
shneorr@post.bgu.ac.il

Abstract. Assembly is an essential step in many industrial manufacturing processes and extensive research has been devoted to robotic assembly over the years. However, the adoption of robots for assembly processes in the industry has been slow and assembly plants are labor-intensive industries. This is due in part to difficulties related to inherent uncertainties in the assembly of deformable objects. In the current chapter, we outline the technologies and methods that can reduce process uncertainty, reduce system susceptibility to uncertainty or, enhance the ability to perceive and react to dynamic changes from the workcell resources, processes, and objective perspectives. We then present a case study of the technologies implemented for reducing the uncertainty in robotic systems developed for the assembly of electrical wire harnesses.

1 Introduction

Industrial robots are widely used, widely deployed, and in general have an increasing market (the worldwide demand for industrial robots dropped by 10% in 2019 reflecting the effects of the Covid 19 pandemic on major industries, but demand has returned to increase in 2020[1]). Assembly is an essential step in many industrial manufacturing processes and extensive research has been devoted to robotic assembly over the years. However, the adoption of robots for assembly processes in the industry has been slow and assembly plants are labor-intensive industries (Shneier et al. 2015; Ho 2018). This is due in part to difficulties related to the assembly of deformable objects.

Essentially, assembly is a sequence of operations conducted so that parts (or subassemblies) are joined together in a particular manner. The operations prescribe a sequence of manipulations of the parts. These manipulations enforce spatio-temporal constraints on the required agent motion of the manipulating entity (human or automation). Tight assembly tolerances, difficult access requirements, delicate interactions, and uncertainties can considerably complicate the assembly manipulations and increase the level of required dexterity. Manipulation of rigid objects determines object pose (position and orientation) along a trajectory. Uncertainties in the position or the orientation of objects in the environment may necessitate trajectory adaptation for successfully accomplishing the required manipulation. An additional source of uncertainty is encountered

[1] International Federation of Robotics. World Robotics 2021 Industrial Robots. https://ifr.org/worldrobotics/

C.-Y. Huang and S. W. Yoon (Eds.): ICPR1 2021, ACES 14, pp. 221–235, 2023.
https://doi.org/10.1007/978-3-031-44373-2_13

when deformable objects are involved since the shape of deformable objects can be affected by the grasp, the motion, or by interactions with other objects. Uncertainties and unintentional changes in object shape during the manipulation may also require motion adaptation or even prohibit object mating. Reducing the impact of the various uncertainties may prescribe gentle motion, i.e., avoiding large forces and moments. Human operators excel in performing gentle motion and in accommodating motion adaptation under the uncertainties typically required in assembly processes. However, gentle handling and motion adaptation is challenging even for modern robotic systems.

Manipulation of deformable objects is a task commonly required both in industrial environments and outside the industrial settings in other areas of human endeavors, such as in domestic activities (e.g., laundry folding, ironing) or in ambient assisted living (e.g., dressing, putting on shoes). Therefore, robotic manipulation of deformable objects is a very active research domain (Chua et al. 2003; Herguedas et al. 2019; Jimenez 2012; Khalil & Payeur 2010; Sanchez et al. 2018). Much progress has been made in the field of deformable object manipulation over the years building on capabilities facilitated by advances in control methods, soft robotics, and perception capabilities. There are clearly significant potential applications for implementing the outcomes of the research on deformable object manipulation in industrial assembly operations.

Assembly processes are typically performed manually in industry, even in industries with highly robotized manufacturing processes, e.g., in the automotive and aerospace industries. The assembly processes which include deformable objects are considered among the most difficult processes for robotization. For example, installing and connecting a car's wire harness has been given as a specific example of a task that humans do better than robots in vehicle assembly in a recent Harvard Business Review article on robotics in the industry (Harbour & Schmidt 2018). The successful translation of the research outcomes and the integration of the robotic systems within factory floor operations requires examining the expected performance of the developed robotic technologies within the wider view of the assembly operation and its integration within the factory floor operations. One dimension to examine, critical for assembly with deformable objects, is the ability of a technology to deal with uncertainty.

A categorization framework that can represent robotic assembly systems can help in providing a coherent understanding of system capabilities and vulnerabilities. There have been several attempts to develop categorization frameworks for robot operations (Shneier et al. 2015; Shneor & Berman 2022). A returning motif in several suggested frameworks is the separation of workcell resources (machine tools, robots, conveyors, etc.) from the manufacturing processes. The vantage point of the manufacturing processes is hierarchically divided into manipulation skills that are associated with a machine or a robot, and production tasks related to the overall production process (Huckaby & Christensen 2022; Pfrommer et al. 2013; Shneor & Berman 2022). An additional framework tier suggested by Shneor & Berman (2022) is the objective tier which addresses process performance measurement. In the current chapter, we will outline the technologies and methods that enhance coping with uncertainty from the workcell resources, processes, and objective perspectives. We will then present a case study of the technologies implemented for reducing the uncertainty in robotic systems developed for the assembly of electrical wire harnesses.

2 Technologies and Methods for Robotic Assembly with Deformable Objects

2.1 Overview

Strategies for coping with assembly process uncertainty can be roughly divided into three basic categories, namely, reducing the uncertainty in the process, reducing the susceptibility of the manufacturing system to the uncertainty, and perceiving and reacting to the dynamic changes related to the uncertainty. Technologies and methods for implementing these strategies with respect to assembly with deformable objects are examined from the workcell resources, processes, and objective tiers (Fig. 1).

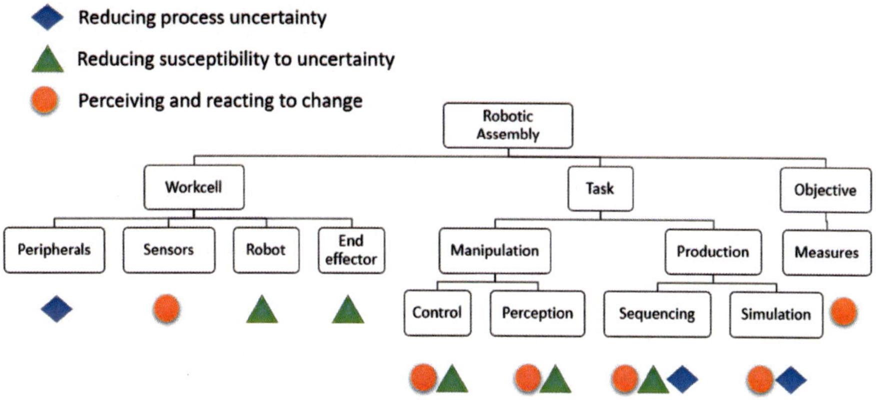

Fig. 1. Overview of the examined technologies and methods for handling assembly process uncertainty and the strategies that can be implemented at each tier.

2.2 Workcell

In the workcell tier, process uncertainty can be reduced by peripheral workstation equipment. Robot and end-effector design can reduce susceptibility to uncertainty and various sensors can facilitate reacting to changes.

2.2.1 Peripheral Workstation Equipment

Various peripheral cell automation devices have been used for reducing process uncertainty in robotic production systems (Groover 2019). In some processes, typically with rigid parts, the uncertainty can be removed altogether, e.g., by fixtures that immobilize a part. Uncertainty can also be partially removed, for example, in part feeders, selectors can be used for permitting only parts in the correct orientation to pass and orientors can reorient the parts that are not in the proper orientation. Such peripheral devices are typically specific to a process and may incur high design and high construction costs.

For deformable objects, solutions reducing position and orientation uncertainty should also prohibit unintentional change to shape during the process. This can be achieved by taking material properties into account when designing the peripheral equipment or by adapting the preprocessing stages of the material. For example, in wiring applications using a dedicated wire feeder connected to the end-effector considerably reduced process uncertainty with respect to using pre-cut wire segments (Hultman & Leijon 2018).

2.2.2 Robot and End-effector

For a given robotic configuration there are directions along which force can be maximally exerted and directions along which it can be most accurately controlled and along the reciprocal directions, velocity can be maximally exerted or most accurately controlled (Chiu 1988). Designing system components, e.g., manipulators and grippers, such that the ability to accurately control forces is high along directions where uncertainty is expected can reduce the susceptibility of the system to uncertainty.

Traditional methods for increasing the ability of the system to handle uncertainty include the kinematic design of the selective compliance assembly robot arm (SCARA) and the remote center of compliance (RCC) wrist adapter (Fig. 2) (Warnecke et al. 1999; Tanie 1999). The RCC wrist adds compliance to lateral forces that may be experienced when inserting a peg into a hole due to miss alignment with the center of the hole causing contact with one side of the hole. Indeed peg-in-hole assembly tasks are very frequent in industry and together with screw operations account for over half of assembly joins (Whitney 2004; Shneier et al. 2015). The kinematic structure of the SCARA robot is tailored for tabletop assembly. The design adds compliance to the horizontal plane but preserves the ability to apply forces in the orthogonal direction.

Fig. 2. Left: An RCC wrist, Right: A SCARA robot

Modern manipulator designs make use of abundant degrees of freedom, e.g., manipulator designs with seven degrees of freedom (unlike the traditional industrial manipulators that have at most six degrees of freedom which are equivalent to the free movements of a rigid body in three-dimensional space). An example of a seven degree of freedom industrial manipulator, in which the self-motion of the elbow is facilitated about a circular path is the KUKA LBR iiwa[2] robot (Kuhlemann et al. 2016). The additional degree of freedom can be used to avoid joint limitations and singularities by elbow positioning, thereby increasing the dexterity of the robot (Zhou & Nguyen 1997).

As for manipulator design, there is also considerable research on designing grippers with abundant degrees of freedom, i.e., flexible grippers and robotic hands. These designs are predominantly anthropomorphic, e.g., the UTAH-MIT hand, the DLR hand, or the shadow hand but non-anthropomorphic, e.g., the Barrett hand (Moralesa et al. 2006) are also suggested. Although much progress has been made both in mechanical design and in hand control algorithms current flexible robotic hands are far from meeting industrial-grade demands for system robustness. An intermediate level between flexible and dedicated grippers, i.e., reconfigurable grippers which include both interchangeable grippers and grippers with interchangeable fingers can be found in industry (Berman and Nof 2011; Ranky 2003). Soft robotics (including soft actuation) is a rapidly growing research field (El-Atab et al. 2020). Soft robots are expected to considerably decrease susceptibility to uncertainty in the future when their control methods and robustness are sufficient for widespread industrial implementation.

Gripper design is an important part of robotic system implementation as grippers are typically designed for the parts handled in a specific application (Raghav et al. 2012; Brown & Brost 1999). In addition to the task constraints, the design of the gripper considers the capabilities of the robotic manipulator to which it is connected, and the capabilities of the system's sensory apparatus (Fantoni et al. 2012; Eizicovits et al. 2016). Gripper designs that are less susceptible to perception errors can increase system robustness to uncertainty. With deformable objects, relevant perception errors may include force and pressure data in addition to position and orientation.

2.2.3 Sensors

Robotic assembly is typically conducted in static settings in which parts are kept fixed by peripheral equipment (Nottensteiner et al. 2021). Sensors have been introduced to assembly environments, especially for small-batch manufacturing to increase system operational flexibility. For assembly with deformable objects, sensors become a crucial component as they can relax fixing precision requirements and can facilitate reaction to dynamic changes. Due to the vast advances in image processing, especially in deep learning-based methods (Wang et al. 2021), vision sensors (including RGB-D sensors) are commonly deployed. For deformable objects, the vision sensors can be integrated with tactile, pressure, or force sensors to convey additionally required data.

[2] https://www.kuka.com/en-de/products/robot-systems/industrial-robots/lbr-iiwa.

2.3 Task Tier

In the task tier, different capabilities can be related to multiple strategies for handling uncertainty. For manipulation tasks, control methods and perception modules can be tuned to react to change, but in addition, the selection of the control method can influence the susceptibility of the system to uncertainty. For production processes operation sequencing methods can contribute to all three strategies of handling uncertainty (reducing process uncertainty, reducing susceptibility to uncertainty, and reacting to change). The use of simulation can lead to reduced process uncertainty and to an increased ability to react to change.

As production becomes more complex and the number of components and the interactions between them increase, optimizing system design and operation becomes more challenging. Collaborative control theory (CCT) is a framework for collaboration between the various players or agents in modern organizations and manufacturing facilities, e.g., products, machines, operators, and controllers (Nof 2007). Collaboration in such highly interconnected environments becomes a necessity for the reliable, timely, and cost-effective achievement of goals.

2.3.1 Manipulation

Robotic assembly of deformable objects requires dexterous manipulation and is subject to multiple uncertainties in duration, exerted forces, induced deformations, etc. Significant progress has been achieved in this domain in recent years (Chua et al. 2003; Herguedas et al. 2019; Jimenez 2012; Khalil & Payeur 2010; Sanchez et al. 2018). Current research centers on achieving high performance and robustness of the required manipulation with special attention to grasping deformable objects (Mu et al. 2019; Tawk et al. 2019).

Advances in autonomous robot control, perception, and in machine learning have led to many control and perception methodologies suitable for handling uncertainty and dynamic changes in the environment. Notable examples include convolutional neural networks (Alzubaidi et al. 2021), deep reinforcement learning modules, and dynamic motion primitives (Nguyen & la 2019; Kroemer et al. 2021). The selection of a control policy or method can influence the susceptibility of the system to uncertainty. For example, impedance control is a control method suitable for tasks that involve contact with the environment. Valency and Zacksenhouse (2000) suggested a practical impedance control method that reduced sensitivity to model inaccuracies and environment uncertainties which is critical for deformable object manipulation. Modeling shape deformation is an additional avenue for reducing uncertainty and improving motion control during the manipulation of deformable objects (Arriola-Rios et al. 2020).

2.3.2 Production

Assembly sequence planning (ASP) has a major impact on establishing and optimizing industrial processes, production performance, and competitiveness (Abdullah et al. 2019; Nahmias & Olsen 2015). ASP is an NP-hard problem and many research projects address assembly sequence planning for robotic manufacturing (e.g., Fakhurldeen et al. 2019; Rodríguez et al. 2020; Tariki et al. 2020). The inherent additional uncertainties in robotic assembly with deformable objects complicate sequence planning which has not yet been widely addressed in the literature.

Many heuristics have been developed for improving ASP outcomes (Nahmias & Olsen 2015). Among the popular heuristics used are soft computing approaches inspired by nature that provide a general solution framework that is suitable for improving optimization problems and can lead to a significant reduction in computation time (Abdullah et al. 2019). Genetic algorithms (GA) is a method loosely based on evolution theory that provides an iterative method for searching the solution space (Marian et al. 2006; Abdullah et al. 2019). GA offers a straightforward way to represent assembly sequences that can be iteratively optimized based on a fitness function. GA enables the incorporation of fitness functions with multiple objectives and flexibility in interpreting constraints (Marian et al. 2006). These, make GA suitable for ASP of robotic operations with deformable objects, which typically require addressing multiple objectives and which have complex constraints. GA are greedy algorithms in terms of execution time and many iterations may be required for optimizing the solution. The initial population largely affects both convergence speed and the quality of the final solution (Lin & Gen 2018).

Zouita et al. (2019) suggest integrating a constraint satisfaction problem (CSP) for the initial GA population generation. However, finding a solution to a CSP is also NP-hard. Several filtering techniques have been developed for reducing the required search space (Li 2017). Among the widely used techniques is Arc consistency (Wang & Yap 2019; Debruyne & Bessiere 1997). The arc consistency3 (AC3) algorithm balances simplicity with efficiency as it maintains only a relatively small number of data structures during search (Zouita et al. 2019; Li 2017). Another way to reduce execution time and improve quality is to look at the similarity to previously found solutions. Such a resemblance can be measured by the longest common subsequence (LCS) index (Bergroth et al. 2000) which LCS examines the relative order of actions yet, it does not mandate their consecutive positioning. A GA with AC3 with constraint satisfaction for initial population generation and a multi-objective fitness function integrating time duration, and LCS was suggested for robotic assembly with deformable objects (Ben-David & Berman 2021). The method was integrated with a database that stores information regarding the assembly process. The database facilitates measuring similarity to sequences of similar products and the generation of feasible solutions based on both product assembly requirements and work cell capabilities.

Simulation can contribute to the analysis of the manufacturing process and to the reduction of uncertainty during run-time. For example, analyzing the influence of perception errors on gripper design using graspability maps was suggested based on a simulation tool to effectively reduce the number of developed physical gripper prototypes and the number of physical experiments (Eizicovits et al. 2016). Ghandi and Masehian (2015) developed a theoretical model for the assembly of deformable objects based on interference relations between parts and compressive stress for assembling them together. In their work, a simulation-based finite element method (FEM) method was used for deformation during assembly.

Simulations are used for expediting various learning processes to improve the ability to react to changes during run-time. For example, reinforcement learning requires many iterations for convergence therefore, most implementations are based, at least to some extent on training in simulation. Many research efforts are underway for improving the ability to transfer simulation results to real-world environments (Zhao 2020). Deep learning algorithms require large training datasets and simulations are also used for creating synthetic datasets that can expedite learning (Nikolenko 2021).

Handling deformable objects carries inherent uncertainties due to material and interaction properties. These influence not only the separate single operation but may carry over across the manufacturing operation sequence. It is therefore important to include a physical simulation within the ASP process. Physics engine simulations, e.g., MuJoCo[3] or Maya[4] have been used in research projects for the development and testing of robotic operations on deformable objects (Li et al. 2018). While providing advantages in terms of simulating realistic scenarios, physics engine simulations typically require extended run times. Robotic simulations are frequently concentrated on modeling the object's behavior (i.e., deformation) during manipulation and rarely (if at all) addresses possible changes in the overall assembly plans (Arriola-Rios et al. 2020; Chang & Padir 2020). A possible approach that can handle process uncertainty and complexity is to use a two-stage ASP method in which assumed high-quality sequences are generated using heuristics followed by testing with a physical simulation. Since simulations with physical engines have non-negligible run-time the thoroughness and quality of the first, heuristic stage is important (Ben-David et al. 2021).

2.4 Objective Tier

Measures and performance indicators can contribute to perceiving and reacting to change. While research projects tend to use basic measures such as completion time or success rate, industrial implementations typically use a large measurement suite with a variety of key performance indicators (KPIs), e.g., production effectiveness or throughput rate (Marvel et al. 2018). Determining the most suitable performance measure set is an important facet in facilitating reaction to change in assembly processes with deformable objects.

3 Robotic wire harness assembly – case study

Investigating wire harness automation is an important research domain with early studies regarding manufacturing strategies (Bertolotti & Griffiths 1987). Wire harness assembly is a frequent and critical task for many electrical and electronic products, in many industries, notably in the aerospace and automotive industries. For example, in the automotive industry, the gaining popularity of hybrid and electric cars is driving wire harness assembly into becoming the central assembly task (Fig. 3). However, wire harness assembly is among the most difficult tasks to accomplish using robotics and is currently predominantly carried out manually (Heisler et al. 2021; Tunstel et al. 2020).

Several research projects on constructing robotic assembly systems for wire harness assembly conducted over the years are reviewed. The research examined was published

[3] https://www.mujoco.org/

[4] https://www.autodesk.com/products/maya/

Fig. 3. Wire harness near connecting to the battery in a hybrid car model (Hyundai Ioniq)

in conferences and journals over a broad timeline (i.e., three decades: the late 90th, the first and second decades of the 21st century). Table 1 presents the analysis of five robotic solutions for wire harness assembly according to the tiers and strategies for coping with assembly process uncertainty considered above (Fig. 1).

The analysis reveals that all three strategies are applied with a higher emphasis on Reducing susceptibility to uncertainty. The progress in technology in the workcell tier

Table 1. Review of wire harness solutions (◆Reducing process uncertainty, ▲ Reducing susceptibility to uncertainty, ●Perceiving and reacting to change)

	Workcell				Task				Objective
	Peripherals	Sensors	Robot	End effector	Manipulation: Control	Manipulation: Perception	Production: Sequencing	Production: Simulation	Measures
(Aguirre et al., 1997)							▲ Avoiding intricate manipulations		● Economic metrics (time, cost)
(Jiang et al., 2010)	◆ Instrumental panel, ten surrounding cameras	● Fixed and on hand cameras, laser range, force sensors	▲ 3 arms Mitsubishi PA10				◆ Divided per robotic arm area	◆ Motion sequence, collisions detections	● Average assembly time
(Yumbla et al., 2019)			▲ 1 arm ABB IRB120	▲ Customized for grasping and aligning wire	▲ Mating process between connectors				● Tolerance between connectors
(Tunstel et al., 2020)	◆ Wire stripper, crimper	● "Eye on hand" approach	▲ 2 arms ABB YuMi IRB 14000	▲ Customized for pinning, heat shrinking	▲ Preprogrammed trajectories	● Based on RGB data			● Average cycle time, first-time yield
(Heisler et al., 2021)	◆ Workspace layout		▲ Human–robot-collaboration (HRC)						● Assessment system for HRC suitability

is considerable. Various oeripherals (e.g., designated jigs) are applied for reducing wire harness uncertainty, advanced sensors are applied for perceiving wire deformations, and designated end-effectors are developed to reduce susceptibility to uncertainty. The focus on the robotic manipulators and end-effectors is also evident in the task tier with more emphasis on manipulation control and perception methods. Production process methods including both sequencing and simulation-based solutions are less investigated. In the objective tier, for over three decades the main metric is assembly time. The conducted analysis indicates that research into the production process methods and the objective measures may lead to improvements in assembly of wire harnesses.

4 Discussion and Conclusions

Robotic assembly with deformable parts is challenging and has attracted much research over the years. However, it is not widespread in the industry, where operations are typically performed manually. The chapter presented various methods for coping with the challenges of robotic assembly with deformable objects in workcell, task, and objective tiers.

The current presentation assumes that the design of the product itself is predefined and therefore focuses on the components of the assembly process. It is well known that product design has a strong influence on product manufacturing and assembly processes and that many aspects straining the production system can be addressed during the product design phase (Redford & Chal 1994). Design for assembly (DFA), or design for manufacturing and assembly is an established active research field. Most of the existing body of work in DFA pertains to rigid parts. More recently researchers have started to develop methods for DFA suitable for deformable parts (Trommnau et al. 2020). Indeed, if technology permits, as suggested by Harbour & Schmidt (2018), it may be best for manufacturing automation to create entirely different products, that do not require the assembly of deformable objects. For example, for products with electrical wire harnesses switching to wireless communication and wireless power transfer can lead to a dramatic change in plausible assembly technology solutions.

The current presentation assumed fully automatic production processes. This is not the case in many assembly scenarios. Collaboration between humans and robots also termed collaborative robots (cobots) in the same physical environment is extensively researched and implemented in multiple assembly scenarios (Malik & Bilberg 2019; Hjorth & Chrysostomou 2022). Indeed, the capabilities of humans to handle uncertainty and the suitability of robots for performing repetitive work with prescribed repeatability can be complementary within a well-planned collaborative process (Malik & Bilberg 2019).

Acknowledgments. This research was supported by the Israel Innovation Authority as part of the Assembly by Robotic Technology (ART) consortium [grant number 67436].

References

Abdullah, M.A., Ab Rashid, M.F.F., Ghazalli, Z.: Optimization of assembly sequence planning using soft computing approaches: a review. Arch. Comput. Meth. Eng. **26**(2), 461–474 (2018). https://doi.org/10.1007/s11831-018-9250-y

Aguirre, E., Ferriere, L., Raucent, B.: Robotic assembly of wire harnesses: economic and technical justification. J. Manuf. Syst. **16**(3), 220–231 (1997). https://doi.org/10.1016/S0278-6125(97)88890-5

Alzubaidi, L., et al.: Review of deep learning: concepts, CNN architectures, challenges, applications, future directions. J. Big Data **8**(1), 1–74 (2021). https://doi.org/10.1186/s40537-021-00444-8

Arriola-Rios, V.E., Guler, P., Ficuciello, F., Kragic, D., Siciliano, B., Wyatt, J.L.: Modeling of deformable objects for robotic manipulation: a tutorial and review. Front. Robot. A **I**, 7 (2020). https://doi.org/10.3389/frobt.2020.00082

Ben-David, S., Shneor, R., Zuler, S., Mann, Z., Greenberg, A., Berman, S.: Simulation-based two stage sequencing of robotic assembly operations with deformable objects. In: 17th IFAC Symposium on Information Control Problems in Manufacturing (INCOM), Budapest (ON-LINE), pp. 7–9 (2021)

Ben-David, S., Berman, S.: A multi-objective fitness function for sequencing robotic assembly operations with deformable objects using a genetic algorithm with constraint satisfaction. In: 26th International Conference on Production Research, Taichung, Taiwan (ON-LINE), July 18–21 (2021)

Bergroth, L., Hakonen, H., Raita, T.: A survey of longest common subsequence algorithms. In: Proceedings Seventh International Symposium on String Processing and Information Retrieval. SPIRE 2000, pp. 39—48 (2000). https://doi.org/10.1109/SPIRE.2000.878178

Berman, S., Nof, S.Y.: Collaborative control theory for robotic systems with reconfigurable end effectors. In: 21st International Conference on Production Research: Innovation in Product and Production, ICPR 2011. July 31-August 4, Stuttgart, Germany (2011)

Bertolotti, G.P., Griffiths, B.J.: A survey of wire harness manufacturing strategies. In: McGoldrick, P.F. (eds.) Advances in Manufacturing Technology II. Boston, MA: Springer, pp. 138–142 (1987). https://doi.org/10.1007/978-1-4615-8524-4_24

Brown, R., Brost, R.: A 3-D modular gripper design tool. IEEE Trans. Robot. Autom. **15**(1), 174–186 (1999). https://doi.org/10.1109/70.744612

Chang, P., Padir, T.: Model-based manipulation of linear flexible objects: task automation in simulation and real world. Machines **8**(3), 46 (2020). https://doi.org/10.3390/machines8030046

Chiu, S.L.: Task compatibility of manipulator postures. Int. J. Robot. Res. **7**(5), 13–21 (1988). https://doi.org/10.1177/027836498800700502

Chua, P.Y., Ilschner, T., Caldwell, D.G.: Robotic manipulation of food products – a review. Ind. Robot Int. J. **30**(4), 345–354 (2003). https://doi.org/10.1108/01439910310479612

Debruyne, R., Bessiere, C.: Some practicable filtering techniques for the constraint satisfaction problem. In: Proceedings of Fifteenth International Joint Conference on Artificial Intelligence (IJCAI'97), Nagoya, Japan August 23–29 (1997)

Eizicovits, D., Van Tuijl, B., Berman, S., Edan, Y.: Integration of perception capabilities in Gripper design using graspability maps. Biosys. Eng. **146**, 98–113 (2016). https://doi.org/10.1016/j.biosystemseng.2015.12.016

El-Atab, N., et al.: Soft actuators for soft robotic applications: a review. Adv. Intell. Syst. **2**, 2000128 (2020). https://doi.org/10.1002/aisy.202000128

Fakhurldeen, H., Dailami, F., Pipe, A.G.: Cara system architecture - a click and assemble robotic assembly system. In: 2019 International Conference on Robotics and Automation (ICRA) (2019). https://doi.org/10.1109/icra.2019.8794114

Fantoni, G., Gabelloni, D., Tilli, J.: How to Design New Grippers by Analogy. Nuclear and Production Engineering, University of Pisa, Report, Department of Mechanical (2012)

Ghandi, S., Masehian, E.: Assembly sequence planning of rigid and flexible parts. J. Manuf. Syst. **36**, 128–146 (2015). https://doi.org/10.1016/j.jmsy.2015.05.002

Groover, M.P.: Automation, Production Systems, and Computer-Integrated Manufacturing, 5th edn. Prentice Hall Int, NJ, USA (2019)

Heisler, P., Utsch, D., Kuhn, M., Franke, J.: Optimization of wire harness assembly using human–robot-collaboration. Procedia CIRP **97**, 260–265 (2021). https://doi.org/10.1016/j.procir.2020.05.235

Huckaby, J.O.D., Christensen, H.I.: A taxonomic framework for task modeling and knowledge transfer in manufacturing robotics. In: Workshops at the Twenty-Sixth AAAI Conference on Artificial Intelligence (2022). http://www.jakehuckaby.com/papers/aaai2012ws_cogrob.pdf

Harbour, R., Schmidt, J.: Tomorrow's factories will need better processes, not just better robots, Harvard Business review (2018). https://hbr.org/2018/05/tomorrows-factories-will-need-better-processes-not-just-better-robots

Herguedas, R., López-Nicolás, G., Aragüés, R., Sagüés, C.: Survey on multi-robot manipulation of deformable objects. In: IEEE International Conference on Emerging Technologies and Factory Automation, ETFA, pp. 977–984 (2019). https://doi.org/10.1109/ETFA.2019.8868987

Hjorth, S., Chrysostomou, D.: Human–robot collaboration in industrial environments: a literature review on non-destructive disassembly. Robot. Comput. –Integr. Manufact. **73**, 102208 (2022). https://doi.org/10.1016/j.rcim.2021.102208

Ho, J.: Cost strategy for product planning under competition. Int. J. Prod. Res. **56**(24), 7444–7457 (2018). https://doi.org/10.1080/00207543.2018.1461273

Hultman, E., Leijon, M.: An updated cable feeder tool design for robotized stator cable winding. Mechatronics **49**, 197–210 (2018). https://doi.org/10.1016/j.mechatronics.2018.01.006

Jiang, X., Koo, K.M., Kikuchi, K., Konno, A., Uchiyama, M.: Robotized assembly of a wire harness in car production line. In: 2010 IEEE/RSJ International Conference on Intelligent Robots and Systems, 2010, pp. 490–495 (2010). https://doi.org/10.1109/IROS.2010.5653133

Jiménez, P.: Survey on model-based manipulation planning of deformable objects. Robot. Comput.-Integr. Manufact. **28**(2), 154–163 (2012). https://doi.org/10.1016/j.rcim.2011.08.002

Khalil, F.F., Payeur, P.: Dexterous robotic manipulation of deformable objects with multi-sensory feedback - a review. In: Jimenez, A., Hadithi, B.M.A. (Eds.), Robot Manipulators Trends and Development, pp. 587–621. InTech. (2010). https://doi.org/10.5772/9183

Kuhlemann, I., Jauer, P., Ernst, F., Schweikard, A.: Robots with seven degrees of freedom: Is the additional DOF worth it?. In: 2016 2nd International Conference on Control, Automation and Robotics (ICCAR) (2016). https://doi.org/10.1109/iccar.2016.7486703

Kroemer, O., Niekum, S., Konidaris, G.: A review of robot learning for manipulation: challenges, representations, and algorithms. J. Mach. Learn. Res. **22**, 1–82 (2021)

Li, H.: Narrowing support searching range in maintaining arc consistency for solving constraint satisfaction problems. IEEE Access **5**, 5798–5803 (2017). https://doi.org/10.1109/access.2017.2690672

Lin, L., Gen, M.: Hybrid evolutionary optimisation with learning for production scheduling: State-of-the-art survey on algorithms and applications. Int. J. Prod. Res. **56**(1–2), 193–223 (2018). https://doi.org/10.1080/00207543.2018.1437288

Malik, A.A., Bilberg, A.: Collaborative robots in assembly: A practical approach for tasks distribution. Procedia CIRP **81**, 665–670 (2019). https://doi.org/10.1016/j.procir.2019.03.173

Marian, R.M., Luong, L.H., Abhary, K.: A genetic algorithm for the optimisation of assembly sequences. Comput. Ind. Eng. **50**(4), 503–527 (2006). https://doi.org/10.1016/j.cie.2005.07.007

Marvel, J.A., Bostelman, R., Falco, J.: Multi-robot assembly strategies and metrics. ACM Comput. Surv. **51**(1), 14 (2018). https://doi.org/10.1145/3150225

Morales, A., Sanz, P.J., Del Pobil, A.P., Fagg, A.H.: Vision-based three-finger grasp synthesis constrained by hand geometry. Robot. Auton. Syst. **54**(6), 496–512 (2006). https://doi.org/10.1016/j.robot.2006.01.002

Mu, X., Xue, Y., Jia, Y.: Robotic cutting: mechanics and control of knife motion. In: International Conference on Robotics and Automation (ICRA), pp. 3066-3072 (2019). https://doi.org/10.1109/icra.2019.8793880

Nahmias, S., Olsen, T.L.: Production and Operations Analysis. IL, USA: Waveland Pr (2015)

Nguyen, H., La, H.: Review of deep reinforcement learning for robot manipulation. In: Third IEEE International Conference on Robotic Computing (IRC), pp. 590–595 (2019). https://doi.org/10.1109/IRC.2019.00120

Nikolenko, S.I.: Synthetic-to-Real Domain Adaptation and Refinement. In: Synthetic Data for Deep Learning. SOIA, vol. 174, pp. 235–268. Springer, Cham (2021). https://doi.org/10.1007/978-3-030-75178-4_10

Nottensteiner, K., Sachtler, A., Albu-Schäffer, A.: Towards autonomous robotic assembly: using combined visual and tactile sensing for adaptive task execution. J. Intell. Rob. Syst. **101**(3), 1–22 (2021). https://doi.org/10.1007/s10846-020-01303-z

Nof, S.: Collaborative control theory for e-work, e-production, and e-service. Annu. Rev. Control. **31**(2), 281–292 (2007). https://doi.org/10.1016/j.arcontrol.2007.08.002

Pfrommer, J., Schleipen, M., Beyerer, J.: PPRS: production skills and their relation to product, process, and resource. In: IEEE 18th Conference on Emerging Technologies and Factory Automation (ETFA), pp. 1–4 (2013). https://doi.org/10.1109/ETFA.2013.6648114

Raghav, V., Kumar, J., Senger, S.S.: Design and optimisation of robotic gripper: a review. In: Proceedings of the National Conference on Trends and Advances in Mechanical Engineering (2012)

Ranky, P.G.: Reconfigurable robot tool designs and integration applications. Ind. Robot Int. J. **30**(4), 338–344 (2003). https://doi.org/10.1108/01439910310479603

Redford, A.H., Chal, J.: Design for Assembly: Principles and Practice. McGraw-Hill, NY, USA (1994)

Rodriguez, I., Nottensteiner, K., Leidner, D., Durner, M., Stulp, F., Albu-Schaffer, A.: Pattern recognition for knowledge transfer in robotic assembly sequence planning. IEEE Robot. Autom. Lett. **5**(2), 3666–3673 (2020). https://doi.org/10.1109/lra.2020.2979622

Sanchez, J., Corrales, J.-A., Bouzgarrou, B.-C., Mezouar, Y.: Robotic manipulation and sensing of deformable objects in domestic and industrial applications: a survey. Int. J. Robot. Res. **37**(7), 688–716 (2018). https://doi.org/10.1177/0278364918779698

Shneier, M. O., Messina, E. R., Schlenoff, C. I., Proctor, F. M., Kramer, T. R., Falco, J. A.: Measuring and representing the performance of manufacturing assembly robots. NISTIR, 8090 (2015). https://doi.org/10.6028/nist.ir.8090

Shneor, R., Berman, S.: Robotic manipulation: an industry-implementation oriented categorization. In: 26th International Conference on Production Research, Taichung, Taiwan (ON-LINE), July 18–21 (2021)

Shneor, R., Berman, S.: The R αβγ categorization framework for dexterous robotic manufacturing processes. Int. J. Product. Res. (2022). https://doi.org/10.1080/00207543.2022.2150907

Tariki, K., Kiyokawa, T., Ricardez, G.A., Takamatsu, J., Ogasawara, T.: 3D model-based assembly sequence optimization using insertionable properties of parts. In: IEEE/SICE International Symposium on System Integration (SII) (2020). https://doi.org/10.1109/sii46433.2020.9026210

Tawk, C., Spinks, G. M., In het Panhuis, M., Alici, G.: 3D printable linear soft vacuum actuators: their modeling, performance quantification and application in soft robotic systems. IEEE/ASME Trans. Mechatron. **24**(5), 2118-2129 (2019). https://doi.org/10.1109/tmech.2019.2933027

Tanie, K.: Robot hands and end-effectors, chapter 7 In: Handbook of Industrial Robotics, Second Edition. S.Y. Nof (Ed.), NY, USA: John Wiley and Sons (1999)

Tunstel, E., et al.: Robotic wire pinning for wire harness assembly automation. In: IEEE/ASME International Conference on Advanced Intelligent Mechatronics (AIM), pp. 1208–1215 (2020). https://doi.org/10.1109/AIM43001.2020.9158905

Trommnau, J., Frommknecht, A., Siegert, J., Wößner, J., Bauernhansl, T.: Design for automatic assembly: a new approach to classify limp components. Procedia CIRP **91**(2020), 49–54 (2020). https://doi.org/10.1016/j.procir.2020.01.136

Valency, T., Zacksenhouse, M.: Instantaneous model impedance control for robots. In: IEEE/RSJ International Conference on Intelligent Robots and Systems (IROS), vol. 1, pp. 757–762 (2000). https://doi.org/10.1109/IROS.2000.894695

Wang, D., Tan, D., Liu, L.: Particle swarm optimization algorithm: an overview. Soft. Comput. **22**(2), 387–408 (2017). https://doi.org/10.1007/s00500-016-2474-6

Wang, R., Yap, R.H.C.: Arc consistency revisited. In: Rousseau, L.-M., Stergiou, K. (eds.) CPAIOR 2019. LNCS, vol. 11494, pp. 599–615. Springer, Cham (2019). https://doi.org/10.1007/978-3-030-19212-9_40

Wang, X.V., Pinter, J.S., Liu, Z., Wang, L.: A machine learning-based image processing approach for robotic assembly system. Procedia CIRP **104**, 906–911 (2021). https://doi.org/10.1016/j.procir.2021.11.152

Warnecke, H.-J, Schraft, R., Hägele, M., Barth, O., Schmierer, G.: Manipulator design. chapter 5. In: Handbook of Industrial Robotics, Second Edition. S.Y. Nof (Ed.), NY, USA: John Wiley and Sons (1999)

Warnecke, H.J., Emmerich, H., Koller, S.: Flexible solution for wiring harness assembly with industrial robots. CIRP Ann. **42**(1), 25–27 (1993). https://doi.org/10.1016/S0007-8506(07)62384-1

Whitney, D.E.: Mechanical Assemblies: Their Design, Manufacture, and Role in Product Development, vol. 1: Oxford University Press (2004)

Yumbla, F., Yi, J., Abayebas, M., Moon, H.: Analysis of the mating process of plug-in cable connectors for the cable harness assembly task. In: 19th International Conference on Control, Automation and Systems (ICCAS), pp. 1074–1079 (2019). https://doi.org/10.23919/ICCAS47443.2019.8971644

Zhou, Z., Nguyen, C.C.: Joint configuration conservation and joint limit avoidance of redundant manipulators. In: Proceedings of International Conference on Robotics and Automation, vol. 3, pp. 2421–2426 (1997). https://doi.org/10.1109/robot.1997.619324

Zhao, W., Queralta, J.P., Westerlund, T.: Sim-to-real transfer in deep reinforcement learning for robotics: a survey. In: IEEE Symposium Series on Computational Intelligence (SSCI), pp. 737–744 (2020). https://doi.org/10.1109/SSCI47803.2020.9308468

Zouita, M., Bouamama, S., Barkaoui, K.: Improving genetic algorithm using arc consistency technic. Procedia Comput. Sci. **159**, 1387–1396 (2019). https://doi.org/10.1016/j.procs.2019.09.309

The Framework and Applications of Anomalous Subsequence Detection in Streaming Data Analysis and Process Monitoring in Intelligent Manufacturing

Hendri Sutrisno[1](✉) and Chao-Lung Yang[2]

[1] Institute of Statistical Science, Academia Sinica, No. 128, Academia Road, Section 2, Nankang, Taipei 11529, Taiwan
hendri@stat.sinica.edu.tw

[2] Department of Industrial Management, National Taiwan University of Science and Technology, No. 43, Section 4, Keelung Rd, Da'an District, Taipei City 106, Taiwan
clyang@mail.ntust.edu.tw

Abstract. In intelligent manufacturing, detecting anomalous signals or time series subsequences and monitoring the process shift are two fundamental tasks to maintain manufacturing quality. Detecting the anomalous signals or process shifting as early as possible is a challenge. This chapter reviews the relevant research works in these two domains and presents the state-of-art methodologies that combine machine learning and data mining techniques to resolve the practical problem in production, named local recurrence rate with robust k-means (LRR-RKMeans) and long short-term memory real-time contrasts control chart (LSTM-RTC). The real-world data in manufacturing were used to evaluate the methods. The methods for anomalous subsequence detection on process monitoring and real-time contrasts control chart for quality control were applied to real-world semiconductor wafer and white wine production cases. The experimental results show that LRR-RKMeans can detect the defective wafer accurately, and LSTM-RTC has a low response delay in noticing process shift in white wine production. The possibility of combining the proposed data analysis framework with an edge-cloud computing system to alleviate the performance boundary in the computational time was further discussed.

Keywords: Anomaly Detection · Process Monitoring · Quality Control · Time Series · Intelligent Manufacturing

C.-Y. Huang and S. W. Yoon (Eds.): ICPR1 2021, ACES 14, pp. 236–250, 2023.
https://doi.org/10.1007/978-3-031-44373-2_14

1 Introduction

Due to the global competition, manufacturers are demanded to be more agile in adjusting their production ability to quickly respond to customers' needs. Indeed, manufacturers need to improve their production process's flexibility, speed, quality, and efficiency to raise their competitiveness in the market. In agile manufacturing, information technology is vital to reduce human error and improve product quality and production efficiency. Based on the conceptual spirit of Industry 4.0, agile manufacturing can be achieved by a more intelligent manufacturing system, where the machines, sensors, and information technology systems are connected as a system. The system allows every element in the production site to collaborate by using an information network to analyze data and adapt to the changes [1]. For example, production machines in the manufacturing site can be integrated through the Internet of Things (IoT) technology to allow the communication between machines (M2M) [2]. Based on the M2M technology, machines can communicate systematically to achieve a more intelligent manufacturing system. For example, once one machine is shut down for repairing, the consecutive machines can be auto-adjusted for different job assignments without human intervention.

A systematic design is essential to store high-frequency sensor data and analyze extensive streaming data simultaneously. Online monitoring and offline batch analysis of the collected data from the production site are crucial for achieving more intelligent manufacturing processes [3]. How to analyze the collected stream data is a challenging task because of the size and the speed of the incoming sensor data [4]. Based on Yang et al.'s work, Fig. 1 shows the data analysis framework representing three significant tasks: processing monitoring, quality control, and data analysis in intelligent manufacturing. The streaming data could be collected from machines and sensors to fulfill the needs of these three tasks.

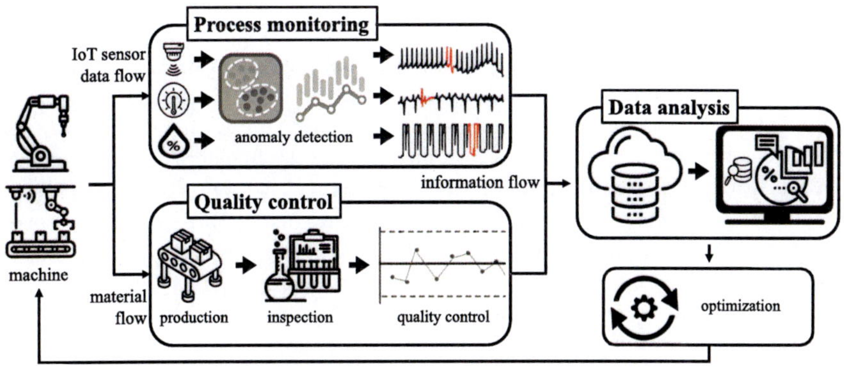

Fig. 1. Data analysis framework of streaming time series in production

Multiple data can be obtained by attaching Internet-of-Thing (IoT) high-frequency sensors to machines. Multiple time series data such as the machine temperature, humidity, pressure, and others can be used to monitor the production process and machine's condition. When unexpected events occur in production, anomalous process or behavior might be observed in the sensor data collected from the system. These anomalies might be in many forms, such as extreme values or unusual time series patterns [5] and might be correlated to a quality defect in production. Indeed, properly analyzing the anomalous process can enable the preventive maintenance of the machines and equipment [6].

The data obtained from the IoT devices are typically highly imbalanced because the anomaly pattern rarely occurs compared to the regular pattern. It is also challenging to understand the anomaly pattern since the obtained data are all coming without labels; thus, most of the developed machine learning methods are hard to be applied to solve the anomaly detection problem [7]. Alternatively, other anomaly detection approaches have been proposed to discover an anomalous pattern that is maximally different from the neighboring patterns without the label of data. These methods are known as the anomaly subsequence (discord) discovery methods, such as the local recurrence rate with robust k-means (LRR-RKMeans) [8] or Matrix Profile (MP) [9]. Discords are the sections of a time series that are anomalous or uniquely presented [10]. Typically, the discord discovery methods find the existing patterns by splitting the data into multiple subsequences and performing a similarity check to assess how close a pattern with its surrounding patterns. The results can be used to indicate the potential signals that might be correlated to system outages or malfunctioning [11].

As mentioned earlier, multiple data can be obtained from the material flow in production, such as those related to the interests in the quality control, for example, product specifications. However, the typical statistical process control (SPC) methods face performance boundaries when the analyzed data is massive. Typically, the limitation is described because of the 4 V's of big data; volume, variety, velocity, and veracity [12].

With the 4 V's challenge, most SPC methods are troubled in the data analysis, and it might lead to longer response delay to the upcoming issues in production. Therefore, catching the process shift as early as possible under the 4 V's challenge is one of the main tasks in intelligent manufacturing. The longer response to issues in production will definitely cost the manufacturers on producing defective products [13], but also on shutting down the production process [14]. Multiple studies also highlight the importance of having a faster reaction to issues in production; in petroleum refining processes [15], in manufacturing processes of compressor's materials [16], in production processes of the plastic button for the clothing industry [17], in wine production processes [18], and in pharmaceutical twin-screw granulation and drying process [19].

Recently, the works on integrating the machine learning technique into SPM to further improve the response delay of SPM have been studied in the literature. Deng et al. [20] introduced the classification concept to the control chart. The monitoring task in SPC was viewed as a classification problem of in-control and out-of-control conditions, named real-time contrasts control chart (RTC). The random forests method was applied as the classifier to their RTC (RF-RTC) in their work. The results show that RF-RTC can alleviate the performance boundary faced by the traditional SPM by achieving far lower response delay. Since then, many works have been reported on integrating other machine learning methods into RTC. Typically, a newer RTC chart is desired to have a lower response delay. Some examples of the new RTC charts are the linear discriminant analysis [21], random forests with multiple voting mechanisms [22], support vector machine method (D-SVM) [23], and D-SVM with the evolutionary algorithm as the optimizer algorithm [24]. The most recent work on RTC aims to be able to train in much more extensive data size (volume in 4 V's), and achieve even lower response delay and at the same time, by applying the long short-term memory (LSTM) as the classifier [18]. The proposed method is called LSTM-RTC, where the stacked-LSTM network was proposed in the architecture. The stacked-LSTM network was applied to catch the streaming data's hierarchical representation and to learn the abnormal patterns on the relatively large dataset with the structure of RTC. As a result, the LSTM-RTC can learn patterns from large datasets and lower the response delay.

Under the proposed data analysis framework in Fig. 1, the operators and managers can quickly react to the abnormal events caused by the malfunction during the production. By integrating LRR-RKMeans for discord discovery and machine learning-based control charts, such as LSTM-RTC for process monitoring, the data analysis process is designed to quickly handle the alarms raised by the anomaly detection and quality inspection from the collected data, respectively. The analyzed results can be utilized to optimize the system setting and configuration of the production machines automatically. In other words, the framework can enable the production machine to self-adjust the system to satisfy the product quality requirements from the market and reduce the downtime in production.

2 Methodology

2.1 LRR-RKMeans for Discord Discovery in Process Monitoring

The importance of discord discovery has been widely discussed in multiple domains, such as in manufacturing [25], medical [26,27], environmental [28], finance [29], sensors data analysis [30], and aircraft engine failure analysis [31]. The discords are usually difficult to detect because their length and location are generally unknown [32]. Despite the importance of the study, the developed methods are less applicable due to the required assumptions and computation complexity.

In the literature, the heuristic ordered time-series symbolic aggregate approximation (HOT-SAX) [10] can be considered one of the most significant discord discovery methods back to the year 2005. The HOT-SAX method is a window-based anomaly discovery technique that utilizes a symbolic approach that reduces data dimensionality by converting the time series data into a string of alphabets. Wang et al. [31], and Linardi et al. [32] noted that the window-based with static window-size anomaly discovery methods, such as HOT-SAX, is less applicable in real-world cases. Most real-world data are multivariate with unpredictable patterns and lengths. Therefore, the need for the predetermined numbers and lengths of the anomaly searching window limits the window-based methods. In addition, the window-based methods do not apply to multivariate time series (MTS) problems where the user has no prior knowledge of the data or the anomalies.

Lately, researchers have extended the discord discovery in multivariate time series (MTS) from window-based to distance matrix-based algorithms. Matrix-based methods converted the MTS into a distance matrix and analyzed the discord based on the matrix. Luo and Gallagher in 2011 [33] proposed the periodicity-based direct search (PBDS) for discovering the discords in MTS with satisfying computational efficiency. Unlike the window-based methods, PBDS does not need the pre-defined number and length of the potential anomalies, which allows it to handle the nearly periodic or quasi-periodic time series. Later in 2013, Luo et al. [30] continued to propose the global discord search (GDS), which does not need the periodicity assumption used in the PBDS method. GDS method directly performs the sampling of the pairwise distance between subsequences to reduce the computational time, especially in a more extensive MTS [30].

In 2019, Hu et al. [34] proposed a more efficient method called the discord search method based on the local recurrence rate (LRRDS). Instead of computing the pairwise distance between all points or subsequences, LRRDS only calculates the distance between a point to its surrounding points as the local recurrence rate (LRR), which indicates the discord's location on a subsequence's distance to the nearby subsequences. Their work showed that LRR could enhance the searching quality and maintain lower computational time than GDS. Yang et al. [35] proposed a new LRRDS method with an automatic time window (LRRDS-ATW) to improve the accuracy of discord further searching. The dynamic window adjusting mechanism essentially selects the adaptive number of surrounding points for comparison in LRR. Their work also extends the matrix representation to contain more local information about the signal variation for improving search accuracy. Although the accuracy of LRRDS-ATW is promising, the method suffers from the extensive computation time due to the complexity of the algorithm. They proved that LRRDS-ATW could discover the discords in short MTS. This advantage is crucial to online discord discovery as the method that can analyze short MTS accurately.

In 2021, Sutrisno and Yang [8] proposed a novel discord discovery method called LRR-RKMeans, which combines LRR [34] with the unsupervised robust k-means clustering (RKMeans) algorithm [36,37] to search out the discords on the periodic MTS data, which is common in a real-world production dataset. The proposed LRR-RKMeans has three advantages: 1) higher searching accuracy, 2) computational efficiency, and 3) simplicity. The proposed method can catch the periodic pattern in the dataset accurately; therefore, it has a promising performance on the production dataset compared to the methods in the literature.

To achieve higher searching accuracy, first, the potential biases caused by the wide data range and the long-term time series trend were filtered out by normalizing the MTS and removing the MTS's long-term trends. Second, the MTS was summarized using LRR to highlight the abnormal non-repeated patterns in a single univariate time series monitoring signal. Third, signal segmentation was performed by converting the monitoring signals into a Boolean sign to differentiate the discords from the expected signals correctly. The segmentation algorithm can avoid splitting an anomalous pattern into several cuts, making the abnormal part clearer. The proposed LRR-RKMeans adopts the dynamic time warping (DTW) [18] as the distance matric to discover the anomalous subsequences from the segmented signals. The RKMeans can escape from calculating two-times distance matrix as in LRRDS and LRRDS-ATW to reduce the computational time. Also, LRR-RKMeans only need to set the lower bound of window size, while LRRDS and LRRDS-ATW need more parameters except for the window size.

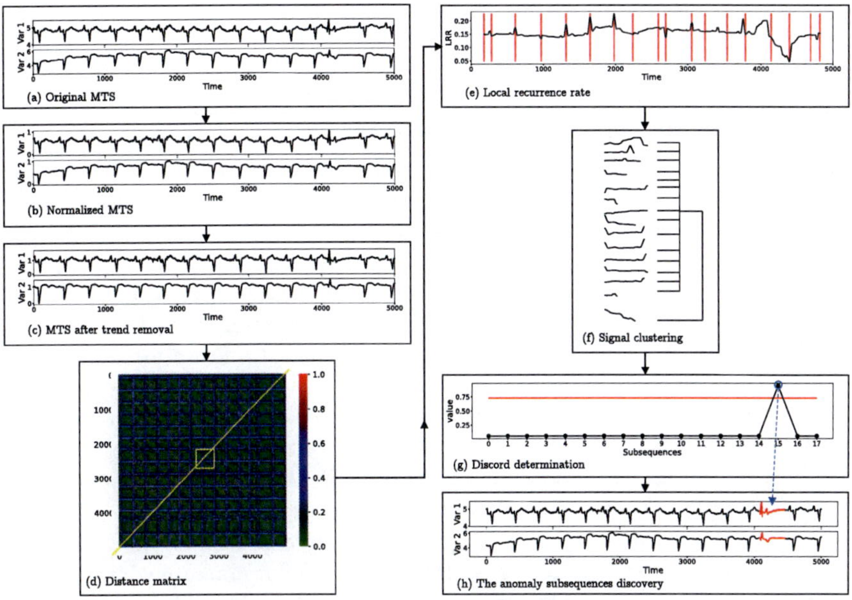

Fig. 2. The flowchart of LRR-RKMeans

Figure 2 illustrates these seven steps of LRR-RKMeans by using the ECG2 5000 [10] dataset with two variables as an example. The MTS data shown on the left-top of Figure 2 are the original MTS. After performing the normalization and trend-removal, the scale of different time series is synchronized, and the long-term trend that might lead to the discord detection bias is removed. Then, the distance matrix among MTS is generated, and LRR can create the monitoring signal. The subsequences of the monitoring signal can be divided based on the proposed segmentation method and clustered by the robust k-means method. The cluster of the subsequences with fewer instances is denoted as discords based on the probability.

2.2 LSTM-RTC for Quality Control

In the literature, applying a real-time contrasts control chart (RTC), one of the popular statistical process monitoring methods, with machine learning method showed the advantages of the combination methods over the traditional statistical methods. Essentially, RTC is a control chart that formulates the process monitoring problem into a classification task. The goal is to distinguish the out-of-control from the in-control processes. For example, distance-based RTC using kernel linear discriminant examination (KLDE) technique overtakes the traditional and classification probability-based control chart in detecting mean process shift [21]. The RTC control chart with weighted voting methods (GWRF, MWRF, and FWRF) and will random forest (RF-RTC) alleviate class imbalance problems and detect process shifts quickly [22]. The other distance-based RTC multivariate process monitoring with a support vector machine (D-SVM) proposed by [23] also reported having a lower response delay than RF-RTC in multiple data distribution. Although the methods mentioned above showed the advantages of applying RTC, they are not suitable for high-dimensional datasets because the data used for training the model is small in their cases.

The LSTM network is a popular time series analysis and processing monitoring method. A novel LSTM ensemble was applied in Choi and Lee's work [38] for time series forecasting by modeling highly nonlinear statistical dependencies. The results show higher forecasting accuracy than the other common forecasting methods. Another study by Ünlü [39] restated the popularity of LSTM model. Ünlü [39] proposed a cost-oriented LSTM model for early detection of error signals in control chart. The results revealed that the LSTM model outperforms SVM and WSVM (weighted support vector machine) in classification and quick abnormal pattern detection accuracy based on signal monitoring in control chart. Therefore, it is interesting to evaluate how the LSTM network can be applied in RTC to lower the response delay in detecting process shifts.

In a report by Yang and Sutrisno [18], a novel chart based on LSTM and RTC was proposed to alleviate the performance boundary on large data streams, named a stacked long short-term memory RTC chart (LSTM-RTC). LSTM-RTC utilizes more extensive training data captured by high-frequency sensors to produce lower response delay. The large LSTM structure in LSTM-RTC can learn the patterns in the streaming time series and RTC for fast process shift detection on various types of data distribution. The network topology in LSTM-RTC was designed to learn the short-term and long-term information and the hierarchical relationship in the streaming data. Thus, in LSTM-RTC, the extensive training data collected by high-frequency sensors were utilized to detect the process shift faster. The experimental outcomes of the synthesized cases and real-world datasets indicated that the LSTM-RTC can surpass the traditional RTC methods in lower response delay by learning very high dimensional data.

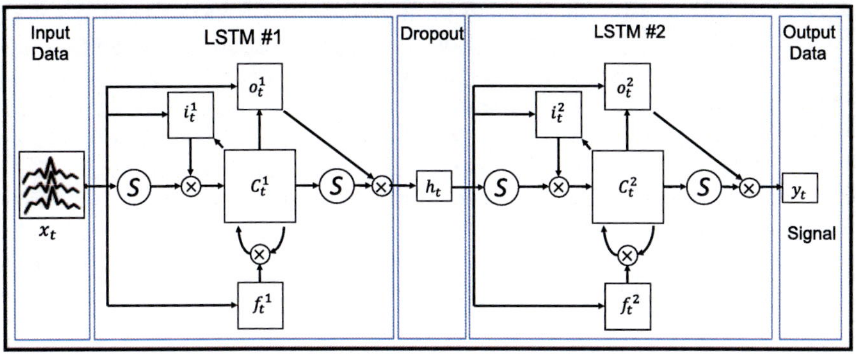

Fig. 3. The network architecture of LSTM-RTC

As in Figure 3, the input data for LSTM-RTC was captured according to the moving window mechanism. Let x_t be a batch of input data at time t. The data was first used to train the first LSTM layer, where x_t is used to update all of the cells altogether. The dimension of data output in the first LSTM layer is the same as x_t. Later on, the first LSTM layer's output will be used to train the second LSTM layer, where a dropout layer connects these LSTM layers. In the second LSTM layer, the hierarchical relationships of the input data were extracted by learning the first LSTM layer's output.

There might be possible for the second LSTM to memorize the pattern in the first LSTM layer instead of learning. This possibility might be due to analyzing the previously processed information from the first LSTM layer. Assume that the result of the first LSTM layer is reliable, then the second LSTM layer might just be memorizing the output pattern by the first LSTM layer. Therefore, to facilitate the model to learn the hierarchical relation of the data, LSTM-RTC applies a dropout layer between the first and second LSTM layers. In the dropout layer, partial information from the first LSTM's output was skipped randomly when used to train the second LSTM layer. Iteratively, in the dropout layer, information was skipped for learning randomly; thus, the dropout layer can prevent the second LSTM layer from memorizing the results from the first LSTM layer. Finally, the output of the second LSTM layer will be used as the monitoring statistics.

The monitoring statistics can be used to classify the in-control and out-of-control states. In LSTM-RTC, the model distinguishes the in-control and out-of-control states based on monitoring statistics. The lower monitoring statistics indicates the in-control state while the higher monitoring statistics is for the out-of-control state. Observations with monitoring statistics lower than the control limit will be classified as in-control and vice versa.

3 Experimental Results

3.1 LRR-RKMeans For Detecting Discord In Wafer Production

This section presents the application of applying LRR-RKMeans on detecting defective items in the wafer production (WP) process dataset in semiconductor industry [40]. The dataset contains a collection of process control measurements from various sensors during the silicon wafers production. The dataset has six monitoring variables: radio frequency forward power (FP), radio frequency reflected power (RP), chamber pressure (CP), 405-nanometer emission (405e), 520-nanometer emission (520e), and direct current bias. The FP and RP variables measure the electrical forward power, and reflected power applied to the plasma, respectively. Variable CP records the pressure of the etching chamber. Variable 405e and 520e record the intensity of light emitted by the plasma on two different wavelengths, respectively. The last variable records the direct current electrical potential difference within the utilized tools.

In this case, we selected the data from batch 1552 to 1569 to simulate the actual production data. We assumed the dataset selected from the batches above can represent the two situations faced in production. First, the number of defective wafers is minimal, and second, the dataset contains abnormal readings (i.e., sensor #2) on the non-defective wafer. Therefore, it is interesting to evaluate how the anomaly detection methods can avoid reporting false alarms. Also, indicating the discord in the WP dataset can help refer the defective product.

In this WP dataset, the number of discords is exactly one. Evaluating the top-5 anomalous subsequences can reveal how sensitive the algorithms tends to raise false alarms. For example, if a method finds three subsequences and one of them is discord, it also means that the method raises two unwanted false alarms. The false alarm issue is critical for assessing the reliability of the methods in real-world applications. Therefore, the discord search method with lower false alarms is desired because it reduces the frequency of checking by the operator due to the alarm raised by the system.

In Table 1, the top-5 discord scores are presented for three methods: LRRDS, LRRDS-ATW, and LRR-RKMeans. The higher the discord score for a particular detection, the higher the confidence in identifying the discord. In the meantime, the lower the discord score for false alarms, the more reliable the method. In Table 1, the black-bolded face indicates if the method successfully detects the discords, while the underlined results show the false alarms.

Table 1. Top 5 discords detection in WP dataset

Method	Score on top-5 detection				
	1	2	3	4	5
LRRDS	**0.98**	<u>0.98</u>	<u>0.98</u>	<u>0.98</u>	0.84
LRRDS-ATW	**0.97**	<u>0.89</u>	<u>0.89</u>	<u>0.89</u>	<u>0.89</u>
LRR-RKMeans	**0.98**	0.02	0.02	0.02	0.02

As shown in Table 1, all models can identify the discords that existed in the WP dataset correctly. However, both LRRDS and LRRDS-ATW generate three and four false alarms, respectively, except for LRR-RKMeans. It also means the propsoed LRR-RKMeans can catch the discord accurately without producing false alarms.

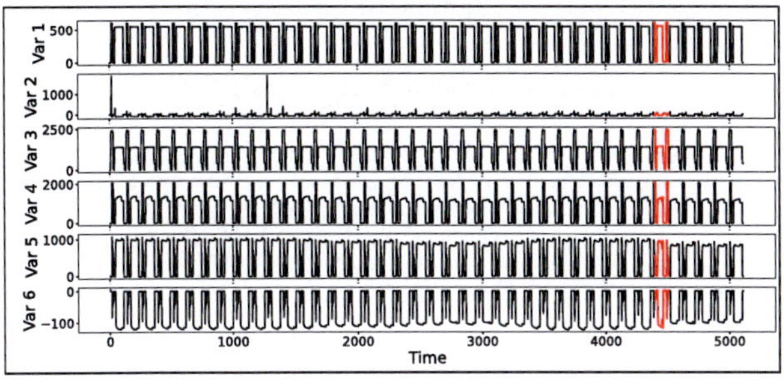

Fig. 4. Discord detection on Wafer Production dataset

Figure 4 visualizes the results of the LRR-RKMeans's discovered discords WP datasets. The red-colored subsequences mark the discovered discords. Consistent with the results in discords in Table 1, these results confirm that LRR-RKMeans is a robust method for discovering discords in MTS.

3.2 LSTM-RTC for Process Shift Detection in Wine Production

This section evaluates multiple RTC charts, such as RF-RTC, FWRF, GWRF, MWRF, D-SVM, and LSTM-RTC on white wine production case [13]. Multiple sensors were attached to monitor the wine specifications in the dataset, such as the acidity, sugars, chlorides, sulfur dioxide, density, and pH. During the production process, the wine quality might be sensitively changed according to the circumstances in the production process. Once the quality of the produced

Table 2. Quality control on the white wine production data [18]

	h	ARL_0	SE	ARL_1	SE
RF-RTC	0.81	195.48	13.38	8.48	1.12
FWRF	0.64	208.46	18.51	6.96	0.45
GWRF	0.64	199.87	15.42	6.13	0.31
MWRF	0.63	209.15	17.41	6.47	0.28
D-SVM	0.85	201.35	8.98	11.73	1.07
LSTM-RTC	0.62	201.16	11.14	**3.91**	0.12

wine is below the standard, the system needs to raise the alarm as quickly as possible so that the operator can react quickly to solve the problem.

Table 2 shows the monitoring results of RTC charts on the white wine production dataset. Columns ARL_0 and ARL_1 show the in-control and out-of-control average run-length (ARL), respectively. As Table 2 reports, the LSTM-RTC has the lowest ARL_1 while the D-SVM chart has the highest ARL_1. It also means that LSTM-RTC is the fastest in detecting process shifts.

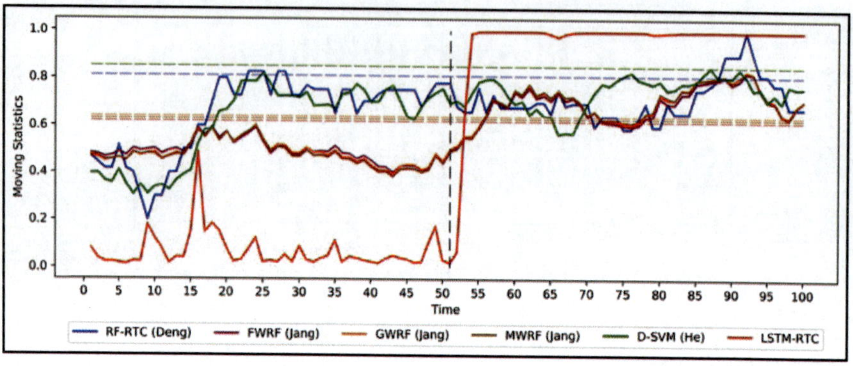

Fig. 5. Monitoring statistics on the white wine production data [18]

Figure 5 highlights the monitoring performance of the tested RTC charts based on the monitoring statistics. Each RTC chart is color-coded: blue, purple, orange, brown, green, and red for RF-RTC, FWRF, GWRF, MWRF, D-SVM, and LSTM-RTC, respectively. The process starts from the in-control state (time 1 to 50) and shifts to the out-of-control state from time 51 to 100. The vertical black dashed line indicated the separation between these two states. The control limit for each chart was shown by the horizontal dashed lines, where the color representation is the same as it in the monitoring statistics.

As Figure 5 illustrates, the monitoring statistics of LSTM-RTC were relatively low in the in-control state and high in the out-of-control state. The differences in monitoring statistics between these two states of the other RTC charts were not so apparent as the differences in LSTM-RTC. These results show that LSTM-RTC has a better classification ability than other RTC charts. The response delay result also restates the advantage of LSTM-RTC over other RTC charts. Compared to the monitoring statistics in other RTC, the monitoring statistics of LSTM-RTC were the fastest in surpassing the control limit. It also means that LSTM-RTC has the lowest response delay.

4 Discussion

A massive amount of data can be captured from production machines in manufacturing sites. Indeed, the size and the frequency of the captured data can trouble real-time data analysis. In the framework in Fig. 1 the streaming data will be monitored based on smaller data batches. The LRR-RKMeans method can be applied to find out anomalous patterns, and these patterns can be monitored by the LSTM-RTC method for process shift detection. The analysis results can be used to notify the workers for faster reaction to process shifts. The collected anomalous patterns can also be constructed as a collection of anomalous patterns. More advanced data analysis can be performed to analyze the correlation between the unique patterns and the quality result of products.

The data processing time is highly crucial in intelligent manufacturing. When the sensor data is obtained, the analysis requires the processing time to determine if the data is anomalous or normal and if the process is in control or out of control. The required time for data analysis is inevitable. However, the analysis time can be reduced by simplifying the data analysis and applying newer data analysis techniques, such as edge-cloud computing. In Edge-and-cloud computing, the data analysis task can be performed directly on the local computers (edge). Only the actual results from the local computers are uploaded to the cloud for computing. The cloud computer can be used to retrain the algorithms and to re-optimize the overall system. However, edge-cloud computing is still an early study, and it needs more studies on multiple production scenarios.

5 Conclusion

Integrating the streaming data analysis framework in a manufacturing system is essential for intelligent manufacturing. Multiple data types can be obtained from either the product or material or the machine itself by the sensor network. The data from the machine can be used to monitor the machine's condition. This chapter introduces one anomalous subsequence detection method called LRR-RKMeans and one short-response-delay process monitoring method called LSTM-RTC for handling streaming data. LRR-RKMeans can detect anomalous and rare patterns in sensor data without labels or prior knowledge. LSTM-RTC method can be applied to notify quickly when the production specification goes out of control. In semiconductor and wine production case studies, applying anomalous subsequence detection and process monitoring shows promising performance.

Monitoring the machine condition and production quality can be further analyzed to re-adjust and re-optimize the production process. It is worth investigating the edge-cloud computing technology in the data analysis framework discussed in Fig. 1. The faster processing time can be achieved by adopting edge-cloud computing. The edge computing for anomalous signals detection can quickly notify the operator once the signals are detected. For the more complex analysis, such as improving the detection accuracy of LRR-RKMeans or the sensitivity of LSTM-RTC, cloud computing is suggested to be applied for data analysis on the recorded data. Our study also shows that the framework can be applied to multiple production processes.

References

1. Zhong, R.Y., Xu, X.W., Klotz, E., Newman, S.T.: Intelligent manufacturing in the context of industry 4.0: a review. Engineering **3**, 616–630 (2017)
2. Chen, K.-C., Lien, S.-Y.: Machine-to-machine communications: technologies and challenges. Ad Hoc Netw. **18**, 3–23 (2014)
3. Nain, G., Pattanaik, K.K., Sharma, G.: Towards edge computing in intelligent manufacturing: past, present and future. J. Manufact. Syst. **62**, 588–611 (2022)
4. Yang, C.L., Sutrisno, H., Lo, N.W., et al.: Streaming data analysis framework for cyberphysical system of metal machining processes. In: 2018 IEEE Industrial Cyber-Physical Systems (ICPS), pp. 546–551 (2018)
5. Wang, Y., Perry, M., Whitlock, D., Sutherland, J.W.: Detecting anomalies in time series data from a manufacturing system using recurrent neural networks. J. Manufactur. Syst. **62**, 823–834 (2022)
6. Wan, J., Tang, S., Li, D., et al.: A manufacturing big data solution for active preventive maintenance. IEEE Trans. Ind. Inform. **13**, 2039–2047 (2017)
7. Zhang, Y.X., Chen, Y., Wang, J., Pan, Z.: Unsupervised deep anomaly detection for multisensor time-series signals. ArXiv, vol. abs/2107.12626 (2021)
8. Sutrisno, H., Yang, C.L.: Discovering defective products based on multivariate sensors data using local recurrence rate and robust k-means clustering. In: The 26th International Conference on Production Research (ICPR-26) (2021)
9. Yeh, C.-C.M., Zhu, Y., Ulanova, L., et al.: Matrix profile I: all pairs similarity joins for time series: a unifying view that includes motifs, discords and shapelets. In: 2016 IEEE 16th International Conference on Data Mining (ICDM), pp. 1317–1322 (2016)
10. Keogh, E.J., Lin, J., Fu, A.W.C.: Hot sax: efficiently finding the most unusual time series subsequence. In: Fifth IEEE International Conference on Data Mining (ICDM'05), vol. 8 (2005)
11. Wang, Z., Pi, D., Gao, Y.: A novel unsupervised time series discord detection algorithm in aircraft engine gearbox. In: ADMA (2018)
12. He, Q.P., Wang, J.: Statistical process monitoring as a big data analytics tool for smart manufacturing. J. Process Control **67**, 35–43 (2018)
13. Cortez, P., Cerdeira, A.L., Almeida, F., Matos, T., Reis, J.: Modeling wine preferences by data mining from physicochemical properties. Decis. Support Syst. **47**, 547–553 (2009)
14. Qu, Y.J., Ming, X.G., Liu, Z., Zhang, X., Hou, Z.: Smart manufacturing systems: state of the art and future trends. Int. J. Adv. Manufact. Technol. 1–18 (2019)
15. Khodabakhsh, A., Ari, I., Bakir, M., Ercan, A.O.: Multivariate sensor data analysis for oil refineries and multi-mode identification of system behavior in real-time. IEEE Access **6**(64), 389–364 405 (2018)
16. El-Midany, T.T., El-baz, M.A., Abd-Elwahed, M.S.: A proposed framework for control chart pattern recognition in multivariate process using artificial neural networks. Expert Syst. Appl. **37**, 1035–1042 (2010)
17. Sentürk, S., Erginel, N., Kaya, I., Kahraman, C.: Fuzzy exponentially weighted moving average control chart for univariate data with a real case application. Appl. Soft Comput. **22**, 1–10 (2014)
18. Yang, C.L., Sutrisno, H.: Reducing response delay in multivariate process monitoring by a stacked long-short term memory network and real-time contrasts. Comput. Ind. Eng. **153**, 107052 (2021)

19. Silva, A.F., Vercruysse, J., Vervaet, C., et al.: Process monitoring and evaluation of a continuous pharmaceutical twin-screw granulation and drying process using multivariate data analysis. Europ. J. Pharm. Biopharm. **128**, 36–47 (2018)
20. Deng, H., Runger, G.C., Tuv, E.: System monitoring with real-time contrasts. J. Qual. Technol. **44**, 27–29 (2012)
21. Wei, Q., Huang, W., Jiang, W., Zhao, W.: Real-time process monitoring using kernel distances. Int. J. Product. Res. **54**, 6563–6578 (2016)
22. Jang, S., Park, S.H., Baek, J.-G.: Real-time contrasts control chart using random forests with weighted voting. Expert Syst. Appl. **71**, 358–369 (2017)
23. He, S., Jiang, W., Deng, H.: A distance-based control chart for monitoring multivariate processes using support vector machines. Ann. Oper. Res. **263**, 191–207 (2018)
24. Wang, F.-K., Bizuneh, B., Cheng, X.-B.: One-sided control chart based on support vector machines with differential evolution algorithm. Qual. Reliab. Eng. Int. **35**, 1634–1645 (2019)
25. Hashemian, H.M., Bean, W.C.: State-of-the-art predictive maintenance techniques*. IEEE Trans. Instrum. Meas. **60**, 3480–3492 (2011)
26. Shoeb, A.H., Guttag, J.V.: Application of machine learning to epileptic seizure detection. In: ICML, pp. 975–982 (2010)
27. Li, G., Bräysy, O., Jiang, L., Wu, Z., Wang, Y.: Finding time series discord based on bit representation clustering. Knowl. Based Syst. **54**, 243–254 (2013)
28. Febrero, M., Galeano, P., Gonzalez Manteiga, W.: Outlier detection in functional data by depth measures, with application to identify abnormal NOx levels. Environmetrics **9**(4), 331–345 (2008)
29. Rahmani, A., Afra, S., Zarour, O., et al.: Graph-based approach for outlier detection in sequential data and its application on stock market and weather data. Knowl. Based Syst. **61**, 89–97 (2014)
30. Luo, W., Gallagher, M.R., Wiles, J.: Parameter-free search of time-series discord. J. Comput. Sci. Technol. **28**, 300–310 (2013)
31. Wang, X., Lin, J., Patel, N., Braun, M.: Exact variable-length anomaly detection algorithm for univariate and multivariate time series. Data Mining Knowl. Discov. **32**(6), 1806–1844 (2018). https://doi.org/10.1007/s10618-018-0569-7
32. Linardi, M., Zhu, Y., Palpanas, T., Keogh, E.J.: Matrix profile goes mad: variable-length motif and discord discovery in data series. Data Mining Knowledge Discov. **34**, 1022–1071 (2020)
33. Luo, W., Gallagher, M.R.: Faster and parameter-free discord search in quasi-periodic time series. In: PAKDD (2011)
34. Hu, M., Feng, X., Ji, Z., Yan, K., Zhou, S.: A novel computational approach for discord search with local recurrence rates in multivariate time series. Inf. Sci. **477**, 220–233 (2019)
35. Yang, C.-L., Darwin, F., Sutrisno, H.: Local recurrence rates with automatic time windows for discord search in multivariate time series. Procedia Manufact. **39**, 1783–1792 (2019)
36. Lei, J., Jiang, T., Wu, K., Du, H., Zhu, G., Wang, Z.: Robust k-means algorithm with automatically splitting and merging clusters and its applications for surveillance data. Multimedia Tools Appl. **75**, 12043–12059 (2016)
37. Yang, C.-L., Sutrisno, H.: A clustering-based symbiotic organisms search algorithm for high-dimensional optimization problems. Appl. Soft Comput. **97**, 106722 (2020)

38. Choi, J.Y., Lee, B.: Combining LSTM network ensemble via adaptive weighting for improved time series forecasting. Math. Prob. Eng. (2018)
39. Ünlü, R.: Cost-oriented LSTM methods for possible expansion of control charting signals. Comput. Ind. Eng. **154**, 107–163 (2021)
40. Olszewski, R.T.: Generalized feature extraction for structural pattern recognition in timeseries data, Report (2001)

PRISM & PGRN Research, Discoveries, and Emerging Challenges [Domains]

On the Optimization of Systems Using AI Metaheuristics and Evolutionary Algorithms

Itshak Tkach(✉) and Tim Blackwell

Department of Computing, Goldsmiths, University of London, New Cross, London SE14 6NW, UK
{itkach001,t.blackwell}@gold.ac.uk

Abstract. In this chapter, evolutionary computation techniques, algorithms and research are presented for the optimization and allocation problems. Several aspects of continuous optimization, systems security and supply networks (SN) are illustrated. The real-life optimization and security problems in systems, automation, SN and law enforcement are NP-hard optimization problems, thus evolutionary algorithms (EA) that employ metaheuristic methods are useful for solving them. EA gain significant interest in recent years, and this chapter summarizes some of the advances in that field and then summarizes their applications for real-life problems. The rest of this chapter is organized as follows. First, the introduction of the developments of nature-inspired EAs and metaheuristics is described. Then the working principles of genetic algorithms (GA), swarm intelligence, and other nature-inspired optimization algorithms are given. Next, the overview of the various applications that were solved and optimized by EAs is presented. The reader of this chapter will be familiar with the following topics: The state-of-the-art AI algorithms and techniques and their working principle. The way to harness AI for optimization and finding optimal solutions. Controlling and optimizing a collaborative system in real-time while addressing several tasks in a complex environment

1 Introduction

Evolutionary algorithms are nature-inspired techniques that are a subfield of computational and artificial intelligence (AI). EA uses an evolutionary process within a computer to address complex engineering problems that require an exhaustive search that traditional algorithms are unable to solve in a finite amount of time. Part of EA is swarm intelligence (SI) algorithms, that employ the emergent collective intelligence of simple agents to solve complex problems (Fig. 1).

EA has been developed since the early 1960s. Evolution strategies (ES) were introduced by Ingo Rechenberg in 1973 [22,23]. Evolutionary programming (EP) was first used by Gary Fogel to use simulated evolution as a learning process

C.-Y. Huang and S. W. Yoon (Eds.): ICPR1 2021, ACES 14, pp. 253–271, 2023.
https://doi.org/10.1007/978-3-031-44373-2_15

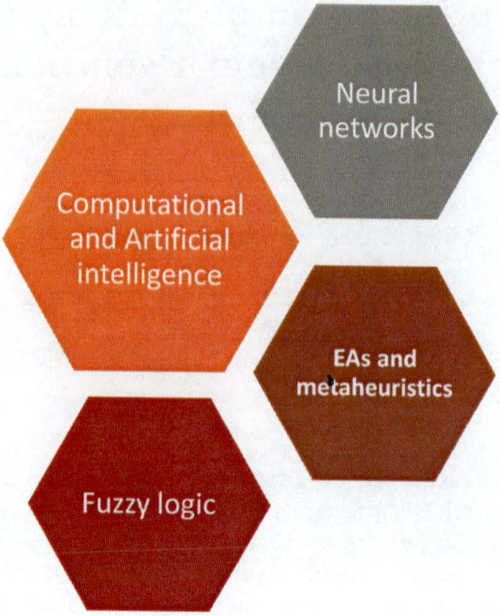

Fig. 1. Computational intelligence and AI disciplines; Neural networks, Fuzzy logic, Evolutionary algorithms and metaheuristics

using finite-state machines [6]. Genetic algorithms (GA) became popular through the work of Holland with studies of cellular automata and formulation of the next generation [10]. Genetic programming (GP) is an extension of GA for program evolution which was introduced by John Koza in 1990 [16]. Simulated annealing (SA) is a metaheuristics that was developed by Kirkpatrick in 1983 [15]. It is a probabilistic algorithm inspired by the annealing of metals in metallurgy. Ant colony optimization (ACO) proposed by Marco Dorigo in 1992 is a probabilistic optimization technique based on the foraging behaviour of ants seeking the shortest path between their colony and a source of food [5]. Kennedy and Eberhart introduced the particle swarm optimization (PSO) metaheuristic in 1995 [14]. PSO is a computational method for optimising problems by using a population of particles moving around the search space according to the principle of particle position and velocity. The bees algorithm (BA), formulated by Pham et al., is based on the foraging behaviour of honey bees [20]. The algorithm exploits global explorative search with local exploitative search. The artificial bee colony (ABC) algorithm is a metaheuristic introduced by Karaboga as an extension of BA, has three types of bee, employed bee, onlooker bee and scout bee [11]. Heterogeneous Distributed Bees Algorithm (HDBA) is a multi-agent metaheuristic algorithm initially introduced by Tkach [30]. It enables solving of combinatorial optimization problems with multiple heterogeneous agents that possess different capabilities and performances. The main AI techniques are summarized in Table 1. Based on the Google Scholar Metrics (2022), the top 6 publications are IEEE Congress on Evolutionary Computation, Swarm and Evolutionary Computation,

Table 1. Timeline of AI techniques

1960	Random Search	
	ES	EP
1970	GA	
1980	SA	Tabu Search
1990	GP	ACO
	PSO	
2000	BA	ABC
2010	HDBA	

Conference on Genetic and Evolutionary Computation, Evolutionary Computation, International Conference on Applications of Evolutionary Computation, International Conference on Genetic and Evolutionary Computing (ICGEC).

2 The Main Algorithms and Their Working Principles

This section describes the main AI evolutionary and metaheuristic algorithms and techniques. The working principles of GA, ACO, PSO, DE and HDBA are illustrated.

2.1 Genetic Algorithm

A genetic algorithm (GA) is an evolutionary algorithm that is used for solving optimization problems with non-polynomial complexity. It was introduced by John Holland in 1975 [10], as a search optimization algorithm based on the mechanics of the natural selection process. The basic concept of this algorithm is to mimic the concept of the survival of the fittest. The evolution usually starts from a population of randomly generated individuals consisting of legitimate candidate solutions, and is an iterative process, with the population in each iteration called a generation. In each generation, the fitness of every individual in the population is evaluated; the fitness is usually the value of the objective function in the optimization problem being solved. The more fit individuals are stochastically selected from the current population, and each individual's genome is modified by crossover (a genetic operator used to combine the information of two individuals to generate a new one) and mutation to form a new generation. The new generation of candidate solutions is then used in the next iteration of the algorithm. Commonly, the algorithm terminates when either a maximum number of generations has been produced, or a satisfactory level of the objective function has been reached.

Algorithm 1. GA pseudocode

```
Initialise a population of n chromosomes
while not end of mission do
    Compute fitness of each chromosome in population
    repeat
        Select a pair of parents with replacement
        Apply crossover and create offspring
        Mutate offspring
        Add offspring to new population
    until n offspring have been created
    Replace the current population with the new population
end while
```

2.2 Ant Colony Optimisation

An ant colony swarm algorithm called Ant Colony Optimization (ACO) was developed [5]. It was inspired by the foraging behaviour of ants in nature which use a pheromone to mark the path on the ground, such as the path from a food source to the nest. Ants can sense the pheromone concentration laid on a path and select the path to the food location with the highest concentration with a higher probability. Similarly, in ACO, each ant is an agent that chooses a solution with a probability that is a function of the performance and the amount of pheromone laid. To force an agent to make legal selections, already visited transitions are disallowed until a cycle is completed (this is controlled by a tabu list); when it completes a cycle, the trail substance on each path visited is updated. After several generations, the algorithm converges to the best path, which represents the optimum or suboptimal solution to the problem.

$$v_i(t+n) = \rho \times v_i(t) + \Delta v_i \geq v_0 \tag{1}$$

where v_i is the intensity of ant i's trail at time t. n is the number of agents (every n iterations - cycle, each ant completed a cycle). ρ is a coefficient such that $(1-\rho)$ represents the evaporation of trail between time t and $t+n$, v_0 is the threshold.

$$\Delta v_i = \sum_{k=1}^{m} \Delta v_i^k \tag{2}$$

where Δv_i is the quantity per unit of trail laid on agent i by the k-th ant between time t and $t+n$; it is given by:

$$p_i^k(t) = \begin{cases} \frac{v_i^{\psi}(t) \times \xi_i^{\omega}}{\sum_{k \in \text{allowed}_k} v_i^{\psi}(t) \times \xi_i^{\omega}} & \text{if } i \in \text{allowed}_k \\ 0 \text{ otherwise} \end{cases} \tag{3}$$

where, visibility ξ_i is quantity S, allowed$_k$ is the set of agents not in tabu list, ψ and ω are parameters that control the relative importance of trail versus visibility.

The stop criterion is given by a threshold that quantifies the desired performance of the agents for a given task. The optimal allocation is given by:

$$V_{OS} = \sum_{k \in \text{ tabu list}} S_j \geq \text{ threshold} \tag{4}$$

where j is the index of agents in the tabu list.

Algorithm 2. ACO pseudocode

```
t = 0
Δv_i^k = 0
Place N ants on agents
while not end of mission do
    Clear tabu list
    Upon task arrival calculate then new task priority f_i
    Update task queue based on priorities
    while termination condition not met do
        if new task arrived then
            break
        end if
        for k = 1 to N do
            Choose the agent i with probability p_i^k(t)
            Insert agent i into tabu list
            Apply trail update
        end for
    end while
end while
```

2.3 Particle Swarm Optimisation

Particle Swarm Optimisation (PSO) is a population metaheuristic algorithm which exhibits a form of swarm intelligence based on social communication. PSO, despite its popular classification as such, is not an evolutionary algorithm.

In its original and most popular formulation, PSO attempts to optimise real-valued objective functions, i.e. to solve

$$\arg\min f(x) \text{ for x } \in X \tag{5}$$

where X is a continuous region of $\mathbb{R}^D$. Swarm individuals are known, by virtue of their simple dynamic properties, as particles. Swarm intelligence is implemented by an inter-particle communication strategy: promising positions in X are communicated via a social network which may be fully connected (the global topology) or by a local topology in which particles only have access to a limited number of neighbours.

Communicated information is historical in the sense that each particle has access to the best position, envisaged as a particle 'memory', ever visited by a social neighbour. The slow spread of information in a local network enhances search during early iterations and, in consequence, local networks are beneficial for complex multi-modal problems. Global communication favours fast convergence and unimodal objective functions [2].

Formally, particles i in a canonical PSO swarm [12,21,25] of N particles have dynamical variables x_i, v_i, representing position and velocity in the search space X, and an internal memory p_i of their best achieved position. The dynamical update rule adjusts particle position by adding an acceleration to velocity. Acceleration is a sum of stochastic attractions to particle memory and to the best memory of a social neighbour, n_i, and friction. The velocity is added to position in order to complete the dynamical update:

$$\begin{aligned} v_i^{t+1} &= wv_i^t + c_1 u_1 \circ (n_i^{t+1} - x_i^t) \\ &+ \ c_2 u_2 \circ (p_i^{t+1} - x_i^t) \\ x_i^{t+1} &= x_i^t + v_i^{t+1} \end{aligned} \tag{6}$$

Here, $u_{1,2} \sim U(0,1)$ are uniform random variables in $[0,1]^D$ and $\circ$ is the Hadamard (entry-wise) product. The 'inertial weight', w, and acceleration coefficients $c_{1,2}$, are two arbitrary (but constrained) positive real parameters chosen to balance convergence and exploration and t labels iteration.

In synchronous updating, iteration $t+1$ begins by updating all memories:

$$p_i^{t+1} = \arg\min^* \left(f(x_i^t), f(p_i^t)\right) \tag{7}$$

where min* returns the first member of the list in the case of non-uniqueness. The iteration is completed by updating all N positions and velocities according to Eq. 6. New information is immediately available to neighbours (asynchronous updating) if Eq. 7 is applied immediately after each position update.

Algorithm specification is completed with a consideration of boundary conditions. Particles are initialised with positions uniform randomly picked in X and with $p_i = x_i$. Initial particle velocity is arbitrary. For example, velocities may be set to zero or chosen as the difference of two random positions in X. The termination condition may be a preset number of function evaluations and/or an acceptable error. Spatial conditions are also arbitrary: particles my be placed on the boundary should they fly outside X, or they might be allowed to continue to move, without evaluation, in $\mathbb{R}^D \setminus X$.

There is no smoothness requirement on f because memory and dynamic update make no reference to function gradient. PSO therefore has a wider applicability than conventional optimisation algorithms which rely on gradient information.

2.4 Differential Evolution

Differential evolution (DE), like PSO, concerns a population of individuals in a region $X \in \mathbb{R}^D$. The aim is again to find the global minimum of a real valued objective function (which need not be smooth). The individuals are not, however, conceived as dynamical particles with velocities and subject to accelerations, but are understood as individuals whose chromosome $x \in X$ is subject to evolution: a cross-over operation generates a new individual ('offspring') from three existing individuals ('parents'). The offspring replaces the primary parent should it prove equal or better ('fitter').

Differential evolution exists in a wide variety of forms. The following specifies the DE/best/1 version. This manifestation is considered competitive and robust [4].

Each DE iteration begins with a determination of the best individual g (similar to the PSO global social topology). Then, for each individual i - the primary parent - two further parents, j and k, are selected such that $i \neq j \neq k$. A random component $r \in \{1, 2 \ldots D\}$ is also selected.

An offspring y_i is generated by applying, for each component $d \in [1, 2, \ldots D]$ of x_i, the conditional point cross-over:

$$\begin{aligned} &\text{if } u \sim U(0,1) < C_R \text{ or } d == r \\ &\quad y_d = g_d + F(x^t_{jd} - x^t_{kd}) \\ &\text{else} \\ &\quad y_d = x^t_{id} \end{aligned} \tag{8}$$

where the parameters $C_R \in [0, 1]$ and $F \in (0, 2]$ are known as the 'cross-over rate' and the 'differential weight'. The random component r ensures that the offspring differs from i. Then, after each component of y has been set, the offspring conditionally replaces i:

$$x^{t+1}_i = \arg\min^* \left(f(y), f(x^t_i)\right)$$

As with PSO, boundary conditions in space and time must be specified.

DE is considered to challenge PSO in optimisation efficacy over common sets of benchmark problems. It has been extensively researched since its inception in 1995 and exists in a bewildering variety of forms [19], including control parameter self-adaptation schemes [7] and local search enhancements for large scale optimisation [18].

2.5 Simulated Annealing

Simulated Annealing (SA) algorithm is a heuristic algorithm, which simulates the process of annealing in metallurgy introduced by Kirkpatrick et al. [15]. The method models the physical process of heating a material and then controlled cooling to alter its physical properties, thus minimizing the system energy. Likewise, SA accepts a worse solution than the current solution with some probability, so it has more possibility to jump out of the local optimal solution and find the global optimal solution. SA starts from some initial solution s and then at each step, a solution $s' \in N(s)$ is generated. If s' improves on s, it is accepted; if s' is worse than s, then s' is accepted with a probability which depends on the difference in the function value $f(s)$ - $f(s')$, and on a parameter T, called temperature. The temperature T is lowered, reducing the probability of accepting solutions worse than the current one. The probability p_{accept} to accept a solution s' is defined as:

$$p_{\text{accept}}(s, s', T) = \begin{cases} 1, \text{ if } f(s') < f(s) \\ \exp \frac{f(s)-f(s')}{T}, \text{ otherwise} \end{cases} \tag{9}$$

Algorithm 3. SA pseudocode

```
Create initial solution s
Initialize temperature T
repeat
while not end of mission do
    Generate a random transition from s to s'
    if f(s') < f(s) or exp((f(s)-f(s'))/T) > random[0,1] then
        s = s'
    end if
end while
Reduce temperature T
until no change in f(s)
return s
```

2.6 Heterogeneous Distributed Bees Algorithm

HDBA is a swarm intelligence algorithm that enables to solve combinatorial optimization problems that include multiple heterogeneous agents that possess different capabilities and performances. It was developed by Tkach et al. [32]. HDBA uses a probabilistic technique taking inspiration from the foraging behaviour of bees.

In this algorithm, each agent is represented as a 'bee', and agent utility, solving p_{ik}, is defined as a probability that the agent k is allocated to the task i. When an agent receives information about the available tasks it calculates its performance for that task. Agents communicate among them selves by broadcasting the position and the priority of the task. The agent's utility function is updated accordingly, and depends on the priority of the task, the distance from the task and the agent's performance on that task:

$$p_i^k(t) = \frac{F_i^\alpha \times \frac{1}{D_{ik}}^\beta \times V_{ik}^\gamma}{\sum_{j=1}^{M} F_j^\alpha \times \frac{1}{D_{jk}}^\beta \times V_{jk}} \text{ if } \Delta_i > 0 \tag{10}$$

where α, β and γ are control parameters that bias the importance of the priority, the distance and the agent's performance respectively (α, β, γ ¿ 0; α, β, $\gamma \in \mathbb{R}$). The probabilities p_{ik} are normalized, and it is easy to show that:

$$\sum_{i=1}^{M} p_{ik} = 1 \tag{11}$$

The HDBA decision-making mechanism uses a wheel-selection rule, where each agent has a probability with which it is allocated to the task from a set of available tasks. Once all the agents' utilities are calculated, each of them selects a task by "spinning the wheel". The HDBA function of agents' utility is developed for the case of heterogeneous agents with different performances. This setting is assumed to improve system performance as it can correlate the agents' utility function with the value of their performances.

Algorithm 4. HDBA pseudocode

```
t = 0
Place N bees on agents
while not end of mission do
    Upon task arrival calculate then new task priority F_i
    Calculate distances of agents from task D_ik
    Calculate performances of agents on tasks V_ik
    for k = 1 to M do
        for k = 1 to N do
            Calculate probabilities for each agent p_ik
        end for
        Apply wheel-selection rule
        Allocate agents according to the selection
    end for
end while
```

3 Summary of Implementation to Optimization Problems

This section presents a brief summary of optimization by AI algorithms conducted on three real-world problems. The presented problems include resource allocation, supply networks security and law enforcement optimization.

3.1 System Resource and Task Allocation

Systems typically consist of multiple resources and tasks that need to be handled with different performances. An example of a sensory system was described in [29]. In this system the performance of the sensor is defined a priori based on the sensor's features, namely detection distance, resolution, and response time. Each sensor can only be allocated to one task at any given time and can be reallocated to another task at any moment. The priority of the task is an application-specific scalar value, where a higher priority value represents a task that has higher importance and must be attended to faster than other tasks. Higher priority tasks also have a higher benefit for completing them. Examples of such tasks include surveillance (gathering information on desired objects), security monitoring (preventing theft of goods and threats, and fire monitoring (forest fire detection and protection), among many others. This system must deal with the real-time detection of unpredictable, unknown tasks arriving at unknown times and locations. The task occurrence is dynamic and unpredictable with different levels of importance of each task and must be detected as fast as possible. The sensors must be allocated to the tasks as fast as possible. The goal is to allocate each sensor an appropriate task at an appropriate time (Fig. 2) and to ensure all tasks are completed in minimum time.

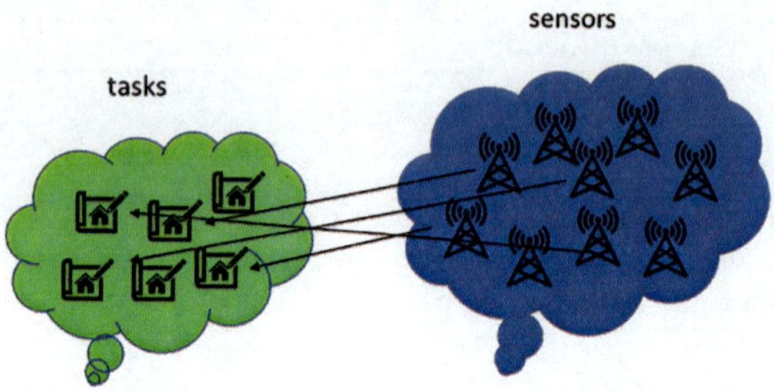

Fig. 2. Distributed multi-sensor system allocation scheme

When there are several tasks that require the same sensors, the allocation depends on the sensors' availability and performance, the physical distance of sensors from the tasks, and the priorities of individual tasks. This is a NP-hard problem that was optimized by AI algorithms. An example of formulating a possible solution to this problem is by using HDBA:

$$p_i^k(t) = \frac{F_i^\alpha \times \frac{1}{D_{ik}}^\beta \times V_{ik}^\gamma}{\sum_{j=1}^{M} F_j^\alpha \times \frac{1}{D_{jk}}^\beta \times V_{jk}} \text{ if } \Delta_i > 0 \tag{12}$$

where F, S and V are the tasks priorities, distances of sensors from the tasks and sensors performances respectively.

3.2 Optimization of Supply Network Security Using Sensory Systems

Supply networks are systems responsible for moving goods, products and services. They consist of a complex network of interrelated entities, including suppliers, manufacturers, retailers, and customers. As part of globalization, supply networks have become more vulnerable to security risks that include facility, information, and cargo security among others. To meet the demands of securing supply networks in today's environment, a comprehensive, technological, and integrated security approach is required to enable monitoring threats of dynamic and unpredictable nature. A sensory system was designed to provide security in supply networks and task administration protocols (TAPs) were used to overcome uncertainties and disturbances (i.e., failures, conflicts in priorities) [31]. TAPs consist of four protocols, one of which applies a bio inspired HDBA for sensors allocation and real time detection of targets for security (Fig. 3). The role of this algorithm was to efficiently allocate high number of sensors to upcoming targets to detect as much targets as fast as possible. Optimal sensors availability related to their monetary cost in the system was achieved by deployment of redundant sensors. Employing TAPs which use HDBA allowed dynamic, real-time allocation of distributed sensors to targets when they occur.

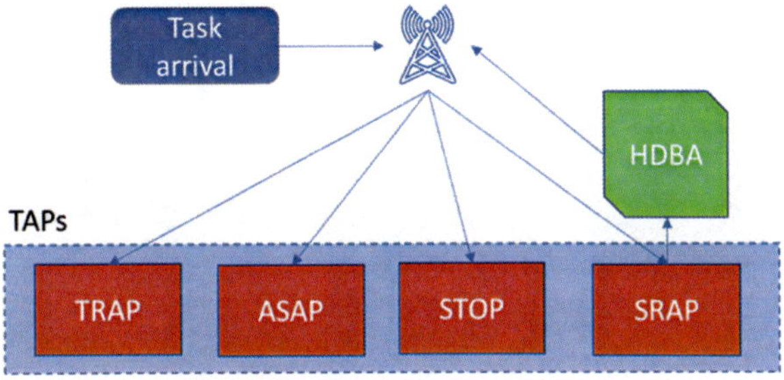

Fig. 3. System procedure scheme

The system was designed as a dual-layer network; a process layer and a monitoring layer. The process layer consists of multiple sensors and is responsible for allocating them to complete tasks. As multi-sensor systems are vulnerable to some risks and problems, the monitoring layer functions at a higher level than the process layer and is used to monitor those problems in the process layer by applying TAPs to handle them. The first problem has to do with tasks that require very long attendance times. These tasks may occupy the sensors allocated to them for long time durations, thus delaying the handling of other tasks by those sensors. In this overloading problem, the sensors are overloaded with a portion of tasks which delays them. A time out policy, which recognizes if a sensor is experiencing a delay while other tasks are waiting for execution, can overcome this problem (STOP). The second problem that the sensory system must overcome relates to tasks that may have much higher priority over other tasks or the same priority as other tasks. A similar problem may arise when, due to malfunction or recognition problems, tasks are perceived by the system to be of a higher priority than they should be. Such tasks may unnecessarily occupy some sensors. In this deception problem, sensors are occupied and delayed by some tasks which may be neither urgent nor important. The system may need to reprioritize those sensors which are unable to perform other tasks that must be handled quickly, ahead of less urgent tasks, due to being close to their deadline

(TRAP). The third problem that the sensory system must overcome is related to a failure of a portion of sensors in the system. In this tampering problem, sensors may fail due to internal properties (e.g. hardware reliability) and external reactions (e.g. weather conditions). The system should be able to monitor these problems to ensure that system performance is unharmed (ASAP). An example of a dual-layer logistic system with distributed sensors to monitor security is illustrated in Fig. 4, and the time-out value could be calculated as follows:

$$to = \mu_{ct} + 2\sigma_{ct} \tag{13}$$

where μ_{ct} is current task mean completion time, and σ_{ct} is the standard deviation of current task completion time.

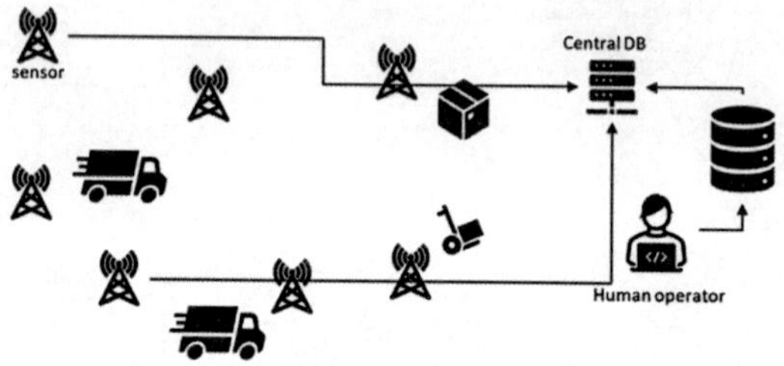

Fig. 4. An illustration of a dual-layer logistic system with distributed sensors to monitor security

This example presented a dual-layer system for applying task allocation algorithm and task administration protocols for efficient target detection for supply networks security that can deal with problems in the allocation process and a protocol for analysing the status of sensors to modify the allocation if necessary. It repeatedly identifies the current state of the system and takes proper actions to deal with allocation problems and improve system performance.

3.3 Optimization of a Law Enforcement Allocation Problem

AI algorithms were implemented by Tkach and Amador [27] for solving a realistic Law enforcement problem by employing HDBA, FMC^{TAH+} and SA. This was a multi-agent problem with heterogeneous skills that work together on tasks to share the workload and improve response time (Fig. 5). The workload associated with each task, indicating the amount of work to be completed for the incident to be processed was different. The simplified total utility derived for performing a task is:

$$U = DL * Cap - Pen \tag{14}$$

where DL is a soft deadline function, Cap is the capability function of the agents performing the task, and Pen is the penalty for task interruption. The algorithms allocated agents to dynamic tasks whose locations, arrival times, and importance levels are unknown a priori. The employed methods were compared to different performance measures that are commonly used by law enforcement authorities. This evaluation was shown to be effective in allocating dynamic tasks to heterogeneous police agents.

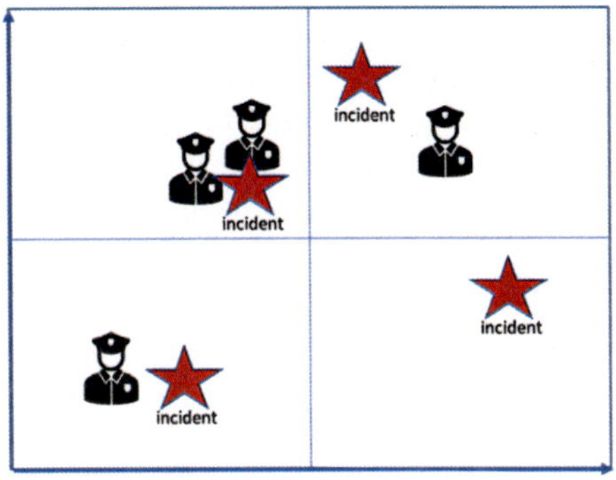

Fig. 5. An example of police officers to incidents/task allocation problem

4 Future Research Perspectives

This section outline three current research directions that aim at producing a controllable laboratory for optimiser study (the barrier tree benchmark), a principled methodology of algorithm comparison (the deployment of separate problem training and tests sets) and a novel swarm intelligence algorithm based on a population of downhill walkers.

4.1 Barrier Tree Benchmark

Optimisation algorithms are frequently assessed in comparative studies. These studies pit algorithm against algorithm over a set of problems that have either been selected or hand-crafted by the authors, or belong to a 'standard' collection known as a benchmark: for example, the CEC and BBOB real-valued benchmarks [9,17]. Although these comparative studies are indicative of algorithm worth, the methodology has several limitations.

Benchmarks, whether standard or assembled by an author from a pool of functions, rely on complex definitions. These functions provide examples of certain problem characteristics such as multi-modality and non-separability, which have been assumed to challenge optimisation algorithms. This approach is valid if the salient problem characteristics are known, if example problems are available for every function class, and if these examples are sufficiently representative.

However current function groupings are arbitrary, test functions often lie in several classes and it is not known if examples are representative of the associated optimisation difficulty of a particular class.

Furthermore, the properties of many exemplar functions extant in the literature, for example the number of local optima, the position and value of the global optimum etc. are often unknown or, occasionally, even in dispute [24].

Despite the division of benchmarks into problem types, algorithm performance is usually assessed by comparative performance over the entire suite. For example, a common performance measure is the number of times a favoured algorithm beats rivals over all problems in the benchmark. Even if trials are subject to rigorous significance testing, the meaning of good performance over an assembly of functions of differing characteristics is unclear: there is no guarantee that cumulative benchmark performance will generalise to real world scenarios.

In summary, current function collections, whether existing in benchmarks or self-assembled, are arbitrary and unstructured. The arbitrariness and mathematical complexity of existing benchmarks nullifies any principled investigation of algorithm performance against problem type and algorithm versus algorithm comparison.

The Barrier Tree Benchmark (BTB) has been proposed as an answer to these objections [28]. The underlying idea is the construction of functions with known topographies. The mathematical properties of BTB member functions is understood and series of controlled algorithm trials over structured test functions can be implemented.

The barrier tree is the fundamental unifying structure of the BTB. A barrier tree is a graphical representation of the critical points (minima and saddles) of a function. This representation arises in evolutionary biology where the objective function is known as a fitness landscape [26]. The fitness landscape, in analogy with a natural landscape of nested valleys, is conceived as a topography of funnels and basins. A basin is a connected region is which all downhill paths (the converse of the adaptive walk in evolutionary biology [8]) starting at any point in the basin lead to a single minimum. (For simplicity, we limit this presentation to landscapes without 'neutral' areas in which a connected region shares a single objective value.) Funnels are connected regions where downhill paths lead to more than one basin.

The barrier tree is a topological representation of landscape topography. It specifies critical values (depths and saddle values) but does not specify basin and funnel extent. The BTB completes the specification by furnishing basins and funnels with unimodal basis functions. These basis functions are, in principle, any unimodal function; for example, symmetrical conical functions or highly conditioned ellipsoids.

The BTB currently exists only as a framework; work is underway to deliver a software package. A preliminary study of the application of the BTB considers one or two basins within a single funnel [28]. Differential evolution was compared to local and global PSO for configurations of varying optima depths and relative sizes. Although these are the simplest conceivable landscapes, the study threw up some unexpected results: optimal DE and PSO parameter settings departed markedly from recommended values in the literature; DE is preferred over PSO for large optimal depths whereas PSO is superior when the basin depths are more similar; PSO performance degrades in dimensions above and DE performance holds steady in all dimensions up to 100.

These findings highlight the ability of the BTB to categorise algorithm performance. The study shows that neither algorithm is an overall winner on the simplest landscapes; a more nuanced matching of algorithm to landscape property has been achieved.

4.2 Parameter Tuning Methodology

The discussion in Sect. 4.1 highlights an overlooked issue concerning the methodology of algorithm comparison. Optimisation algorithms are parameterised. In the absence of any theory, parameter settings are arbitrary. For example, PSO swarm size and DE cross-over rate must be set by the researcher. Values are often extracted from the literature. In comparative tests where the merits of a novel algorithm A are being promoted in comparative trials, the parameters of A are typically tuned to provide good performance on the test set (which may have been selected by the researcher) or a specific benchmark. These trials are clearly unfair unless the comparison algorithms have also been optimised on the chosen test set. And even if optimal versions of the comparison algorithms are lined up, there is a further issue of generalisation. There is no guarantee that an optimal algorithm on a test set of, say, 30 functions, will continue to be superior on a 31st function. Moreover, there is no guarantee that good performance on any artificial test set will extend to success on real world problems.

The generalisation failure of tuned algorithms has been recognised in the machine learning (ML) community. The danger of overfitting a machine learning statistical model to a single test set is countered by preparing training, verification and test sets. Model hyperparameters are tuned by comparing trained models with performance on the verification set; the optimised model is then tested once on the reserved test set. The unseen test set represents future unknown problem instances. Even this process is not infallible because information leaks from the validation set into the model during the training-validation cycle [3].

The ML methodology does not directly translate to the problem of algorithm parameter tuning because ML training, validation and test set data is similar whereas there is a huge diversity of mathematical functions. For example, a data set of cat and dog images or a set of digitised handwritten digits consists of relatively heterogeneous data but the functions in the CEC benchmark range from simple unimodal problems to complex multi-function compositions. Future images of a cat or dog, or a handwritten digit, are not expected to be too dissimilar to those in the the original data set but objective functions can take many forms. And, although image space is extremely large, models are usually trained with tens of thousands exemplars. ML succeeds because the objectives are narrow: a model trained on pet images is not expected to classify household appliances. It would not be feasible, with current computational resources, to tune optimisation algorithms on similarly sized training sets.

The barrier tree benchmark suggest ways of transferring the ML methodology to optimisation algorithm tuning because the benchmark is structured. Although the BTB consists of a vast set of functions, limited only by floating point precision, principled subdivisions into problem classes can be achieved. For example, the set of all single funnel functions with one hundred or fewer optima, or the set of all bimodal functions with depths and basin radii in defined limits. Representative training and test examples can then be picked from these restricted classes. Algorithm parameters would be optimised on the training set and fair comparisons can be performed with the unseen test set.

To make these remarks more precise, consider tuning PSO on bimodal topographies.

A BTB bimodal function f for an D-dimensional search space $X = \mathcal{F}_{12} \cup \mathcal{B}_1 \cup \mathcal{B}_2$, comprising a funnel $\mathcal{F}$ and two basins $\mathcal{B}_{1,2}$, is defined:

$$f(x) = \begin{cases} f_{1,2}(x) & \text{if } x \in \mathcal{B}_1 \text{ or } \mathcal{B}_2 \\ f_{12}(x) & \text{otherwise } (x \in \mathcal{F}) \end{cases}$$

where $f_{1,2}$ are functions describing the topography of the optimal basins (for example, cone functions, $f_{1,2} = |x - x_{1,2}| + d_{1,2}$, of depths $d_{1,2}$ and optimal positions $x_{1,2}$) and a unimodal function f_{12} is chosen for $\mathcal{F}$ such that (i) $f_{12}(\mathcal{F}) > f_{1,2}(B_{1,2})$ and (ii) f_{12} decreases throughout $\mathcal{F}$. $B_{1,2}$ can be chosen to be D-balls of radii $r_{1,2}$. A specific instance of f is specified by picking $r_{1,2}, d_{1,2}$ and $x_{1,2}$. Suppose a set of distinct BTB bimodal f-instances has been prepared. PSO parameters N (swarm size), w (inertia weight) and $c_{1,2}$ (acceleration parameters), Eq. 6, would then be tuned on 80% of the instance set and tested for generalisation on the remainder.

4.3 Swarm Walker

Basins and funnels are defined by downhill paths - all points in a basin lead to the basin optimum, and all points in a funnel lead to two or more basins. The concept is suggestive: can we exploit downhill paths in a novel optimisation algorithm? The downhill path is continuous, so we would have to imagine a *downhill walker* that progresses by steps of a finite length. A single walker would move downhill until it reached a basin bottom which may be sub-optimal. A single walker would not be expected to succeed on multimodal landscape with many basins, but a population of walkers who collaborate to find promising paths might provide the basis for a feasible optimisation algorithm. This idea is the basis of a proposed Swarm Walker algorithm.

Another driver behind the development of the Swarm Walker algorithm is the need to find a clean model that isolates the role of swarm intelligence from the rule that specifies individual movement. PSO is notoriously hard to analyse because the dynamics, which combines random attractive forces, particle memory and a social network, is involved and complicated. There have been attempts to simplify PSO particle movement, for example the so-called bare bones particle swarm [13] but performance had to be enhanced with extra mechanisms (e.g. [1]). A downhill walker, on the other hand, does not need intricate dynamic or evolutionary mechanisms. The walker could simply sample the local terrain until it finds a downhill spot, and then move there. Swarm intelligence can be invoked by biasing a walker's search towards neighbours (social or spatial) without the invocation of a memory or particle velocity.

A basic Swarm Walker algorithm is in development and initial results on single funnel-many basin landscapes reveal comparable performance to PSO and DE, and better performance on bi-modal landscapes where the optima values are close. The algorithm implements a swarm of N downhill walkers. Each walker attempts to move downhill with an adjustable step length λ by picking a point from a hypersphere of radius λ centred on a point between the walker and its best neighbour; the walker moves to the trial position if the trial is downhill from its current position. In order to adapt λ to the local terrain, the step length is reduced by a factor in $(0, 1)$ after a preset number of unsuccessful trials.

The algorithm is relatively unadorned and several refinements await exploration. As with PSO, global and local social networks can be instigated, or spatial neighbours, as in bird flocks, could be tested. A downhill walker might occasionally walk uphill with a rule similar to simulated annealing: an uphill step is chosen with decreasing probability. Walkers might (in analogy to ant colony optimisation) be attracted to other walkers paths. It is hoped that by progressively combining features of the main paradigms such as PSO (social network), DE (spatial position) and ACO (stigmergy as implemented by evaporating pheromone trails), an understanding might be reached on the interactions between mechanisms and their role in an 'intelligent' population. The development of a powerful optimiser would be a useful spin-off.

5 Summary

AI techniques that include evolutionary algorithms and metaheuristics were presented and explained in regard to systems collaboration, integration and optimization. The methods and systems described in this chapter provide several examples for the possible practical use of AI. The examples include optimization tools, multi-sensory systems and supply network security methods with the ability to optimize complex systems and allocate agents to tasks in real-time. The structure of presented algorithms and protocols can be used for different scenarios and problems, but the parameters should be adjusted and tuned for a specific case study and be optimized based on the objective of the specific system for optimal results. System designers can determine the best method for the specific problem which is essential for systems that require continuous operation, e.g., monitoring systems, lifesaving systems, complex systems with different types of agents and complex formations. Finally, some future research perspectives for optimization were outlined.

References

1. Blackwell, T.: A study of collapse in bare bones particle swarm optimization. IEEE Trans. Evol. Comput. **16**(3), 354–372 (2011)
2. Blackwell, T., Kennedy, J.: Impact of communication topology in particle swarm optimization. IEEE Trans. Evol. Comput. **23**(4), 689–702 (2019)
3. Chollet, F.: Deep Learning with Python. Simon and Schuster, New York (2021)
4. Das, S., Suganthan, P.N.: Differential evolution: a survey of the state-of-the-art. IEEE Trans. Evol. Comput. **15**(1), 4–31 (2011). https://doi.org/10.1109/TEVC.2010.2059031
5. Dorigo, M.: Optimization, learning and natural algorithms. Ph.D. thesis, Politecnico di Italy, Milano (1992)
6. Fogel, L.J.: Biotechnology: Concepts and Applications. Prentice-Hall, Upper Saddle River (1963)
7. Georgioudakis, M., Plevris, V.: A comparative study of differential evolution variants in constrained structural optimization. Frontiers in Built Environment **6**, 102 (2020)
8. Gillespie, J.H.: Molecular evolution over the mutational landscape. In: Evolution, pp. 1116–1129 (1984)
9. Hansen, N., Auger, A., Ros, R., Mersmann, O., Tušar, T., Brockhoff, D.: Coco: a platform for comparing continuous optimizers in a black-box setting. Optim. Methods Softw. **36**(1), 114–144 (2021)
10. Holland, J.: Adaptation in Natural and Artificial Systems. University of Michigan Press, Ann Arbor (1975)
11. Karaboga, D.: Artificial bee colony algorithm. Scholarpedia **5**(3), 6915 (2010)
12. Kennedy, J.: Small worlds and mega-minds: effects of neighborhood topology on particle swarm performance. In: In: Proceedings of the 1999, Congress of Evolutionary Computation, vol. 3, pp. 1931–1938. IEEE Press (1999)
13. Kennedy, J.: Bare bones particle swarms. In: Proceedings of the 2003 IEEE Swarm Intelligence Symposium, SIS 2003 (Cat. No. 03EX706), pp. 80–87. IEEE (2003)
14. Kennedy, J., Eberhart, R.: Particle swarm optimization. In: Proceedings of the 1995 IEEE International Conference on Neural Networks, pp. 1942–1948 (1995)
15. Kirkpatrick, S.: Optimization by simulated annealing. Science **220**(4598), 671–680 (1983)
16. Koza, J.R.: Genetic programming: A paradigm for genetically breeding populations of computer programs to solve problems, vol. 34. Stanford University, Department of Computer Science Stanford, CA (1990)
17. Liang, J.J., Qu, B., Suganthan, P.N., Hernández-Díaz, A.G.: Problem definitions and evaluation criteria for the CEC 2013 special session on real-parameter optimization. Comput. Intell. Lab. Zhengzhou Univ. Zhengzhou, China Nanyang Technol. Univ. Singapore, Tech. Rep. **201212**(34), 281–295 (2013)
18. Molina, D., LaTorre, A., Herrera, F.: Shade with iterative local search for large-scale global optimization. In: 2018 IEEE Congress on Evolutionary Computation (CEC), pp. 1–8. IEEE (2018)
19. Pant, M., Zaheer, H., Garcia-Hernandez, L., Abraham, A., et al.: Differential evolution: a review of more than two decades of research. Eng. Appl. Artif. Intell. **90**, 103479 (2020)
20. Pham, D., Ghanbarzadeh, A., Koc, E., Otri, S., Rahim, S., Zaidi, M.: The bees algorithm. Technical Note, Manufacturing Engineering Centre, Cardiff University, UK, pp. 44–48 (2005)

21. Poli, R., Kennedy, J., Blackwell, T.: Particle swarm optimization: an overview. Swarm Intelll. **1**, 33–57 (2007)
22. Rechenberg, I.: Evolutionsstrategie – optimierung technischer systeme nach prinzipien der biologischen evolution. Ph.D. thesis (1973)
23. Schwefel, H.P.: Numerische optimierung von computer-modellen mittels der evolutionsstrategie.(Teil 1, Kap. 1-5). Birkhäuser (1977)
24. Shang, Y.W., Qiu, Y.H.: A note on the extended rosenbrock function. Evol. Comput. **14**(1), 119–126 (2006)
25. Shi, Y., Eberhart, R.: A modified particle swarm optimizer. In: Congress on Evolutionary Computation, pp. 69–73 (1998)
26. Stadler, P.F.: Fitness landscapes. In: Lässig, M., Valleriani, A. (eds.) Biological Evolution and Statistical Physics. Lecture Notes in Physics, pp. 183–204. Springer, Heidelberg (2002). https://doi.org/10.1007/3-540-45692-9_10
27. Tkach, I., Amador, S.: Towards addressing dynamic multi-agent task allocation in law enforcement. Auton. Agents Multi-Agent Syst. **35**(1), 1–18 (2021). https://doi.org/10.1007/s10458-021-09494-x
28. Tkach, I., Blackwell, T.: Measuring algorithm performance on a conical barrier tree benchmark. In: GECCO 2022: Genetic and Evolutionary Computation Conference (2022)
29. Tkach, I., Edan, Y.: Distributed Heterogeneous Multi Sensor Task Allocation Systems. Springer, Heidelberg (2020). https://doi.org/10.1007/978-3-030-34735-2
30. Tkach, I., Edan, Y., Jevtic, A., Nof, S.Y.: Automatic multi-sensor task allocation using modified distributed bees algorithm. In: 2013 IEEE International Conference on Systems, Man, and Cybernetics, pp. 1401–1406. IEEE (2013)
31. Tkach, I., Edan, Y., Nof, S.Y.: Multi-sensor task allocation framework for supply networks security using task administration protocols. Int. J. Prod. Res. **55**(18), 5202–5224 (2017)
32. Tkach, I., Jevtić, A., Nof, S.Y., Edan, Y.: A modified distributed bees algorithm for multi-sensor task allocation. Sensors **18**(3), 759 (2018)

Systematic Review of Inclusive Design via Neuroergonomics as Assistance for Atypical Neurology

John Biechele-Speziale, William Raymer(✉), and Vincent G. Duffy

Purdue University, West Lafayette, IN 47906, USA
{jbiechel,wraymer,duffy}@purdue.edu

Abstract. This study explores the application of neuroergonomics to improve the efficacy of human computer interfaces in the domain of cognitive work, with a specific focus on individuals with mild cognitive impairments. We applied various data analysis tools to perform a systematic literature review on this topic. A bibliometric analysis-style approach is used for the systematic review. The metadata used for this study was extracted from three databases: Scopus, Google Scholar, and Web of Science; said data was analyzed using VosViewer, MaxQDA, Google nGram Viewer, CiteSpace, and custom R scripts to generate usable insights into the field and to substantiate our review. Mendeley was used to generate a bibliography which was crossreferenced via LATEX. This study reviews the application of neuroergonomic techniques to design jobs or assistive technologies to improve operators' cognitive work capabilities in light of their limitations when they interact with work systems. This article demonstrates a current gap in the literature and briefly discusses improving job design via unique neuroergonomic analysis to reduce the impact of poorly designed jobs on neurodivergent individuals, and the likely concomitant economic impact of reducing the friction of cognitive work.

Keywords: Neuroergonomics · Job Design · Human Factors · Bibliometric Analysis · Cognitive Work · FaceReader · Human-Computer Interfaces · Brain-Computer Interfaces · MaxQDA · CiteSpace · Mendeley · Machine Learning

1 Introduction

For most of human history, work has encompassed primarily physical labor. However, with the boom of the industrial revolution, then the subsequent data and information age, cognitive work has supplanted physical labor as the primary value generation mechanism of the modern world (Hancock et al., 2021). Hence, it stands to reason that the finite cognitive resources we have available should be maximally leveraged to generate more wealth, and maximize overall well-being. However, there exists an uneven distribution of cognitive resources, and some tasks may prove difficult if not impossible for individuals with certain cognitive impairments or atypical neurology, and even worse, many neurodivergent individuals are actively discriminated against in the workplace (Zolyomi

C.-Y. Huang and S. W. Yoon (Eds.): ICPR1 2021, ACES 14, pp. 272–284, 2023.
https://doi.org/10.1007/978-3-031-44373-2_16

and Snyder, 2021). As it currently stands, these individuals may not be as effective in achieving the same throughput or quality of work as their neruotypical peers, as jobs are usually designed for the mean or median worker, not the extremes, which often leads to disability as a consequence of noninclusive design practices (Vanderheiden et al., 2021).

To accommodate for this, improve efficacy of job designs, and increase the value any individual can generate through cognitive work, it is crucial to consider cognitive design patterns that are maximally neuroergonomic to the worker, and this has been woefully under-considered until only recently. Inclusive job design was mentioned in the Handbook of Human Factors and Ergonomics (Vanderheiden et al., 2021), but primarily focused on physical disabilities, with less exposition on mental ones. The focus that is provided is also primarily given to individuals with low IQs, but this is not always a consequence of atypical neurology.

Recent advancements in autonomous vehicles have demonstrated the ability reduce vehicle accidents caused by attention fatigue in ADHD drivers (Ebadi et al., 2021). However, the ultimate goal of autonomous vehicle development is to fully automate driving and mitigate or eliminate human interaction with the vehicle piloting process, also known as L4 driving (Rajasekhar and Jaswal, 2015). This near full automation of the task, while excellent for improving safety, does not address the issue of allowing workers the ability to better engage with their task. Instead, this design philosophy simply replaces them with autonomous systems and removes them from consideration altogether.

Instead, we propose a hybrid approach of varying autonomy levels which depends on, and closely integrates with, the users' dynamic neural profile in a collaborative feedback system. Such a system acts more like a crutch in the most basic sense, or can act like a mechanised exoskeleton in its most sophisticated implementation. The idea is to augment human capacity to maximize "cognitive leverage" that is, allow users to lift maximal cognitive loads while exerting themselves only as much as necessary, with the augmented system aiding with the parts of the task the user struggles with. A vehicle related example of this sort of augmentation is autonomous braking, which still allows users the chance to control the vehicle, and only take over when deemed absolutely necessary to prevent injury (Ebadi et al., 2021). Most current systems work by reading the environment, but neglect the user in their calculus, except during the course of research. This is clear from the explanation of the various self-driving levels, as none of them claim to aid the user in integrating with the system more effectively (Rajasekhar and Jaswal, 2015). However, by integrating such partially autonomous solutions with non-invasive brain-measuring devices, or even with less sophisticated technology such as facial and emotional recognition analysis, it may be possible to advance this technology to improve prediction accuracy based on cognitive markers and autonomously adjust the level of aid required based on the immediate needs of the user.

Articles demonstrating the use of bibliometric analysis methods used in this study include Cross-Cultural Design in Consumer Vehicles to Improve Safety: A Systematic Review, The Reaches of Crowdsourcing: A Systematic Review and Data Mining Methodology in Support of a Systematic Review of Human Aspects of Cybersecurity (Koratpallikar and Duffy 2021; Dishman and Duffy, 2021; Duffy and Duffy 2020a). Applications considering construction and an error prevention context are shown in the systematic review related to the human aspects of environmental mishaps (Duffy and Duffy 2020b).

While this sort of transhumanistic idea has long been a part of science fiction with regards to physical augmentation, and has recently seen advances due to increased funding, cognitive enhancements have been neglected until recently due to the variety of challenges associated with developing high resolution BCIs. Some examples include producing valid standards for the signal preprocessing steps, producing participant independent classification of mental states, and removing nonaffective states from the data (Mu¨hl et al., 2014). However, with the machine learning revolution that's currently taking place, it seems apparent that now is the ideal time to increase funding to begin increasing the volume of work in this field with the eventual goal of enabling as many people as possible to engage more effectively with their work, and improve the quality of their personal lives. Indeed, there are already some steps being taken in automating the BCI customization process, but this still have substantial ground to cover until ready for less critical applications (Vidaurre et al., 2011). The authors of this article believe that neurodivergent individuals would be excellent to evaluate these emerging technologies with, as they stand to benefit immensely from such research, and the ability to develop novel theranostic tools for cognitive conditions with clearly defined evaluation criteria could reveal opportunities to provide personalized cognitive aid (Fig. 1).

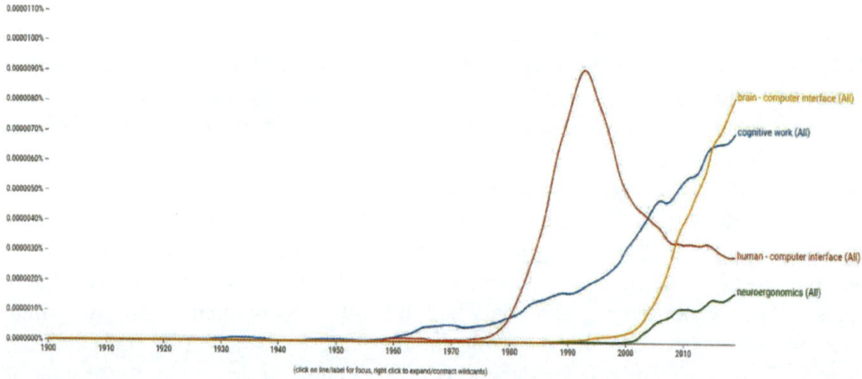

Fig. 1. This figure illustrates the result of the Google nGram analysis of our four topics: "cognitive work", "neuroergonomics", "brain-computer interface", and "human-computer interface", limited from 1900 to 2019.

As shown, the topic of cognitive work started drawing attention around the 1960's and has continued in an upward trajectory since, similar to brain-computer interface which got its start in the early 1990s. Neuroergonomics, by contrast, has not been addressed much and appears relatively new, first appearing around the year 2000. Finally, there has been an apparent decrease in interest with respect to human-computer interfaces as compared to the 1990s.

2 Purpose of Study

This study aims to conduct a systematic literature review regarding the current work in the Neuroergonomics, HCI/BCI, and cognitive work space with respect to neurodivergence. While much prior work has been done within the overlap of neuroergonomics and computer interfaces, our review has revealed an apparent lack of work on individuals with atypical neurology. In doing this review, we hope to draw attention to this apparent gap in the literature, and aim to encourage experts in these spaces to allocate resources to further explore the intersection of these fields. Hopefully, as a result of these efforts, alternative therapeutics will be developed to assist neurodivergent individuals with the unique challenges of cognitive work they face in today's market. To this aim, literature metadata and relevant analysis tools were employed, including CiteSpace (Chen), VosViewer (Nees Jan van Eck), MaxQDA (VerbiSoftware), and more (Fig. 2).

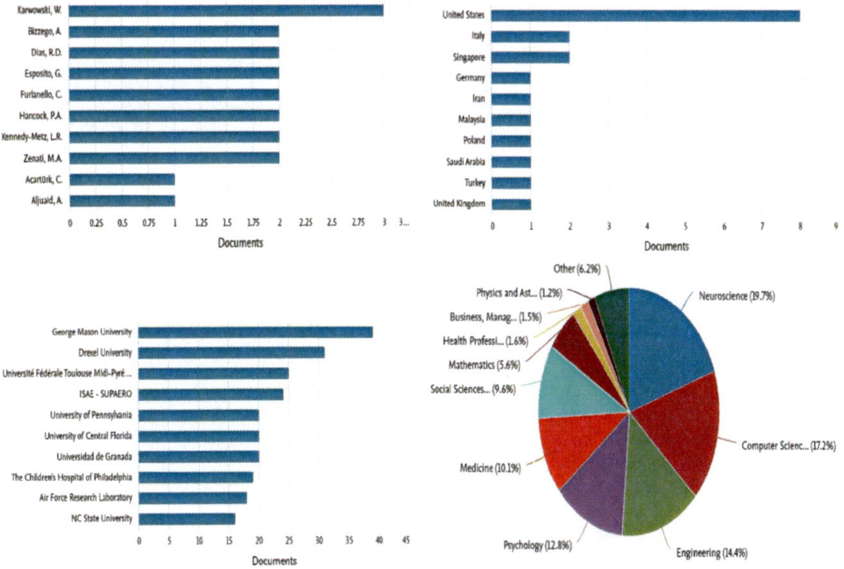

Fig. 2. These figures were obtained from Scopus and shows the number of published articles using cognitive work on the topics of neuroergonomics, and cognitive task analysis. These leader figures show leaders by author (top-left) country (top-right), institution (bottom-left) and field(bottom-right).

3 Research Methodology and Preliminary Analysis

3.1 Data Collection

The review began by extracting metadata related to our topics of interest across multiple databases. We selected Google Scholar (Google), Scopus (Elsevier), and Web of Science (Clarivate). Google Scholar was chosen as it has the largest number of sources by far, as it includes patents, articles, conference proceedings, dissertations and more. Scopus and Web of Science were selected both as more conservative alternatives, and were used as compliments of each other, as a contingency in the event their databases contained exclusive results.

All bibliometric data was filtered down to consider academic articles only. Results of our search terms for both individual and joint queries are enumerated in Table 1. After these searches were performed, the metadata of the articles was exported and used in statistical analyses in further sections. Based on the peak in Google nGrams, and the $\sim$ 3 times larger corpus of work for the Brain-Computer Interface as Compared to the Human-Computer Interface, we chose to focus on the former for the remainder of this article, even though the intersection of topics is smaller; this actually helps us demonstrate the main point of our introduction: there is insufficient research generally within this intersection of topics, and we were unable to find any article that focused on atypical cognition.

Table 1. This table lists the keywords searched within each database of interest, as well as the number of results for each term and combination of terms. We only included academic articles in our search results. As shown, there is only a small fraction of interest in either intersection of these topics, even though the topics themselves are heavily connected.

Keywords	Google Scholar	Scopus	Web of Science
"Neuroergonomics"	398	162	239
"Cognitive Work"	1550	508	501
"Human-Computer Interface"	1950	2191	1365
"Brain-Computer Interface"	4170	8985	6356
Terms 1–3	155	10	3
Terms 1,2,4	30	0	0

3.2 Publication Trend Analysis

After doing our primary searches, we investigated the number of publications per year from 2017 until the present, to see if any recent trends were identifiable both among the databases and the topics themselves. This is illustrated in Fig. 3. This precursor data was also subject to an ANOVA to determine which variables had a statistically significant impact on the publication count over the last 5.5 years (results in Table 2).

As we can see within Table 2, both the topic and database, have individual and combined statistically significant effects on the number of results returned for each topic across the last 5.5 years. More interestingly, we see a super-linear increase in Fig. 3 in the BCI field, which indicates an accelerating interest in that field, but an apparent neglect or lack of interest in the related fields of cognitive work, human-computer interfaces, and neuroergonomics.

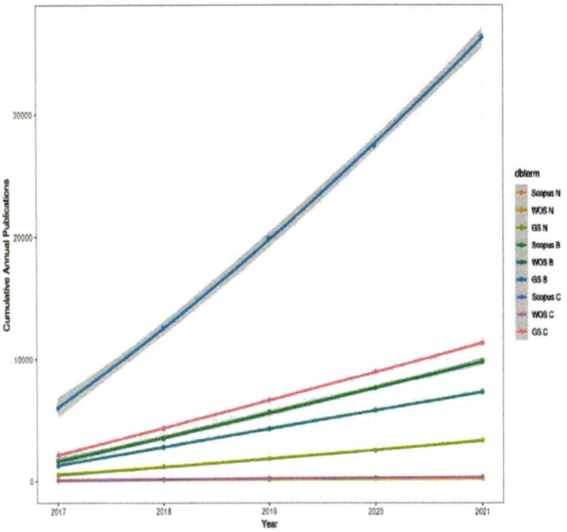

Fig. 3. Regression analysis of the cumulative publication values for different database term combinations. As shown, all but the Google Scholar + Brain Computer Interface combination grow linearly, but the outlier grew super-linearly, indicating an acceleration of interest in BCI research as compared to the relative constant interest in our other topics.

Table 2. ANOVA results on our search terms from Table 1 over the last 5.5 years (20172022). As shown, both the topic, the database, and the combined variables have statistically significant impacts on the number of articles found across our search terms.

	Effect	DF_n	DF_d	F	p	$p \leq .05$	ges
1	Topic	2	45	71.288	1.12e-14	T	0.760
2	Database	2	45	59.441	2.35e-13	T	0.725
3	Both	4	45	16.462	2.23e-08	T	0.594

3.3 Popular Engagement Measure: Twitter Trend Analysis

Not only did we do frequency and trend analysis across formal article databases, we also examined term frequency and relative popularity on the modern day public square: Twitter. To accomplish this, we leveraged Vicinitas to mine the data related to the engagement level across our different topics. As shown in Table 1, we used four key terms: cognitive work, neuroergonomics, HCI and BCI, respectively. Vicinitas queried them across Twitter to produce both word clouds and the cumulative engagement and post timeline for each topic. As we can see, it appears as though the first two terms are logarithmic in nature, with large initial growth that has tapered off over time, and the latter two appear almost exponential; however, there appears to be substantial overlap between them, which could be caused by the less precise nature of speech on Twitter as compared to academic journals.

4 Results

4.1 MaxQDA Content Analysis

Subsequent to our trend analysis, many of the top articles and book chapters on neuroergonomics were identified and used to view potential overlap with our other topics of interest. This was accomplished through text mining and word frequency analysis via the MaxQDA software. Figure 5, illustrates the top 30 informative keywords across the various neuroergonomics papers. Many of these should overlap with the other topics such as EEG, cognitive, task, performance, workload, etc. However, when we performed the remainder of our analyses, we did not find the level of correlation between these topics that we expected, nor did we find any articles focused on individuals with solely mental disabilities. Instead, we found a focus on improving the performance of already high performing individuals. We find this trend interesting, as it stands to reason that applying these cognitive enhancement methods to top performers will engender diminishing returns much faster than if they were used on lower performers first, though this is purely speculation at this stage (Fig. 4).

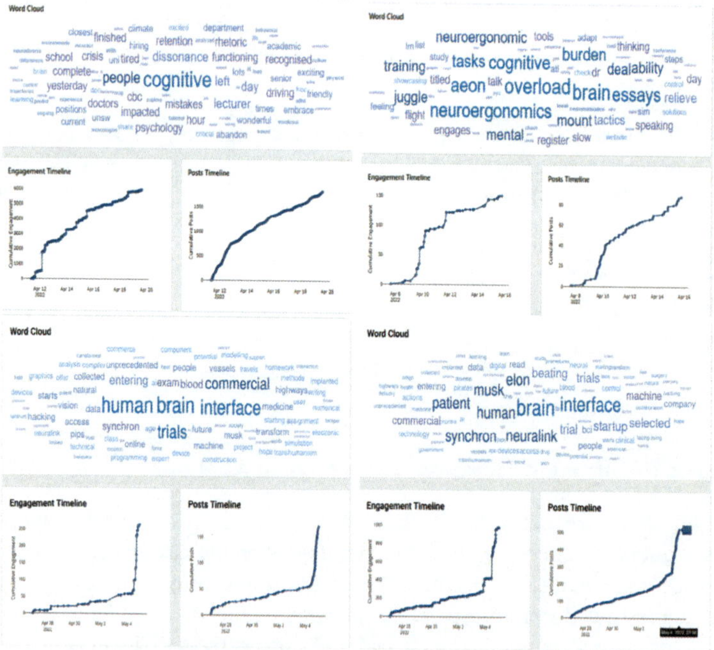

Fig. 4. Popular trend analyses for our 4 topics mentioned in Table 1. The keywords from top left to bottom right are: cognitive work, neuroergonomics, human computer interface, and brain computer interface. All sub-figures were obtained from Vicinitas.

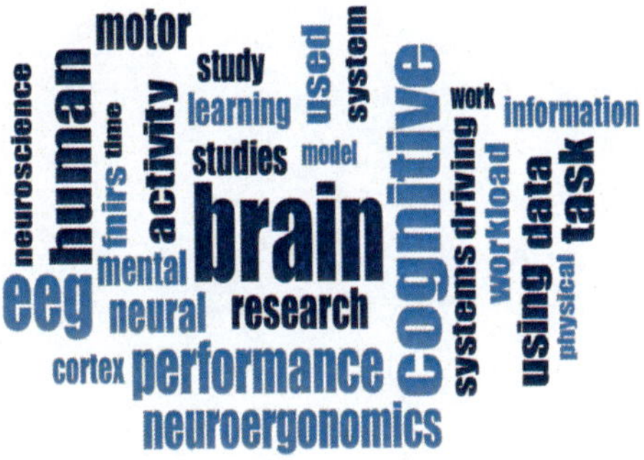

Fig. 5. This figure illustrates some of the main topics from many of the articles from the search term "neuroergonomics". 8 of the most cited articles, along with Chapter 31 in the Handbook of Human Factors and Ergonomics, along with some articles from the VosViewer analysis were compiled into a project to do this word frequency visualization.

4.2 VosViewer Co-citation and Content Analysis

A co-citation is defined as the event when two articles have been cited together in a third article. This information allows us to construct a graph to analyze the degree distribution of the different articles. VosViewer is an excellent tool for this task: metadata extracted from Scopus and Web of Science in the RIS formats were given to VosViewer as new projects. These extracted files had a total of 1221 unique articles. For statistical rigor, we limited articles to a minimum citation value of 3, anything below this cutoff was not considered for the analysis. The resulting clusters for our topics, and the associated content analysis are shown in Fig. 6.

The link strength was set to 2 as a minimum number of citations of cited references to generate a large number of articles and then the strength 5 as a minimum number of citations of cited references to generate the most co-cited articles. As a result, a total of 72 articles and 3 articles for the respective strengths remained after the constraints were applied. These words were then mapped to the graph in Fig. 6. After our initial analysis, VosViewer recommended the top articles with the strongest link strength, many of which were referenced during the preparation of this article, and are shown in Fig. 7.

4.3 CiteSpace Cluster Analysis

The co-citation analysis in VOSviewer has limitations like producing a citations burst. These limitations can be overcome by Citespace. CiteSpace is a software that can be used to perform co-citation analysis and extract labels for each cluster. It can also be used to generate a citation burst diagram that indicates the period during which a particular article was cited the most. To perform the analysis in CiteSpace, a keyword search was carried out in web of science which yielded 161 results. These results were exported along with cited references in text format. This folder was opened in CiteSpace, and co-citation analysis was carried out. Labels for the clusters were extracted using keywords. The results are shown in Fig. 8. The names of the clusters represent the articles in a particular cluster. This helps identify various sub-topics within a topic area. It also helps identify articles within a sub-topic.

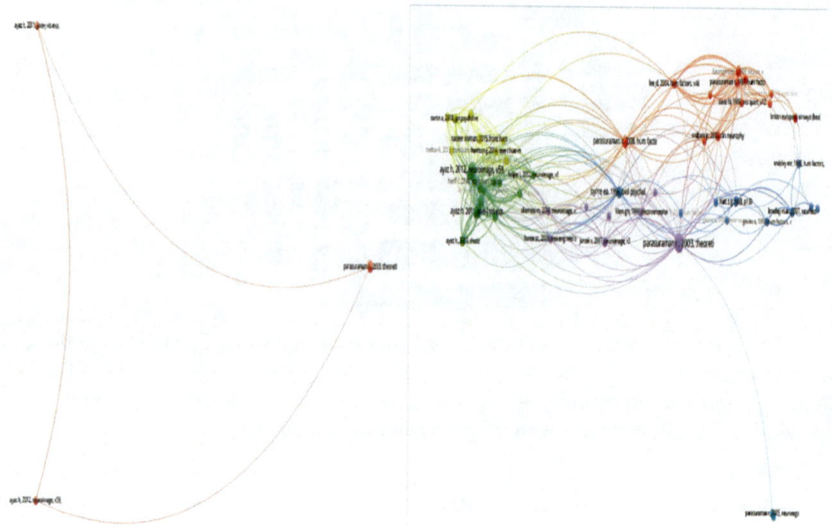

Fig. 6. This figure was obtained from VosViewer by using chapters 7, 31, and 13 from the Handbook of Human Factors and Ergonomics and 6 other articles published on the subject of neuroergonomics and human-computer interface. We used 2 as the parameter for the value references in common and the co-citation as another parameter. VosViewer produced 72 co-cited articles which produced the figure above. This figure shows the most co-cited articles and the least co-cited articles. This Co-citation analysis method is used to find articles that have been cited together in another article which provides information regarding the degree of connectivity between articles.

Selected	Cited reference	Citations	Total link strength
✓	ayaz h, 2012, neuroimage, v59, p36, doi 10.1016/j...	6	50
✓	ayaz h, 2011, jove-j vis exp, doi 10.3791/3443	5	44
✓	parasuraman r., 2003, theoretical issues e, v4, p1, ...	10	40
✓	ayaz h, 2013, front hum neurosci, v7, doi 10.3389/...	4	37
✓	jobsis ff, 1977, science, v198, p1264, doi 10.1126/s...	3	29
✓	byrne ea, 1996, biol psychol, v42, p249, doi 10.101...	4	28
✓	herff c, 2014, front hum neurosci, v7, doi 10.3389/...	3	26
✓	parasuraman r, 2008, hum factors, v50, p511, doi ...	4	26
✓	naseer noman, 2015, front hum neurosci, v9, p3, ...	3	25
✓	izzetoglu k, 2004, int j hum-comput int, v17, p211,...	3	24
✓	ayaz h, 2010, ieee eng med bio, p6567, doi 10.1109...	3	23
✓	fishburn fa, 2014, front hum neurosci, v8, doi 10.3...	2	23
✓	liu yc, 2017, sci rep-uk, v7, doi 10.1038/srep43293	2	23
✓	mckendrick r, 2014, neuroimage, v85, p1014, doi 1...	2	23
✓	lee jd, 2004, hum factors, v46, p50, doi 10.1518/hf...	3	22
✓	okamoto m, 2004, neuroimage, v21, p99, doi 10.1...	2	22
✓	harrison j, 2014, ieee t hum-mach syst, v44, p429, ...	2	20
✓	parasuraman r, 1997, hum factors, v39, p230, doi ...	3	20
✓	baumgartner t, 2008, neuron, v58, p639, doi 10.10...	2	17

Selected	Cited reference	Citations	Total link strength
✓	holper l, 2012, neuroimage, v63, p212, doi 10.1016...	2	17
✓	madhavan p, 2007, theor iss ergon sci, v8, p277, d...	2	17
✓	mayer rc, 1995, acad manage rev, v20, p709, doi 1...	2	17
✓	peck e.m.m., 2013, p sigchi c hum fact, p473, doi ...	2	17
✓	wickens cd, 2007, theor iss ergon sci, v8, p201, doi...	2	17
✓	derosiere g, 2013, front hum neurosci, v7, doi 10.3...	2	16
✓	dimoka a, 2010, mis quart, v34, p373	2	16
✓	gateau t, 2015, plos one, v10, doi 10.1371/journal....	3	16
✓	hart s g, 1988, p139	3	16
✓	riedl r, 2010, mis quart, v34, p397	2	16
✓	riedl r, 2012, j neurosci psychol e, v5, p63, doi 10.1...	2	16
✓	wolpaw jr, 2002, clin neurophysiol, v113, p767, doi...	3	16
✓	daly jj, 2008, lancet neurol, v7, p1032, doi 10.1016/...	2	15
✓	bunce sc, 2006, ieee eng med biol, v25, p54, doi 1...	2	14
✓	klem g h, 1999, electroencephalogr clin neurophy...	2	14
✓	parasuraman r, 2009, mil psychol, v21, p270, doi 1...	2	14
✓	ayaz h., 2010, thesis	2	13
✓	izzetoglu m, 2005, ieee t neur sys reh, v13, p153, d...	2	13
✓	mehta rk, 2013, front hum neurosci, v7, doi 10.338...	2	13

Fig. 7. This figure shows the co-cited articles resulted from VosViewer, the cited references, the number of times each article has been cited, and the total link strength, which measures the degree of that node in the graph.

Top 14 References with the Strongest Citation Bursts

References	Year	Strength	Begin	End	1996 - 2003
*AM PSYCH ASS, 1994, DIAGN STAT MAN MENT, V0, P0	1994	3.01	**1996**	1999	
ENDERS W, 1993, AM POLIT SCI REV, V87, P829	1993	2.46	**1996**	1997	
MALLONEE S, 1996, JAMA-J AM MED ASSOC, V276, P382	1996	4.69	**1997**	2000	
TUCKER JB, 1997, JAMA-J AM MED ASSOC, V278, P362	1997	3.49	**1999**	2001	
SIMON JD, 1997, JAMA-J AM MED ASSOC, V278, P428	1997	3.49	**1999**	2001	
FRANZ DR, 1997, JAMA-J AM MED ASSOC, V278, P399	1997	3.22	**1999**	2001	
CARTER A, 1998, FOREIGN AFF, V77, P80	1998	2.97	**1999**	2001	
OKUMURA T, 1996, ANN EMERG MED, V28, P129	1996	2.59	**1999**	2001	
ZILINSKAS RA, 1997, JAMA-J AM MED ASSOC, V278, P418	1997	2.36	**1999**	2001	
TOROK TJ, 1997, JAMA-J AM MED ASSOC, V278, P389	1997	3.62	**2000**	2001	
RICHARDS CF, 1999, ANN EMERG MED, V34, P183	1999	3.21	**2000**	2001	
HENDERSON DA, 1999, SCIENCE, V283, P1279	1999	2.66	**2000**	2001	
INGLESBY TV, 1999, JAMA-J AM MED ASSOC, V281, P1735	1999	2.65	**2000**	2001	
KEIM M, 1999, ANN EMERG MED, V34, P177	1999	2.26	**2000**	2001	

Fig. 8. This figure represents the results obtained from CiteSpace showing the articles obtained from Web of Science. The top sub-figure shows the cluster of the most co-cited articles with the author as the label of each article. In the second part of the figure, we identified the articles with the most citation bursts in this field. As shown, nearly all the most influential papers were published over 20 years ago, indicating a reduced interest in the topic. From this analysis, we can select a number of articles for further review where Mendeley would be used to generate a reference list (Elsevier, a). This reference list would then be used to perform a co-citation analysis using VosViewer to determine the connectivity between these articles.

5 Discussion

As shown in the results section, we can clearly identify that despite the heavy clustering and dense networking of these topics, the overall number of papers that intersect with all three of these topics and provide insights on individuals with atypical cognition is lacking. This is unfortunate, as the economic impacts on neurodiverse (specifically autistic) individuals has been explored: autistic individuals are less likely to take particular jobs either out of lack of enjoyment, or the inability to cope with particular working requirements due to sensory or social over-stimulation. This limitation of potential job prospects, and the reduced ability to fill roles impacts an estimated 70 million people worldwide (Cowen, 2011). Individuals with ADHD may either be unwilling or unable to accept jobs with increased attentional demands without substantive assistance to limit distracting stimuli, and are estimated at 5.29% of the population or approximately 370,300,000 people worldwide (Polanczyk et al., 2007).

Hence, as a rough estimate, we can expect approximately 6% of the global population to have reduced job prospects, or reduced job efficacy directly caused by their atypical cognition alone. Taking to heart the ideas of accessibility through design from (Vanderheiden et al., 2021), a rough estimate of economic improvement given assistive devices or more inclusive job design can be performed. If we assume that the proposed hybrid cognitive interventions will only impact the neurodiverse population with autism and ADHD (unlikely), and that these interventions will only increase their maximum productive throughput by 40%, we can determine the impact this would have on global GDP. In 2020, global GDP was 80 trillion dollars, thus our estimate predicts it would have increased by 40% of 6% or by an extra 192 billion dollars (Bank, 2021). However, the development of cognitive assistive devices are not limited to individuals with mild cognitive impairments or atypical neurology, which could open the doors to marked economic productivity for a majority of the population as well, as was seen during the personal computer and web revolution of the 1970s to the present.

Given the possibility of impact on the individual and global economic scales that such hybrid, cognitive interventions may have, the authors believe this field should be funded with a particular focus on developing aid for neurodiverse individuals which may be generalized as research and development continues.

6 Conclusion

In summary, we conducted a systematic review for exploring the application of neuroergonomics to improve the efficacy of human computer interfaces in the domain of cognitive work, with a specific focus on individuals with atypical neurology. We extracted the metadata from three different databases: Scopus, Google Scholar, and Web of Science. From Scopus we observed the US has had the most number of publications compared to the remaining countries that published on these topics and observed the topic has gained increasing attention since 2017. We also used VosViewer, Google nGram Viewer, Vicinitas, and CiteSpace to run the cluster and trend analyses on our bibliometric metadata. From both Google nGram Viewer and Vicinitas we observed an upward trend on BCI and cognitive work, which is encouraging.

However, in absolute numbers, there are only about 10 published articles from Scopus, 155 articles from Google Scholar, and 3 articles from Web of Science, but none of them have a primary focus on our target demographic, illustrating a substantive gap in the literature. We outlined a back-of-the-envelope style estimate of the negative economic impact that continued lack of work in this area may have, and hope this article will convince both principal investigators and funding agencies to consider work in this area as a worthwhile pursuit.

7 Future Work

Finally, it seems less than ideal to end on such a bleak picture of the current body of work. Instead, we illustrate some of the astounding possibilities that can be achieved if more funding is placed into the appropriate areas, and our biases around disabilities are appropriately accounted for. A recent grant awarded by the NSF has shown some incredible promise with some of the work performed with the funding. Li and Nam showed that an non-invasive BCI device, known as a SSVEPbased collaborative BCI, allowed individuals with ALS to control a robot along pre-defined paths with their brain signals. While there was no direct improvement in their motor ability, the device may be able to return some level of autonomy to these individuals, especially if the BCI can be leveraged with partially autonomous exoskeletons. More work like this has been performed recently, such as the use of invasive and non-invasive BCI devices to allow individuals with locked-in syndrome to communicate with the outside world (Birbaumer, 2006).

While this work is an excellent step forward among many that focus on physical disabilities (Jackson and Mappus, 2010), the authors of this bibliometric analysis believe the lack of focus on hidden disabilities, such as cognitive impairments, sensory processing, and/or attentional issues, is neglecting job design for an invisible minority who are still impacted by their atypical neurology regardless of how visible it is to researchers, job designers, or funding agencies. Considering how much of a role cognitive work is playing in our economy, the authors hope that drawing attention to this gap in the literature will allow funding resources to be leveraged to not only improve the quality of life and job satisfaction of these neurodivergent individuals, but that there will a commensurate, measurable economic benefit as a result.

Acknowledgements. The National Science Foundation (Award Number 2128970 and 2128867) are thanked for supporting the research reported in this paper. Any opinions, findings, conclusions, and/or recommendations expressed in this material are those of the authors and do not necessarily reflect the views of the NSF.

References

Bank, W.: Gross domestic product 2020 (2021)

Birbaumer, N.: Breaking the silence: Brain–computer interfaces (bci) for communication and motor control. Psychophysiology **43**(6), 517–532 (2006)

Chen, C. Citespace. https://citespace.podia.com/faq

Clarivate. Web of science. https://www.webofscience.com

Cowen, T.: An economic and rational choice approach to the autism spectrum and human neurodiversity. SSRN Scholarly Paper 1975809. Social Science Research Network, Rochester, NY (2011, December)

Dishman, S., Duffy, V.G.: The Reaches of Crowdsourcing: A Systematic Literature Review. In: International Conference on Human-Computer Interaction, pp. 229–248. Springer, Cham (2021, July)

Duffy, B.M., Duffy, V.G.: Data mining methodology in support of a systematic review of human aspects of cybersecurity. In: International Conference on Human-Computer Interaction, pp. 242–253. Springer, Cham (2020a, July)

Duffy, G.A., Duffy, V.G.: Systematic literature review on the effect of human error in environmental pollution. In: International Conference on Human-Computer Interaction, pp. 228–241. Springer, Cham (2020b, July)

Ebadi, Y., et al.: Impact of l2 automated systems on hazard anticipation and mitigation behavior of young drivers with varying levels of attention deficit hyperactivity disorder symptomatology. Accident Analysis Prevention **159**, 106292 (2021)

Elsevier. Mendeley. https://www.eslevier.com/solutions/scopus

Elsevier. Scopus. https://www.eslevier.com/solutions/scopus

Google. Google scholar. https://scholar.google.com
Hancock, G., Longo, L., Young, M., Hancock, P.: Handbook of Human Factors and Ergonomics, 5th edn. John Wiley and Sons, Chapter Mental Workload (2021)
Jackson, M.M., Mappus, R.: Brain-Computer Interfaces: Applying our Minds to Human-Computer Interaction, Chapter Applications for Brain-Computer Interfaces. Springer, London, London (2010)
Koratpallikar, P., Duffy, V.G.: Cross-cultural design in consumer vehicles to improve safety: a systematic literature review. In: International Conference on Human-Computer Interaction, pp. 539–553. Springer, Cham (2021, July)
Li, Y., Nam, C.S.: A collaborative brain-computer interface (bci) for als patients. Proceedings of the Human Factors and Ergonomics Society Annual Meeting **59**(1), 716–720 (2015)
Mu¨hl, C., Allison, B., Nijholt, A., Chanel, G.: A survey of affective brain computer interfaces: principles, state-of-the-art, and challenges. Brain-Computer Interfaces **1**(2), 66–84 (2014)
Nees Jan van Eck, L. W. Vosviewer
Polanczyk, G., de Lima, M.S., Horta, B.L., Biederman, J., Rohde, L.A.: The worldwide prevalence of ADHD: a systematic review and metaregression analysis. American Journal of Psychiatry **164**(6), 942–948 (2007, June)
Rajasekhar, M.V., Jaswal, A.K.: Autonomous vehicles: The future of automobiles. In: 2015 IEEE International Transportation Electrification Conference (ITEC), pp. 1–6 (2015)
Vanderheiden, G.C., Jordan, J.B., Lazar, J.: Handbook of Human Factors and Ergonomics, 5th edn. John Wiley and Sons, Chapter Designing for People Experiencing Functional Limitations (2021)
VerbiSoftware. Maxqda 2022. https://www.maxqda.com; Software. 2021
Vidaurre, C., Kawanabe, M., von Bünau, P., Blankertz, B., Müller, K.R.: Toward unsupervised adaptation of lda for brain–computer interfaces. IEEE Transactions on Biomedical Engineering **58**(3), 587–597 (2011)
Zolyomi, A., Snyder, J.: Social-emotional-sensory design map for affective computing informed by neurodivergent experiences. Proceedings of the ACM on Human-Computer Interaction **5**(CSCW1), 77:1–77:37 (2021, April)

Product and Corporate Culture Diffusion via Twitter Analytics: A Case of Japanese Automobile Manufactures

Yuta Kitano[1], Shogo Matsuno[2(✉)], Tetsuo Yamada[1], and Kim Hua Tan[3]

[1] The University of Electro-Communications, Chofu, Japan
[2] Gunma University, Maebashi, Japan
s.matsuno@gunma-u.ac.jp
[3] University of Nottingham, Nottingham, England

Abstract. Instilling a company's products and culture in consumers is important not only for the effective introduction of new product development but also for strategically connecting to the next innovation by capturing product feedback more effectively and automated. This paper proposes a computer-supported decision analysis by text mining of the official Twitter accounts of five Japanese automobile manufacturers conducted to search for effective Social Networking Service (SNS) operation strategies for companies. The number of retweets for posts of the official account is one of the most important indicators when considering the use of Twitter for corporate public relations. This study analyzes how the occurrence of retweets is related to other controllable variables for 1) posting timing, 2) elements in Twitter and 3) tweet text. Specifically, trends are investigated by collecting words contained in tweets posted by official accounts over a certain period of time and counting their occurrence frequency. Moreover, it analyzes the impact of additional information such as when the tweet was posted, the URL contained in the tweet, hashtags, mentions, and images. As a result, it was found that the influence of the posting timing was the largest, and the influence of the feature words contained in the text was larger than the length of the text. This suggests that for corporate accounts to increase the numbers of retweets for their posts, it is effective to post tweets containing words that are characteristic of the community to which the corporate account belongs during a time zone when the average number of retweets is high.

Keywords: Social Media · Text Analytics · eWOM · Owned Media · Information Diffusion

1 Introduction

People can see numerous information on daily life with easy access in today's information society. Additionally, they have obtained more and more information from their screens with the spread of smartphones [1]. These social changes have also affected corporate

PRISM 30 Special Sessions

C.-Y. Huang and S. W. Yoon (Eds.): ICPR1 2021, ACES 14, pp. 285–294, 2023.
https://doi.org/10.1007/978-3-031-44373-2_17

advertising efforts. In the past decades, TV commercial was one of the most famous advertising ways. However, while corporate TV commerce cost is decreased year by year though advertising costs in Japan are increased [2]. On the other hand, Social Networking Service (SNS) or social media has been the way to obtain multiple information for the people due to spread of smartphones [3].

Table 1 shows the literature review on social media analysis and text mining. Originally, the main purpose of SNS was promoting connections between people. As time went on, it has become a source of new information for people [4]. Companies have also been influenced by the spread of SNS and have developed advertising strategies for smartphones as well. One of the strategies is the opening of their own SNS accounts. This is now used as an inexpensive advertising tool for companies [5, 6]. For instance, most of Japanese companies have their own SNS accounts, and there is various usage by companies. Most of Japanese companies have their own SNS accounts, and there are various usages by each company. In particular, Twitter has a characteristic which spread of tweets across communities has the effect of increasing the final spread of tweets [7]. People who see the tweet, called followers, can LIKE or Retweet (RT) the tweet, which means spreading the tweet to their own followers. Companies use Twitter to communicate their activities and their products to users or potential customers. Therefore, analytics of Twitter usage by companies will clarify methods of advertising strategies in the future information age. In terms of Twitter advertising phases, the first phase is to operate a corporate account on Twitter. This is a way of using Twitter as a destination for your company's advertising activities. The second phase is to improve the quantity and quality of user generated contents (UGC) such as electronic word of mouth (eWOM) [8] in order to improve corporate branding. The third phase is to use Twitter advertising where sending out tweets for advertising are reached to users who are not following you. Although there has been text mining research on Twitter for companies in recent years [9, 10], when, how, and what to tweet to obtain more RT for a company account has not been established. Ma et al. [7] used the subject of corporate scandal case to investigate the impact of Twitter. They looked at the accounts of the companies that caused the scandal and the tweets of numbers of public users to determine the impact. However, they insisted on the use of Twitter from a risk management perspective, focusing on the negative aspects of companies and not on the promotion of corporate activities. Kitano et al. [11] focused on the content of tweets and used a technique called text mining to identify commonalities in the content of tweets. However, they only did word frequency analysis and network analysis. No specific managerial Twitter strategy has been proposed. Han et al. [12] on promoting user engagements in the Twitter platform has been provided insights in comparing multiple industry sectors. Like this, although there are several studies on corporate Twitter accounts, they have not conducted any research on managerial strategies focused on the statistical numbers of RT.

This study focuses on RT as an indicator to adapt the Japanese corporate Twitter account and analyzes using text mining its relationship for below: 1) tweet post timing, which 2) controllable variables (URL, hashtag, mentions, attached images), and 3) tweet content.

Table 1. A summary of literature review on social media analysis and text mining.

Authors	Target Social Media		Target Language	Using Characteristics for Text Mining			Purpose
	By Consumers	By Companies		Word counts	Sentiment	Network	
Tse et al (2016) [9]	*		English	*	*	*	Verification on how TM can provide companies with better decision-making abilities in crisis management
Ma et al (2019) [7]	*		English	*	*	*	Identifying the sentiment towards the overbook incident
Jeong et al (2019) [10]	*		N/A	*	*		The approach proposal is able to evaluate the potential opportunity available for the product topics to improve
Kitano et al (2021) [11]		*	Japanese	*		*	Showing the strategic differences that exist between Twitter and technical reports in Japanese manufacturing industry
This study		*	Japanese	*		*	Proposing a Twitter strategy based on increasing RT for corporate accounts

2 Methodology

2.1 Research Procedure

Figure 1 shows the experimental designs and constructing procedure of datasets in this study. Before the analysis, it is necessary to obtain the target tweets data from Twitter. In this study, five Japanese automotive companies are selected as examples since it is an industry with high advertising costs. One of the reasons is that they have Twitter data enough followers for possessing and products to advertise with. Another reason why the Japanese automotive industries are treated in this study is that these industries have higher advertising costs than others industries [13], and needs a more effective advertising method. The most recent year data, 2019, was used as the target time span from 1st April 2019 to 30th March 2020. Table 2 shows the corporate Twitter accounts detail included in our collected dataset in this study. The number of followers was taken and described during the dates of data extraction. Since the numbers of RTs are highly dependent on followers, and it is difficult to com-pare the data among companies. At first, the difference between the largest and smallest numbers of tweets over the year was about 1,000 in this data set. These tweets data include tweets in response to an account called Replies.

Next, the numbers of followers who see the tweets on the corporate account all exceeded 100,000. There is a large difference in the maximum numbers of RT relative to the numbers of followers. Specially, Subaru has more followers than Mazda, however, Mazda has more maxi-mum RT than Subaru. It can be said that the maximum numbers of RT on a company account do not necessarily depend on the numbers of followers. Moreover, Mazda's standard deviation (SD) is the largest in the set of analyzed companies' due to the maximum RT of about 17,000. This means that the tweets have been spread more than the usual tweets in Mazda. Next, in order to find out when tweets should be posted, the timing of tweets is analyzed for month, day of the week and hour. The number of tweets posted and the average number of RT are calculated for each time scale, and the correlation between them is used to find the right time to post. Next, it is examined which four controllable variables, hashtag, mention, URL and photo, Twitter characteristics should be included in the tweet. RT averages are compared with and without the four controllable variables.

Furthermore, text mining [10, 14] is conducted to analyze the words inside the tweets. Word counts analysis [15] is used to examine trends in tweets. After that, network analysis, which visualizes the network of co-occurring words, is used to divide the tweets into several groups. Additionally, RT for each of those groups are related to know which tweet categories are good impression to followers. Finally, the results will be discussed and how they can be used in actual operations.

2.2 Text Mining (TM) and Corporate Account Examples

Text Mining [7, 9] is a type of data mining for character strings. Sentence data is firstly divided into words. Next, phrases and the frequency, word correlation, and time series are analyzed. Since TM is an analytical method for natural languages, it depends on the type of language used. The tweet data written in Japanese language is treated in this

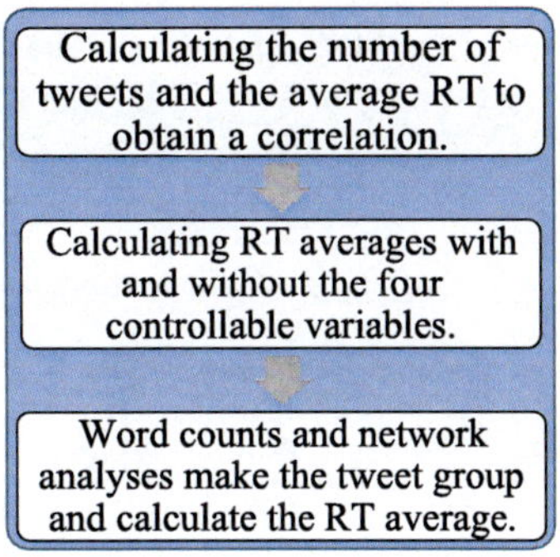

Fig. 1. Experimental designs and constructing procedure of datasets in this study

Table 2. The corporate accounts' status and RT averages.

Companies	Year	Followers (04.Sep.2020)	Number of Tweets	Max of RT	Mean of RT	Median of RT	SD of RT
Mazda	2019	118532	332	17954	253.7	128.0	1016.9
Honda	2019	235609	1383	4019	105.9	37.0	261.2
Toyota	2019	267030	965	4116	104.7	48.0	238.8
Nissan	2019	277714	792	7743	127.2	40.0	477.4
Subaru	2019	166667	335	3532	183.1	76.0	313.7

study because in order to understand the strength of Japanese companies, it is necessary to analyze Japanese language. Japanese has linguistic properties that are very different from English. A first major difference is that words are not separated by spaces. Another one is the problem that double-byte characters are usually used in Japanese descriptions over computers. Therefore, it is necessary to pre-processing such as normalization and tokenization peculiar to Japanese in order to create a bag of words (BoW) and measure the number of words. For the above reasons, this study uses the Text Mining Studio (TMS) of NTT DATA Mathematical Systems Inc. [16] to process the Japanese texts. In addition, to analyze a question: "What kind of sentences are spread by users?", two methods of analysis are used in the TM: word counts analysis and network analysis.

3 Results

3.1 Analysis of Controllable Variables in Twitter

This section provides an analysis and discussion of observable factors other than the content of tweets on Twitter. Firstly, the elements are analyzed by "Hashtag (#)", "Mention (@)", "URL" and "Photo". Table 3 shows the average RT for each element. As the overall trend, more than half of the tweets contain "Hashtag", "URL" and "Photo". It is found that the tweets containing "Hashtag" and "Photo" have a higher average RT than total average RT. However, the tweets contain "URL" had a lower average RT than the overall average.

Table 3. Average of RT for each element.

Companies	Hashtag#		Mention@		URL		Photo	
	Without	With	Without	With	Without	With	Without	With
Mazda	102.1	262.9	261.1	107.8	323.4	183.0	195.3	292.7
Honda	111.3	104.2	97.5	181.3	168.0	88.7	64.7	112.1
Toyota	65.9	108.0	107.0	75.7	120.5	98.4	58.0	109.8
Nissan	98.0	133.0	131.2	64.0	185.1	109.5	67.9	141.8
Subaru	181.1	185.0	185.5	106.1	204.2	167.0	56.5	256.5

3.2 Analysis of the Relationship Between RT and Timing of Tweets Posted

In order to know when a tweet is sent out, the relationship between data on the time a tweet was posted and RT is examined by "Month", "Day of the week", and "Hour". At first, the correlation between the numbers of tweets posted and average RT are shown in Fig. 2. In the top 30 RT, there was a strong correlation from 0.5 to 0.9 with hours. About 80% of posting timing were between 8 am and 6 pm, which means business hours. Regarding the posting timing, the highest number of posts and RT averages are recorded at 11 am and 3 pm for Nissan and 11 am and 5 pm for Mazda.

3.3 Analysis of Tweet Sentence Using TM

In this analysis, using TM word counts analysis and network analysis, tweets were examined on a word basis and on a group basis in a network based on co-occurring words to know what kind of tweet content is appeared. However, there are also many proper nouns in the Twitter feed of the automotive industry such as product name and event name. Thus, proper nouns related to products, races and events are replaced to find meaning in proper nouns.

An example of a network diagram where the replacement has been carried out is shown in Fig. 3. Overall, product-related tweets have a higher RT average than ones in the other groups as shown in Table 4. In terms of tweet categories, tweets containing product names tend to be more likely to be retweeted than event or race related tweets. However, in some groups the RT average was lower than the overall average. Additionally, these are event-related tweets, which are common in the automotive industry. Therefore, when the announcement and promotion is conducted, the content of the event could be improved with photo as mentioned in Sect. 3.1 and product topics.

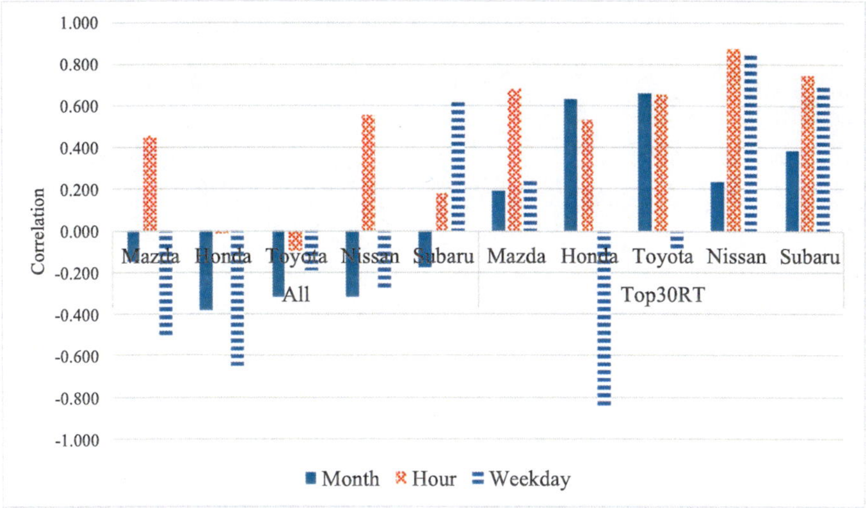

Fig. 2. Correlation between the numbers of tweets posted and average RT

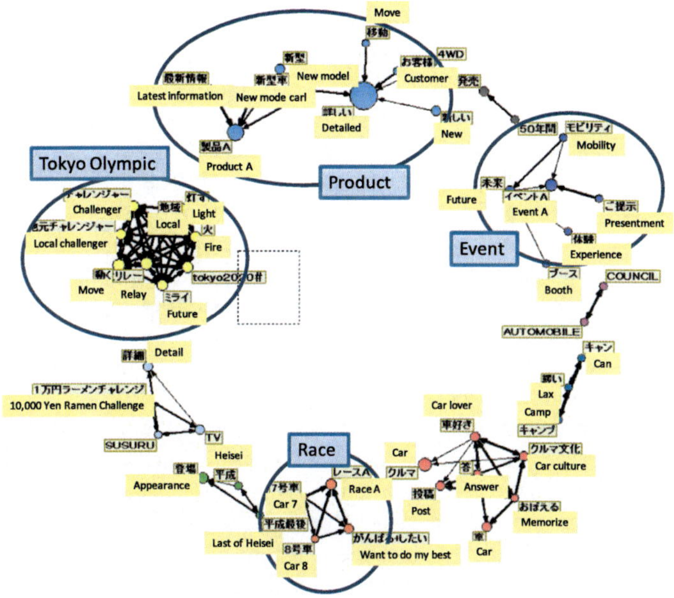

Fig. 3. Network diagram after proper noun substitution (Toyota)

3.4 Discussion

The relationship between timing of tweet posts and RT trends was examined. In terms of the timing of postings, the numbers of postings were increased during the months when there were events, weekdays are more common than holidays throughout the company,

Table 4. RT average of the three groups in 5 companies

	Race		Product		Event	
Companies	Percentage of relevant tweets	RT average	Number of relevant tweets	RT average	Number of relevant tweets	RT average
Mazda	6.02	157.95	33.73	205.99	14.46	167.67
Honda	20.68	65.93	7.52	150.73	12.36	59.93
Toyota	16.87	162.76	15.41	148.68	6.48	62.57
Nissan	1.52	83.50	18.81	178.14	8.33	62.65
Subaru	13.73	136.43	37.31	263.78	21.79	199.73

and postings time zones are concentrated between 8 am and 6 pm, which means business hours. Regarding the timing of RT, there were no trends in RT that were characteristic of the month or day of the week. However, there were some trends in the time of posting. The highest numbers of posts and average RT are recorded at 11 am and 3 pm for Nissan and 11 am and 5 pm for Mazda. Some studies found that Twitter users tend to have more RT at around 6 am, 12 am and 6 pm on weekdays than any other time [17]. In other words, it is recommended to tweet before those times, namely peak hours. This suggests that the strategy of biasing the time of posting to certain times, such as Subaru and Mazda, is also effective. However, even if they do not bias the numbers of tweet they post such as Nissan, some companies are still producing more RT during the peak hours. Another strategy for companies with sparse tweet times is to tweet at the peak times. In this analysis, peak times varied slightly by each company. The reason could be that it depended on the attributes of the followers and the content of the tweets. Therefore, it is important to determine the difference between their own peak times. At another focus point about RT timing relationship, there was a negative correlation for some accounts. Honda and Toyota, for which a negative correlation was observed, hardly changed by day of the week, so it is expected that such a result was obtained. However, it is difficult to make a reliable discussion with only this data, and this is a limitation of this study and becomes one of future works.

Moreover, the overall trend of the relationship between Twitter variables and RT propensity was observed where more than half of the tweets contained "Hashtag", "URL" and "Image". The advantage of these tweets over text-only tweets is that they can be more informative. The tweets containing "Hashtag" and "Image" have higher average RT than total average RT. However, the tweets containing "URL" had a lower average RT than the overall average. This suggests that Twitter citizens tend to be more likely to RT hashtags that indicate important keywords and photos that visually convey information. Indeed, an article pointed out that more than half of people who use timeline-based social media such as Twitter only read the header and do not understand the content of the URL [18]. Therefore, tweets tend to obtain more RT if they include a summary of the URL content and a photo of the content that company would like to convey.

In addition, the relationship between the content of the tweets posted and the propensity to RT was examined. Overall, product-related tweets have a higher average RT than the other groups. However, none of the company-specific groups derived from the network analysis had a high correlation with the average RT. In other words, these analyses did not lead to the identification of groups with high RT. On the other hand, in some groups the average RT was lower than the overall average. The group is event-related tweets, which are common in the automotive industry. Therefore, the announcement and promotion for the content of the event needs to be improved with better choice of photo and product topics. In terms of tweet categories, tweets containing product names tend to be more likely to be retweeted than event or race related tweets. However, top tweets from each company did not necessarily include names of products, and there were a few minor categories that were not grouped together. Thus, it suggests that there are numbers of other factors that contribute to RT.

In this study, three separate aspects of Twitter are analyzed, and it found strong associations with RT in each. However, it did not take into account their crossover or other factors as a limitation. Therefore, there are numbers of factors that may contribute to this RT process, including the psychological factors of Twitter users. Meanwhile, the question is whether these findings, such as hashtags and photo attachments, are also common in Instagram and Facebook. Twitter is superior in terms of spreading information, while Instagram and Facebook differ in terms of the age of their users and the proportion of images posted.

4 Conclusions

This study analyzed its relationship for tweet post timing, which controllable variables, and tweet content using text mining. In this analysis, it turns out that one of the most important factors in tweeting on a corporate account is the time of day, rather than the season when you post. One of the reasons is that some of the companies tweeted without regard to time. The other reason was that the companies that tweeted more at certain times also had higher average RT than others. Moreover, this study focused on observable Twitter elements and found that the average RT was higher when "Hashtag" and "Photo" were present. Additionally, all five companies included URL in more than about half of their tweets. However, the results showed that ones without URL received more RT. These results will be useful information for actual tweeting strategies. In addition, the content of the tweets was text-mined, and experiments were conducted on word-based and group-based clustering. The results showed that the average RT was higher than one in the group of tweets containing the product name. However, the average RT is not observed as high as those containing products.

Further studies should examine whether the factors and characteristics found to be effective in this study are related to be used in RT, and provide information that will be useful in actual management. In addition, we would like to examine in detail how the positive or negative of the correlation with RT timing affects the situation of diffusion.

Acknowledgments. This research was partially supported by the Japan Society for the Promotion Science (JSPS), KAKENHI, Grant-in-Aid for Scientific Research (A), JP18H03824, from 2018 to 2023. The authors would like to thank Hottolink Inc. For providing the Twitter analyzing software.

References

1. Ministry of Internal Affairs and Communications: Survey on time and information behavior of information and communication media. https://www.soumu.go.jp/iicp/research/results/media_usage-time.html. (in Japanese). Accessed on 28 Apr. 2020
2. Dentsu: 2018 Advertising costs in Japan. http://www.dentsu.co.jp/news/sp/release/2019/0228-009767.html. Accessed on 12 Oct. 2020
3. Han, S., Min, J., Lee, H.: Antecedents of social presence and gratification of social connection needs in SNS: a study of twitter users and their mobile and non-mobile usage. Int. J. Inf. Manage. **34**(4), 459–471 (2015)
4. Omuka, I.: The History of SNS. The Institute of Electronics. Info. Commu. Eng. (IEICE) **9**(2), 70–75 (2015)
5. Idota, H., Bunno, T., Tsuji, M.: How social media enhances product innovation in Japanese firms. Multidisciplinary Social Networks Research **540**, 236–248 (2015)
6. Moe, W.W., Schweidel, D.A.: Opportunities for innovation in social media analytics. J. Prod. Innov. Manage. **34**(5), 697–702 (2017)
7. Ma, J., Tse, Y.K., Wang, X., Zhang, M.: Examining customer perception and behavior through social media research: an empirical study of the united airlines overbooking crisis. Transp. Res. Part E **127**, 192–205 (2019)
8. Hennig-Thurau, T., Gwinner, K.P., Walsh, G., Gremler, D.D.: Electronic word-of-mouth via consumer-opinion platforms: what motivates consumers to articulate themselves on the internet? J. Interact. Mark. **18**(1), 38–52 (2004)
9. Tse, Y.K., Zhang, M., Doherty, B., Chappell, P., Garnett, P.: Insight from the horsemeat scandal exploring the consumers' opinion of tweets toward tesco. Ind. Manag. Data Syst. **116**(6), 1178–1200 (2016)
10. Jeong, B., Yoon, J., Lee, J.M.: Social media mining for product planning: A product opportunity mining approach based on topic modeling and sentiment analysis. Int. J. Info. Manage. **48**(C), 280–290 (2019)
11. Kitano, Y., Yamada, T., Tan, K.H.: Technological innovation, new solutions, branding, and promotion: twitter and technical report use in japanese's companies. Enterp. Info. Sys. **15**(10), 1683–1712 (2021)
12. Han, X., Gu, X., Peng, S.: Analysis of Tweet Form's effect on users' engagement on Twitter. Cogent Business & Management **6**(1), (2019)
13. TOYOKEIZAI ONLINE: Ranking of the top 300 companies with the most advertising costs. https://toyokeizai.net/articles/-/187757. Accessed on 29 Oct. 2020
14. NTT DATA Mathematical Systems Inc.: Text Mining Studio. https://www.msi.co.jp/tmstudio/. Accessed on 24 Jan. 2021
15. Cohen, A.M., Hersh, W.R.: A survey of current work in biomedical text mining. Brief. Bioinform. **6**(1), 57–71 (2005)
16. NTT DATA Mathematical Systems Inc.: Text Mining Studio Manual ver. 6.0. (2018)
17. Social Media Trend: What is the best time of day to post. https://social-dog.net/trend/a-2. Accessed on 19 Jan. 2021
18. Maksym, G., Arthi, R., Augustin, C., Arnaud, L.: Social clicks: what and who gets read on twitter?. In: Proc. of the 2016 ACM SIGMETRICS/IFIP international Conf. on Measurement and Modeling of Computer Science, pp. 179–192 (2016)

Reflow Oven Recipe Optimization Approaches Based on Data-Driven Simulation

Zhenxuan Zhang, Yuanyuan Li, Sang Won Yoon, and Daehan Won(✉)

State University of the New York at Binghamton, Binghamton, NY, US
dhwon@binghamton.edu

Abstract. The temperature settings for the reflow oven chamber (i.e., the recipe) are critical to the quality of a Printed Circuit Board (PCB) in the surface mount technology because solder joints are formed on the boards with the placed components during the reflow. Inappropriate temperature profiles cause various defects, such as cracks, bridging, and delamination. Solder paste manufacturers generally provide the ideal thermal profile (i.e., target profile), and PCB manufacturers have attempted to meet the given profile by fine-tuning the oven's recipe. The conventional method tunes the recipe to gather thermal data with a thermal measurement device and adjusts the profile through trial-and-error. That method takes a lot of time and effort, and it cannot guarantee consistent product quality because it depends on the engineers' skills. In this paper, two approaches are introduced. The first uses a Random Forest Regression (RFR) model to generate the defect metric (DM) with different recipe inputs. DM is a customized measure calculated from the post-Automatic Optical Inspection (AOI). The RFR is trained with empirical, experimental data and serves as the objective function of an optimization model. The optimization model adopts an Evolution Strategy (ES) with an adaptive search region and identifies the best recipe. The proposed model has essential significance for the solder reflow process (SRP). The second approach is adopting a Backpropagation Neural Network to simulate the air temperature from the stage-based (ramp, soak, and reflow) input data segmentation to boost the computational efficiency and optimize the recipe settings according to the simulation. According to the requirements of Industry 4.0, the machine learning method is applied in this research to explore more information from the data to build an efficient simulation model. The application of the simulation model makes the optimization process efficient while saving a lot of experimental materials and time. The experimental results prove the effectiveness of the entire model. Specifically, the identified recipe reduces the defects by 54% compared with the original recipe. The model is consistent with the actual experimental results in the first approach, and the second method identified recipe shows 99% fitness in terms of R^2 to the targeted profile within 10 min of starting the experiment.

Keywords: Soldering Reflow Process · Simulation · Stage-based Segmentation · Optimization · Backpropagation Neural Network

C.-Y. Huang and S. W. Yoon (Eds.): ICPR1 2021, ACES 14, pp. 295–314, 2023.
https://doi.org/10.1007/978-3-031-44373-2_18

1 Introduction

Surface mount technology (SMT) has been widely applied in the electronics industry's printed circuit board (PCB) assembly systems for the past two decades. An SMT assembly system consists of several stages: screen printing, chip mounting, reflow, and inspection. Screen printing deposits solder paste into the apertures of a stencil mounted on the printed circuit board (PCB). A squeegee is then used to clear excess solder from the stencil, a process intended to leave the desired amount of solder paste on the PCB's pads [1]. In the chip mounting stage, PCBs usually travel on a conveyer belt through a line of placement machines [2]. The soldering reflow process (SRP) is the last step in a surface mount technology (SMT) line to attach components to a PCB, which is the focus of this study [3]. For the industry 4.0 standard, AI-based data-driven analytics, prediction, simulation, optimization, and control methods have been widely studied in recent years.

As recipes are the most important factor of SMT and SRP, many studies have aimed at finding the right recipe or building dynamic systems that adjust themselves to limit defects. Generally, the recipe is optimized for the SRP based on the solder paste thermal profile, which is the temperature range of the paste during its transit through the oven. [17, 18]. The solder paste manufacturer will provide a target thermal profile and some recommended parameters according to the solder paste composition. Usually, a well-shaped thermal profile from a proper recipe will generate high-quality PCB products. Thus, minimizing the gap between the target profile and the real thermal profile is the most common way to optimize the recipe and has been studied widely in recent years [17, 18]. For example, regression analysis and optimization methods are applied in one research to obtain the heating factor, along with a thermal profile measure, successfully achieve the target profile with their optimized recipe [18]. Some other literature optimizes the recipe based on the defect level. For instance, the grey-based Taguchi method is used to optimize the parameters during SRP to minimize defects such as bridging and spattering. The recommended optimal condition performs better than the initial condition [14]. Most studies focus on mechanical defects (i.e., voids, non-wetting, solder balls, etc.). Limited research is conducted on the solder fillet and pad overhang inspection. So, these three defects are studied in this research. As a popular tool in industry-related research, simulation has become the main trend in SRP studies to save the cost of experiments. Computational fluid dynamics (CFD), finite element (FE), and finite difference (FD) are all widely applied in thermal studies [14, 17, 19]. The numerical simulation model is used in some research to predict the thermal profile [18]. This research applies multiple machine learning (ML) methods to simulate the defect performance under different recipe settings. Specifically, a defect metric (DM) is designed based on the actual inspection tested value. Then, the ML-based simulation model is introduced as the objective function of the optimization model. A customized Evolution Strategy (ES) with a dynamic search region is proposed to minimize the defect metric.

In the target thermal profile, multiple key features are highly correlated with the solder joint quality: the ramping slope, the soaking time, the peak temperature, and the time above liquidus (TAL). In comparable studies of peak temperature and TAL, the sizes and spacing of the secondary precipitates Sn_3Ag_3 and Cu_6Sn_5 generated during reflow affected the quality of the solder joints [27]. As SAC solder joints form, hotter peak temperatures or longer TALs increase the number of secondary precipitates, decreasing the spacing. As SAC solder joints form, hotter peak temperatures or longer TALs increase the number of secondary precipitates, reduce the dislocation of components. With hotter peak temperatures or longer TALs, the thicker intermetallic compounds (IMC) layer forms as the solder paste melts. The layer is a compound formed from the copper substrate reacting with the tin in the solder paste [27]. Based on comparative research, a thin layer of IMC is good for the stability of the solder joint, but as thicker IMC layers are low ductile, they can result in brittle failures [28].

This research identifies the optimized reflow recipe settings to fit the target thermal profile. It is the simplest, most direct, and most effective method to predict the quality of a PCB when comparing the experimental profile to the target profile. The observed profile is obtained from the k-type thermocouples attached to the solder joints. A non-contact prediction model proposed by the previous research [3] is used for predicting the solder joint temperature to improve testing efficiency. It reduces the redundancy of the experiment of the predicted thermal profile and the target profile. The result can be regarded as an evaluation method of the oven status in real time for quality control.

The rest of this paper is organized as follows: Sect. 2 introduces some literature related to recipe optimization and defect study; Sect. 3 illustrates the methods applied in this research; Sect. 4 presents the experiment material, parameter settings, and results analysis; conclusion and future works are discussed in Sect. 5.

2 Background

For SMT, the smart manufacturing study trend uses online learning by connecting the machines through all stages to produce models that can be used to improve the final product. Those studies focus on the increase of the throughput and the quality of the products. To achieve the goals, multiple studies were done to optimize machine settings for the machines and adjust inspection thresholds.

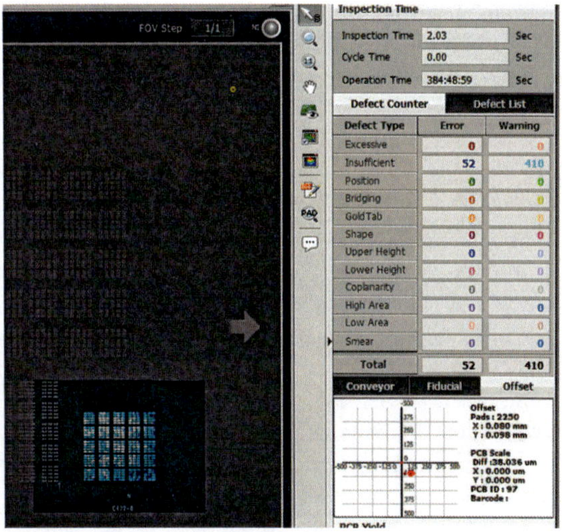

Fig. 1. Snapshot of the solder pastes inspection result interface for the Koh Young "aSPIre 3" SPI machine

In solder screen printing, multiple parameters need to be controlled. Data-driven machine learning algorithms are needed to control and optimize SMT because of the variety of solder pastes, the quality of the stencils, and the printing settings, which include printing pressure, the printing speed, the cleaning cycle, and the cleaning frequency. In recent studies, multiple AI-based approaches were used in the solder screen printing process. One of the reinforcement learning approaches, Q-learning, established an online control model for printing parameters to increase the throughputs [1]. Using machine learning algorithms, including regression tree (RT) and support vector regression (SVR), researchers studied the relationship between printing parameters and the solder paste volume. They used mixed-integer non-linear programming (MINLP) to optimize the parameters for any paste volume [4]. Researchers then developed a multi-objective optimization model with evolutionary strategies (ES) [5]. In 2019, a dynamic predictive model for volume detection with real-time memory updates in the printing process was developed and achieved over 92% R^2 coefficient [6]. The optimization model was developed with a guided evolutionary search optimization model for solder printing [7]. AI-based algorithms were used for solder printing anomaly diagnosis and prognosis with ensemble learning [8]. For all the stencil cleaning profiles developed, a classification was proposed with the approach of a convolutional neural network (CNN) [9]. Besides the printing parameters, the cleaning cycle studies were necessary. The AI-based approaches include applying a recurrent neural network (RNN) for predicting the cleaning cycle for the stencil [10] and then developing it into a boosting-based intelligent model for a stencil-cleaning prediction model, which can provide more efficient solutions [11]. The solder screen printing quality can be checked with the solder printing inspection (SPI) machines. For this study, the solder printing quality has been checked using the Koh Young "aSPIre 3" SPI machine, and the interface for the inspection results for the SPI machine is shown in Fig. 1.

A heuristic algorithm was proposed to speed up the mounting process in 2017 [2]. Multi-phase heuristics were proposed for optimizing dual-delivery placement on an assembly line to balance the workload of multiple mounters [12]. In 2018, an adaptive clustering-based genetic algorithm (GA) was proposed for optimizing the dual-gantry mounter machine, which resulted in reducing the total gantry moving distance by more than 5% [13].

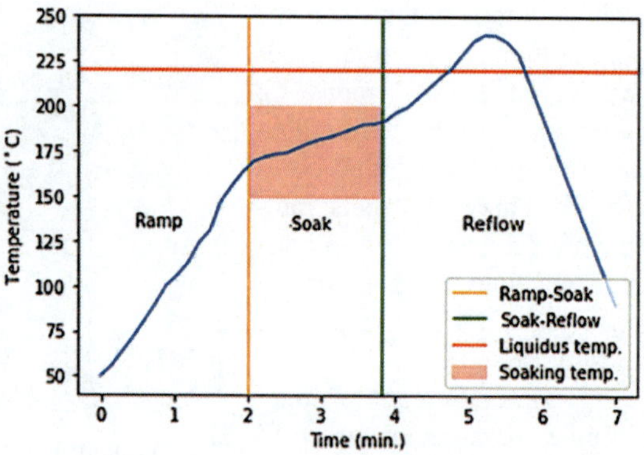

Fig. 2. Target profile of Indium 8.9HF SAC305 Pb-free paste [solder 36]

A forced convection reflow oven is widely adopted for SRP because of its high throughput and heat distribution. A typically forced convection reflow oven consists of several connected zones. Each zone has different temperatures. A PCB will go through all the zones on a conveyor system, and the solder paste between the board and component changes from liquid to solid to make permanent connections. The solder joint (the solid phase of solder paste) generation is controlled and affected by the temperature inside the oven [14]. Solder joint performance is also determining the PCB quality. An investigation showed that more than 80% of PCB defects are related to solder joints [15]. Also, the cost per defect during SRP is greater than the defects from other processes (i.e., stencil printing and component pick-and-placement processes) [16]. Therefore, the reflow oven temperature setting (i.e., recipe), which is highly related to product quality, is a critical factor of SMT. There are four temperature stages during SRP: preheating, soaking, reflowing, and cooling. The wrong parameter setting for any stage will lead to defects. For example, an improper ramp-up rate during preheating and reflowing stage will cause tombstoning, bridging, and void defects [16]. Thus, a proper recipe is necessary for SRP to avoid defects. In this study, based on the manufacturer's datasheet, the target TAL is 60 s, with a recommended range of 45–90 and an acceptable range of 30–120. The target peak temperature is 240 °C, as shown in Fig. 2, with a recommended range of 230–250 °C and an acceptable range of 220–260 °C.

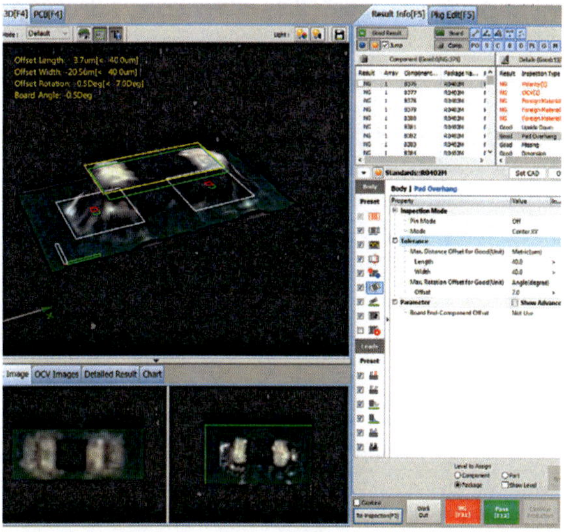

Fig. 3. Automatic Optical Inspection (AOI) result interface for the Koh Young "Zenith" AOI machine

The primary determinant of a thermal profile is the environment inside the reflow oven, which includes, but is not limited to, the temperature settings, blow rates, and conveyor speed. Reflow ovens with forced convection are widely used in SMT assembly lines. It is possible to handle high throughput with this reflow oven, and the heat is evenly distributed across the PCBs. The instruments used for this project are the Heller 1700W and the Heller 1707MKEV forced convection reflow oven, which contains six and seven heating zones, respectively, followed by one cooling zone. Only one conveyor connects different zones inside the chambers in those reflow ovens. Heat is transferred from the heated air to the board, component, and solder paste inside the oven chamber. The heat transfer efficiency determines the thermal profile. The temperature performance of forced convection reflow ovens during SRP has been extensively studied. Based on test results, one of the studies shows that heat transfer coefficients differ between periods [29]. The heat transfer coefficient should be obtained and applied to the prediction model for evaluation purposes. This study calculates the heat transfer coefficients for each zone separately. The prediction and optimization model is used in stages with a thermal profile segmented by the periods in the reflow process.

Automatic Optical Inspection (AOI) is widely adopted in the SMT line to inspect the component performance by obtaining PCB images from an optical apparatus such as a camera [20]. The AOI machines can inspect the quality of the solder joints, in the lab for conducting our experiments is the Koh Young "Zenith" AOI machines for the pre-and post-AOI inspections. The interface for the inspection results for the AOI machine is shown in Fig. 3. Pre-AOI is used to inspect the pick-and-place performance before SRP. Post-AOI checks the component status after SRP. The AOI machine used in this research is from Koh Young Technology Inc. Multiple checkpoints can be tested during the inspection. Examples include component missing inspection, pad overhang inspection, coplanarity inspection, upside-down component inspection, and solder fillet and offset inspection. According to previous experiments, the pad overhang and solder fillet are the focus of this research because of the high frequency of these defects. The location of the components is checked during the pad overhang inspection. The solder fillet inspection checks the shape of each solder joint. AOI will provide the inspection results, such as offset value for every pad on the board. According to the actual data measured by the AOI machine and the target value of each inspection, the DM can be obtained. Thereby, the simulation model can be trained, and the final optimal solution can be obtained from the optimization model. The proposed recipe optimization framework has a low requirement for the experiment because of the usage of the simulation model. With the efficient optimization model, product quality will be improved, and throughput will also be increased. More importantly, the application of AOI machine and ML methods satisfied the requirement of Industry 4.0 with high efficiency and automation degree. The proposed model can be widely applied to the SMT domain for recipe optimization.

Various research is conducted on optimizing reflow oven recipes because of the significance of an SMT line. Optimization based on the thermal profile shape is one crucial direction. For example, FE was applied to simulate the Ball Grid Array (BGA) thermal profile under different recipes. The simulation model was developed with ANSYS, which is a commonly used simulation software in thermal studies. A first-order optimization algorithm coded in ANSYS was used to optimize the SRP-related parameters, including maximum temperature and temperature ramp-up rate. The optimal solution showed an xx percent reduction in temperature- and stress-related defects from the initial parameters, which proves the high efficiency of the simulation and optimization model [19]. Another research adopted regression as the thermal profile simulation model. Different heat factor values were investigated to determine the best candidate to minimize the gap between the target thermal profile and the practical one. The optimized setting achieved the target thermal profile and was suggested to be applied to similar products [20]. The thermal stress distribution is another popular direction that other researchers study in SMT related research. The FE method remains widely adopted in much research in the thermal stress domain. It was used in a study to simulate the thermal distribution and provided information for the gray-based Taguchi design to pursue a better thermal stress distribution. Various related factors were investigated, such as board density, cooling temperature, inlet velocity, and conveyor speed. According to the Analysis of Variance (ANOVA) results, inlet velocity significantly influences the thermal distribution. The best setting that will lead to an even thermal distribution was successfully identified [21].

Defect minimization is an essential branch of SRP study. The common defect and corresponding reasons are discussed in [22], which provides insights for this research. An

RSP optimization model was developed in a recent study. CFD simulates the product quality of different parameters (i.e., soak temperature, peak temperature, etc.). Then, Taguchi design and ANOVA are applied to determine the optimal parameter settings. Experimental results show that the performance characteristics improved significantly compared to initial settings [14]. Shear force and warpage were discussed in the literature [23]. An Artificial Neural Network (ANN) was established to predict the shear force with actual Design of Experiment (DOE) data. Then, an optimization model was developed to maximize the shear force. The confirmation experiment results proved that the optimal settings increase the shear force significantly. The recipe was optimized to improve the solder joint reliability in assessments such as fatigue test performance and obtained promising results [24]. However, limited research studied the pad overhang and solder fillet issues. So, these defects are discussed in this research.

The temperature has been widely studied because it is the most critical factor in the SRP. The two major research directions are simulation-based and experiment-based. Experiment-based studies produced many significant conclusions. The comparable research projects show that the heat transfer effects on characteristics of the PCB boards and components change with the board material, board size, board thickness, component thickness, and the density of the components on the board. In one of the comparable research projects, the results show that the time to reach the melting point on the surface has a linear relationship with the thickness of the board. The temperature difference between the surface and middle plane when the PCBs reach peak temperature is under 10 °C, which can be considered negligible [30]. In research on PCBs with different sizes, thicknesses, and heights, the results show the surface sizes of the boards do not significantly affect heat transfer as different-sized boards have close results for thermal profiles. The results show that boards traveling higher in the oven are exposed to tighter thermal profiles. The temperatures of thinner boards have larger heating factors and can be heated and cooled faster with a higher peak temperature under the same recipe settings. Thicker boards have smaller heating factors and slower passive cooling speeds, but the temperature falls more significantly in the cooling zone [31].

The comparative studies used ANN, NLP, and GA for the machine learning optimization approaches in the reflow setting optimization studies [32, 33]. From the comparative studies, the heating factor Q_n is presented as a comprehensive formulation of the peak temperature T_p and TAL [32]. The backpropagation neural network (BPNN), one of the ANN approaches, was introduced to describe the non-linear relationship between the reflow settings and the thermal reflow profiles. With each period's upper and lower bound constraints, the problem can be formulated as an NLP and solved to get sets of optimal solutions. The GA is widely used to find the global optimal solution among the optimized reflow settings.

Machine learning and artificial intelligence are widely applied in many domains to achieve classification and prediction functions in the big data era. By inputting factors such as soak time, reflow time, and peak temperature in the SMT domain, ANNs were applied to predict the shear force tolerance of the reflowed solder joint. Good accuracy was obtained when comparing the prediction results with the experimental results [19]. ANN has many advantages; for example, it is very good at handling non-linear data with high generalization capability. A neural network model fits this research well because of the nonlinearity of the data and the need to apply the proposed model to unknown data. The comparative studies show that the thermal profile can be well-predicted from the reflow settings, which means the optimized reflow parameters can be well-optimized from the ideal target thermal profile. This study proposes a multi-stage BPNN model to predict the zone air temperature from the target thermal profile.

3 Methodology

3.1 Defect Metric

A customized DM is proposed in this research to quantify the performance of the AOI. Because the AOI machine only provides the inspection result with "good" or "no good" for a PCB, a DM could illustrate the detailed status needed to improve the performance. The DM is calculated with the following equation:

$$DM = \alpha \frac{1}{n} \frac{1}{m} \sum_{i=1,j=1}^{i=m,j=n} \frac{\left|U_{ij} - X_{ij}\right|}{U_{ij}} + P_{ij}, \tag{1}$$

, where α is the weight for the different checkpoints, m is the number of checkpoints, n is the number of pads on the board, U_{ij} represents the target value, X_{ij} is the actual tested value, P_{ij} is the penalty value to increase the DM when the pre-AOI shows worse results than the post-AOI on the same checkpoint. There are three steps to obtain the penalty value:

1. Calculate all the gaps between the inspected value and target value from pre-AOI results.
2. Sort the data, divide it into four groups from small to large, and then calculate each group's average for each component type.
3. Check the rank of the component, and the corresponding group average value is the penalty value that will be added to DM.

3.2 Simulation Model

A simulation model can be established from the DM. In this research, five machine learning models are tested: Decision Tree Regression (DTR), Support Vector Regression (SVR), Adaboosting Regression (ABR), Gradient Boosting Regression (GBR), and Random Forest Regression (RFR). Those regression models are widely applied in industry, health care, manufacturing, and agriculture. More details and applications of regression models used in this research can be found in [25]. The five regression models tested with different numbers of training data will be discussed in Sect. 4.

3.3 Optimization Model

Evolution strategy (ES) is a biological evolution-inspired optimization method. The concept of ES is to implement a particular process of stochastic variation multiple times with the following rules. New offspring are generated from parents in each generation, and the fitness is evaluated to identify the best offspring. The best offspring will be selected as the parents of the next generation. Here, the offspring can be understood as a candidate solution, and parents are the candidate solutions that have been checked [26]. More detail can be found in [26]. In this research, ES is selected as the main optimization model. During optimization, the search region is updated dynamically according to the current candidate solution, which is the recipe setting in this research. Because of the high dimension of the recipe, the convergency efficiency is low, which harms the local optimal solution. The proposed method with an adaptive search region helps limit and determine the search region, improving the convergence speed. The proposed Adaptive Evolution Strategy Search (AESS) details are presented in Algorithm 1.

3.4 Recipe Initialization Method (RIM)

The recipe initialization method (RIM) is used to obtain the initial recipe for collecting the data to train the model. In manufacturing lines, the initial recipe is usually designed by engineers with expertise in the SMT assembly line, especially those experienced in the reflow process. With the knowledge of thermal conduction and convection during the reflow process, along with the experience and results from the previous studies, the conveyor speed of 35 cm/minute is determined by calculation of the measured total length of the reflow oven, along with the time recommended in the target thermal profile. The blowing rate of the blowers is set as the default setting, 100%.

The conveyor speed is constant in the reflow oven, and the peak temperature on the board is obtained when the board leaves the last heating zone so that the time axis can coordinate with the reflow oven's length. According to the location of the board in the reflow oven, the target profile can be segmented accordingly to match the corresponding zones, and the maximum temperature of the corresponding thermal profile for each zone will be set as the recipe for each zone; for the zones corresponding to the reflow period, zone 6 and 7, 20 °C are added to the maximum temperature based on the experience gained from previous studies. The initial recipe is shown in Table 1.

Table 1. Initial recipe from RIM for the seven-zone oven.

Zone 1 (°C)	Zone 2 (°C)	Zone 3 (°C)	Zone 4 (°C)	Zone 5 (°C)	Zone 6 (°C)	Zone 7 (°C)
90	131	172	182	190	237	260

3.5 Stage-Based Segment Data Collection

After determining the initial recipe settings, the next stage is to prepare the data. For the model proposed, only one experiment is required to collect the data. In this study, ten replicates were performed, and after eliminating noise from the data collection phase, the average was used for the model training. The data was collected from boards traveling through the oven under the initial recipe, the RIM.

The data obtained from the experiment include the board and the zone air temperature above the board. The board temperature and the zone air temperature obtained were split into segments according to the stages of the corresponding heating zones in the reflow oven. The data was divided into seven zones, and the zones in the same period, i.e., ramping, soaking, and reflowing, were combined. In the end, the data were split into five segments with five stages, namely (1) room temperature to ramping corresponding to zone 1; (2) ramping corresponding to zone 2; (3) ramping to soaking corresponding to zone 3; (4) soaking corresponding to zones 4 and 5; and (5) reflowing corresponding to zones 6 and 7. The data segments corresponding to the five stages were applied to the model. The solder joint thermal profile is the model's input, and the simulated zone air temperature is the model's output. The center point of each zone's predicted air temperature has been set as the reflow recipe of the heating zones. The process is shown in Fig. 4.

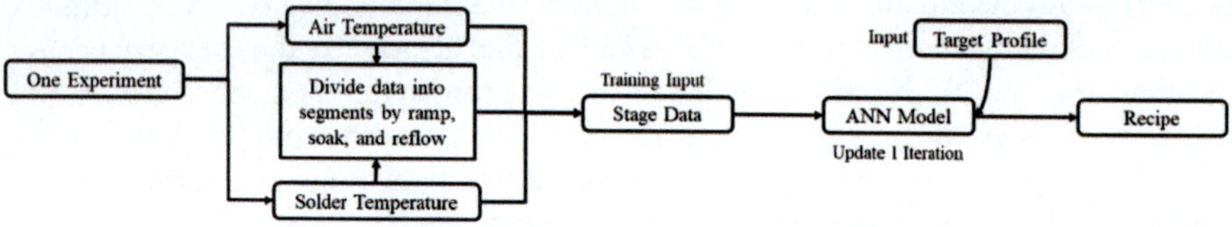

Fig. 4. The flow of the proposed reflow parameter optimization model [36]

3.6 Zone Air Temperatures Prediction Model

For the stage-based segmentation approach, with the segmented data, a multi-stage BPNN model with five layers was constructed and trained in Python. The model takes the joints' thermal profiles as the input, passing through the three fully connected hidden layers. The predicted zone air temperature is the output of the model. According to the previous subsection, the optimized reflow recipe is obtained accordingly. Each hidden layer has 100 neurons. The activation functions used in the model are rectified linear units (ReLu) for each of the hidden layers, and the linear activation function is used in the output layer. As for the optimizer of the multi-stage BPNN, the adaptive moment (Adam) estimation is used. The framework of the model is shown in Fig. 5.

The 3 hidden layers are fully connected, meaning that each node in the first hidden layer is directly connected to every node in the second hidden layer. Each node in the second hidden layer is directly connected to every node in the third hidden layer. Because the hidden layers are fully connected, the input data would be processed through every node during every iteration, and the weight would be updated after each iteration. With the stage-segmented data as inputs, a 3-hidden-layer construction results in a promising outcome compared with other model constructions. Meanwhile, the computing time is more than ten times faster than complicated constructed neural network models.

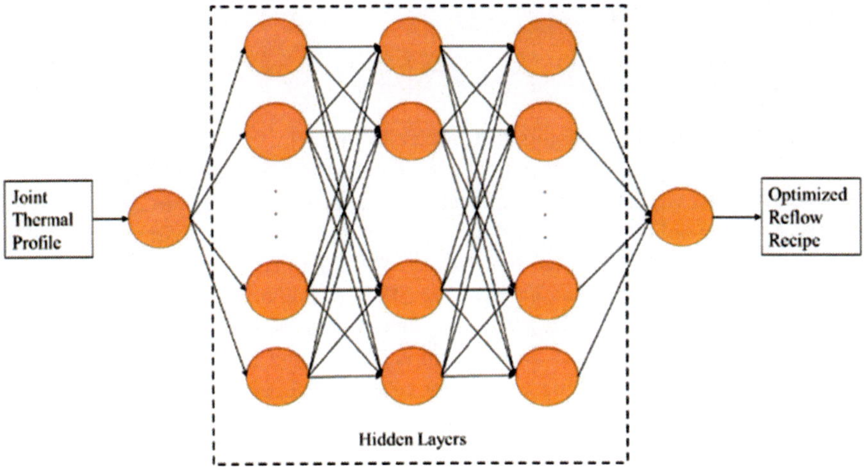

Fig. 5. Multi-stage BPNN five-layer framework of the proposed reflow parameter optimization model [36]

3.7 Recipe Estimation Method

The final stage is the test data's optimization model for the reflow recipe. The test data input is the manufacturer's recommended target thermal profile because its recipe aims to optimize the reflow setting. Hence, the actual profile fits the recommended profile as much as possible. In this research, the multi-stage BPNN optimization model is trained for each of the five stages sequentially as the stages flow in the RSP. The well-trained model was applied to the segmented target thermal profile corresponding to the five stages to predict the zone air temperature in each stage. With the predicted zone air temperature for each stage, the zone air temperature in the reflow oven can be estimated. Because the estimated zone air temperature in the reflow oven is time-series data, the optimized reflow recipe settings $\hat{R}$ can be estimated according to Eq. (2). t_s indicates when a point on the board, called the time-measuring point, enters the zone. t_e indicates when the time measuring point leaves the zone. $\hat{T}_a(t)$ indicates the estimated zone air temperature where the measuring point enters the reflow oven at time t.

$$\hat{R} = \frac{1}{t_e - t_s} \sum_{t_s}^{t_e} \hat{T}_a(t), \tag{2}$$

In this research, the experiments are performed, and the optimization model is applied to the solder joint temperature under the passive components. According to a comparative study, the solder joints' temperature of the passive components is almost the same as the board temperature [34]. The components used in this research are passive capacitors and resistors with sizes of 0.4 × 0.2 mm, 0.6 × 0.3 mm, and 1 × 0.5 mm. The experiment results and the validating results of the optimized reflow recipe will be discussed in Sect. 4.

4 Experiment and Results Analysis

4.1 Design of Experiments for Different Recipes

To obtain the training and validation data, several experiments were conducted. The Design of Experiments (DOE) covered 9 combinations of the recipe. Experiments with four combinations were conducted to create testing data to validate the simulation. In this research, to avoid the thermal profile shape change of preheating part and reflowing parts, only the middle 4 zones are selected. $L_9(3^4)$ is applied to design an experiment with 4 factors and 3 levels for each factor. The DOE settings and practical experimental results are summarized in Table 2. The testing data and results are summarized in Table 3.

The reflow oven used for the data collection and optimization model experiment is a Heller 1700W with six heating zones, one cooling zone, and a temperature control accuracy of $\pm$ 3(°C). As mentioned, the AOI machine to collect the inspection data is from Koh Young Technology Inc. The testing PCB is a $15cm \times 16cm$ FR-4 glass epoxy board. Three capacitors with different sizes are on the board, which are 0402, 0603, and 1005. The number of each component is 750 pieces.

Table 2. $L_9\left(3^4\right)$ DOE results of different recipes.

Zone 2 (°C)	Zone 3 (°C)	Zone 4 (°C)	Zone 5 (°C)	DM
140	155	170	190	2.96
140	165	180	200	2.88
140	175	190	210	2.50
150	155	180	210	2.30
150	165	190	190	1.20
150	175	170	200	2.55
160	155	190	200	2.35
160	165	170	210	1.75
160	175	180	190	1.43

Table 3. Testing data.

Zone 2 (°C)	Zone 3 (°C)	Zone 4 (°C)	Zone 5 (°C)	DM
145	160	180	195	2.43
150	175	175	200	2.15
160	170	185	200	1.86
165	170	185	210	1.11

Algorithm 1 Adaptive Evolution Strategy Search (AESS)

Initialization: Define a P population size, G generations, M iterations, and current solution of ith input CS_i, initial search region $SR_{1_{Low}}$ and $SR_{1_{Up}}$, and search region limitations of ith input $SR_{i_{Low}}$ and $SR_{i_{Up}}$.
for $i < M$ **do**
  for $\theta = 1, 2, \ldots, G$ **do**
    for $\rho = 1, 2, \ldots, P$ **do**
      Select a subset of individuals from P
      Generate the offspring of the selected individuals within $SR_{i_{Low}}$ and $SR_{i_{Up}}$
      Implement mutation for the offspring
      Evaluate the fitness of the new offspring and select the best offspring
    end for
  end for
  if $i > 3$ and the maximum gap among CS_{i-2}, CS_{i-1}, and CS_i is less than 10 **then**
    $SR_{i+1_{Low}} = Min(CS_{i-2}, CS_{i-1}, CS_i)$; $SR_{i+1_{Up}} = Max(CS_{i-2}, CS_{i-1}, CS_i)$,
  end if
end for

The stage-based segment data collection and recipe optimization model experiment is conducted on the Heller 1707MKEV reflow oven with a temperature control accuracy of $\pm$ 3(°C). Three components, 0402M, 0603M, and 1005M, were soldered to the testing board, and each component had 250 pieces. The temperature is measured by the Mega MOLE with 20-channel K-type thermocouples. As shown in Fig. 6 the thermocouples are attached at the 4 corners and the center of the board, and an additional thermocouple that stands vertically 1 *cm* above the board measures the air temperature.

Fig. 6. Thermocouples with Mega M.O.L.E profiler attached on testing PCB

The experiment was conducted with the initial recipe mentioned in Sect. 3.2. After the data is collected from the investigation, the temperature of the solder joints and the zone air temperature collected above the measured solder joints are used for training the model. In actual production lines, fewer experiment data provides multiple advantages, including faster obtaining of the optimized result, less material waste, and less labor. This AI-based approach requires only the sample data from one experiment for training the model. After training, the model was tested with 7 different profiles as "target profiles." The optimized recipe obtained for each profile was validated with 1 experiment. Cross-validation was performed using the experimental data from the different recipes as training input and validated with test cases.

After training the model, the optimization can be conducted using the target thermal profile as the input data. The optimized reflow recipe was validated with one more experiment, and the performance is evaluated based on the *R*2, and the root mean square error (RMSE). The sum squared regression (SSR) shows the variance explained in the regression can be calculated with Eq. (3), and the total sum of squares (SST) can be calculated with Eq. (4), R^2 is an indicator to show the fitness of the thermal profile from the optimized recipe compared to the target profile and can be calculated with Eq. (5), and the RMSE is a measurement that indicates the difference between the optimized recipe result to the target profile and can be calculated with Eq. (6). $\hat{y}_i$ represents the predicted value, and $\overline{y}_i$ represents the mean value. Based on the recommended and acceptable range of the peak temperature from the manufacturer's datasheet, the reference lever of the RMSE is 10, which means if the RMSE exceeds 10, the optimized recipe is not acceptable.

$$SSR = \sum_{i=1}^{i=n} (\hat{y}_i - y_i)^2, \tag{3}$$

$$SST = \sum_{i=1}^{i=n} (\overline{y}_i - y_i)^2, \tag{4}$$

$$R^2 = 1 - \frac{SSR}{SST} \tag{5}$$

$$RMSE = \sqrt{\frac{1}{n} \sum_{i=1}^{i=n} (\hat{y}_i - y_i)^2}, \tag{6}$$

4.2 Simulation Model Selection and Parameter Setting

Five regression models are tested with different numbers of training data. The size is from 6 to 9, and the training data are randomly selected from Table 2. To be specific, when testing data size is 6, 7, or 8, four combinations of the different number of runs are randomly selected. However, for the condition when the data size equals 9, only one combination, which includes all the data, is used. The results of the average gap between

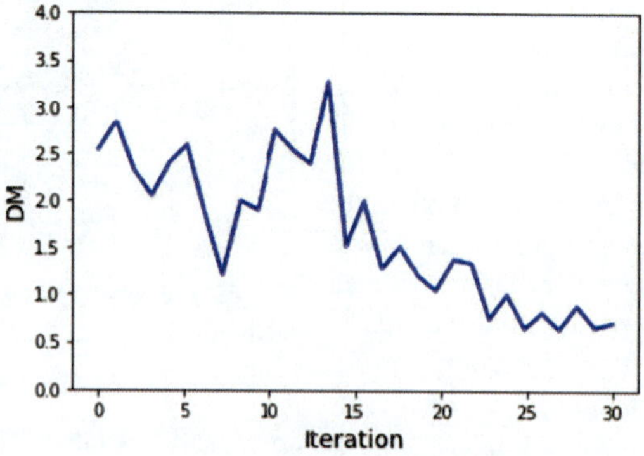

Fig. 7. Optimization results [35]

the target value and prediction results are summarized in Table 4. According to the prediction results, RFR outperforms other regression models. Also, the gap value obtains the lowest value when there are six input data. So, RFR with six input data is determined as the simulation model in this research to predict the defect level with different RTRs input and will be further used in the optimization model, which will be discussed in the later chapter.

Table 4. The average DM gap between the target value and prediction result.

Input data	DTR	SVR	ABR	GBR	RFR
6	0.32	0.63	0.54	0.61	**0.05**
7	0.83	0.40	0.83	0.91	**0.37**
8	0.92	0.44	0.28	1.02	**0.29**
9	0.84	1.01	**0.54**	0.74	0.59

4.3 Optimization and Results Analysis

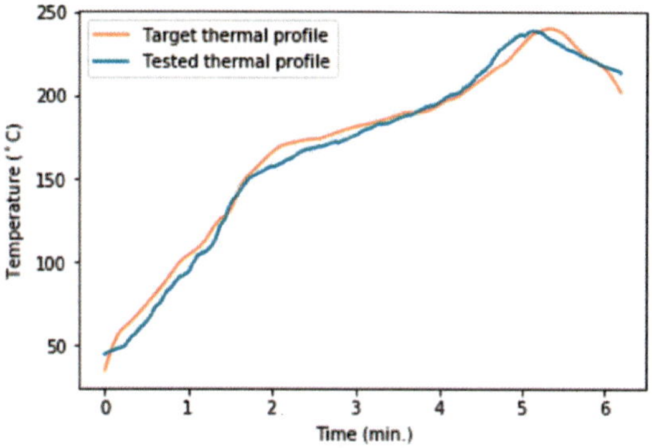

Fig. 8. Thermal profile of optimized [35] recipe.

The objective of the optimization model is to minimize DM. RFR will generate the DM according to the recipe provided by the optimization model. The parameters of the optimization models are presented in Table 5. The parameter α, which is the weight information, is defined as 1 in this research. The value can be changed in a future study with more information on the checkpoint significance. The optimization result of 30 iterations is shown in Fig. 7. The final optimized DM is 0.65, and the corresponding recipe settings are 157, 165, 193, and 210. Table 6 presents three iterations with the solution and corresponding DM value. The values are similar to the final optimal solution with some small gaps, which shows the reliability of the optimization model. One confirmation experiment was conducted and produced a 0.62 DM. The thermal profile is tested with a practical experiment to validate the optimal recipe's performance further. According to Fig. 8, the tested thermal profile fits the target profile well, and the value is 0.97, proving the optimized recipe's reliability.

Table 5. Parameter setting.

Parameter	Value
α	1
P	100
G	50
M	30
$SR1_{Low}$	110, 125,140, 160
$SR1_{Up}$	190, 205, 220, 240

Table 6. Optimization process examples.

Iteration	DM	Zone 2 (°C)	Zone 3 (°C)	Zone 4 (°C)	Zone 5 (°C)
25	0.65	154	170	193	210
27	0.64	157	170	190	208
29	0.66	158	166	192	210

4.4 Stage-Based Segment Model Result

The experimental results are shown in Figs. 9 and 10. The R^2 fitness score increased from 0.92 to 0.99. The RMSE was reduced by 65.2%. The optimization model requires only one iteration to obtain the optimized recipe, and the calculation time for this model is less than 5 s. The optimized recipe can be brought with one experiment and validated with one more experiment.

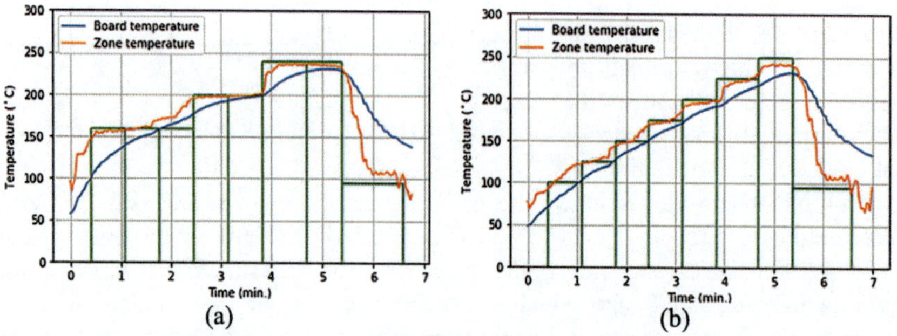

Fig. 9. (a) Initial recipe result ($R^2 = 0.92$, RMSE = 14.09), (b) Random Initial Recipe generated from arithmetic series [36]

Because the model used a random recipe from an identical product for the initial training data, it is generally applicable to any recipe. Two different initial recipes were used for validating the performance, and the model obtained an optimized recipe setting with the same performance. Even with the limited experience in the field of the SMT assembly line, the optimized recipe can be obtained within 10 min, including the experiment time. Also, in the same reflow oven and production (boards and components), with different target thermal profiles provided, the optimization can be performed with the data from one experiment. Another advantage of this multi-stage reflow parameter optimization model is separating the zones by the corresponding stages. This model can be extended to different reflow ovens with a different number of zone and designs.

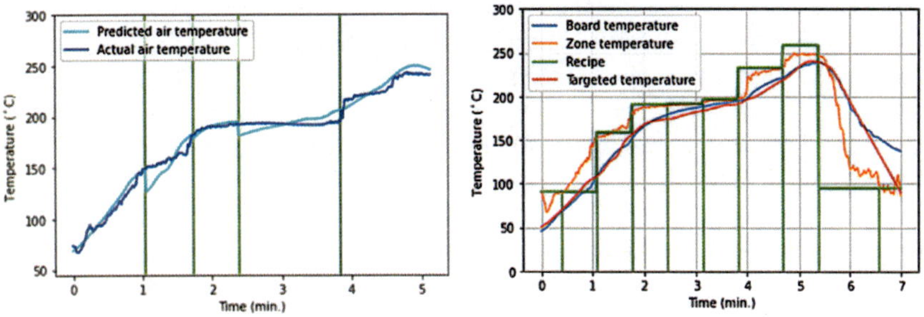

(a) (b)

Fig. 10. (a) predicted and validated air temp., (b) optimized recipe ($R^2 = 0.99$, RMSE = 4.91) [36]

5 Conclusion and Future Work

A recipe optimization model is proposed in this research. Specifically, a DM is designed to represent the quality of the PCB. Then, the RFR model is applied as the simulation model to predict the DM with different recipe inputs. Also, the simulation model is the objective function of the optimization model. AESS is proposed in this research to improve convergency efficiency. The optimized recipe setting reduces the DM value by 54%. The recipe is tested with the confirmation experiment to validate the performance of the optimal solution. The experiment result keeps consistent with the optimization model by showing a gap of 0.03. Experimental and optimization results show that defects are minimized significantly, which proves the promising ability of the RFR-based simulation model and the AESS-based optimization model.

With the multi-stage BPNN optimization model, the optimized result can be obtained within 10 min. The training process takes the actual solder joint temperature as the input. In this study, the actual thermal profile of the passive components solder joints has been proved close to the board temperature, which is easy to obtain. The optimizing process takes the target thermal profile as the solder paste manufacturer's input. From comparing the initial and optimized recipe results, the R^2 and RMSE has been improved by 7.6% and 65.2%, respectively. In addition, multiple advantages can be found with the proposed model. For instance, the optimization process with the multi-stage BPNN model does not require any SMT-related field experience for subjective judgment, which increases the automation and fulfillment of the industry 4.0 requirements. In addition, compared to taking the complete profile data as the model's input, the 5-staged segmented data leads to a significantly shorter computation time.

However, there are some limitations to this research. The proposed model mainly applies to passive components on the PCBs. For the larger components and packages, the temperature of the solder joints underneath the package could have some gap with the passive component solder joints. Therefore, an adaptive optimization model for the reflow recipe should be proposed that satisfies both the passive components and the large-sized packages (e.g., BGAs) to be close to the target thermal profile. Moreover, because the solder joint temperature underneath the bigger-sized packages is hard to measure, a prediction model should be based on the size and thickness of the large-sized components. That research would increase the possibility of understanding the relationship between the thermal profiles of the passive components and oversized packages and propose a model that can eventually provide the optimal solution to satisfy all the components on the same board.

Acknowledgment. This work was partly supported by the Integrated Electronics Engineering Center for Advanced Technology in Electronics Packaging of Binghamton University.

References

1. Khader, N., Yoon, S.W.: Online control of stencil printing parameters using reinforcement learning approach. Procedia Manufacturing **17**, 94–101 (2018)
2. He, T., Li, D., Yoon, S.W.: A heuristic algorithm to balance workloads of high-speed SMT machines in a PCB assembly line. Procedia Manufacturing **11**, 1790–1797 (2017)
3. Li, Y., He, J., Won, D., Yoon, S.W.: Noncontact reflow oven thermal profile prediction based on artificial neural network. IEEE Trans. Compon. Pack. Manuf. Technol. **11**(12), 2229–2237 (2021)
4. Khader, N., Lee, J., Lee, D., Yoon, S.W., Yang, H.: Multi-objective optimization approach to enhance the stencil printing quality. Procedia Manufacturing **38**, 163–170 (2019)
5. Khader, N., Yoon, S.W.: Stencil printing process optimization to control solder paste volume transfer efficiency. IEEE Trans. Compon. Pack. Manuf. Technol. **8**(9), 1686–1694 (2018)
6. Lu, H., Wang, H., Yoon, S.W., Won, D., Park, S.: Dynamic predictive modeling of solder paste volume with real time memory update in a stencil printing process. Procedia Manufacturing **38**, 108–116 (2019)
7. Lu, H., He, J., Won, D., Yoon, S.W.: A guided evolutionary search approach for real-time stencil printing optimization. IEEE Trans. Compon. Pack. Manuf. Technol. **11**(2), 333–341 (2020)
8. Alelaumi, S., Wang, H., Lu, H., Yoon, S.W.: A predictive abnormality detection model using ensemble learning in stencil printing process. IEEE Trans. Compon. Pack. Manuf. Technol. **10**(9), 1560–1568 (2020)
9. Alelaumi, S., He, J., Li, Y., Khader, N., Yoon, S.W.: Cleaning Profile Classification Using Convolutional Neural Network in Stencil Printing. IEEE Trans. Compon. Pack. Manuf. Technol. **11**(11), 2003–2011 (2021)
10. Wang, H., He, T., Yoon, S.W.: Recurrent neural network-based stencil cleaning cycle predictive modeling. Procedia Manufacturing **17**, 86–93 (2018)
11. Wang, H., Lu, H., Won, D., Yoon, S.W., Srihari, K.: A boosting-based intelligent model for stencil cleaning prediction in surface mount technology. Procedia Manufacturing **38**, 447–454 (2019)

12. He, T., Li, D., Yoon, S.W.: A multi-phase planning heuristic for a dual-delivery SMT placement machine optimization. Robo. Comp.-Integr. Manuf. **47**, 85–94 (2017)
13. He, T., Li, D., Yoon, S.W.: An adaptive clustering-based genetic algorithm for the dual-gantry pick-and-place machine optimization. Adv. Eng. Inform. **37**, 66–78 (2018)
14. Lau, C.-S., Abdullah, M.Z., Khor, C.Y.: Optimization of the reflow soldering process with multiple quality characteristics in ball grid array packaging by using the grey-based Taguchi method. Microelectronics International (2013)
15. Kong, F.-H.: A new method of inspection based on shape from shading. In: 2008 Congress on Image and Signal Processing, vol. 2, pp. 291–294. IEEE (2008)
16. Tsai, T.-N.: Modeling and optimization of reflow thermal profiling operation: a comparative study. J. Chinese Inst. Indu. Eng. **26**(6), 480–492 (2009)
17. Gong, Y., Li, Q., Yang, D.G.: The optimization of reflow soldering temperature profile based on simulation. In: 2006 7th International Conference on Electronic Packaging Technology, pp. 1–4. IEEE (2006)
18. Gao, J., Wu, Y., Ding, H.: Optimization of a reflow soldering process based on the heating factor. Soldering & Surface Mount Technology (2007)
19. Whalley, D.C.: A simplified model of the reflow soldering process. In: ITherm 2002. Eighth Intersociety Conference on Thermal and Thermomechanical Phenomena in Electronic Systems (Cat. No. 02CH37258), pp. 840–847. IEEE (2002)
20. Song, J.-D., Kim, Y.-G., Park, T.-H.: SMT defect classification by feature extraction region optimization and machine learning. The Int. J. Adv. Manuf. Technol. **101**(5), 1303–1313 (2019)
21. Lau, C.-S., Abdullah, M.Z., Che Ani, F.: Optimization modeling of the cooling stage of reflow soldering process for ball grid array package using the gray-based Taguchi method. Microelectronics Reliability **52**(6), 1143–1152 (2012)
22. Rajewski, K.: SMT process recommendations. Defect minimization methods for a no-clean SMT process. In: IEEE Technical Applications Conference and Workshops. Northcon/95. Conference Record, pp. 354–362. IEEE (1995)
23. Lin, Y.-H., Deng, W.-J., Shie, J.-R., Yang, Y.-K.: Optimization of reflow soldering process for BGA packages by artificial neural network. Microelectronics International (2007)
24. Geczy, A., Szőke, P., Zsolt, I.-V., Miklos, R., Radu, B.: Soldering profile optimization for vapour phase reflow technology. In: 2011 IEEE 17th International Symposium for Design and Technology in Electronic Packaging (SIITME), pp. 149–152. IEEE (2011)
25. Rathore, S.S., Kumar, S.: A decision tree regression based approach for the number of software faults prediction. ACM SIGSOFT Software Engineering Notes **41**(1), 1–6 (2016)
26. Hansen, N., Arnold, D.V., Auger, A.: Evolution strategies. In: Springer handbook of computational intelligence, pp. 871–898. Springer, Berlin, Heidelberg (2015)
27. Wentlent, L.A., Genanu, M., Alghoul, T.: Effects of laser selective reflow on solder joint microstructure and reliability. In: 2018 IEEE 68th Electronic Components and Technology Conference (ECTC), pp. 425–433. IEEE (2018, May)
28. Pan, J., Chou, T.C., Bath, J., Willie, D., Toleno, B.J.: Effects of reflow profile and thermal conditioning on intermetallic compound thickness for SnAgCu soldered joints. Soldering & Surface Mount Technology (2009)
29. Illés, B.: Distribution of the heat transfer coefficient in convection reflow oven. Appl. Thermal Eng. **30**, 1523–1530 (2010)
30. Straubinger, D., Bozsóki, I., Bušek, D., Illés, B., Géczy, A.: Modelling of temperature distribution along pcb thickness in different substrates during reflow. Circuit World (2019a)
31. Alaya, M.A., Géczy, A.: Effect of pcb thickness and height position during heat level type vapour phase reflow soldering. In: In 2019 42nd International Spring Seminar on Electronics Technology (ISSE), pp. 1–6 (2019)

32. Tsai, T.N.: Thermal parameters optimization of a reflow soldering profile in printed circuit board assembly: A comparative study. Appl. Soft Comput. **12**(8), 2601–2613 (2012)
33. Tsai, T.N.: TCW014. Development of a closed-loop diagnosis system for reflow soldering using neural networks and support vector regression. International Journal of Industrial Engineering 21(1)
34. Schüßler, F.K.D., Franke, J.: Influences on the reflow soldering process by components with specific thermal properties. Circuit World **35**(3), 35–42 (2009)
35. Li, Y., Won, D., Yoon, S.W.: Reflow Oven Recipe Optimization Based on Simulation. PRISM 30 Special Sessions (2021)
36. Zhang, Z., Li, Y., Yoon, S.W., Won, D.: Reflow Thermal Recipe Segment Optimization Model Based on Artificial Neural Network Approach. Lecture Notes in Mechanical Engineering (2022)

Optimization in Pharmacy Automation System

Nieqing Cao[1], Mohammad Sa'eed Alattar[1], Yu Jin[1](✉), Soongeol Kwon[2], and Sang Won Yoon[1]

[1] Department of Systems Science and Industrial Engineering, Binghamton University, State University of New York, 4400 Vestal Pkwy E, Binghamton, NY 13902, US
yjin@binghamton.edu

[2] Department of Industrial Engineering, Yonsei University, 50 Yonsei-ro, Seodaemun-gu, Seoul 03722, South Korea

Abstract. Prescription demand and the complexity of patients' pharmaceutical protocols have significantly increased during the last decade. To achieve greater effectiveness of the overall prescription fulfillment process, the development and deployment of modern pharmacy automation systems, known as mail order pharmacy (MOP) or central fill pharmacy (CFP) systems, have been accelerated in recent years. Such advanced systems adopted automated robotic dispensing systems (RDS) and collation systems that can prepare more than tens of thousands of prescriptions per day. Designing and operating large-scale pharmacy systems are very complicated and expensive to ensure their expected throughputs and patient safety consideration. Therefore, a thorough system evaluation and investigation for potential improvement regarding the performance and operational efficiency should be conducted. This chapter aims to provide the detailed working mechanisms of pharmacy automation systems and introduce five important optimization problems in pharmacy automation, which include the RDS planogram design optimization, RDS replenishment optimization, collation system analysis, order scheduling optimization, and pharmacy database mining. To better demonstrate the optimization modeling in the context of pharmacy automation, a case study of the RDS replenishment process optimization based on modeling and simulation approaches is presented. The chapter also provides several research and development directions, which can potentially facilitate the realization of smart pharmacy automation solutions in the era of Industrial 4.0.

Keywords: Central Fill Pharmacy · Robotic Dispensing System · Replenishment Process · Modeling and Simulation

1 Introduction

Healthcare spending in the United States has dramatically increased during the recent decade, especially for prescription drugs. According to the Centers for Medicare and Medicaid Services estimation, prescription drug spending, which may reach $580 billion to $610 billion through 2021, will continue to grow in 10 years in ten years [1, 2]. The release of new drugs and the expanded use of high-priced drugs are the two key factors

C.-Y. Huang and S. W. Yoon (Eds.): ICPR1 2021, ACES 14, pp. 315–337, 2023.
https://doi.org/10.1007/978-3-031-44373-2_19

that drive the prescription drug spending growth. In addition, the 2019 coronavirus disease (COVID-19) outbreak has further increased prescription demand for treatments and medications. When the World Health Organization declared the rare COVID-19 infection to be a worldwide pandemic, there was a substantial rise in positive cases [3]. As of July 2022, there were approximately 570 million confirmed cases worldwide. As the number of verified cases rises, so will the need for prescription medications [4]. Figure 1 demonstrates that the worldwide usage of medications has increased during the previous decade.

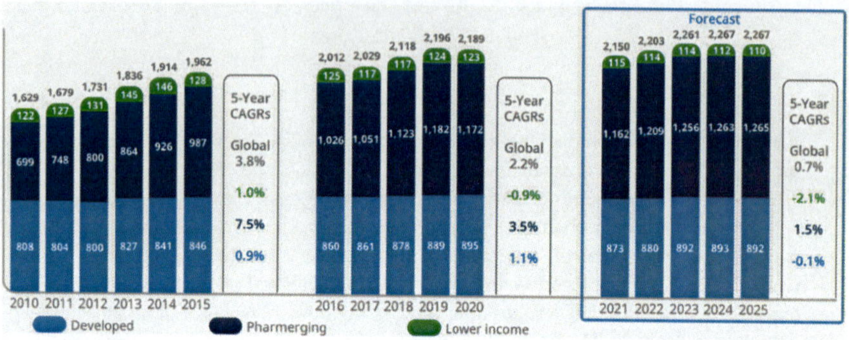

Fig. 1. Historical and projected use of medicine by segment, 2010–2025, defined daily doses (DDD) in billions [5].

To fulfill the increased prescription demand, the traditional pharmacy systems have evolved by adopting widespread technological advancements in robotics and automation that permit not only high productivity but also pharmaceutical safety. Before the automation engaged in pharmacy, the following items were prevalent [6]:

- Time wasted searching for a patient's prescription
- Having long queues where the patients wait until their prescriptions get ready
- Hiring additional staff, whether permanent or temporary, when volume demand increases
- Filling wrong drug name, quantity, or dosage strength
- Losing track of pharmacy throughput, staff productivity, inventory levels, and customer satisfaction

The use of robots, dispensing systems, automated bar-code scanning, and an intelligent enterprise pharmacy fulfillment software platform improves the pharmaceutical ordering and dispensing process, which thereby improves medication administration and inventory management operations [7]. Automation techniques also liberate pharmacists and technicians from time-consuming, repetitive manual tasks so that they can focus more on clinical tasks and improve the quality of patient care. In this case, pharmacists can spend more time in consultation to better understand patients' concerns and deliver more dedicated healthcare services [8]. For example, Beard and Smith quantify the benefits of using robotic dispensing machines in a hospital by calculating dispensing errors and staff efficiencies [9]. The results show that drug safety can be guaranteed by linking electronic prescriptions and a robot dispensing machine through the elimination of dispensing errors. Additionally, four hospital workers can be released and focus more on patients instead of drug dispensing, in which the filling efficiency enhancement results.

A typical type of modern pharmacy automation system, which is known as the Central Fill Pharmacy (CFP) system, has been actively deployed in recent years, such as Amazon Pharmacy [10]. A CFP system is a facility that processes and fills a large volume of prescription orders received from multiple retail pharmacies [11] in off-site locations. Completed prescriptions are shipped to the affiliated retail pharmacies and then delivered to the consumers at the retail stores. The central fill concept may be used by both big and small retail pharmacies. Due to the high cost and operational challenges of owning and operating a central fill facility, large pharmacy chains can build their own central fill facility, while smaller retail chains can opt for central fill as a service by hiring a third-party partner to assist with the dispensing and distribution of prescription drugs.

A CFP system consists of various workstations to perform the filling, order collation, packing, and sortation [12]. The auto-fill workstations are used to automatically put countable pills into bottles with prescriptions. After dispensing the medication, the bottle is put on the conveyor to be transported to the subsequent workstations. The manual-fill stations are for prescriptions that cannot be filled by a robot and must be filled by a technician, who then sets the completed prescription in a tote and on a conveyor. If a single order contains several bottle items for the same client, each bottle will go to the collation station until the whole order has been collated. The customer's purchase would be packed at packing stations and dispatched after being compiled at sortation stations. A conveyor system links all workstations and transports prescriptions to their final destination until they are ready to be dispatched to the consumer. Utilizing automated dispensers, robots, conveyors, and imaging technologies, CFP facilities may fill prescriptions at an accelerated pace. Figure 2 shows a general process of CFP systems. Figure 3 is an illustration of a typical CFP system layout arrangement. Different retail stores may adopt different layout solutions according to local demand and facility limitations.

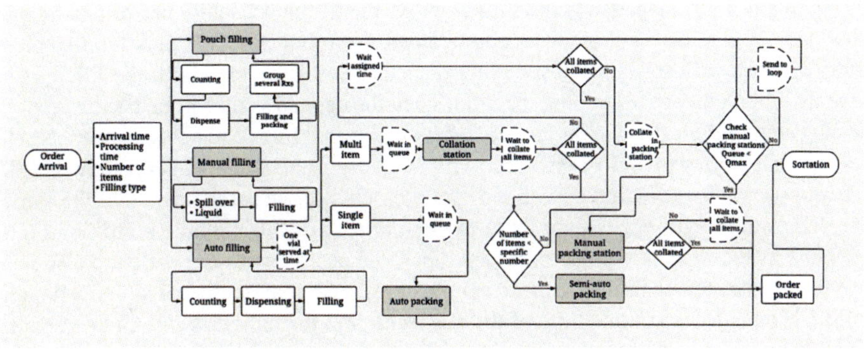

Fig. 2. General CFP systems process flow diagram.

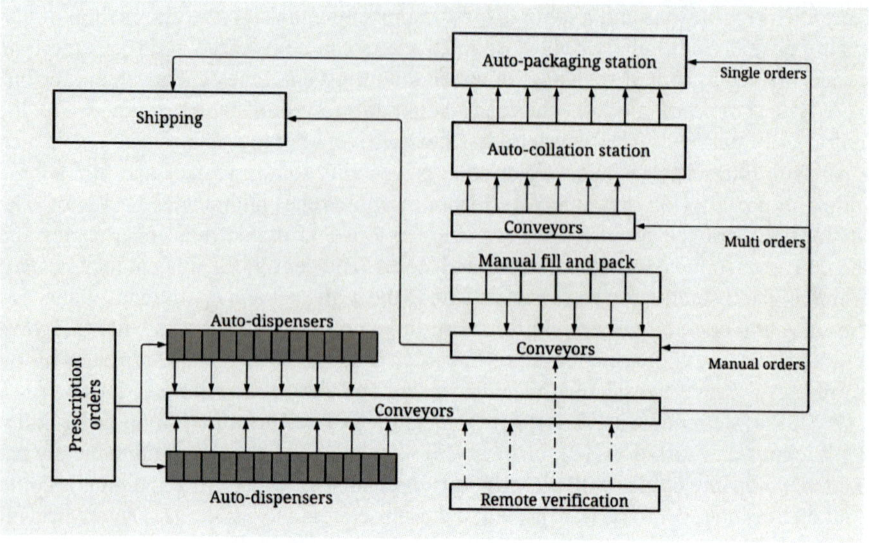

Fig. 3. Typical CFP layout diagram [13].

Among all these stations, the robotic dispensing system (RDS) for auto-dispensing and collation robot units are two critical automated machines that enable pharmacy automation with high productivity to fill over 30,000 patient prescriptions per work shift [14]. RDSs allow the CFP systems to store, count, and release the most commonly prescribed medications, which ensures convenience, accuracy, and agility in the auto-dispensing process. A typical RDS consists of a robot arm, imaging systems, and multiple shelves of dispensers. Each dispenser is specific for one type of medication classified by the National Drug Code (NDC) and employs a software-controlled counting strategy to ensure accuracy and convenience in the dispensing process. The auto-dispensing process has two main steps, dispensing and filling, where the filling is usually performed by the robot arm. When the RDS receives a new order, the robot arm will pick up a labeled empty vial, move it to the corresponding dispenser, wait for the vial to be filled, and place the filled vial at the capping location. Then the capped vials will be transported to the next station after the weight and the label are verified by the imaging system. Next, the vials that belong to multi-item orders will be transported to the collation station [15]. The robot arm of a collation station will put the items of the same order into the same tube. The station will not release them to the tote underneath the tube matrix until all the items of the order are gathered at the tube. The order governing software also has predefined rules to constrain the maximum waiting time of all the items at the collation station. Figure 4 depicts one type of the RDS unit and the collation robot unit used in the CFP system.

To make sure automated stations can work efficiently and effectively to fulfill the large-volume and highly customized orders, the operation and design at both station and system levels should be optimized and validated. In the next section, a detailed discussion of four important optimization research problems will be provided. Then a case study is

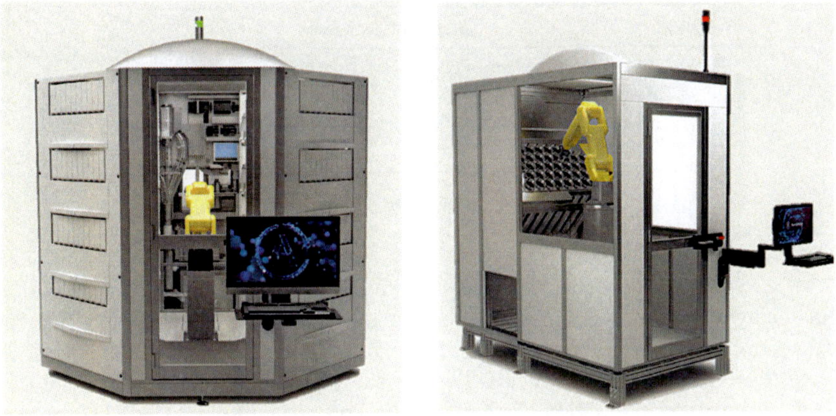

(a) RDS unit (b) Collation robot unit

Fig. 4. RDS unit and collation robot unit [16].

demonstrated to show how simulation modeling and analysis can be conducted to solve the proposed optimization problems, followed by the directions for future research and potential innovation in this domain.

2 Optimization Problems and State-of-the-Art Technologies in Pharmacy Automation

To efficiently manage the CFP system and keep the productivity high in pharmacy automation, the design and operation of key automated machines and CFP systems should be well evaluated and improved. In this section, five optimization problems, as well as the corresponding modeling and analysis approaches, will be introduced to summarize the research efforts that have been done or need to be done for the RDS planogram design, medication replenishment, collation improvement, order scheduling, and pharmacy database mining.

2.1 RDS Planogram Design Optimization

The RDS throughput is mainly determined by the travel time needed for the robot arm to travel to the dispenser, which depends highly on the dispenser location within the RDS. Therefore, different dispenser allocation strategies significantly affect the system throughput [17]. In addition, if medications of a multi-item prescription order are assigned in the same RDS, the order cannot be processed in parallel on different RDSs and order collation delay is more likely to increase [17]. To improve the efficiency of RDSs and downstream stations, a comprehensive and sophisticated dispenser allocation strategy is desired to minimize robot arm travel distance and separate associated medications into different RDS units. Figure 5 shows an example of the priority distribution for dispenser allocation in the RDS based on robot arm travel time.

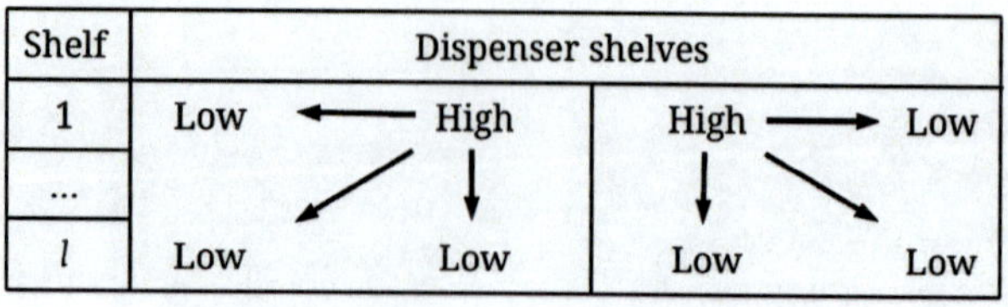

Fig. 5. Shelf slot priority distribution [17].

An RDS planogram provides a mechanism for the allocation of pharmaceuticals inside a single RDS unit and their distribution among many RDS units [17]. Sundaramurthy proposed a novel RDS planogram design to assign medications that are frequently ordered together [11] to different RDSs. This study provided a summary of the basic procedures involved in the RDS planogram design procedure for a single RDS as shown in Fig. 6. After obtaining transactional data and system design rules, the acquired data is checked and cleaned such that only demand data that contain countable pills that the RDS can process are retained. Based on the cleaned transactional data, demand patterns and the association of different medications are studied. Then the dispenser position and the number of dispensers necessary for each medication can be determined based on the extracted demand pattern and medication association. By studying the associations between various prescriptions and the distribution of pharmaceuticals to various units, the CFP system performance can be enhanced by lowering the collation delay and completion time of each order [18].

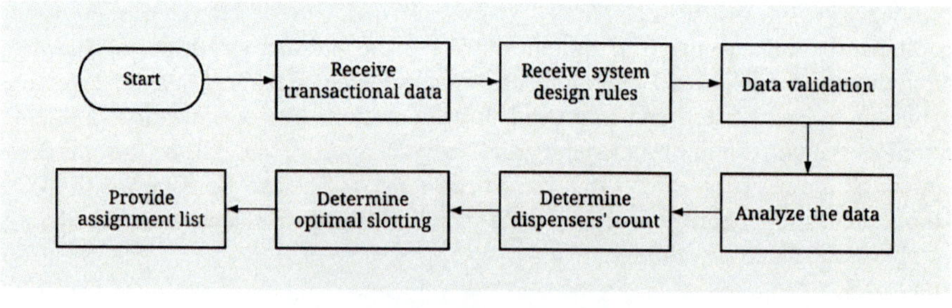

Fig. 6. RDS Planogram Process [11].

Wang et al. proposed a novel framework to optimize the parallel robotic dispensing planogram, as shown in Fig. 7 [17]. Specifically, the transaction database of the CFP facility is utilized to extract the drug association by applying the association rule mining method. The extracted association would be exploited by a multi-objective optimization model to achieve 1) minimization of association among medications in each RDS unit, 2) workload balance among RDS units, and 3) minimization of robot arm travel distance. It is anticipated that the output of the model would produce an optimal planogram for each RDS with a balanced workload among RDS units. In this study [17], the medication association is quantified by using Association Rule Mining (ARM), which is integrated into the optimization problem for the RDS planogram design to ensure the separation of NDCs that are commonly ordered together among the various RDS units. Figure 8 shows group-based and graph-based association rule results.

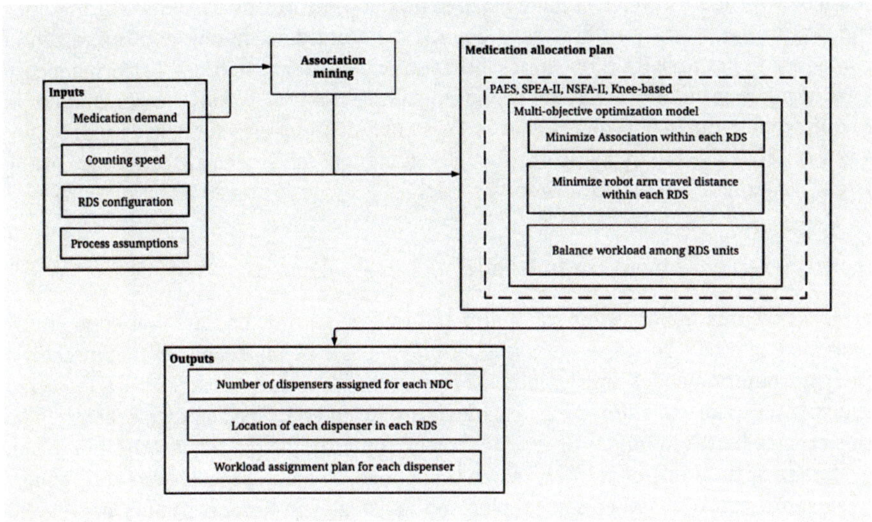

Fig. 7. Multi-objective parallel robotic dispensing planogram optimization framework [17].

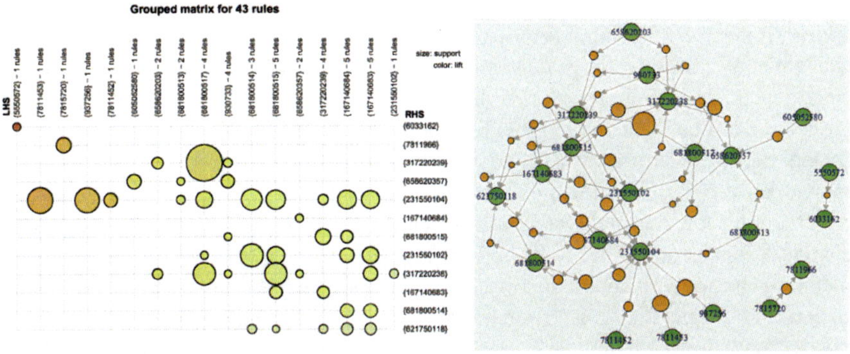

(a) Grouped matrix-based visualization.

(b) Graph-based visualization.

Fig. 8. Association rules visualization for support ≥ 0.0005 [17].

After mining the associations between medications, a mathematical model has been constructed, and three objectives are to be minimized: 1) the total support, which represents the association level between NDCs within each RDS unit; 2) maximum dispenser workload, which is used to balance the workload between RDS units; and 3) the total robot arm travel distance, which is the distance the robot arm must travel multiplied by the demand. Due to the fact that planogram design is an NP-hard issue that becomes unsolvable as the number of product categories increases [19], heuristic approaches are used to address such complicated challenges. To solve the mathematical problem, the study [17] employs the evolutionary algorithm.

In summary, the RDS planogram design optimization is an important problem that should be well addressed to maintain the high productivity of CFP systems. The methods proposed in the existing studies have applied different data mining and optimization approaches to improve RDS machines utilization downstream stations' performance. In terms of potential future work, the proposed studies assume that all the demand could be completed in the given work period and system downtime was not considered. Future work may consider the dynamic change in demand and the stochastic properties of the system components in the optimization model.

2.2 RDS Replenishment Optimization

An RDS machine includes one robot arm and hundreds of dispensers that contain different medications. The dispenser itself has only a basic storage capacity that might not satisfy the requirement of high-volume CFP systems. Therefore, a canister is attached to the dispenser as an extra storage backup device to extend the basic storage capacity. The canisters can be taken down for replenishment while the dispenser is working. When the storage of the canister is lower than a predefined threshold, the system will request replenishment. If operators do not replenish the canister in time, a "rundry error" will be generated and reported to the system. To fulfill the increased demand volumes, efficient inventory supply for the dispensers is one of the key challenges in CFP production planning.

The replenishment process contains five critical components, including operators, working strategy, replenishing priority, extra backup canisters, and replenishment carts, which are configurable settings that can be considered to define different scenarios. The general replenishment process is shown in Fig. 9. When there exist empty canisters and there is no extra canister in the replenishment station, the cart operator retrieves them from the RDS according to priority and sends them to the replenishment station. The stock clerk works to obtain specific stock bottles filled with the corresponding medications for each empty canister in the replenishment station. The replenishment technician fills the empty canisters using the stock bottle and then puts the filled canisters at the pickup window in the replenishment station. Then, the cart operator sends back the refilled canisters and installs the canisters on the relative dispensers. However, if extra canisters are in the replenishment station, they can be immediately filled by the stock clerk and replenishment technician and sent to the RDS first.

Several studies have analyzed the replenishment process and improved the throughput of RDS. Wang and Yoon introduced the detailed machine configuration and working mode of the RDS and optimized multiple variables (i.e., the number of backup canisters and reorder point of each medication) in the dispenser replenishment mechanism [14]. Dauod et al. proposed a receding horizon control strategy, a real-time optimization approach, to boost the RDS replenishment decisions in the CFP system [20, 21]. There exists minimal literature that considers human operations in the replenishment process of the RDS. Instead of focusing only on the RDS settings, O'Connor et al. considered the number of operators and staff costs in the continuous-time Markov Chain

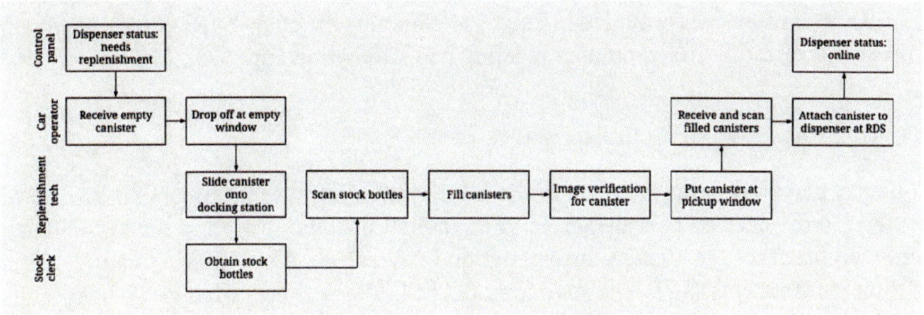

Fig. 9. Replenishment process overview.

Table 1. Summary of relevant studies on RDS replenishment process.

Paper		Wang and Yoon [14]	Dauod et al. [20]	Dauod et al. [21]	O'Connor et al. [22]	Case study in Sect. 3
Configurations of concern	Reorder point	√		√	√	√
	Canister size				√	√
	Num. of backup canisters	√	√	√		√
	Num. of operators				√	√
Objective		Min cost	Min cost	Min cost	Min cost	Provide operator strategies
Methodology		Mixed integer programming	Mixed integer quadratic programming	Mixed integer programming	Markov chain	Discrete-event simulation

model, which is exploited to show the inventory status of dispensers [22]. Table 1 summarizes the consideration of configurations and methodologies when making replenishment decisions.

Although RDS is highly automated, manual operations performed by operators are still crucial to achieve the desired throughput. The replenishment process can be affected by undesired factors and complex interactions between automated systems and operators, which is difficult to be formally formulated. Therefore, there is an urgent need to adopt a systematic approach that is uniquely designed to model the replenishment process, which includes manual operations, and investigate proper staffing and resources ensuring desired performance by reflecting real-world practice. In this regard, the modeling and simulation approach can be a potential methodology to emulate real-world practice, analyze system performance under various scenarios, and design system settings and operation strategies to efficiently manage and control the overall replenishment process

[15, 23]. A case study is provided in Sect. 3 to elaborate the study of RDS replenishment process optimization based on the modeling and simulation approach.

2.3 Collation System Optimization

Collation plays an important part in the CFP system because all orders consisting of multiple items need to be transported to the collation station; waiting time of collated items might affect the system throughput and makespan. A collation system consists of three components [15]. The first component is the scanner, which will trigger the conveyor and robot to assign an item to a tube when it passes through. Tube, as the second component, will hold multiple items belonging to the same order. When a tube is full, it will stop accepting new items even if the coming items are from the same order as the accepted ones. When all tubes of the collation station are filled, the system will be blocked. The maximum number of objects allowed in a single tube and the number of tubes in each station are determined by the collation station design specifications. The exit system is the third component. After collecting all items, the tubes will release all items in a tote and then pass the tote to the exit. Figure 10 depicts the operation of this system under the assumption that the capacity of each tube is five.

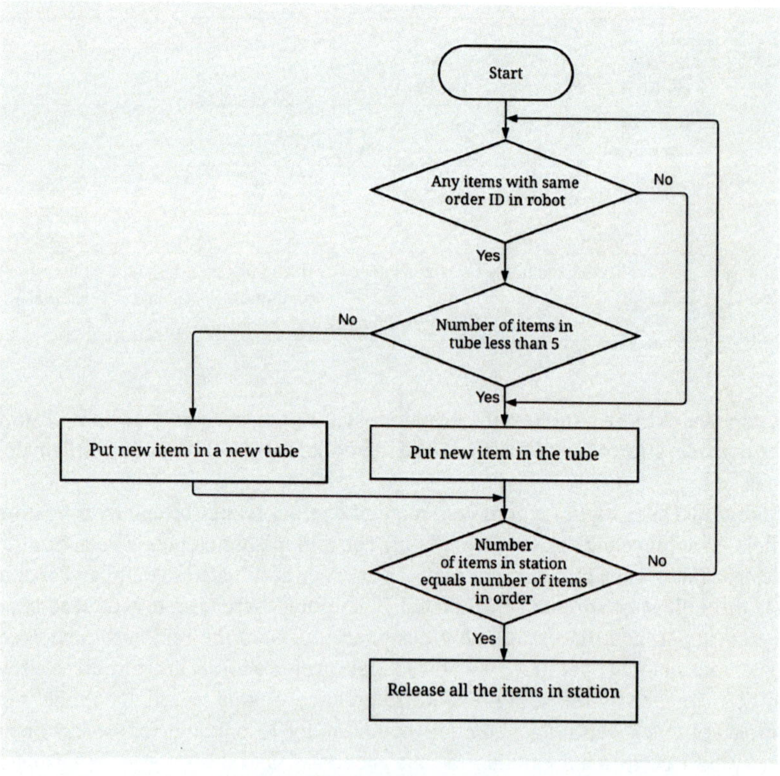

Fig. 10. Collation robot process [15].

Several performance metrics can be used to evaluate the collation process. Collation delay, which is the waiting time for an order from the arrival of the first item at the collation station until the arrival of the last item [24], is widely used to evaluate the system efficiency. Li and Yoon demonstrated using a factorial multivariate analysis of variance and simulation findings [25] that the collation delay is a crucial element for the CFP system throughput. The large collation delay indicates that many items are looping on the conveyor and waiting to be collated. When the number of medications exceeds the conveyor capacity, the system will be blocked and cannot process any orders without manual intervention. However, it is not sufficient to consider reducing the collation delay without investigating its adverse impact on the makespan. There is a trade-off between reducing the collation delay and makespan, particularly when receiving a large number of multi-item orders. Therefore, a multi-objective optimization problem should be established to minimize both collation delay and makespan. Three genetic algorithms were used to study multi-objective optimization in the CFP system: vector evaluated genetic algorithm (VEGA); multi-objective genetic algorithm (MOGA); and non-dominated sorted genetic algorithm-II (NSGA-II) [24]. Besides the collation delay, the number of tubes utilized in one collation station is another key measurement to evaluate the collation utilization during a working shift. Lower tube utilization indicates the redundant space in the collation machine and higher equipment cost, operation cost, and maintenance cost. However, full utilization without any buffer area indicates the non-flexible or non-adaptable status of the collation system. Therefore, the utilization of the collation station can also be considered in the optimization model to minimize the collation delay and improve the CFP system throughput while balancing the tube utilization of collation machines.

Several studies investigate the collation component of CFP systems. Collation delay may be influenced by the dispensing sequence [26]. Intuitively, first-enter & first-dispense can save more collation delay than other rules [27]. However, considering the flexibility of filling machines, other heuristic algorithms can provide more efficient solutions. According to the testing results of several algorithms, the genetic algorithm—the best among others—can reduce the collation delay by 96% [28]. In addition, two collation system designs with or without robots is compared by using the discrete event simulation methodology to model complicated systems and the interplay of items and environment [15]. The results show that the system with a robot can provide better performance, especially in reducing the collation delay when the majority of the orders include more than five items or have a long fill time window. However, considering the construction cost and the performance of the downstream packing station, more studies should be conducted to compare different collation designs and strategies.

2.4 Order Scheduling Optimization

The complex system dynamics and highly customized demand structure distinguish CFP systems from other production lines. It is also challenging to handle such large quantities of orders efficiently and safely by coordinating all the auto or manual stations in a CFP system. Different orders may require different sets of processes. To improve the overall system performance throughout a work shift, the received orders should be "smartly" dispatched by considering resource utilization and order preparation efficiency. Four common order scheduling rules are widely used in the CFP system to determine the processing sequence and priority of each order [27].

- *First Come, First Serve*: Each drug category is assigned a sequence ID based on the order arrival time. The items are filled according to their order arrival time. The earlier an order is received, the higher priority that the order gets to be processed. The early received orders get higher priority to be processed. The later processes will also rely on the sequence ID.
- *First Enter Prioritize*: Once a medication (item) from an order group is released to system (start counting), the whole group's priority will be increased. The later process will also rely on the new priority, higher priority, first serve. This strategy dynamically changes the original order sequencing, and the priority update process is also different from traditional scheduling, because the sequence is generated during operation process.
- *First Enter Prioritize, Count Finish First*: The first step of this strategy is the same as First Enter Prioritize; order entering the system also relies on priority. However, in the later processes, instead of considering priority, only those early finished orders will be released first to the next process. This strategy combines the first two strategies with the difference in updating the priority of other items in one group. Also, only order entering process relies on priority ID; the later processes are based on which order is first finished in the previous process.
- *First Enter Prioritize & First Dispense Prioritize*: In addition to increasing the group priority at the order entering point, the order group priority will be further increased when any order in the group starts dispensing. The later processes also rely on the new priority, higher priority, first serve.

The order scheduling problem in CFP systems can be solved by a multi-objective optimization model to consider multiple performance indicators simultaneously. One example is discussed in Sect. 2.3 where an integer programming mathematical model is proposed to minimize collation delay or makespan of the order [13]. In this study, the fill time window (FTW) [25], which is defined as the time difference between the first and last dispensed medications of a prescription order, is considered as the objective function and the makespan as the constraint for this multi-objective optimization. A unique adaptive parallel tabu search technique is also given to solve this NP-hard order scheduling issue effectively. Four different heuristic methods, vector evaluated genetic algorithm (VEGA), multi-objective genetic algorithm (MOGA), non-dominated sorted genetic algorithm-II (NSGA-II), and longest processing time (LPT), are applied to solve the optimization model. The results indicate that the NSGA-II provided the best frontier in larger scales of work [24].

Based on the conventional order scheduling methods and the previous optimization models, a new threshold- and priority-based dispatching strategy is proposed [5] to dynamically adjust the order sequence based on the real-time system performance as shown in Fig. 11. The proposed order dispatching strategy is implemented by using discrete-event simulation, where the key performance indicators of the system will be evaluated iteratively. Specifically, once the system receives the order, the system will assign the order a priority level based on the type of the order. Then the system will process the order with the highest priority first. After the order enters the system, the performance of different stations will be evaluated and compared to a predetermined threshold. If a specific indicator reaches its threshold, the system will adjust the priority of the subsequent orders/items. Because the strategy is implemented in a simulation model as a virtual replica of the real system, all the thresholds and priority rules can be easily tested and adjusted based on the engineers' needs.

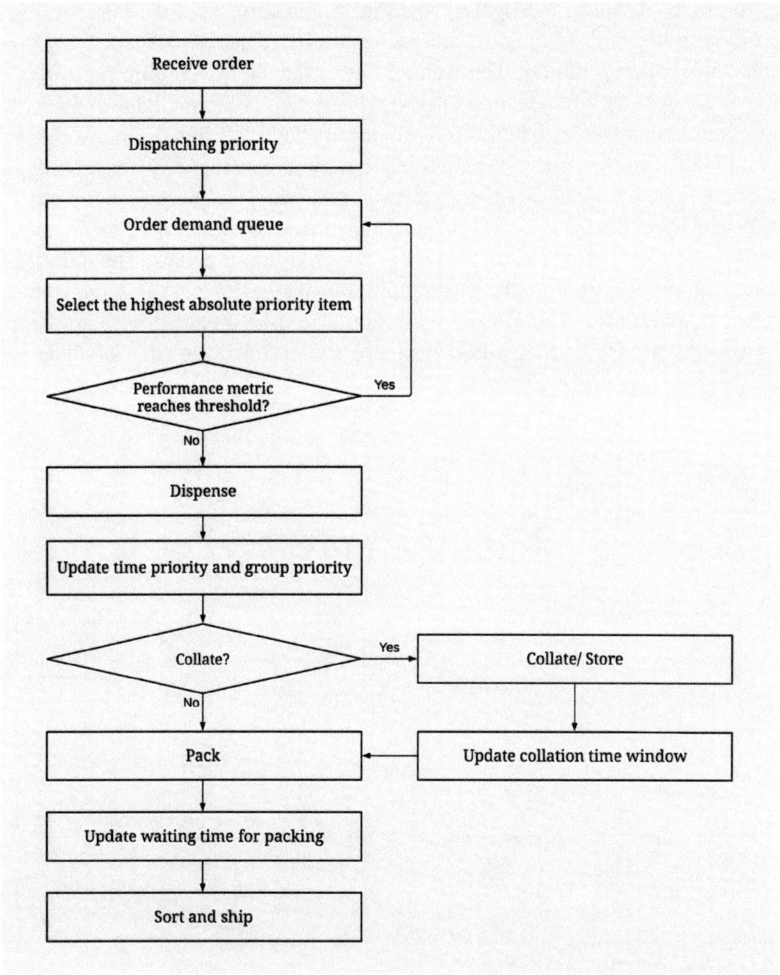

Fig. 11. Threshold- and priority-based dispatching rule workflow [5].

2.5 Pharmacy Database Mining

Currently, automated pharmacies, particularly CFP systems that fulfill prescriptions sent by retail pharmacies, are confronted with a substantial increase in prescription order volume. Every day, an enormous quantity of transactional data is created at each facility. The previous section discussed different approaches to improve the system performance, but the proposed methods or models rely highly on the assumptions or simplified prior knowledge that did not catch the complex system dynamics and demand variety. Therefore, it is desirable to retrieve detailed information from the prescription database to extract more efficient design guidelines and inventory management techniques to improve the current automation solutions.

For instance, Khader investigates potential medication regimen interactions for a variety of individuals [6]. Mining the pre-prescription data assists in identifying the most frequently prescribed products. The frequent item sets are prescription groupings that regularly occur in a single transaction for several patients. The knowledge of five common item combinations provides insight into the pharmaceuticals that are more likely to be ordered and delivered together. The RDS planogram is considerably aided by capturing these associations and detecting the patterns in transactional prescription data. This is accomplished by optimizing the dispensers' allocation to the various medications and by ordering and distributing the dispensers among numerous robots. The duration from the moment an order is put into the system until it is packaged and validated is the entire cycle time for that order. Therefore, minimizing the time involved with automatically filling prescriptions for orders would decrease the entire cycle time of these orders,

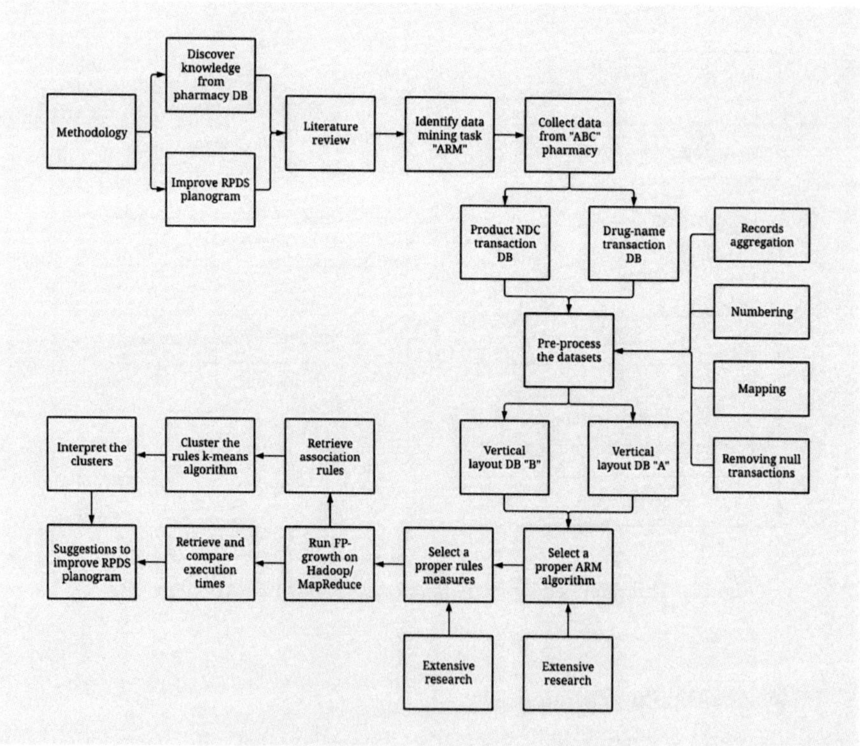

Fig. 12. Process map of methodology [6].

which results in an improvement in throughput because the system is able to accept more orders. In addition to these advantages, the primary objective of the project is to raise awareness of the significance of mining transactional prescription data to identify drug trends and connections. In addition, the project is to value the knowledge obtained from the underused pharmaceutical databases that have been kept. Figure 12 depicts how to locate database rules using the Frequent Pattern Growth method.

3 Case Study: Study of RDS Replenishment Process Based on Modeling and Simulation Approach

This section presents a case study of RDS replenishment process conducted by the authors to show how simulation modeling and analysis can be applied to evaluate the system/machine performance via a sensitivity analysis. The results can provide insights for practitioners to design and manage efficient replenishment operations with limited resources.

3.1 Replenishment Process

As introduced in Sect. 2.2, the replenishment process contains five critical components, which include operators, working strategy, replenishing priority, extra backup canisters, and replenishment carts, that can be configured for to define different analysis scenarios and identify the best replenishment. The operators work to transport canisters between replenishment stations and the RDS using a replenishment cart and replenish the empty canisters in the replenishment station. The cart operator, stock clerk, and replenishment technician are three different types of operators and are responsible for different tasks as demonstrated in Fig. 9. Important criteria such as the number of operators, its working strategy, and replenishment priority can have a significant impact on the processing efficiency and the occurrence of rundry errors. To test different system configurations, the complex and continuous interaction among the components should be considered and modeled to simulate the replenishment process before further evaluation and analysis. In this study, an RDS machine that includes one robot arm and 80 dispensers with different medications is simulated and analyzed.

3.2 Methodology: 3D Discrete-Event Simulation Model

A promising methodology to imitate the systems' performance without physical considerations is the Discrete Event Simulation (DES), which will be employed to simulate CFP systems to experimentally analyze the proposed replenishment performance under different operational conditions and constraints. In this study, a 3D simulation model is developed based on DES principles. Important system performance indicators, such as makespan, rundry error occurrence, and resource utilization, are studied to assess manual operations. In addition, the visualization of human operations can help track the specific replenishment process at any time, which reflects the real human performance in the pharmacy.

The list of assumptions made while developing the simulation model is given below, and assumptions on processing time related to the replenishment process are shown in Table 2. In the replenishment simulation model, there are a total number of 4,559 vials that need to be filled with medications. The demand data is uniformly generated during a nine-hour shift from two RDS machines. There are 160 types of medications, which are uniformly assigned to the demand order. Because the inventory capacities of canisters and dispensers are volume, the basic unit "Quantity/100cc" is utilized to transform volume and quantity.

- There are two RDS machines with 160 dispensers that contain 160 different medications.
- The dispenser cannot work if it runs out of drugs (rundry error exists), and the system will deal with the next order because other dispensers can work. The skipped order will be handled after the dispenser has been replenished.
- Canisters (500cc capacity) and dispensers (800cc capacity) are considered in the baseline model.
- Reorder point is the 100cc inventory level in the dispenser. When the dispenser's inventory level is lower than the low-level sensor (100cc), the canister will release its chamber (500cc).
- Each dispenser has one attached canister at the RDS. No extra canisters are considered in the baseline model.
- One cart operator with one cart, one stock clerk, and one replenishment technician are in the baseline model.

Table 2. Simulation model assumptions for processing time relating replenishment process.

Machine & Staff	Task	Processing time
RDS	Fill vials	Tri[12.9,13.3,13.7] sec/vial
Replenishment technician	Replenish canisters	Tri[220.0,227.0,234.0] sec/canister
	Stock filled canisters at pickup window	Tri[13.1,13.5,13.9] sec/canister
Stock clerk	Bottle operation	Tri[90.3,93.1,95.9] sec/canister
Cart operator	Unload carts	Tri[7.8,8.0,8.3] sec/canister
	Load carts	Tri[8.7,9.0,9.2] sec/canister
	Replenish dispensers with canisters	Tri[220.0,227.0,234.0] sec/canister
	Travel time between the RDS and workstation	Tri[45.0,75.0,105.0] sec

- At the same priority level, each empty canister's priority is set according to the timestamp.
- The replenishment cart can carry at most 24 canisters at once.

The replenishment simulation model is developed with Demo3D, which is an industry-leading simulation software with advanced visualization and capability in DES. A simulation demo that shows the replenishment process layout is provided in Fig. 13(c). In the constructed simulation model, the system is able to monitor and update the dispenser's and canister's status in real time by following the flow chart depicted in Fig. 13(a). The detailed replenishment steps implemented in the simulation model are presented in Fig. 13(b). When there are empty canisters, the cart operator will pick at most 24 empty canisters to the replenishment station. The stock clerk and replenishment technician will find the corresponding medication bottles and refill the canisters accordingly one by one. After all those canisters are refilled, the cart operator will send them back and attach them to the dispenser.

3.3 Experiments and Results

Based on the established simulation model, six scenarios with different configuration settings of five replenishment components are proposed to evaluate the performance of the replenishment process. Table 3 provides the detailed scenario settings where Scenario 1 is considered as the baseline model. It is worth noting that the backup canisters shown in the table mean the additional canisters in the replenishment station, which exclude the one attached to the dispenser.

Table 4 presents the simulation results of six tested scenarios in terms of the five evaluation perspectives indicators to assess the replenishment under different settings. Makespan of RDS machines and replenishment work reflect the influence of delay caused by rundry error and replenishment process in a nine-hour shift. The total number of replenished canisters and rundry errors indicates the efficiency of the replenishment process in terms of canister allocation and rundry error reduction. Moreover, the actual staff utilization is calculated to justify the potential headroom and bottleneck to improve the process.

It is observed that the RDS machine makespan obtained by the baseline model (Scenario 1) is far beyond a nine-hour shift, indicating/which indicates that the system cannot fulfill the expected demand within the shift time due to the pending rundry errors. This is because the cart operators cannot work synchronously with the other two types of operators in the baseline case. Changing the staff arrangements or operations (Scenarios 2 & 3) or using extra backup canisters (Scenarios 4 & 5) can help the system to clear up earlier due to the reduced delay resulting/that results from the rundry errors. The replenishment efficiency improvement is contributed by the increased staff utilization. For the cart operators, if they can pick up the filled canisters with high priority rather than all filled canisters, then the RDS machines can provide timely responses to deal with the rundry error even though the cart operators may spend more time traveling. However, simply increasing the number of cart operators (Scenario 6) may not achieve the same expected improvement due to the time wasted waiting the canisters to be filled in front of the pickup window. In summary, more flexible staff arrangement or more sufficient backup canisters can help the RDS machines to reduce the rundry errors significantly.

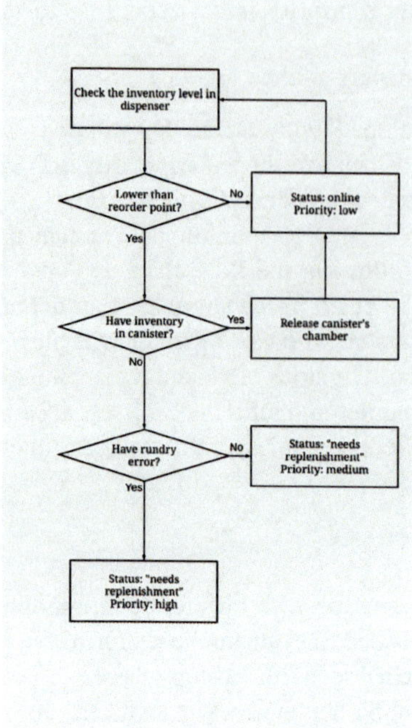

(a) Dispenser update process.

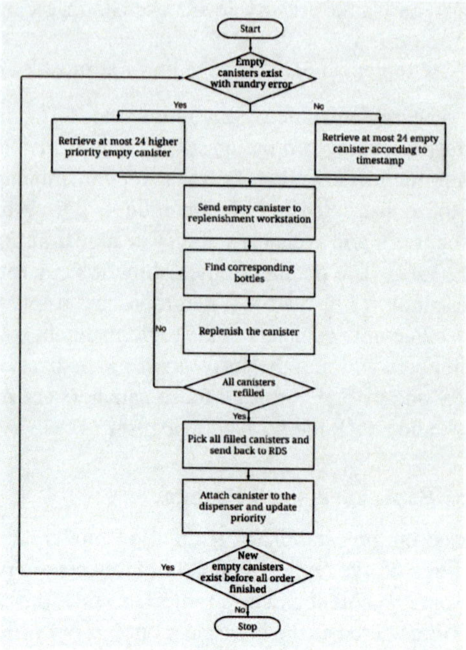

(b) Replenishment simulation model flowchart.

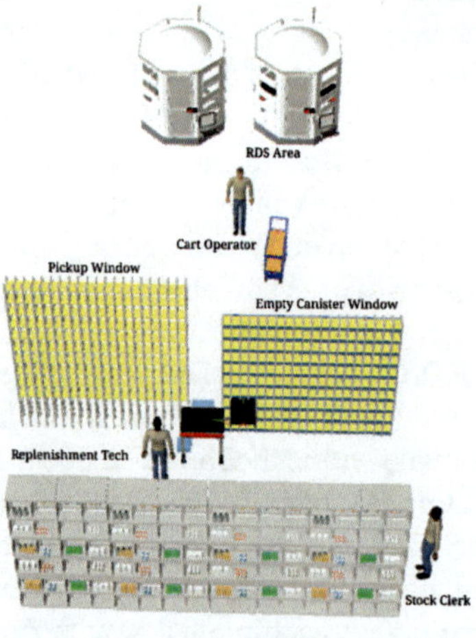

(c) Replenishment simulation model layout.

Fig. 13. Replenishment simulation model.

This case study provides a simple example of how to use simulation modeling and analysis to evaluate different system designs and operation strategies. Similar frameworks can be adopted and integrated with the proposed optimization models discussed

Table 3. Scenario settings for numerical experiment.

Settings	Scenario 1	Scenario 2	Scenario 3	Scenario 4	Scenario 5	Scenario 6
Num. of operators	3	3	3	3	3	4 (2 cart operators)
Backup canisters	-	-	-	Another one for each dispenser		-
Replenishment process	Described in Section 3	Described in Sect. 3		Canisters can be filled in the station first. Cart operator will first delivery filled canisters to RDS		Described in Section 3
Staff arrangement	Cart operator waits when doing replenishment	Cart operator can go back to the RDS when the canisters are doing replenishment		Cart operator waits when the canisters are doing replenishment		Cart operator waits when doing replenishment
Cart operator operations in workstation	Pick all filled canisters	Pick filled canisters with high priority first	Pick all filled canisters	Pick filled canisters with high priority first	Pick all filled canisters	Pick all filled canisters

Table 4. Summary of experimental results.

Evaluation indicators		Scenario 1	Scenario 2	Scenario 3	Scenario 4	Scenario 5	Scenario 6
Makespan of RDS machines		14.5 h	**9.8 h**	12.7 h	**9.9 h**	12.7 h	14.3 h
Makespan of replenishment work		29.4 h	17.1 h	**16.5 h**	**16.6 h**	**16.4 h**	23.6 h
Total num. of replenished canisters		133	112	100	100	**96**	125
Total num. of rundry errors		18	10	**7**	**6**	**8**	20
Actual utilization	Stock clerk	18.7%	32.1%	34.2%	33.9%	34.4%	23.4%
	Replenishment tech	48.3%	83.3%	88.7%	87.6%	88.8%	60.5%
	Cart operator	53.3%	99.1%	99.2%	99.2%	99.2%	39.3% & 28.0%

in Sect. 2 to implement heuristic algorithms dynamically as the process progress or validate the optimization results without physical implementation. Moreover, because the 3D simulation model constructed by the Demo3D provides real-time dashboards, the system/machine performance not only can be evaluated by the end of the working shift but also can be monitored in real time to track the status of each component.

4 Discussion: Towards Smart Pharmacy Automation System

With the advancement of contemporary information and computer technologies, it is possible to identify many opportunities to evolve the current pharmacy automation solutions into more intelligent and reliable smart manufacturing systems. Different from other manufacturing applications, confidential protection, medication safety, and other domain-specific constraints should be considered when adopting the new techniques into pharmacy automation systems.

One promising direction is to introduce data-driven analytical approaches to facilitate the development of a data-supported decision-making process. Deep mining and analyzing the massive amounts of transactional data collected in CFP systems can help the company to improve the system design based on the patients' and the market's needs. In addition, operation management, such as inventory control, can also be improved by reducing inventory holding costs and preventing out-of-stock exceptions. Moreover, the automated machines can be improved based on the identified bottlenecks and headroom by optimizing the robot arm trajectory and medication allocation in dispensers.

The applications of Artificial Intelligence (AI) have also attracted great attention in the pharmaceutical automation industry. Despite the fact that automation stations, such as RDSs and collation systems, are now commonly used in the CFP system, the software cannot respond to unexpected demand changes or operation interruptions. For instance, the current replenishment strategy involves operators who restock the canister based on the reported inventory shortage. With AI-based methods, it is possible to predict the potential shortage of a specific medication based on the historical record and then plan the replenishment operation precisely and accurately ahead of time. Another potential application of AI is to propose vision-based real-time route planning to reduce the potential congestion and improve the conveyor utilization of CFP systems.

Another innovation that can be introduced in the industry is to establish a more advanced simulation model, which is referred as a Digital Twin or machine emulator in the context of Industrial 4.0, to establish a virtual replica of physical systems via database synchronization, PLC controlling, network proctoring and communication, and simulation modeling. A virtual environment can help engineers to test pharmacy automation solutions safely and improve the quality control of operations. The overall software testing/maintenance cost and time can be reduced using the virtual software testing environment where virtual machine emulators or Digital Twins will replace actual machines, workstations, or subsystems. Moreover, all the data analytical methods and AI-based models can be integrated into the Digital Twin to plan future production, reduce manufacturing exceptions, and improve productivity. It can also provide a virtual workspace for training staff and customers and promoting company sales and marketing initiatives, which can improve customer relations by providing a virtual "hands-on" experience.

5 Conclusions

CFP systems, as advanced automation systems, have been actively deployed in the pharmacy industry in recent years to satisfy the drastically increasing prescription demand and the complexity of patients' pharmaceutical protocols. RDSs and robot-based collation systems are essential automated facilities for pharmacy automation. Numerous studies concentrating/that concentrate on optimization problems at both machine and system levels have been conducted to improve the system design and operational efficiency. This chapter provides an overview of five reprehensive optimization problems, including/which include RDS planogram design optimization, RDS replenishment optimization, collation system analysis, order scheduling problem, and pharmaceutical database mining. As a result of the fact that many optimization models are NP-hard problems, heuristic approaches are often utilized to solve the proposed optimization models. A case study is provided to illustrate how simulation modeling and analysis can assist in the system evaluation, which can be useful to validate the optimization results and capture the system dynamics that cannot be easily formalized. Due to the complexity of the CFP systems, there are still many challenging questions that need to be addressed. Potential future research directions that apply the new techniques of Industrial 4.0 have been discussed to inspire multidisciplinary collaboration for the realization of smart pharmacy automation systems.

Acknowledgements. This study was supported by the Watson Institute of Systems Excellence (WISE) at Binghamton University and by iA. The authors would like to thank the anonymous reviews for their valuable comments in improving the quality of this manuscript.

References

1. Conti, R.M., Turner, A., Hughes-Cromwick, P.: Projections of US prescription drug spending and key policy implications. JAMA Health Forum **2**(1), e201613 (2021). https://doi.org/10.1001/jamahealthforum.2020.1613
2. U.S. prescription drug spending as high as $610 billion by 2021: report. https://www.reuters.com/article/us-usa-drugspending-quintilesims/u-s-prescription-drug-spending-as-high-as-610-billion-by-2021-report-idUSKBN1800BU
3. WHO: World Health Organization. (2019). Coronavirus disease (COVID-19) outbreak situation
4. Ayati, N., Saiyarsarai, P., Nikfar, S.: Short and long term impacts of COVID-19 on the pharmaceutical sector. DARU J. Pharmace. Sci. **28**(2), 799–805 (2020)
5. Yang, Y.: Threshold-and Priority-Based Dispatching Rule in Mail-Order Pharmacy Automation Systems (Doctoral dissertation, State University of New York at Binghamton) (2021)
6. Khader, N.: Frequent pattern mining in a pharmacy database through the use of Hadoop. State University of New York at Binghamton (2014)
7. Angelo, L.B., Christensen, D.B., Ferreri, S.P.: Impact of community pharmacy automation on workflow, workload, and patient interaction. J. American Pharmac. Ass. JAPhA **45**(2), 138–144 (2005)
8. Tan, W.S., Chua, S.L., Yong, K.W., Wu, T.S.: Impact of pharmacy automation on patient waiting time: An application of computer simulation. Annals Academy of Medicine Singapore **38**(6), 501 (2009)

9. Beard, R.J., Smith, P.: Integrated electronic prescribing and robotic dispensing: a case study. Springerplus **2**(1), 1–7 (2013)
10. Shaya, F.T., Eddington, N.D.: Disruptive innovation in pharmacy: Lessons from the amazon frontier. In: JAMA Health Forum, vol. 1 (American Medical Association, 2020), pp. e200, 038–e200, 038 (2020)
11. Sundaramurthy, S.S.: Mining Frequent Itemsets of a Central Fill Pharmacy Transaction Database to Enhance the Planogram of Robotic Dispensing System (Doctoral dissertation, State University of New York at Binghamton) (2018)
12. O'Connor, R.: Minimizing Replenishment Cost in a Central Fill Pharmacy Using a Markov Chain (Doctoral dissertation, State University of New York at Binghamton) (2020)
13. Li, D., Yoon, S.W.: A novel fill-time window minimisation problem and adaptive parallel tabu search algorithm in mail-order pharmacy automation system. Int. J. Prod. Res. **53**(14), 4189–4205 (2015)
14. Wang, H., Yoon, S.W.: Drug dispenser replenishment optimization via mixed integer programming in central fill pharmacy systems. In: 2016 Industrial and Systems Engineering Research Conference, ISERC 2016 (2016)
15. Li, Y., Zhang, Q., Yoon, S.W.: Discrete event simulation-based collation system analysis in mail-order pharmacy automation system. In: IIE Annual Conference. Proceedings, pp. 828–833. Institute of Industrial and Systems Engineers (IISE) (2019)
16. Leading-edge pharmacy automation solutions. Retrieved September 18, 2022, from: https://iarx.com/
17. Wang, H., Dauod, H., Khader, N., Yoon, S.W., Srihari, K.: Multi-objective parallel robotic dispensing planogram optimisation using association rule mining and evolutionary algorithms. Int. J. Comput. Integr. Manuf. **31**(8), 799–814 (2018)
18. Khader, N., Lashier, A., Yoon, S.W.: Pharmacy robotic dispensing and planogram analysis using association rule mining with prescription data. Expert Syst. Appl. **57**, 296–310 (2016). https://doi.org/10.1016/j.eswa.2016.02.045
19. Hansen, J.M., Raut, S., Swami, S.: Retail shelf allocation: a comparative analysis of heuristic and meta-heuristic approaches. J. Retail. **86**(1), 94–105 (2010). https://doi.org/10.1016/j.jretai.2010.01.004
20. Dauod, H., Serhan, D., Wang, H., Khader, N., Yoon, S.W., Srihari, K.: Robust receding horizon control strategy for replenishment planning of pharmacy robotic dispensing systems. Robo. Comp.-Integr. Manuf. **59**, 177–188 (2019). https://doi.org/10.1016/j.rcim.2019.04.001
21. Dauod, H., Wang, H., Khader, N., Yoon, S.W., Srihari, K.: Real-time dispenser replenishment optimization based on receding horizon control. Procedia Manufacturing **11**, 1782–1789 (2017). https://doi.org/10.1016/j.promfg.2017.07.313
22. O'Connor, R., Yoon, S.W., Kwon, S.: Analysis and optimization of replenishment process for robotic dispensing system in a central fill pharmacy. Comp. Two Collartio Ind. Eng. **154**, 107116 (2021). https://doi.org/10.1016/j.cie.2021.107116
23. Alhaag, M.H., Aziz, T., Alharkan, I.M.: A queuing model for health care pharmacy using software Arena. In: 2015 International Conference on Industrial Engineering and Operations Management, IEEE 2015, pp. 1–11 (2015)
24. Mei, K., Li, D., Yoon, S.W., Ryu, J.H.: Multi-objective optimization of collation delay and makespan in mail-order pharmacy automated distribution system. The Int. J. Adv. Manuf. Technol. **83**(14), 475–488 (2016)
25. Li, D., Yoon, S.W.: Simulation Based MANOVA Analysis of Pharmaceutical Automation System in Central Fill Pharmacy. IEEE InternationalConference on Industrial Engineering and Engineering Management, pp. 1647–1651 (Dec. 2012)
26. Wang, H., Yoon, S.W.: Evaluation and optimization of automatic drug dispensing/filling system. Proceedings of the 3rd Annual World Conference of the Society for Industrial and Systems Engineering (2014)

27. Wang, H., Serhan, D.M., Yoon, S.W.: Collation delay optimization using discrete event simulation in mail-order pharmacy automation systems. In: Proceedings of the 2016 Industrial and Systems Engineering Research Conference (2016)
28. Dauod, H., Li, D., Yoon, S.W., Srihari, K.: Multi-objective optimization of the order scheduling problem in mail-order pharmacy automation systems. The Int. J. Advan. Manuf. Technol. 1–11 (2016)

Managing a Retail Store and the Associated Warehouse with a Knowledge-Driven Approach

Pisut Koomsap[1], Chih-Fan Tan[1,2], Yu-Ju Lin[2(✉)], and Chin-Yin Huang[3]

[1] Asian Institute of Technology, Pathum Thani, Thailand
[2] Tunghai University, Taichung, Taiwan
yujulin@thu.edu.tw
[3] Industrial Engineering and Enterprise Information, Tunghai University, Taichung, Taiwan

Abstract. For today's retail stores that directly serving final customers' demand have some special characteristics and restrictions, such as varieties of goods yet limited space in a retail store, a customer order includes various products but each of them is with a small number of quantities, products with different features like seasonal, etc. The competitiveness of the retail store relies on the utilization of the space and fulfillment of the customer demands. To increase the competitiveness, the retail store is normally linked with an associated (small) warehouse. The intelligence of managing the associated warehouse for the retail store has an extraordinary meaning in today's dynamic environment.

This research use data-driven approach to analyze historical customer order to unearth the relevance, customer buying behavior, time series, seasonal, etc. Then by using knowledge-driven approach (ontology) to define the classes and properties, relations between retail store, warehouse, and products, the knowledge can be shared, reused, and communicate in the company. Finally, via inference engine with Semantic Query-enhanced Web Rule Language (SQWRL) to infer the correlations from the ontology and find out the knowledge and logics. The research helps the retail store quickly delivery the products and lets the retail store and its warehouse possess the ability to respond to the various changes.

Keywords: Data-Driven Approach · Knowledge-Driven Approach · Retail Store · Semantic Query-enhanced Web Rule Language · Warehouse

1 Introduction

It is important for a retail store and its associated warehouse (usually small) to integrate the advancement of the technologies to move towards automation and intelligence. Data-driven is an important step toward identifying users' activities [1]. Retail can use data-driven approach to analyze the correlation data and create opportunities. However, it is challenge for retailer to search valid information from such an enormous data. Knowledge-driven approach is used as a source of domain knowledge. By using ontology, meaningful information is retrieved from the database to help in making decisions [2].

For the retail store, through these technologies, it can grasp accurately the market trends and increase the added value of products and services. With these actions, retail

C.-Y. Huang and S. W. Yoon (Eds.): ICPR1 2021, ACES 14, pp. 338–354, 2023.
https://doi.org/10.1007/978-3-031-44373-2_20

can use it to make some sales strategy, and it may affect customer buying behavior. The application of intelligent warehousing ensures the speed and accuracy of input in all aspects of warehouse management, ensures that the enterprise can grasp the real data of the inventory in a timely and accurate manner, and controls the enterprise inventory. It also can match the characteristics of different commodities, storage conditions, and even import and export patterns. According to the intelligent warehouse and retail store, it can provide customers for better service.

This research focuses on studying the operations of receiving and storage for the retail store and the warehouse. In this research, the retail stores use data-driven approach to find the insight from tens of thousands of order data. The knowledge-driven approaches present knowledge representation tools to model activities and exploits logical reasoning for activity inference [3]. Using knowledge-driven approach to integrate the information of the retail store, warehouse, products, and the insight from order data, it can grasp the information accurately for the retail store. Finally, this research uses Semantic Query-enhanced Web Rule Language (SQWRL) to infer the rules for the operational decisions in the retail store and warehouse. These decisions can help the warehouse to have the suitable storage capacity, smooth traffic flow, and storage-allocation. This research integrates these methods to help retailers manage their warehouse, solve those problems and restrictions in the retail, and provide better service to attract consumers.

2 Literature Review

With the rapid changes in customer buying preferences, the retail industry is growing very fast, and the global retail competition structure has also changed. Whether the retail industry can gain a place in the market depends on the performance of its supply chain, a balance between responsive and efficient warehouse operation, and its ability to respond quickly to customer orders and high efficiency [4].

Orders placed by retailers generate demands at the warehouse, which acts as the source of supply for the retailers. The warehouse replenishes its inventory from an external supplier. There is a holding cost charged against each unit of inventory per unit time at the retailers and the warehouse and a corresponding set-up cost charged for each order placed at the warehouse and each retailer [5]. As marketing becomes more customer-centric, the accuracy of decisions with regards to which potential customers to engage in relationships with is becoming more important [6]. When setting up warehouse operations, the challenge for retailers directly facing customers is the strict delivery time, which has led to shortened picking time. Each order contains only a small number of items, but with various types [7].

There are some different problems in the retail store and its associated warehouse. Previously, researchers applied the following distinct approaches to solve them:

(1) Giannikas et al. use a warehouse management system (WMS) and distributed intelligence approach to maintain the flexibility of being responsive to short-term changes in customer demands and maintain the service level [8].
(2) Saleheen et al. use modern information technology and secondary data to analyze and improve facility layout to achieve a higher level of productivity in the warehouse management [4].

(3) Zhang et al. study the warehousing construction model by considering the market growth rate and the number of active users. They make use of big data and improve the distribution system to explore the transformation of e-commerce logistics under the third party logistics [9].
(4) Weidinger et al. [7] believe scattered storage should be used as a storage placement method for small orders with a small number of items.

Another issue related to storage layout problems is the dynamic nature of customer demand order as well as the way to group and handle/store products in a warehouse, and the demand for items always varies with seasons. When considering the dynamic nature of customer demand orders, the manager needs to periodically review the characteristics of order demand and modify the stock location accordingly [10].

3 Intelligent Management System

With the rapidly changing consumer needs and with diverse consumption patterns, it will cause a heavy load for the retail store and its warehouse. There are varieties of goods yet limited space in a retail store. Usually, a customer order of a retail store includes various products; each of them is with a small number of quantities. In the current retailer, the emphasis is on fast, fast replenishment and fast shipment to meet customer needs. However, in a fast-changing environment, how to control the retail store and its warehouse without making mistakes is a big challenge.

In retail store or warehouse, managers often need to make decisions and choices about products storage, product placement, and sale strategy. In the past warehousing systems, it is difficult to express the logic and to construct the relations between warehouses, retail store, products, and customer behaviors. The knowledge framework established by the knowledge-drive approach allows the same knowledge field to have common language to communicate, and the meaning of the data and the concept can be understood during the process of inference.

The competitiveness of the retail store relies on the utilization of the space and fulfillment on the customer demands. Hence, how the retail store and warehouse can become intelligent by using historical shopping data of the customers for the operational decisions is a challenge.

The framework of this research is shown in Fig. 1. First of all, the retail store needs to find the characteristics, insight, and tacit knowledge, including relevance, customer buying behavior, time series and seasonal. Then the manager can use these characteristics to help the retail stores. Our study suggests using data mining to find the features from the historical customer orders. Afterward, by integrating the information form internal information of the retail store, warehouse, products, ontology is applied to construct the knowledge model. Based on the ontology knowledge, inference rules are developed to retrieve the answers from the managerial questions. The inference rule is a way that can infer the correlation from a large amount of knowledge. This step is for the manager that can find out the knowledge and logic between concepts and attributes hidden in the ontology knowledge model. Finally, the system integrates the three decision processes to deliver a storage decision.

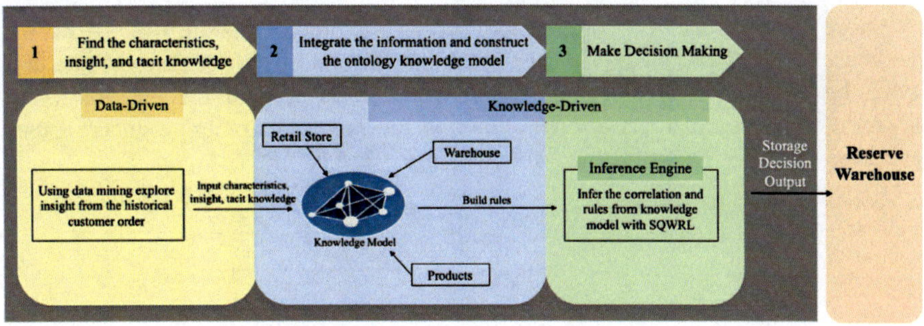

Fig. 1. The Warehouse of Retail Store Framework Overview

Step 1: Using Data Mining to Find Tacit Knowledge

This research uses apriori algorithm as the data mining technique. Apriori algorithm is for frequent item mining and association rule learning on relational databases. Apriori can be used to determine association rules, including applying in market basket analysis [11]. Market basket analysis is an algorithm that can find out the relevance of products in customer purchase. The found relevance may change the product arrangements on the shelfs to increase the customer's convenience of shopping; thus increasing the store's revenue.

Step 2: Using Ontology to Construct Knowledge Model

This study uses Protégé [12] to implement ontology. Ontology can represent semantic knowledge of the environment and provide activities modeling and logical reasoning to realize knowledge conceptualization. This research uses ontology to integrate the information of the retail store, warehouse, products, and the features from historical customer order to construct the knowledge model.

Step 3: Using Inference to Make Decisions

Then we use inference to make decisions. Inference can infer the correlation from a large amount of knowledge. Fudholi et al. [13] indicated Semantic Query-enhanced Web Rule Language (SQWRL) can be used to perform a query in the result of menu recommendation. SQWRL provides Structured Query Language (SQL) like operations to format knowledge retrieved from OWL ontology. SQWRL takes a standard Semantic Web Rule Language (SWRL) rule as the antecedent and effectively treats it as a pattern specification for a query [14]. In this research, after constructing an ontology knowledge model, the manager can according to the scenario setting to make the inference rules, and find out the knowledge and logic which between concepts and attributes hidden in the ontology knowledge model.

4 Implementation

Suppose there is a retail store and have a small warehouse near the retail store. Each product has different characteristics and costumers have different shopping behaviors. This study is focused on the warehouse in the retail store, so the information about the retail store is simpler and less complicated. Scenario setting for this study in Fig. 2 is the warehouse received a batch of products, and the manager needs to decide if the products have to be replenished directly in the retail store or to be stored in the warehouse? Do the products have special characteristics or have relevance with other products? If they have to be stored in the warehouse, which storage space is suitable?

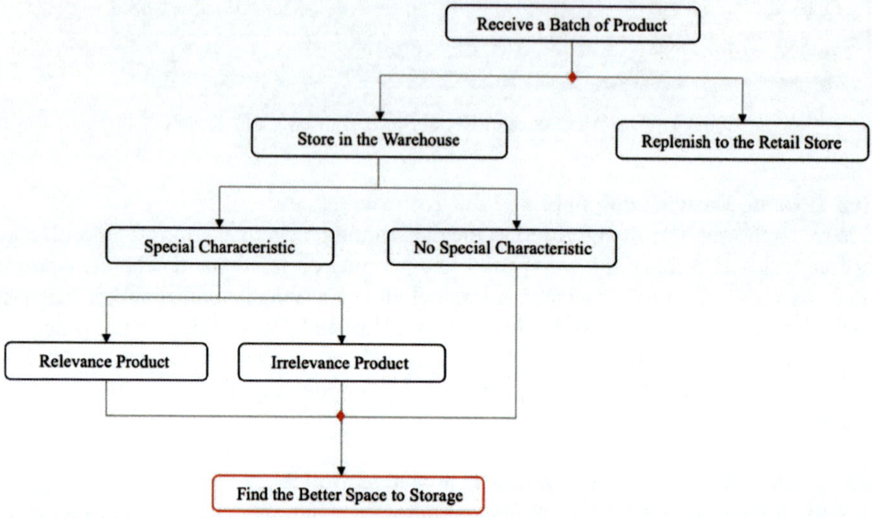

Fig. 2. Scenario Setting

4.1 Case Description

Description of the warehouse and retail store environment is as follows. The retail store that shown in Fig. 3 (a) has 3 aisle, 6 rows of the shelf, and each row has 5 shelves. The warehouse near the retail store that shown in Fig. 3 (b) has 2 aisles, 4 rows of the shelf and each row has 2 shelves. The shelf size of the retail store and the warehouse are shown in Fig. 4. The size of the shelf in the retail store has measures approximately 100 cm wide × 50 cm deep × 70 cm high. The size of the shelf in the warehouse has measures approximately 200 cm wide × 100 cm deep × 100 cm high.

The products' information uses the open dataset from UCI ML repository. It has 18537 customer orders and 44860 items. After cleaning the data, only 55 items were chosen to construct the model to verify the feasibility of the knowledge model.

4.2 Development of the Knowledge Model

The process of developing the ontology knowledge model is shown in Fig. 5. First, this study uses apriori algorithm of market basket analysis to find the tacit knowledge of the products in the customer order. Then construct the ontology knowledge model, including

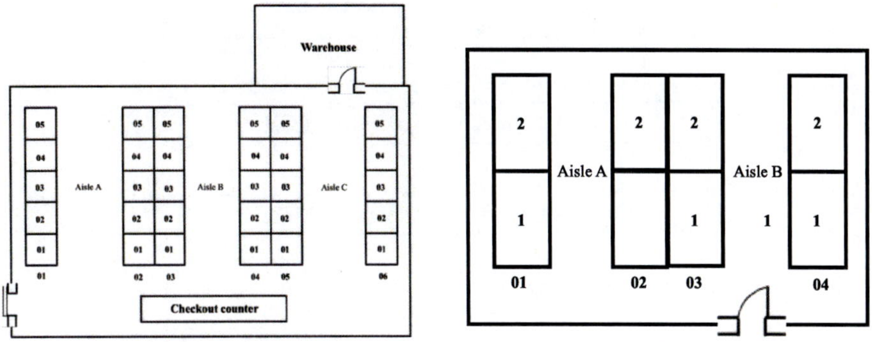

(a) Layout of the retail store (b) Layout of the warehouse

Fig. 3. Layouts of the retail store and the associated warehouse

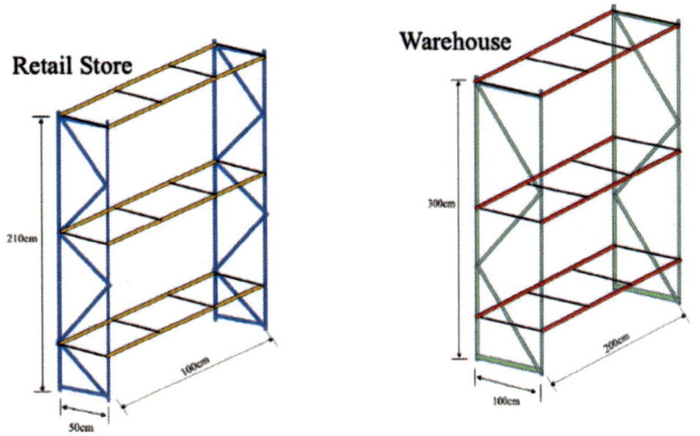

Fig. 4. The Shelf Size of the Retail Store and the Warehouse

explicit knowledge and tacit knowledge of the products, warehouse, and retail store. It is to let all the knowledge can be shared and reused in the company. Finally based on the environment scenario settings, it can establish inference rules in accordance with the ontology knowledge model.

4.2.1 Construction of Products Knowledge Model

The market basket analysis gave the values of 1% support and 70% confidence for people buying maximal 3 items at once. The definitions of support, confidence, and lift are specified in the Appendix A. The analysis generated 36 rules in Appendix B. For the 36 rules of the analysis, all of values of the lift indicator are larger than one which means the products have a positive correlation in each basket.

According to the result of the data mining, this study separates the products into 4 different kinds of products, as follows.

(1) Staple Merchandise: Staple merchandise is the product with the largest proportion of sales, and the main source of turnover and profit in the retail store. According to

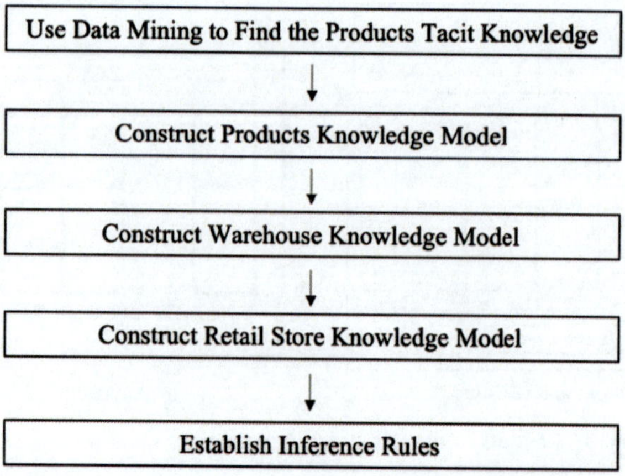

Fig. 5. Process of Develop Ontology Knowledge Model

the summary of the baskets in the data mining, this research took top 15 products to be the staple merchandise and do the knowledge model.

(2) Related Merchandise: Related merchandise has a strong correlation with the staple merchandise. Related merchandise is usually purchased with the staple merchandise. It can increase the sales volume of staple merchandise. According to the 36 rules of the data mining result, six rules show the relationship with the staple merchandise. Eight products were related merchandise from the six rules.
(3) Series of Merchandise: Series of merchandise indicates the group of two or three products that were purchased together usually. In this research, according to the 36 rules of the data mining result, it has 30 rules shows the relationship in the series of merchandise. Twenty-two products were series of merchandise from the 30 rules.
(4) Other Merchandise: Other merchandise is the product that is not so prominent in retail stores, and do not have relationship with other products. In this research, 10 products are other merchandise.

In the retail store, there are varieties of products, and its warehouse has thousands of different products. A product has many different characteristics. To build complete product information needs to understand all the knowledge of the product, including explicit knowledge and tacit knowledge. By applying Protégé [12] to build the ontological knowledge model: four merchandise classes, 15 individuals in the staple merchandise, 8 individuals in the related merchandise, 22 individuals in the series of merchandise, and 10 individuals in the other merchandise (Fig. 6). Different kinds of products have different characteristics and must be managed in different ways.

There are 5 data properties in the products, including priority, relationship, selling well or not, and the number in stock or on the shelf (Fig. 7). Due to space limitation, the details are not addressed for each data property. Those data properties of the products provide information to the manager for inventory replenishment decision on the shelf of the retail store or in the warehouse.

This research took the product namely alarm clock bakelike red of staple merchandise to be the example, and the information is shown in Fig. 8. Alarm clock bakelike red is

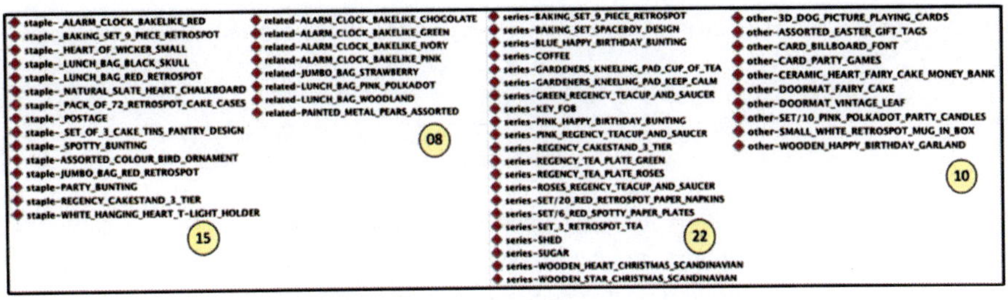

Fig. 6. Individuals in Four Kinds of Products

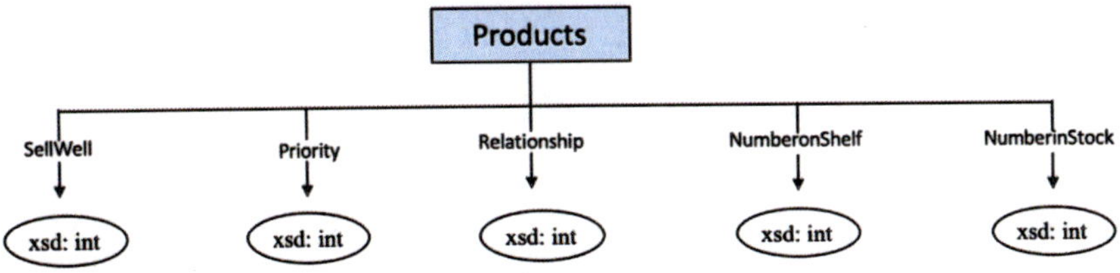

Fig. 7. Product's Data Properties

staple merchandise. According to its importance, the priority is set value 1. This product is selling well and has a relationship with other products, so for the relationship and selling well were both set value 1. And alarm clock bakelike red has 500 in the stock and 1000 on the shelf.

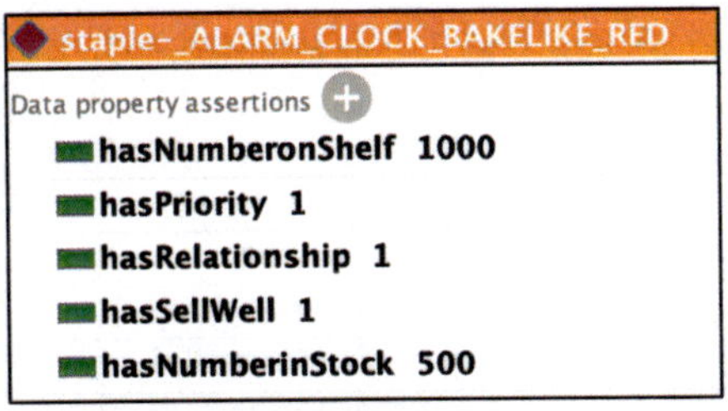

Fig. 8. Example of Staple Merchandise Data Properties

For object properties, there are four types, including the product shows (1) having a relationship with which product, (2) been supported by which product, (3) been stored in which shelf in the warehouse, and (4) with a display on which shelf in the retail store. Here, taking the related merchandise- lunch bag pink polka dot as the example, the object properties are shown in Fig. 9. Lunch bag pink polka dot is related to the product jumbo bag strawberry and lunch bag woodland, and it also supports the product jumbo bag red retros pot and lunch bag red retros pot. By the information, the manager can know the lunch bag pink polka dot that can support jumbo bag red retros pot and lunch bag red retros pot to increase the sales volume. In this case, they can be put together on the shelf to stimulate purchases or store it together in the stock to make replenishment easier and faster.

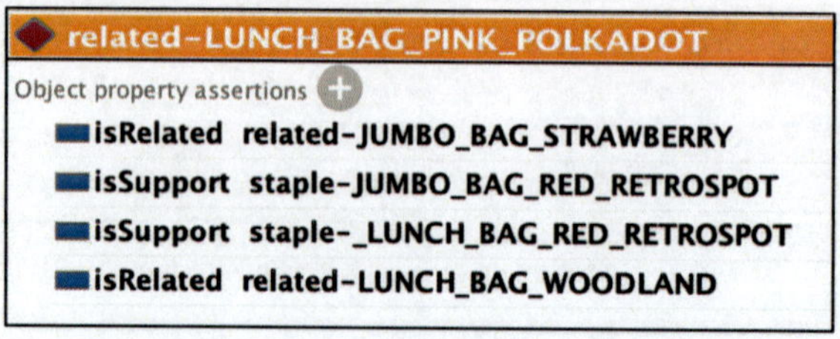

Fig. 9. Example of Object Properties for a Related Merchandise

4.2.2 Construction of Warehouse Knowledge Model

The operation of a warehouse affects the entire retail store. It is associated with the speed of replenishment, customer desire to buy the product, sales volume, etc. In addition, it may influence the supply chain, including out of stock problem, the capacity of the warehouse, miscalculated sales to order too many products, and so on. Warehouse knowledge model represents a thoughtful consideration in operational decision of the warehouse.

As was mentioned in Sect. 4.1, it describes the detailed of the warehouse that it has 2 aisles, 4 rows of the shelf and each row has 2 shelves. Each shelf has three layers, namely lower, medium, and higher. According to these descriptions, classes and individuals of the ontology are shown in Fig. 10 and Fig. 11. In Fig. 11, Wshelf01-1H means the shelf in the warehouse and the storage space is locates at the aisle A on the row 01 of the first shelf in the higher layer. Due to space limitation, the properties of the warehouse are not addressed.

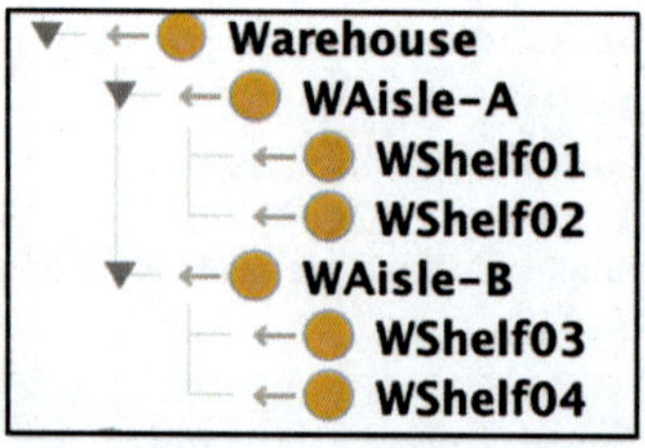

Fig. 10. Warehouse Situational in Ontology Knowledge Model

Figure 12 demonstrates data properties of Wshelf03-1L and Wshelf03-1M as examples, where the shelves are on row 03 but in different layer. These two shelves have priority one in the warehouse, because they are near the door. The shelf in the lower layer has warehouse space; it has 2000 capacity, but it only has 1000 products in the stock. On the other hand, the shelf in the medium layer has not warehouse space; it also has 2000 capacity, but it has already 2000 products in it.

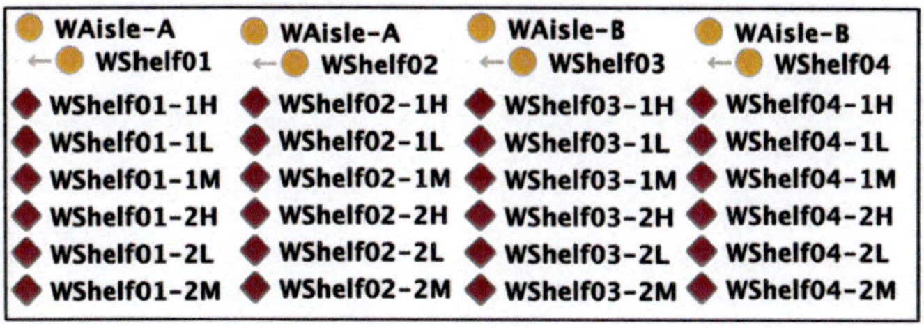

Fig. 11. Individuals of the Warehouse in Ontology Knowledge Model

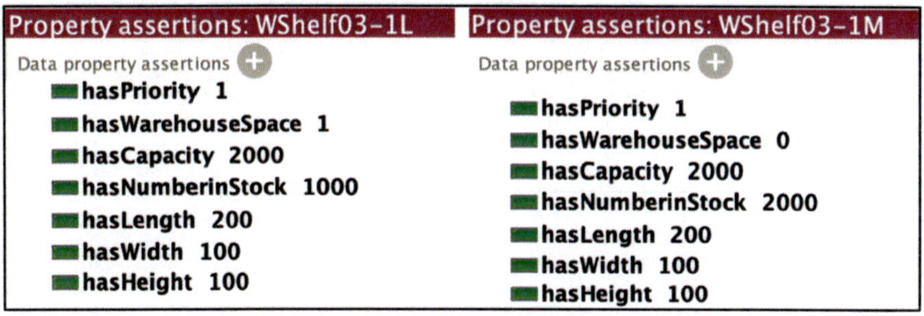

Fig. 12. Example of Two Shelves in the Warehouse

4.2.3 Construction of Retail Store Knowledge Model

Retail stores are front-line to face the customers, attracting customers to buy products. Therefore, the merchandise placement and moving line of a retail store are crucial. Section 4.1 describes the detailed of the retail store that has 3 aisles, 6 rows of the shelf, and each row has 5 shelves. Each shelf has three-layers, namely lower, medium, and higher. According to these descriptions, the ontology model is developed and shown in Fig. 13. By taking an example in Fig. 13, Rshelf01-01H means the shelf in the retail store is located at the aisle A at the row 01 of first shelf in the higher layer. Due to space limitation, the object properties and data properties are not addressed.

4.3 Inference Rules

After building the above knowledge models, the designer can through the envisaged situation to develop influence rules. With the knowledge models, the logical and causal relationship between classes and properties is to find the tacit knowledge and logical rules.

4.3.1 Finding Products Information in the Warehouse

For the retail warehouse, sometimes it may be messy, and does not have any standard storage rules in the warehouse. According to this knowledge model, robots if equipped can through the computer and the inference rule find the product information in the

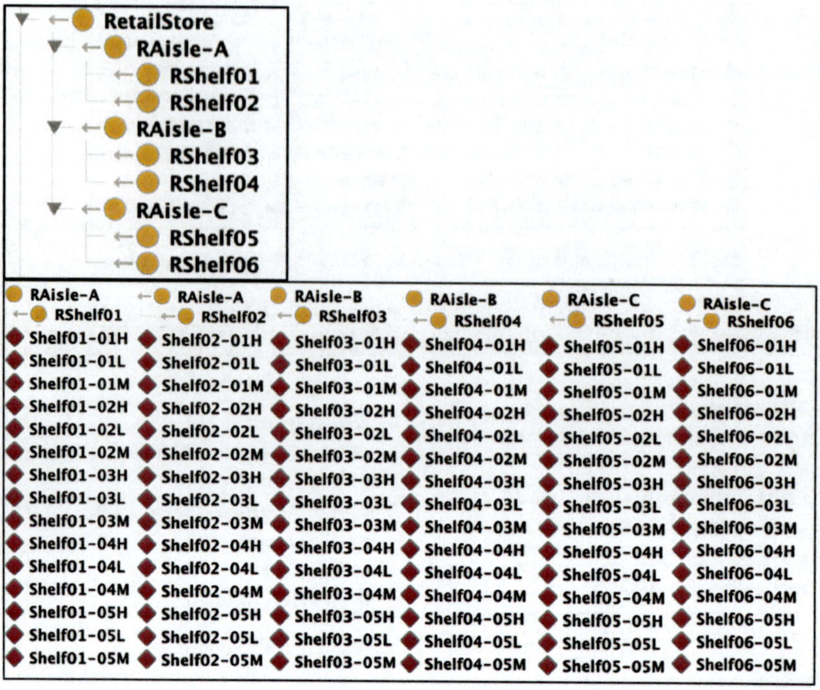

Fig. 13. Classes (top) and Individuals (bottom) of the Retail Store in Ontology Knowledge Model

warehouse. In this research, to show the application reality, the inference rule of SQWRL that can find all the products stored in the warehouse is listed as follows:

```
Rule 01 Products(?_products) ^ hasPriority(?_products, ?_priority)
    ^ hasNumberonShelf(?_products, ?_Noonshelf)
    ^ hasNumberinStock(?_products, ?_Noinstock)
    ^ isStoragein(?_products, ?_warehouse)
    ^ Warehouse(?_warehouse) ^ hasCapacity(?_warehouse, ?_capacity)
    -> sqwrl:select(?_products, ?_warehouse, ?_priority, ?_capacity, ?_Noinstock, ?_Noonshelf)
```

The above rule of SQWRL give a query to the knowledge model for questions:

(1) Which warehouse space that the product storage in?
(2) Is there any space to store other product?

The result can show the capacity of the warehouse, the number of the products in stock of the warehouse, and the number of products on shelf of the retail store. The example result in Fig. 14 shows they all in the staple merchandise with a highest priority. Manager need to pay caution for those products.

_products	_warehouse	_priority	_capacity	_Noinstock	_Noonshelf
:staple-WHITE_HANGING_HEART_T-LIGHT_H...	:WShelf03-1M	1	2000	500	1000
:staple-_NATURAL_SLATE_HEART_CHALKBOA...	:WShelf04-1L	1	2000	500	1000
:staple-_SET_OF_3_CAKE_TINS_PANTRY_DESI...	:WShelf03-1L	1	2000	500	1000
:staple-_HEART_OF_WICKER_SMALL	:WShelf04-1M	1	2000	500	1000
:staple-PARTY_BUNTING	:WShelf03-1M	1	2000	500	1000
:staple-_PACK_OF_72_RETROSPOT_CAKE_CA...	:WShelf04-1M	1	2000	500	1000
:staple-_ALARM_CLOCK_BAKELIKE_RED	:WShelf03-1H	1	2000	500	1000
:staple-_LUNCH_BAG_RED_RETROSPOT	:WShelf03-1H	1	2000	500	1000
:staple-REGENCY_CAKESTAND_3_TIER	:WShelf03-1M	1	2000	500	1000
:staple-ASSORTED_COLOUR_BIRD_ORNAMENT	:WShelf03-1L	1	2000	500	1000
:staple-_POSTAGE	:WShelf04-1M	1	2000	500	1000
:staple-_SPOTTY_BUNTING	:WShelf04-1H	1	2000	500	1000
:staple-JUMBO_BAG_RED_RETROSPOT	:WShelf03-1M	1	2000	500	1000
:staple-_LUNCH_BAG_BLACK_SKULL	:WShelf04-1L	1	2000	500	1000
:staple-_BAKING_SET_9_PIECE_RETROSPOT	:WShelf04-1H	1	2000	500	1000

Fig. 14. Result of Inference Rule

4.3.2 Finding the Relationship of Products

Through the relationship of the products, mangers can change the sales method to attract the customer buying the goods, so the sales volume can be increased. There are three rules to describe the relationship of products as follows:

Rule 02 Products(?_products) ^ isRelated(?_products, ?_relatedproducts)
-> sqwrl:select(?_products, ?_relatedproducts)
Rule 03 Products(?_products) ^ isSupport(?_products, ?_supportproducts)
-> sqwrl:select(?_products, ?_supportproducts)
Rule 04 Products(?_products) ^ isSupport(?_products, ?_supportproducts)
^ isRelated(?_products, ?_relatedproducts)
-> sqwrl:select(?_products, ?_supportproducts, ?_relatedproducts)

For Rule 02, it can retrieve the related products that are usually in the customer's basket with the other one or two products. The manager can use this information to put these products together on the shelfs to ease the searching for the products. For Rule 03, it can retrieve the support products that may increase the sales volume of staple merchandise and expand the scope of target customers. The manager can use this information to put these products together in a prominent place to attract customers. When applying to the warehouse, the products can be place in a location so the replenish time is minimal. Rule 04 is a combination of Rule 02 and Rule 03. Finding the products by the rule can help the manager to improve the turnover of the retail store and attract the customer by placing the products in the prominent shelf. In addition, the products with high turnover can also be placed in the convenient locations of the warehouse.

4.3.3 Finding the Better Storage Space in the Warehouse

The operation of a warehouse affects the entire retail store and the supply chain. Two rules are developed to help to find a better space to store the product in the warehouse; hence reducing time of handling distance and replenishing operations.

Rule 05 Warehouse(?_warehouse) ^ hasWarehouseSpace(?_warehouse, ?_warehousespace)^ wrlb:equal(?_warehousespace, 1)^ hasPriority(?_warehouse, ?_priority)
-> sqwrl:select(?_warehouse, ?_warehousespace, ?_priority)^ sqwrl:orderBy(?_priority)

Rule 06 Warehouse(?_warehouse) ^ hasProducts(?_warehouse, ?_products) ^ hasCapacity(?_warehouse, ?_WarehouseCapacity) ^ Products(?_products) ^ hasNumberinStock(?_products, ?_NoinStock)
-> sqwrl:select(?_warehouse, ?_products, ?_WarehouseCapacity, ?_NoinStock)

For Rule 05, it help find the warehouse space, ordered in accordance with the priority. This rule is especially helpful for the staple merchandise due to its frequent moving in and out. The results of Rule 06 provide a cross-reference warehouse information based on product properties and relevance.

This section describes how to utilize knowledge modeling by apriori algorithm for the customer basket and ontology for the product, retail store, and warehouse to support the manager in warehousing and shelf management for the products within limited spaces of the retail store and the associated warehouse.

5 Conclusions and Future Research

Today's retail stores and their associated warehouse have unique characteristics, including limit storage capacity, seasonal merchandise, huge number of various small-quantity orders, and so on. The method to manage the warehouse is also different from the past, because more characteristics need to be considered. In addition to the need to arrange for products to be stored in a suitable location, and to be quickly replenished to the shelf, managers should quickly grasp the information and status of warehousing, retailers and products.

This research used data-driven approach to assist knowledge-driven approach to intelligently manage the retail store and its associated warehouse with the following features:

(1) This research verifies that knowledge driven method can be applied in the retail store. It allows us not only to describe the product information, but also to consider the location of the shelf in the warehouse and retail store.
(2) In the retail store and warehouse, many things need to be considered, like the characteristic of the products, capacity of the warehouse, customer behavior, and so on. This research uses knowledge-driven approach (ontology) that presents the knowledge of the system.
(3) The inference rules support intelligently the manager's decisions or daily operations on the products, the retail store, and the warehouse.

The ontology knowledge model and the inference rules in this research are not complete. In the future, attempts to complete the following three works are necessary.

(1) Considering the weights/dimensions of the goods and modify the SQWRL about the weight limit.
(2) Expanding the knowledge model (ontology and inference rules), including customers, suppliers, logistics, operational queries/decisions, and so on..

(3) Connecting the ontology knowledge model to the databases of the retail store to be consistently support the decisions and operations timely.

Acknowledgements. Grant project No. 107-2221-E-029-022-MY2 from Ministry of Science and Technology, Taiwan.

Appendix A: Definitions of Support, Confidence, and Lift

The following definitions are from Wikipedia [15].

Support is an indication of how frequently the itemset appears in the dataset. It is defined as:

support(A ∩ B) = (# of transactions in the dataset containing A and B)/(total # of transactions in the dataset).

Confidence is the percentage of all transactions satisfying A that also satisfy B. It is defined as:

Conf(A = > B) = support (A ∩ B)/support(A).

The lift of a rule is defined as:

Lift(A = > B) = support (A ∩ B)/(support(A) * support(B)).

Appendix B: Market Basket Analysis Result (36 Rules)

	Left hand side	Right hand side	support	confidence	lift
1	{SUGAR}	{SET 3 RETROSPOT TEA}	0.0108	1.0000	92.2239
2	{SET 3 RETROSPOT TEA}	{SUGAR}	0.0108	1.0000	92.2239
3	{COFFEE, SUGAR}	{SET 3 RETROSPOT TEA}	0.0108	1.0000	92.2239
4	{COFFEE, SET 3 RETROSPOT TEA}	{SUGAR}	0.0108	1.0000	92.2239
5	{SUGAR}	{COFFEE}	0.0108	1.0000	67.9011
6	{COFFEE}	{SUGAR}	0.0108	0.7363	67.9011
7	{SET 3 RETROSPOT TEA}	{COFFEE}	0.0108	1.0000	67.9011
8	{COFFEE}	{SET 3 RETROSPOT TEA}	0.0108	1.0000	67.9011
9	{SET 3 RETROSPOT TEA, SUGAR}	{COFFEE}	0.0108	0.7363	67.9011
10	{SHED}	{KEY FOB}	0.0111	1.0000	63.2662
11	{KEY FOB}	{SHED}	0.0111	0.7031	63.2662

(*continued*)

(*continued*)

	Left hand side	Right hand side	support	confidence	lift
12	{REGENCY TEA PLATE GREEN}	{REGENCY TEA PLATE ROSES}	0.0103	0.8377	56.4684
13	{SET/6 RED SPOTTY PAPER CUPS}	{SET/6 RED SPOTTY PAPER PLATES}	0.0106	0.8174	56.1209
14	{SET/6 RED SPOTTY PAPER PLATES}	{SET/6 RED SPOTTY PAPER CUPS}	0.0106	0.7296	56.1209
15	{WOODEN STAR CHRISTMAS SCANDINAVIAN}	{WOODEN HEART CHRISTMAS SCANDINAVIAN}	0.0114	0.7439	44.0541
16	{PINK HAPPY BIRTHDAY BUNTING}	{BLUE HAPPY BIRTHDAY BUNTING}	0.0116	0.7072	38.5590
17	{GREEN REGENCY TEACUP AND SAUCER, ROSES REGENCY TEACUP AND SAUCER}	{PINK REGENCY TEACUP AND SAUCER}	0.0172	0.7185	27.7464
18	{GREEN REGENCY TEACUP AND SAUCER, REGENCY CAKESTAND 3 TIER}	{PINK REGENCY TEACUP AND SAUCER}	0.0121	0.7166	27.6727
19	{PINK REGENCY TEACUP AND SAUCER, ROSES REGENCY TEACUP AND SAUCER}	{GREEN REGENCY TEACUP AND SAUCER}	0.0172	0.8788	26.3168
20	{PINK REGENCY TEACUP AND SAUCER, REGENCY CAKESTAND 3 TIER}	{GREEN REGENCY TEACUP AND SAUCER}	0.0121	0.8721	26.1163
21	{PINK REGENCY TEACUP AND SAUCER}	{GREEN REGENCY TEACUP AND SAUCER}	0.0207	0.7979	23.8950
22	{PINK REGENCY TEACUP AND SAUCER, REGENCY CAKESTAND 3 TIER}	{ROSES REGENCY TEACUP AND SAUCER}	0.0118	0.8450	23.8403
23	{GREEN REGENCY TEACUP AND SAUCER, PINK REGENCY TEACUP AND SAUCER}	{ROSES REGENCY TEACUP AND SAUCER}	0.0172	0.8329	23.4999

(*continued*)

(*continued*)

	Left hand side	Right hand side	support	confidence	lift
24	{GREEN REGENCY TEACUP AND SAUCER, REGENCY CAKESTAND 3 TIER}	{ROSES REGENCY TEACUP AND SAUCER}	0.0140	0.8248	23.2726
25	{REGENCY CAKESTAND 3 TIER, ROSES REGENCY TEACUP AND SAUCER}	{GREEN REGENCY TEACUP AND SAUCER}	0.0140	0.7507	22.4817
26	{SET/6 RED SPOTTY PAPER PLATES}	{SET/20 RED RETROSPOT PAPER NAPKINS}	0.0104	0.7148	21.8655
27	{PINK REGENCY TEACUP AND SAUCER}	{ROSES REGENCY TEACUP AND SAUCER}	0.0196	0.7563	21.3373
28	{GARDENERS KNEELING PAD CUP OF TEA}	{GARDENERS KNEELING PAD KEEP CALM}	0.0220	0.7286	20.2786
29	{GREEN REGENCY TEACUP AND SAUCER}	{ROSES REGENCY TEACUP AND SAUCER}	0.0240	0.7173	20.2379
30	{ALARM CLOCK BAKELIKE GREEN, ALARM CLOCK BAKELIKE PINK}	{ALARM CLOCK BAKELIKE RED}	0.0141	0.7844	16.9476
31	{ALARM CLOCK BAKELIKE GREEN, ALARM CLOCK BAKELIKE IVORY}	{ALARM CLOCK BAKELIKE RED}	0.0112	0.7820	16.8941
32	{BAKING SET SPACEBOY DESIGN}	{BAKING SET 9 PIECE RETROSPOT}	0.0165	0.7321	16.5490
33	{ALARM CLOCK BAKELIKE CHOCOLATE}	{ALARM CLOCK BAKELIKE RED}	0.0120	0.7070	15.2748
34	{LUNCH BAG PINK POLKADOT, LUNCH BAG WOODLAND}	{LUNCH BAG RED RETROSPOT}	0.0106	0.7323	12.7469
35	{LUNCH BAG PINK POLKADOT, JUMBO BAG STRAWBERRY}	{JUMBO BAG RED RETROSPOT}	0.0101	0.7866	10.9717

(*continued*)

(*continued*)

	Left hand side	Right hand side	support	confidence	lift
36	{PAINTED METAL PEARS ASSORTED}	{ASSORTED COLOUR BIRD ORNAMENT}	0.0118	0.7204	9.9434

References

1. Azkune, G., Almeida, A., López-de-Ipiña, D., Chen, L.: Extending knowledge-driven activity models through data-driven learning techniques. Expert Syst. Appl. **42**(6), 3115–3128 (2015)
2. Girase, A., Patil, S.: Devloping knowledge driven ontology for decision making (2016)
3. Rodríguez, N.D., Cuéllar, M.P., Lilius, J., Calvo-Flores, M.D.: A fuzzy ontology for semantic modelling and recognition of human behaviour. Knowl.-Based Syst. **66**, 46–60 (2014)
4. Saleheen, F., Miraz, M.H., Habib, M.M., Hanafi, Z.: Challenges of warehouse operations: A case study in retail supermarket. Int. J. Supp. Chain Manage. **3**(4), 63–67 (2014)
5. Teo, C.-P., Shu, J.: Warehouse-retailer network design problem. Oper. Res. **52**(3), 396–408 (2004)
6. Fazel Zarandi, M.: A retail ontology: formal semantics and efficient implementation. Master Thesis, University of Toronto (2007)
7. Weidinger, F., Boysen, N., Schneider, M.: Picker routing in the mixed-shelves warehouses of e-commerce retailers. Eur. J. Oper. Res. **274**(2), 501–515 (2019)
8. Giannikas, V., Lu, W., McFarlane, D., Hyde, J.: Product intelligence in warehouse management: a case study. In: Industrial Applications of Holonic and Multi-Agent Systems, pp. 224–235. Springer (2013)
9. Zhang, Z., Shi, X., Xing, M.: Research on e-commerce logistics warehousing mode under the new retail. In: Proceedings of the 2018 International Conference on Information Management & Management Science, pp. 48–52 (2018)
10. Liu, C.-M.: Optimal storage layout and order picking for warehousing. Int. J. Oper. Res. **1**(1), 37–46 (2004)
11. Agrawal, R., Srikant, R.: Fast algorithms for mining association rules. In: Proc. 20th int. conf. very large data bases, VLDB 1994, pp. 487–499. Citeseer
12. Musen, M.A.: The Protégé project: A look back and a look forward. AI Matters **1**(4), 4–12 (2015)
13. Fudholi, D.H., Maneerat, N., Varakulsiripunth, R., Kato, Y.: Application of Protégé, SWRL and SQWRL in fuzzy ontology-based menu recommendation. In: 2009 International Symposium on Intelligent Signal Processing and Communication Systems (ISPACS), pp. 631–634. IEEE (2009)
14. O'Connor, M.J., Das, A.K.: SQWRL: a query language for OWL. In: OWLED 2009, vol. 2009
15. Wikipedia: Association rule learning (2022)

Crop Plants Stress Monitoring with Bayesian Network Inference in Cyber-Physical System

Win P. V. Nguyen[1(✉)], Puwadol Oak Dusadeerungsikul[2], and Shimon Y. Nof[3]

[1] Grado Department of Industrial and Systems Engineering, Virginia Tech, Blacksburg, VA, USA
winn@vt.edu

[2] Department of Industrial Engineering, Chulalongkorn University, Bangkok, Thailand

[3] PRISM Center and School of Industrial Engineering, Purdue University, West Lafayette, IN, USA

Abstract. Propagating crop plant stresses and diseases risk severe losses of agricultural produce. Without effective monitoring and treatment procedures, stresses and diseases can propagate to the surrounding plants and eventually create irreparable damage. To address the problem, the Agricultural Robotic System for Plant Stress Propagation Detection (ARS/PSPD), working as a cyber-physical system is developed. In this cyber-physical system, the robot agents are assigned scanning tasks to detect stresses in greenhouse crop plants. The cyber-physical network modeling, which provides better situation awareness and augments advanced collaborative scanning protocols, is utilized to develop the crop plant stress propagation detection model. Three collaborative scanning protocols (baseline, disruption propagation network analysis, and Bayesian network inference) are designed, implemented, and validated in this study. The scanning protocols minimize errors and conflicts in scanning task allocation and enable better crop plant stress detection. Computer numerical experiments have been performed to validate the designed protocols. Results show that the scanning protocol utilizing Bayesian network inference significantly outperforms other protocols: It results in fewer undetected crop plant stresses, and fewer redundant scans.

Keywords: Agricultural Robotic System · Collaborative Scanning Protocols · Crop Stress Detection · Disruption Propagation · Situation Awareness

1 Introduction and Background

Agricultural crops are vulnerable to anomalous stress situations, even in a greenhouse environment [1]. Such stress situations include sudden changes in temperature, water levels, humidity levels, diseases, and pests. Timely and reliable stress/disease detection is required lest irreversible damage occurs. Such damage can be as high as 40% of food production [2]. This is complicated by the large-scale nature of agricultural food production activities, which renders exhaustive scanning and detection for plant stresses and diseases costly or even infeasible. Manual plant stress detection activities performed by

C.-Y. Huang and S. W. Yoon (Eds.): ICPR1 2021, ACES 14, pp. 355–362, 2023.
https://doi.org/10.1007/978-3-031-44373-2_21

human operators involve the operators walking into the plot and inspecting the plants one by one, taking up as much as 20 km per day [3, 4]. However, undetected stresses/diseases can spread and propagate to nearby plants, due to the proximity between the plants. While such propagation can be damaging, the propagation pattern can be inferred and applied to guide the next scanning decisions. For example, if a small area of plants is found to be stress-free or disease-free, the probability of nearby areas having stresses/diseases can be assumed to be lower, and vice versa.

Early detection of stress and disease in plants is important because it can ensure agricultural yields and quality with minimal cost. With the advancement in robotic and communication technologies as well as knowledge-based information, early detection of stress/disease in plants becomes possible [5]. Agricultural cyber-physical systems enable real-time communication and control, information sharing between agents, and real-time reaction to detected plant stresses. Agricultural robotics enables precise and repetitive plant monitoring tasks by remote operators. In addition, plant stress/disease detection technologies [6] (e.g., sensors and hyperspectral cameras) support the monitoring process as they help localize stress/disease locations in the field. All mentioned technologies, however, require an Agricultural Robotic System (ARS) with protocols to enable collaboration between agents, algorithms, and knowledge-based information effectively.

In this work, the Agricultural Robotic System for Plant Stress Propagation Detection (ARS/PSPD) is developed to guide the scanning decisions of robot agents to detect stresses in greenhouse crop plants. ARS/PSPD utilizes and combines the knowledge of plant stress propagation [4], disruption propagation network modeling [7, 8], and Bayesian network statistical inference [9] to guide the scanning decisions. The system is an extension of the agricultural robotics framework [1, 4], and adapts the collaborative response to disruption propagation (CRDP) framework [7] to the agricultural plant stress/disease setting. The ARS defines the problem setting, the agents involved, and the scanning tasks, whereas the PSPD formulation captures the plant stress occurrence and propagation mechanisms. Both functions enable better situation awareness and augment the development of advanced scanning protocols. Furthermore, Bayesian network (BayesNet) statistical inference is applied and utilized to guide the inference of stress/disease propagation and thus improves the scanning decisions. In this study, three collaborative scanning protocols are implemented and validated. The first scanning protocol is the baseline protocol, selecting scanning locations randomly. The second scanning protocol, which is the adaptive network scanning protocol, is developed based on the ARS adaptive scanning protocol [4], prioritizing scanning plants next to known stresses/diseases. The third scanning protocol expands further by constructing and using a BayesNet to infer and update the probability grid of the entire greenhouse based on all observations made thus far. Then, numerical experiments are conducted to validate ARS/PSPD and its three scanning protocols. The results show that the scanning protocol developed based on BayesNet outperforms the adaptive network scanning protocol, which in turn outperforms the baseline random scanning protocol.

The remainder of the article is organized as follows. Section 2 presents the ARS/PSPD model, the PSPD problem formulation, and the three collaborative scanning protocols. Section 3 presents the numerical experiments and the results. Section 4 presents the conclusions and discusses the future work directions.

2 Methodology

In this section, the Agricultural Robotic System for Plant Stress Propagation Detection (ARS/PSPD) model is presented. This section specifies the agents and tasks involved in the ARS. Then, the mathematical formulation of the PSPD problem is presented and explained. The three collaborative scanning protocols are then introduced. An illustration of the ARS/PSPD model is presented in Fig. 1.

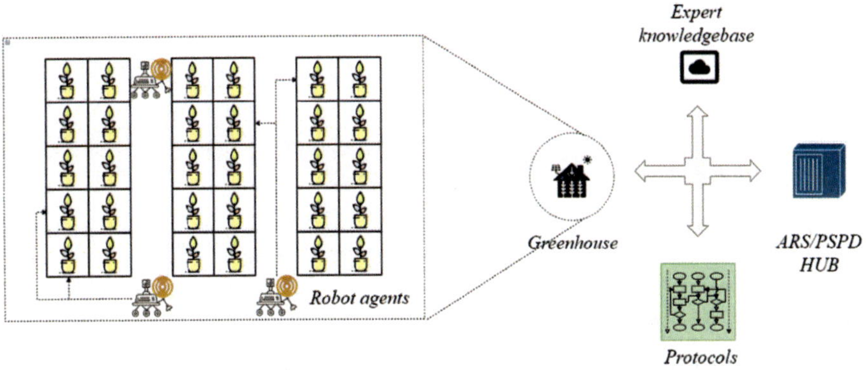

Fig. 1. ARS/PSPD illustration

The ARS/PSPD model is the expansion of the ARS model [1, 4, 10]. The ARS is a multi-agent system developed for monitoring and early detecting stress/disease in plants. The system comprises three main types of agents (i.e., human, robot, and sensors) with protocols, algorithms, and knowledge-based information. In ARS, human operators are not responsible for manually inspecting the plant, but they are the decision-makers who control important parameters and solve unexpected real-time problems. Instead, the robot mounted with multiple sensors is the monitoring agent moving into the greenhouse to inspect the plant at the assigned locations. The robot will communicate with the expert knowledgebase to decide the plant's status (either stressed or not stressed). The stress status of the current plant will lead to the next scanning decision.

The formulation of the plant stress propagation detection (PSPD) problem is as follows. The set of plants is a rectangular $L \times W$ grid G (L: length, W : width, L, $W \in \mathbb{Z}^{+}$), with each plant occupying each discrete cell and can be subjected to stresses. The stresses can propagate once to a set of predetermined directions. Within the scope of this work, the directions include up, down, left, and right. Applying the CRDP [7, 11] to this problem, each plant is modeled as a node $n \in N$, and each possible propagation direction is modeled as a directed edge $e = \left(n_i, n_j\right) \in E$. An illustration is given in Fig. 2.

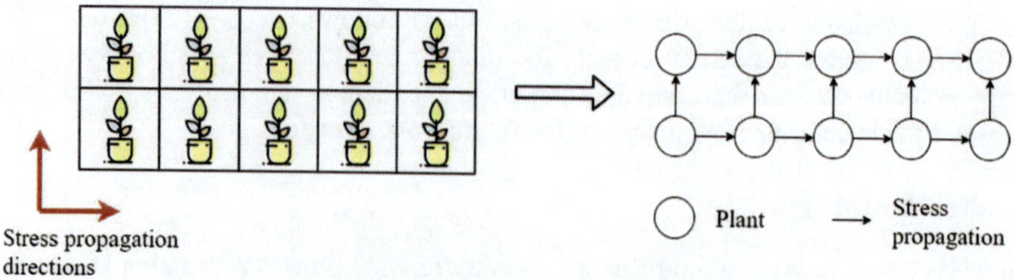

Fig. 2. Network modeling of stress propagation example.

To model stress propagation, two variables associated with each plant n is defined. The base stress status $S_0(n) \in \{0, 1\}$ is defined, with $S_0(n) = 1$ denoting active/positive stress status, and $S_0(n) = 0$ denoting no base stress at plant n. The probability $p_0 \in [0, 1]$ is defined as the probability that each plant has a source of stress that can induce stress on the plant itself and propagate along the directed edges originating from the plant. Formally,

$$S_0(n) = \begin{cases} 1, & \text{with probability } p_0 \\ 0, & \text{otherwise} \end{cases} \tag{1}$$

The parameter p_0 and the stress propagation directions can be selected based on previous knowledge. Then, the propagation of stresses is defined with the final stress status $S(n) \in \{0, 1\}$, which is set to 1 if either $S_0(n) = 1$ or there is a directed edge that originates from another node n_i, pointing into n, and $S_0(n_i) = 1$. Formally,

$$S(n) = \begin{cases} 1, & \text{if } S_0(n) = 1 \text{ or } \exists e = (n_i, n_j) \in E : (S_0(n_i) = 1 \text{ and } n_j \equiv n) \\ 0, & \text{otherwise} \end{cases} \tag{2}$$

The ARS does not know about $S_0(n)$ and $S(n)$ until plant n is scanned, and each scanning task reveals to the value of $S(n)$ to the ARS. The scanning/observation status $O(n) \in \{0, 1\}$ of a plant n is defined, with $O(n) = 1$ if a plant has been observed by the ARS, and $O(n) = 0$ otherwise. However, there is a limited number of scanning that can be performed by the ARS, with the scanning budget $O_b \in \mathbb{Z}^+$ defined as the maximum number of observations allowed by the ARS.

$$O_b \geq \sum_{n \in N} O(n) \tag{3}$$

To measure the scanning performance, two performance metrics with minimization goals are defined. The first performance metric $M_1 \in \mathbb{Z}$ is the total number of undetected stresses, or alternatively, the total number of stressed plants that were not scanned. Formally,

$$M_1 = |\{n \in N : S(n) = 1 \text{and} O(n) = 1\}| = \sum_{n \in N} \max(0, S(n) - O(n)) \tag{4}$$

The second performance metric $M_2 \in \mathbb{Z}$ is the total number of redundant scans, or alternatively, the total number of scanned plants that were not stressed. Formally,

$$M_2 = |\{n \in N : O(n) = 1 \text{and} S(n) = 0\}| = \sum_{n \in N} \max(0, O(n) - S(n)) \tag{5}$$

The ARS/PSPD simulation logic is as follows.

ARS step 1: InitialiseN, E, p_0, O_b, O_p.

ARS step 2: Initialise origin stress : $\forall n \in N$: if $\text{unif}(0, 1) < p_0$: $S_0(n) \leftarrow 1$, else $S_0(n) \leftarrow 0$

ARS step 3: Propagate stress : $\forall n \in N$: set the values of $S(n)$ per definition.

ARS step 4: Allocate scanning :

For $i := 0$ to O_b : Decide $O(n)$ according to scanning protocol, resulting in n_{next}, then $O(n_{next}) \leftarrow 1$.

ARS step 5: Compute M_1, M_2

To guide the scanning/observation decisions, different collaborative scanning protocols are developed and employed. The protocols require collaborative information sharing to eliminate redundant scanning, which are scanning by different agents of the same plant in a short period of time. The first scanning protocol, random sampling protocol, is defined as:

$$\text{Protocol}_1 : n_{next} \leftarrow \text{randomly select from}(n \in N : O(n) = 0\} \tag{6}$$

This protocol allows scanning coverage but does not consider the stress propagation pattern.

The second protocol, adaptive network scanning protocol [4], is defined as:

$$\text{Protocol}_2 : n_{\text{next}} \leftarrow \underset{n \in N : O(n)=0}{\text{argmax}} \left\{ \left| \left\{ n_i \in N^{in}(n) : O(n_i) = 1 \text{and} S(n_i) = 1 \right\} \right| \right\} \tag{7}$$

The third protocol, BayesNet scanning protocol, is developed based on Bayesian network inference. To utilize this protocol, a Bayesian network [12, 13] is constructed based on N and E. A BayesNet node is created for each variable $S_0(n)$ and $S(n)$ for all plants $n \in N$. For each plant $n \in N$ separately, a BayesNet directed edge is created between from $S_0(n)$ to $S(n)$ of that specific plant. And there, for each edge $e = \left(n_i, n_j\right) \in E$, a BayesNet directed edge is created between $S_0(n_i)$ to $S\left(n_j\right)$. For each BayesNet node associated with $S_0(n)$, the probability table is based on p_0. And for each BayesNet node associated with $S(n)$, the conditional probability table is constructed so that the value of $S(n)$ is the logical OR function of all the incoming edges into the BayesNet node.

After each observation, the ARS feeds the status $S(n^*)$ of all observed plants n^* into this Bayesian network, which then infers the stress status of all unobserved plants. Even though $S_0(n)$ is not observed by the ARS, the Bayesian network can infer these probabilities, and computes the probabilities $P(S(n)|\{S(n^*) : O(n^*) = 1, \forall n^* \in N\})$ for all remaining plants n that have not been scanned. Then, at each observation:

$$n_{next} \leftarrow \underset{n \in N : O(n)=0}{\text{argmax}} \{P(S(n)|\{S(n^*) : O(n^*) = 1, \forall n^* \in N\})\} \tag{8}$$

3 Experiments and Results

In this section, the ARS/PSPD model, the PSPD problem formulation, and the three scanning protocols are validated with numerical experiments. The set of nodes N is a 10-by-10 grid of 100 nodes in total, and all combinations of cardinal propagation directions (up, down, left, and right) are investigated. Due to the symmetry of the grid, only five combinations are investigated: 1-direction, 2-direction-opposite, 2-direction-orthogonal, 3-direction, and 4-direction. The p_0 value is selected to be 0.15, and the scanning budget O_b is 50, at which point each simulation run is terminated. The three scanning protocols are applied. The numerical experiments are conducted on Python 3, and the BayesNet calculations are provided by the library Pomegranate [14, 15]. The performance metrics M_1 and M_2, both with minimization objectives, are reported. A total of 100 replications (randomizing $S_0(n)$ for all nodes $n \in N$) are simulated for each factorial combination. This amounts to 1,500 simulation runs in total.

The performance comparison between the three scanning protocols, grouped by stress propagation direction, (with 95% confidence interval bars) is provided in Fig. 3 and Fig. 4.

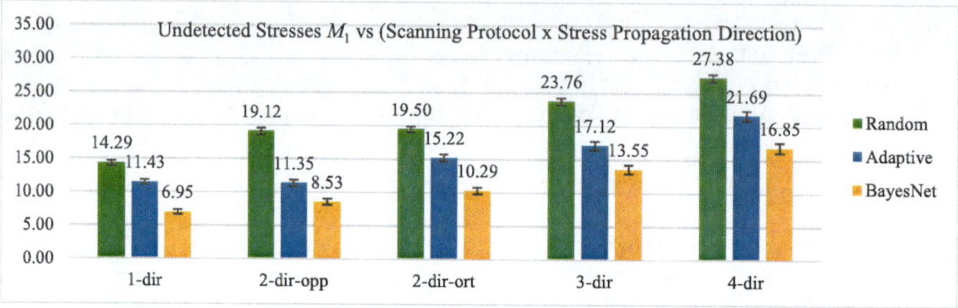

Fig. 3. Comparison of undetected stresses between scanning protocols, grouped by stress propagation directions.

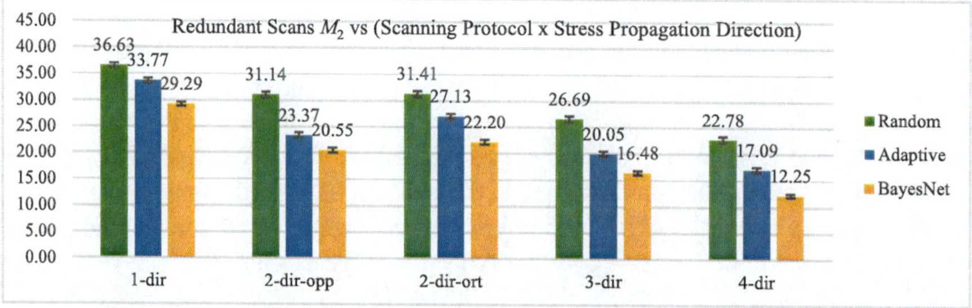

Fig. 4. Comparison of redundant scans between scanning protocols, grouped by stress propagation directions.

The results presented in Fig. 3 and Fig. 4 are averaged across all 100 replications of each factorial combination. In all five cases of stress propagation directions, the BayesNet scanning protocol outperforms the other two protocols in both M_1 and M_2, followed by the adaptive scanning protocol. With respects to M_1, the BayesNet protocol outperforms the adaptive scanning protocol by 26.87% and the random scanning protocol by 46.02% on average. Across all give cases of stress propagation directions, the BayesNet protocol outperforms the adaptive scanning protocol by 20.85% to 39.19%, and the random scanning protocol by 38.45% to 51.36%. With respects to M_2, the BayesNet protocol outperforms the adaptive scanning protocol by 17% and the random scanning protocol by 32.21% on average. Across all give cases of stress propagation directions, the BayesNet protocol outperforms the adaptive scanning protocol by 12.07% to 28.32%, and the random scanning protocol by 20.04% to 46.22%. The experiment results indicate that the BayesNet protocol outperforms the adaptive scanning protocol and the random scanning protocol in both M_1 and M_2, thus have the superior plant stress detection performance.

4 Conclusions and Future Work

This paper presents the Agricultural Robotic System for Plant Stress Propagation Detection (ARS/PSPD) model, which consists of the agricultural robotic system designed for plant stress detection. The plant stress propagation detection problem is formulated by applying the CRDP framework to the agricultural setting, which allows for better situation awareness and scanning task allocation. Three collaborative scanning protocols are developed: random scanning protocol (as baseline), adaptive scanning protocol, and Bayesian network scanning protocol. By leveraging the knowledge of plant stress propagation, a Bayesian network can be constructed to infer the stress status of unobserved plant locations, allowing effective use of scanning resources to detect plant stresses. Numerical experiments are conducted to validate the ARS/PSPD model and its scanning protocols. The experiment results show that the BayesNet scanning protocol provides superior detection performance compared to the other scanning protocols implemented. The experiment results indicate that the ARS/PSPD model can be applied to the plant stress detection and monitoring problems, and Bayesian network inference can be utilized to improve scanning performance.

This work can be further expanded into several challenging research directions: (1) applying computational statistics and/or machine learning techniques to infer the plant stress occurrence probabilities and the plant stress propagation directions; (2) considering different types of plant stresses with different occurrence and propagation characteristics; (3) including human-in-the-loop design, which allows human intervention in task allocation and decision-making, when it may be needed; (4) considering different scanning operation characteristics, budget, and time limits.

Acknowledgments. The research presented here is supported by the PRISM Center for Production, Robotics, and Integration Software for Manufacturing & Management at Purdue University. In addition, the research on ARS is supported by BARD project Grant# IS-4886-16R, "Development of a Robotic Inspection System for Early Identification and Locating of Biotic and Abiotic Stresses in Greenhouse Crops;" and the research on human-robot workflow by NSF project Grant# 1839971, "FW-HTF: Collaborative Research: Pre-Skilling Workers, Understanding Labor Force Implications and Designing Future Factory Human-Robot Workflows Using a Physical Simulation Platform".

References

1. Guo, P., Dusadeerungsikul, P.O., Nof, S.Y.: Agricultural cyber physical system collaboration for greenhouse stress management. Comput. Electron. Agric. **150**, 439–454 (2018). https://doi.org/10.1016/j.compag.2018.05.022

2. Oerke, E.-C., Dehne, H.-W.: Safeguarding production—losses in major crops and the role of crop protection. Crop Prot. **23**, 275–285 (2004). https://doi.org/10.1016/J.CROPRO.2003.10.001
3. Khan, A., Martin, P., Hardiman, P.: Expanded production of labor-intensive crops increases agricultural employment. Calif. Agric. **58**, 35–39 (2004). https://doi.org/10.3733/ca.v058n01p35
4. Dusadeerungsikul, P.O., Nof, S.Y.: A collaborative control protocol for agricultural robot routing with online adaptation. Comput. Ind. Eng. **135**, 456–466 (2019). https://doi.org/10.1016/j.cie.2019.06.037
5. Dusadeerungsikul, P.O., Liakos, V., Morari, F., Nof, S.Y., Bechar, A.: Smart action. In: Agricultural Internet of Things and Decision Support for Precision Smart Farming, pp. 225–277. Elsevier (2020)
6. Wang, D., et al.: Early Tomato Spotted Wilt Virus Detection using Hyperspectral Imaging Technique and Outlier Removal Auxiliary Classifier Generative Adversarial Nets (OR-AC-GAN), (2018). http://elibrary.asabe.org/abstract.asp?aid=49283&t=5
7. Nguyen, W.P.V., Nof, S.Y.: Collaborative response to disruption propagation (CRDP) in cyber-physical systems and complex networks. Decis. Support Syst. **117**, 1–13 (2019). https://doi.org/10.1016/j.dss.2018.11.005
8. Zhong, H., Nof, S.Y.: The dynamic lines of collaboration model: collaborative disruption response in cyber–physical systems. Comput. Ind. Eng. **87**, 370–382 (2015). https://doi.org/10.1016/j.cie.2015.05.019
9. Gibson, G.J., Otten, W., Filipe, J.A.N., Cook, A., Marion, G., Gilligan, C.A.: Bayesian estimation for percolation models of disease spread in plant populations. Stat. Comput. **16**, 391–402 (2006). https://doi.org/10.1007/s11222-006-0019-z
10. Dusadeerungsikul, P.O., Nof, S.Y., Bechar, A.: Detecting stresses in crops early by collaborative robot-sensors-human system automation. In: IISE Annual Conference and Expo 2018, pp. 1084–1089. Institute of Industrial and Systems Engineers, IISE, Orlando, FL, United states (2018)
11. Nguyen, W.P.V.: Collaborative Response to Disruption Propagation (CRDP), PhD Dissertation, School of IE, Purdue University, W. Lafayette IN, USA (2020)
12. Pearl, J.: Bayesian networks a model of self-activated memory for evidential reasoning. In: Proceedings of the 7th Conference of the Cognitive Science Society, pp. 329–334 (1985)
13. Pearl, J., Paz, A.: Graphoids: a graph-based logic for reasoning about relevance relations. In: Du Boulay, B., et al. (eds.). University of California (Los Angeles). Computer Science Department (1987)
14. Schreiber, J.: pomegranate. GitHub Repos (2014)
15. Schreiber, J.: Pomegranate: fast and flexible probabilistic modeling in python. J. Mach. Learn. Res. **18**, 5992–5997 (2017)

Future Challenges in Systems Collaboration and Integration

Augmenting Human-Machine Teaming Through Industrial AR: Trends and Challenges

Mohsen Moghaddam(✉)

Department of Mechanical and Industrial Engineering and Khoury College of Computer Sciences, Northeastern University, Boston, MA 02115, USA
m.moghaddam@northeastern.edu

Abstract. Industrial Augmented Reality (AR) is an emerging spatial computing technology which involves the use of head-mounted displays or hand-held devices such as tablets or smartphones to superimpose digital content onto the worker's physical to foster their productivity, learning, and interactions with machines, tools, and other workers. Industrial AR has been adopted in many industries such as manufacturing, healthcare, aerospace, and defense, predominantly for training or remote assistance purposes. Yet, several technical and technological challenges remain to be addressed for industrial AR to evolve from a spatial visualization tool to a more intelligent and adaptive assistive tool that not only augments the spatial and causal reasoning of workers but can also provide them with just-in-time training and support on the job. This chapter provides some technical background on industrial AR and underscores several research and development directions which can potentially materialize this vision.

Keywords: Augmented reality · Collaboration · Assistive technology · Training

1 Introduction

The rapid growth of Artificial Intelligence (AI) and spatial computing technologies such as AR are transforming the landscape of work and human-machine interaction in several industries. These technologies are increasingly adopted by many companies to complement human work and upskill workers [1]. This is also in part due to the shortage of skilled workers, workforce aging and retirement, shifting skill requirements, and the increasing complexity of industrial technology. In manufacturing industry, for example, most companies have predicted a steady demand for workers over the next few years [2], despite shedding nearly 5 million workers between 2000 and 2016 [3], as COVID-19 has increased the need to produce more goods domestically [4]. Yet, about 26% of industrial workers in the United States are retiring [5] and finding skilled workers is more challenging that ever [6]. It is anticipated that near 2.4 million manufacturing jobs will be left unfilled by 2030, which is likely to incur a cost of $2.5 trillion to the U.S. manufacturing GDP [7].

The skills gap in industry is due to the need for complex, career-spanning expertise that are hard to automate in the foreseeable future [8]. Some companies are gradually

C.-Y. Huang and S. W. Yoon (Eds.): ICPR1 2021, ACES 14, pp. 365–385, 2023.
https://doi.org/10.1007/978-3-031-44373-2_22

adopting AR as an experiential training and assistive technology to train and upskill their workers faster [9]. Boeing was one of the early adopters of industrial AR for wire assembly of aircrafts, which led to a 25% reduction in their cycle time and nearly eliminated all the errors that used to occur during the assembly process [10]. Similar early results have been reported by EU-funded STARMATE [11, 12] and SKILLS [13–15] projects, as well as companies such as Honeywell [16], Porsche [17], and Mercedes-Benz [18]. More recent studies also point to the affordances of industrial AR for fostering performance and learning on tasks such as assembly [19–21], maintenance [22, 23], and inspection [24–26].

Extant approaches to industrial AR are mostly concerned around AR authoring [22, 27–31], object tracking and registration [20, 32, 33], comparative analysis of various AR hardware (e.g., headset, tablet, projector, haptic) [19, 25, 34, 35], and remote assistance [36]. The affordances of industrial AR for intelligent, adaptive, and personalized teaming and collaboration between humans and machines are yet to be discovered. Specifically, it is necessary to understand how AR coupled with AI technologies can enhance learning and adaptability of workers on the job through intelligent human-machine teaming, while avoiding potential risks associated with over-dependence on technology and stifled innovation. This chapter first provides a background summary of industrial AR followed by a discussion on the following fundamental research topics and directions for future research and development in this domain.

A. Delivering a given task instruction to a worker through AR can be done in a variety of mode such as text, alert, image, animation, or video. However, each mode is likely to have a different impact on the worker's efficiency, error rate, learning, independence, and cognitive load. It is therefore necessary to explore different modes of instruction delivery through AR and their impact the worker and work. Understanding the usability and limitations of different modes can potentially inform more optimized design of AR user experiences tailored to worker needs and specific task requirements.
B. AR can be used as a training tool prior to task execution or as an assistive tool during task execution. It is important to make such distinction to delineate training scenarios, where AR support is removed after training, from assistive scenarios where AR support is used on a just-in-time basis. Deciding which route to take depends partly on the worker choice complexity [37], novelty and extent of task components, procedures, and functional attributes [38], and the required level of reasoning and decision-making.
C. Learning sciences research underscores the necessity of scaffolding and fading mechanisms that align with the learner's attention and cognitive processes to help them construct knowledge. One-size-fits-all delivery of task information through AR must be replaced with an intelligent system that dynamically scaffolds instructions to the subject matters that workers need information on. Previous research underscores the necessity of devising scaffolding mechanisms that align AR instructions with the learner's attention and cognitive processes to help them construct knowledge [39–41]. Hence, it is necessary to understand the nature of the scaffolding that AR affords, and how to design it in the most effective way for the ongoing success of individual workers through intelligent worker-AR teaming.

D. Extant methods are mainly concerned with the provision of procedural knowledge [42] through AR—the knowledge related to performing sequences of actions. However, this approach merely helps a worker learn "how" to perform a given task without effectively learning the "why" behind work instructions, quality assurance guidelines or specifications, and informal job knowhow. Only by understanding the deeper causal relationships behind the procedural instructions can workers develop the cognitive agility to solve new problems and adapt to new circumstances.

The remainder of this chapter is organized as follows. Section 2 provides a brief overview of the state-of-the-art in industrial AR. Section 3 presents a case study on industrial AR in manufacturing that aims at illuminating research topics A-D. Section 4 discusses several research challenges and research directions within the scope of topics A-D.

2 State-of-the-Art in Industrial AR

A comprehensive review of industrial AR in manufacturing and assembly is provided by [43], which highlights the technical features, characteristics, and industrial applications of AR. The review article categorizes the AR applications in the assembly domain into training, design and planning, and guidance. The main research challenges identified include tracking and registration, collaborative AR interfaces, 3D workspace scene capture, and context-aware knowledge representation. Similar surveys [44, 45] have been conducted on industrial AR applications in maintenance, which emphasize operation-specific applications, AR hardware and development platform comparison, visualization methods, tracking, and authoring solutions. The key technical challenges identified by these articles include automated authoring, context-aware adaptation, and human-AR interactions. Other studies also point to similar challenges in the areas of technology (e.g., tracking/registration, authoring, UI, ergonomics, processing speed), organization (e.g., user acceptance, privacy, cost), and environment (e.g., industry standards for AR, employment protection, external support) [46, 47].

The rest of this section discusses some of the prior work associated with research topics A-D presented in Sect. 1. Topic A seeks to explore the impact of various modes of task instruction delivery through AR on the skill acquisition of workers. Topic A has been addressed by many studies from different angles. For example, a comparison between the effects of verbal, paper-based, and AR instructions on manufacturing workers' productivity, quality, stress, help-seeking behavior, perceived task complexity, effort, and frustration was conducted by [19]. A field study on AR-assisted assembly [48] shows progress in physical and temporal demands, effort, and task completion time. A study on paper-based and head-mounted AR instructions for assembly [49] indicate significant reductions in error rates and task completion times through AR. The effects of an AR fault diagnosis app on the performance of novices with AR support and experts with no AR support were studies by [50]. Results showed that AR-supported novices outperformed experts with no AR support in terms of completion time, accuracy, and cognitive load. The effects of an AR app compared to pictures on inspection task performance were studies by [25], which showed improved task completion time, error rate, gaze shifts, and cognitive load. The effectiveness of different modes of AR information delivery and their measured impact on various task performance metrics on a real-life manufacturing task were recently investigated by the author his team [51], which is reported in Sect. 3.

Topic B presented in Sect. 1 aims at understanding of the affordances of industrial AR as a preliminary training tool versus a just-in-time assistive tool. Many recent studies have partially addressed this topic by investigating the usability, acceptability, and organizational challenges of industrial AR. Field interviews were conducted by [35] to understand the perspectives and acceptance of AR as an assistive tool among assembly workers, which reported generally positive attitude about AR by many workers. A field experiment was also conducted by [52] to study the organizational and technological challenges of industrial AR for industrial workforce training, specifically hardware and software limitations, user acceptance, ergonomics, usability, cost, and integration into shop floor processes. The study concluded that there is a lack of sufficient research on organizational issues, especially on user acceptance and integration. Another study [53] explored the impact of a quiz mode in AR where the user must successfully complete part selection quizzes in addition to AR training prior to task performance. It was shown that the number of errors in new assembly tasks can be reduced by 79% compared to baseline AR training. The usability of AR as an assistive tool for maintenance workers was studied by [54], which reported that a relatively high usability of their AR app compared to traditional modes of instruction. The conditions under which AR can be most effective as an assistive tool versus a training tool are yet to be determined. Section 4 reports some of those conditions based on the findings from a recent study by the author and his team [51].

Topics C and D aim to generate new insights on the potential for AR coupled with AI to enable effective human-technology teaming in industrial workplaces. Extant literature in industrial AR already reports many studies on intelligent context-aware AR apps for industrial applications. A cognition-based interactive AR assembly guidance systems was developed by [33, 55], which leverages tracking and registration techniques for context-aware delivery of task instructions. Another study [56] integrates an intelligent tutoring system comprised of domain knowledge, student models, and pedagogical models into AR for personalized learning. A comprehensive review of research on AI-enabled AR systems was done by [57], which mainly addresses vision system calibration, object tracking and detection, pose estimation, rendering, registration, and virtual object creation in AR. Despite these remarkable efforts, several knowledge gaps related to Topics C and D remain to be addressed. First, learning sciences research underscores the necessity of scaffolding and fading mechanisms [58–61] that align with the learner's attention and cognitive processes to help them construct knowledge [39, 41]. More research is needed on transitioning from "one-size-fits-all" instructions towards personalized and adaptive teaming between workers and industrial machinery through AR. Not addressing this need can lead to overdependence on technology, lack of innovation by workers, and limited industry adoption, among other potential unintended consequences. Second, extant methods are mainly concerned with the provision of procedural knowledge [42] through AR; i.e., the knowledge related to performing sequences of actions. This approach may only help workers learn "how" to perform a given task without effectively learning the "why" behind work instructions, quality assurance guidelines/specifications, and informal shop floor knowledge. Only by understanding the deeper causal relationships behind the procedural instructions can workers develop the cognitive agility to solve new problems and adapt to new circumstances. The remainder of this chapter provides insights into these challenging yet transformative research topics in industrial AR.

3 Case Study: Marine Engine Assembly

This section presents a case study conducted by the author in cooperation with a marine engine manufacturer in Massachusetts and Massachusetts Manufacturing Extension Partnership (MassMEP) to explore some aspects of topics A-D presented in Sect. 1. Details of this study and experimental results are available in [51]. The task involves assembling fuel cell modules of custom-made marine engines, which require following different procedures for each engine model. Traditionally, the task is performed by an experienced assembler who using standard hand/power tools to assemble the fuel cell following instructions provided on a one-page instruction sheet that includes technical drawings and bill of materials (Fig. 1). The main challenge of this manufacturer is that training their novice workers who usually come from mechanic or machinist backgrounds often takes several weeks or months and is done by their experienced workers. This is costly for the manufacturer and often leads to reduced productivity of their existing workers and high scrap or rework rates attributed to their new workers. The core objective of the study was to understand if and how AR can help address this challenge.

3.1 Design of Experiments

Participants. 20 engineering students from Northeastern University participated in the study, including 11 undergraduate students and 9 graduate students, 6 females and 14 males, 4 freshmen, 3 sophomores, 2 juniors, 5 seniors, 5 masters, and one PhD. All participants had an average to high level of familiarity with electro-mechanical assembly using simple tools, and little or no prior experience with AR. A questionnaire was provided to the participants in the beginning to collect their demographics and prior related experiences and use the data to counterbalance the experimental groups. The participants received a brief introduction to the assembly tasks and tools prior to the experiments. They were also briefly trained on using HoloLens 2 (AR headsets) for browsing through the AR app, steps, and different modes of instructions.

Task and Apparatus. The experiments involved electro-mechanical assembly of a fuel cell module for marine engines (Fig. 2, top left), a representative and relatively complex assembly task. The bill of materials contains 26 groups of components that must be assembled over 13 steps using standard tools such as open-ended wrench and Allen socket and rachet. The components were placed on a numbered grid on the worktable in front of the participants (Fig. 2, top right). The participants were divided into two groups of AR-based and paper-based instruction. The AR app was developed using Unity and Mixed Reality Toolkit (Fig. 2, bottom left). The app provides the instructions associated with each step in three different modes (see Fig. 3): (a) expert capture videos with vocal cues generated by mounting a GoPro on the forehead of an expert worker and recording

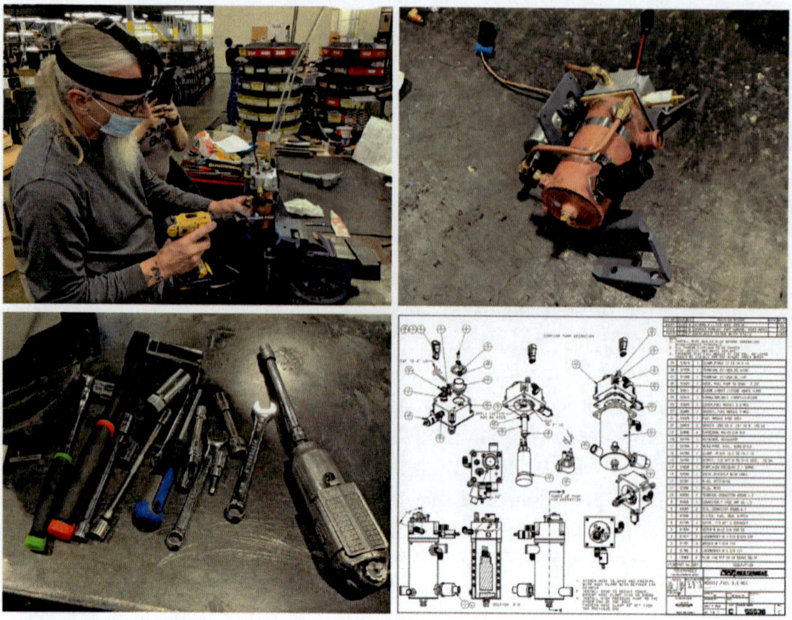

Fig. 1. *Top left*: An expert worker performing the fuel cell assembly task. *Top right*: The fuel cell assembly module. *Bottom left*: Hand and power tools used for assembly. *Bottom right*: Technical drawing and bill of materials.

their task performance (Fig. 2, bottom right), (b) textual descriptions of assembly guides and information for each step (*e.g.*, part numbers, tools, procedures) along with images of the parts to be assembled in that particular step, and (c) interactive 3D CAD animations that allow users to view, rotate, and replay a holographic animation of the assembly step. The AR hardware used for the experiments were HoloLens 2 headsets.

Procedure. A between-subject experiment design was used where the participants were divided into two groups (Fig. 4): Group 1 (AR) and Group 2 (paper). The paper guides include written step-by-step task instructions along with the technical documentation and bill of materials (Fig. 1, bottom right). Each participant performed three assembly cycles on three consecutive days and then returned after a few days to perform a final assembly cycle using the opposite means of instruction (*i.e.*, paper for Group 1 and AR for Group 2). At the end of each round of experiments, both groups of participants filled out a NASA Task Load Index (TLX) questionnaire [62], and the participants who used AR also responded to the following questions:

- In a few words, explain your opinion about the use of AR as a training or assistive tool for manufacturing workers.
- Tell us about your experience with HoloLens (*scale*: very low, low, neutral, high, very high).
  - How do you rate the level of comfort/fit of HoloLens?

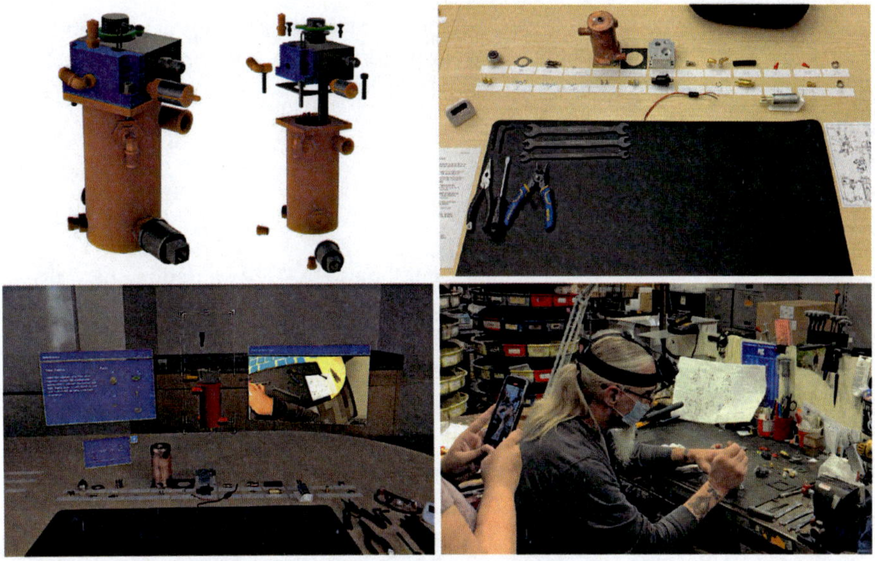

Fig. 2. *Top left:* The assembly part CAD model and exploded view. *Top right*: Worktable setup and tools used for assembly. *Bottom left*: The AR app interface for one assembly step, including textual descriptions and part images, interactive CAD animation, and expert capture video. *Bottom right*: Recording expert capture videos. From [51].

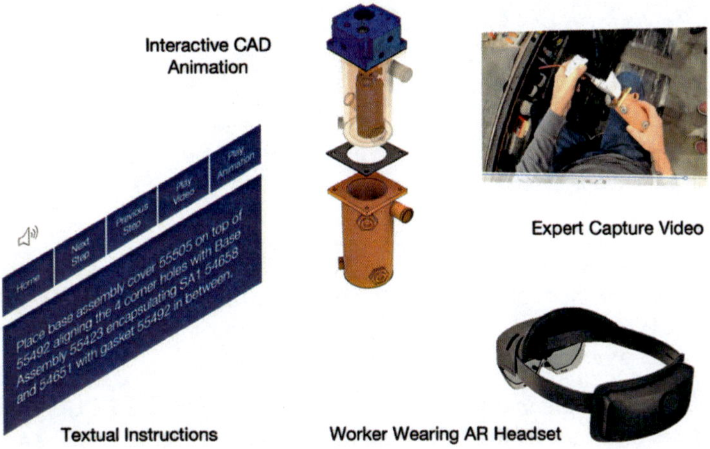

Fig. 3. A schematic of an assembly step instruction in the AR app.

- How satisfied are you with the job you did?
- How do you rate your knowledge of the process to do it without the HoloLens?
- How much do you prefer to learn from a person rather than the HoloLens?
- How distracting or cumbersome do you find HoloLens?

- How do you rate the impact of different modes of AR instructions on your ability to learn the assembly task and improve your performance? (*scale*: not at all, very little, somewhat, quite a bit, a great deal).
  - Text and images
  - Expert capture videos
  - Interactive 3D animations
- In a few words, explain your opinion about the use of AR as a training or assistive tool for manufacturing workers.
- What new, potentially interactive features would you recommend being incorporated in the AR guides?

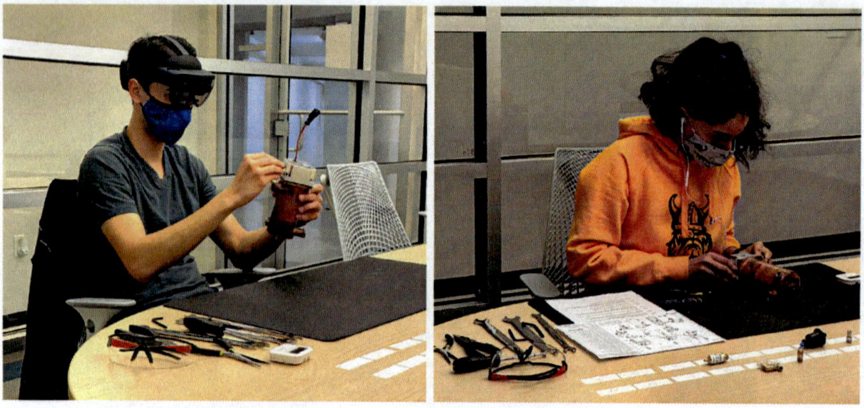

Fig. 4. *Left*: Participant using AR-based task information (Group 1). *Right*: Participant using paper-based task information (Group 2).

The following data was also collected by observing each individual experiment: round of experiment, mode of guide, time to completion (min), frequency of help-seeking behavior (*i.e.*, number of questions asked during assembly), the types of questions asked (if any), number of errors, and the types of errors made (if any).

Metrics. *Time to completion*: The time needed or taken by the participant to complete the task. It was measured by timing the assembly cycle. *Number of errors*: The number of errors made during each assembly cycle. It was measured by counting the number of errors per cycle and recording the type of error(s) for further analysis. *Help seeking behavior* [19]: The number of times help is requested by the participant per assembly cycle. It was measured by counting the number of times help is requested per cycle and recording the question for further analysis. *Learning curve*: The degree of competence to which the acquired assembly skill is retained through the passage of time. It was measured by recording the amount of improvement in time-to-completion, number of errors, and help seeking behavior over time over temporally separated rounds of experiments on a given task. *Independence from AR*: The ability of AR-trained workers to accomplish the same task without AR, and the impacts of AR on task performance of traditionally trained workers. It was measured by bringing the participants in after a few days to perform the assembly task with the opposite means of task information delivery, record and compare their time-to-completion, number of errors, and help-seeking behavior against their best recorded performance prior to the gap. *Cognitive load*: The amount of working memory used to complete the task following the instructions. It was measured using the NASA-TLX questionnaire. *Utility of different AR modes*: The usefulness of different modes of AR information delivery for learning a task. It was measured using the qualitative questionnaire mentioned above.

Hypotheses. *AR significantly improves (H1) time-to-completion, (H2) number of errors, (H3) help-seeking behavior, (H4) learning curve, (H5) retention, and (H6) cognitive load of workers compared to paper-based instructions.*

3.2 Results

The main results of the experiments are presented in Fig. 5. The key findings of the study were as follows: AR reduces the number of errors by 31–84%. The task completion times of the two groups are about the same; however, that was partly due to the unfamiliarity of participants with AR and some technical issues. Further, most participants reported absolute independence from AR after two/three cycles, which points to the effectiveness of AR in improving task competency, and yet its low utility as an "assistive tool" for routine tasks. Further, several participants suggested devising interactive help and voice command systems.

Time-to-Completion. Statistical analysis of the results of experiments (Fig. 5) indicates a significant difference between the mean time-to-completion achieved by participants in Groups 1 and 2 in Rounds 2 and 3. Group 2 (paper) significantly outperformed Group 1 (AR) in the second and third rounds of experiments in terms of task completion time, even though Group 1 showed a slightly better performance in Round 1. Hypothesis H1 was therefore rejected. It is speculated that Group 2 participants gradually transitioned from following the task information to using their memory to complete the task, while Group 1 participants still went through the AR Instructions. The relatively poor performance of the AR group in terms of completion time was also partly because of their unfamiliarity with the AR headset.

Number of Errors. Statistical analyses on the mean number of errors presented in Fig. 5 show that Group 2 made a significantly higher mean number of errors compared to Group 1 in Round 3 of the experiments. These results are also partly due to a significant reduction in the number of errors made by the participants in Group 1, while the other group maintained an almost steady and relatively higher number of errors throughout Rounds 1–3. With these findings, hypothesis H2 can be accepted, which indicates the significant impact of spatiotemporal alignment of task information and visual/vocal cues with experience on the number of errors made during task performance.

Help-Seeking Behavior. Only two participants from each group sought help related to AR app, part orientation, and sequence of assembly. This observation was consistent

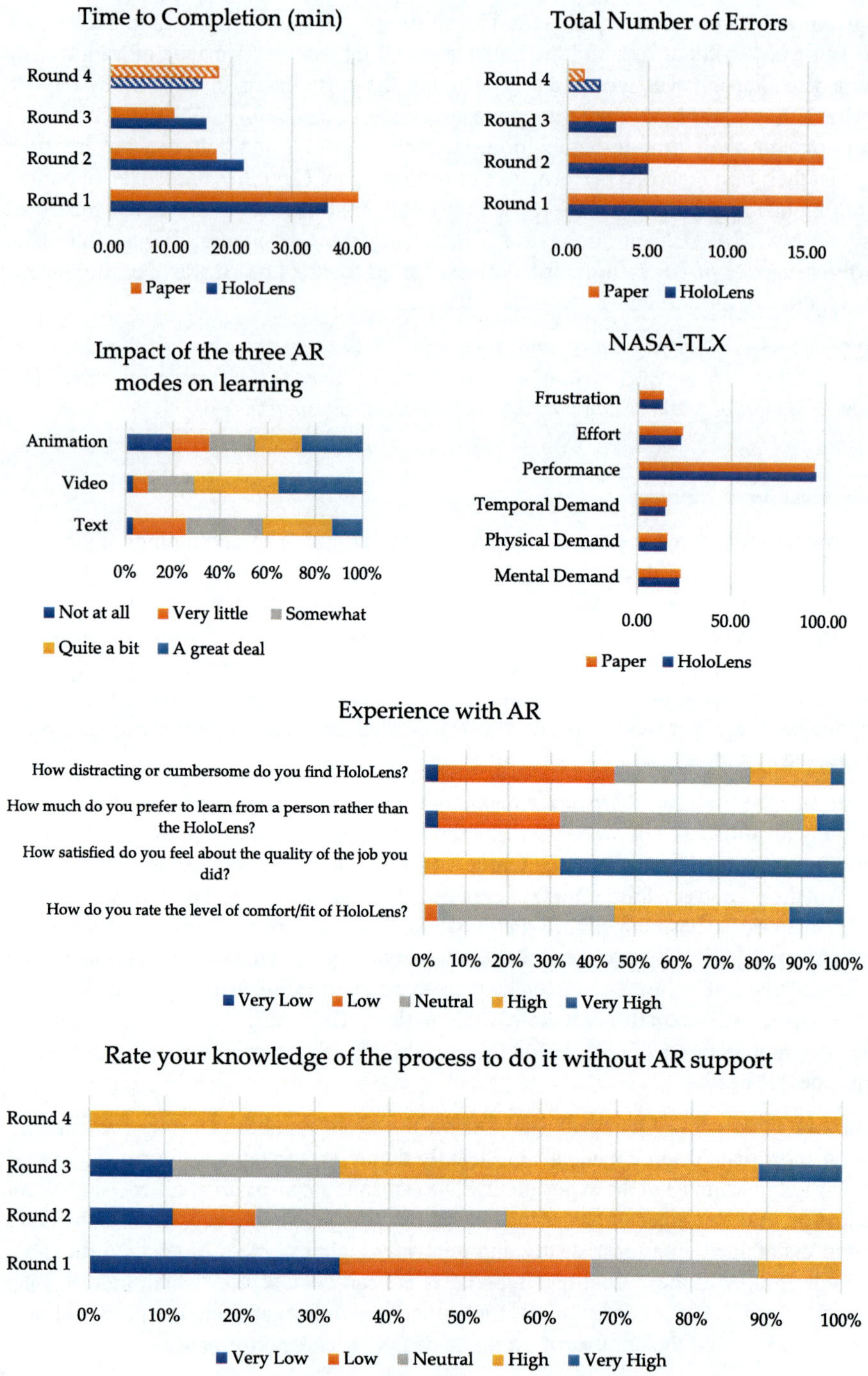

Fig. 5. Results of the human subject experiments.

with the comments made by the plant manager of the marine engine manufacturer that workers often tend to overthink and not reach out for help due to fear of embarrassment or overconfidence, which may lead to failures down the line. This is yet another motivation to create AR training and assistance systems that can address the needs and struggles of workers without them having to reach out (or not reach out due to the feelings or fear or embarrassment) to their coworkers or supervisors for help. These findings reject hypothesis H3.

Learning Curve. Statistical analyses on the differences between mean time-to-completion and mean number of errors of each group between Rounds 1 and 2, Rounds 2 and 3, and Rounds 1 and 3 indicated significant reductions in mean time-to-completion between each round for both groups. It can therefore be stated that the means of instruction (*i.e.*, paper versus AR) does not have any noticeable impact on task completion time. However, observations also showed that although Group 2 made no improvement in the number of errors made during assembly, Group 1 participants were able to significantly reduce the number of errors between Rounds 1 and 3. It is thus concluded that not only the use of AR leads to fewer errors, but it also helps workers significantly reduce the number of errors in subsequent rounds of operation. Hypothesis H4 is therefore accepted.

Independence from AR. The mean time-to-completion of each group in Round 3 and Round 4 (see Fig. 5) comprised two interesting observations. First, Group 1 who switched from AR guides in Round 3 to paper guides in Round 4 could complete their task even slightly faster in Round 4 than in Round 3. Here is a quote from a Group 1 participant after completing Round 4: "*It was less cumbersome to assemble the components without the AR headset on, but the paper drawings were much harder to interpret. I much prefer the CAD animations; I imagine if I were to have started first with the paper-based instructions and drawings, it would have taken me much longer to complete the task initially. I suspect the only reason it took me around the same time to complete the task with paper-based instructions is simply because I had assembled the component three times already.*" Second, Group 2 demonstrated significantly longer completion times in Round 4 using AR than in Round 3 using paper, which is to some extent due to their lack of prior experience with AR app. Moreover, Group 1 maintained their relatively lower mean number of errors in Round 4 even after a few days, while the mean number of errors by Group 2 was significantly reduced in Round 4. This indicates the role of AR in accelerating workers' learning and competency, its usefulness for traditionally trained workers in avoiding more errors during task performance, and better memory retention than paper-based instructions which results in a significantly lower number of errors even after AR support is removed. Hypothesis H5 was therefore accepted.

Cognitive Load. Results of the NASA-TLX questionnaire (Fig. 5) indicate almost identical levels of mental demand, physical demand, temporal demand, perceived performance, effort, and frustration for both groups. These observations therefore reject hypothesis H6. The assembly task was perceived as not too challenging by most participants.

Modes of Instruction Delivery. The collected responses to the questions about the impacts of different modes of AR (*i.e.*, text, 3D animation, video) on task performance, independence from AR, and experience with AR varied among the participants. Some found text instructions and 3D CAD animations more helpful: *"The text instructions and CAD animation together provided a great deal of detail about how to complete the current step. Reading the instructions and visualizing the task through the animation provided clarity on how to complete the task and what the subassembly should look like afterward. Although helpful, the video was not completely necessary, and I skipped it for most of the steps." "Being able to rotate and view the CAD model was super helpful during assembly. It allowed me to easily understand how all the parts fit together. The other two were useful, but tended to get in the way as I was putting physical pieces together."* Some other participants, however, found videos with vocal cues along with textual instructions more helpful: *"The CAD animation was somewhat useful, but I preferred the video, as the instructor assembled the part at about the same speed that I was. Additionally, there were little comments that helped, which a silent CAD animation didn't include." "I tended to listen to the verbal cues from the video, occasionally checking the text to confirm part numbers, and only once or twice double-checking with the CAD animation."*

Independence from AR. Group 1 were initially highly dependent on AR: 70% of the participants reported high or very high dependence on AR. This dependency, however, gradually declined as only 20% of Group 1 were highly or very highly dependent on AR by the end of Round 3. On the contrary, 90% of the participants from Group 2, who switched from paper to AR in Round 4, stated that they are highly independent from AR. Nevertheless, switching to AR helped this group significantly reduce the number of errors made during assembly. Moreover, only 10% of the participants expressed a preference to be trained by a person rather than by the AR app. These number are expected to vary for actual manufacturing workers who belong to different age groups and educational levels/backgrounds.

AR for Just-In-Time Assistance. The participants were asked to comment on the use of AR as an assistive tool for workers. Text-to-speech features were recommended to read out the textual instructions. Some also suggested the use of voice commands for hands-free interaction with the AR content. Menu-based, non-procedural provision of task information were suggested by some participants so the user can call certain instructions on demand, rather than having to go through a fixed sequence of steps. It was suggested to include more interactive and personalizable layout design for the AR app so the user can use the layout they feel most comfortable in. Some participants suggested help options that allow the user can get assistance when something goes wrong or if they have a question about the task.

4 Discussion: Towards Intelligent AR Systems

Industrial AR has the potential to transform workplace-based learning for future workers and thus bridge the labor market mismatch and enhance the productivity and/or quality of future work. Yet, the technology is still evolving, and several key challenges associated with technology development, socioeconomic impacts, and human factors are yet to be addressed. The following section discusses several directions in line with research topics A-D discussed in Sect. 1 with the vision of turning AR into an intelligent, adaptive, and personalized assistant for incumbent and new workers across different industries.

State-of-the-art industrial AR systems still offer limited personalized interactions between workers and AR, and primarily offer procedural, one-size-fits-all instructions with minimal attention to the individual worker's needs and knowledge. This may lead to potential unintended consequences such as overdependence on technology and stifled innovation [13, 15, 63], and also hinder industry adoption. The provision of procedural knowledge [42] through AR—the knowledge related to performing sequences of actions—merely helps workers learn "how" to perform a given task without effectively learning the "why" behind work instructions. That prevents workers from developing a deep understanding of the underlying causal relationships behind the procedural instructions can workers develop the cognitive agility to solve new problems and adapt to new circumstances, especially in tasks that involve complex reasoning and decision-making. Future research must explore how AR can intelligently tailor scaffolds to the specific needs of workers to enhance not only their performance efficiency but their complex reasoning skills for solving novel problems and adapting to changing work environments. It is therefore critical to build new methods at the nexus of AI, AR, and human-machine interaction to understand how various sources of multimodal data captured by AR devices, industrial machinery, and any other smart device or sensor can be harnessed to interpret, predict, and guide the behavior of industrial workers and enable intuitive human-machine teaming.

Future research must build new inference engines into AR for interactive and personalized task assistance informed by multimodal context data, user intent, and multi-dimensional digital data. The outputs of the inference engine can be generated either automatically or in response to user inputs (*e.g.*, buttons, menus, dialog, hand gestures, gaze). The automated outputs can be more focused on critical task information (*e.g.*, safety features, warnings) while the on-demand outputs may involve declarative knowledge or task assistance. Such inference engines can progressively tailor the instructions to the specific needs of users based on the collected data on their performance. That is, they can capture data on the history of user interactions with the auto-generated content (e.g., interactions with menus, questions asked, or "gaze-time") for each content, and gradually remove contents below a certain usage threshold. A set of logical rules can be applied for the inference engine to generate proper automated or on-demand system outputs in any of the following or similar forms (Fig. 6):

- **Spatially registered 3D visualizations.** Given the identified 6D poses of objects, AR systems can superimpose task guidance and digital information on physical objects (*e.g.*, part, robot, controller), along with textual instructions, and visualizations to boost the spatial reasoning abilities of users.
- **Notifications.** Considering user intent and task status, AR systems can generate visual or auditory notifications, based on job sheets and operations manuals, to ensure the user is aware of important safety and operational features.
- **Recommendations.** AR systems can leverage multidimensional digital data (*e.g.*, GD&T, 3D annotations, material specifications, and process notes) to assist workers with reasoning about observed task progression.
- **Spatial/causal reasoning animations.** AR systems can include detailed 3D animations of the process for users to visualize what cannot be seen during operation.

- **Expert capture videos.** AR systems. Can provide users with audiovisual "expert stories", captured using GoPro or over-the-shoulder cameras, which can be activated using voice command, menus/buttons, or hand gestures on demand.
- **Remote assistance.** AR systems can also allow users to video call remote experts from the app and share their screen to get immediate assistance.

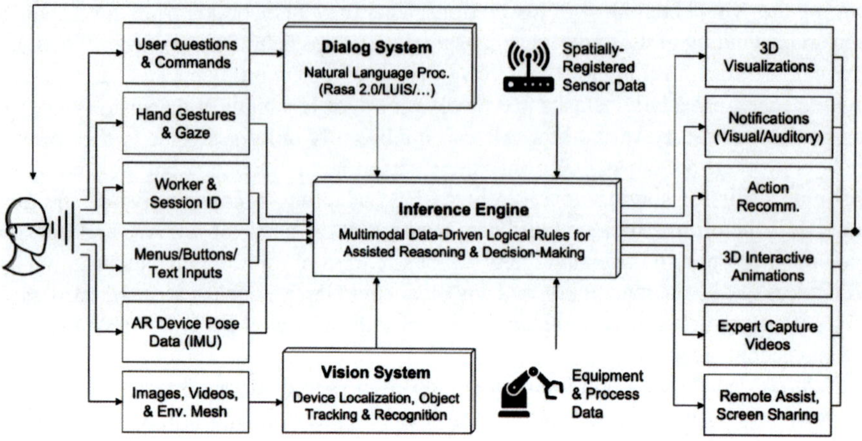

Fig. 6. Schematic of an intelligent AR system with an inference engine for context-aware human-machine teaming in industrial settings.

Consider, for example, human-robot collaboration in industry where workers and collaborative robots synchronously process the same task in a shared physical workspace. This is an increasingly common scenario in many industrial settings such as factories, warehouses, and distribution centers, especially given the fact that human-robot teaming is known to reduce idle time by 85% compared to when the task is performed by all human teams [64]. Yet, the integration of collaborative robots into factories is currently limited to structured operations with known, minimal, and fully predictable interactions with humans. That is, collaborative robots are currently being used in factories as an advanced, automated tool rather than an active and intelligent coworker for the human operator. This "black-and-white" approach to automation in factories has reached its limits, because many manufacturing tasks such as machine tending, assembly, inspection, and part transfer are not 100% repetitive and involve many variations that cannot be handled by exiting robots or require time-consuming reprogramming. This makes today's robotic coworkers inadequate assistants to human workers and impedes human-robot teamwork.

AR coupled with AI capabilities can bridge this gap by functioning as a mediator to enable the worker to preview and modify the programs and policies taught to the robot, which will in turn lead to the progressive adaptation and convergence of shared mental models between them (Fig. 7). AR can also facilitate the communication of the worker's intent to the robot through both explicit (*e.g.*, hand gestures, gaze) and implicit (*e.g.*, eye/head gaze, wearables) modes of interaction. AR can also enhance robot-to-worker interaction through multimodal communication channels such as 3D visualization, haptics, and/or auditory signals. Such an intelligent AR mediator system can potentially advance the understanding of how transparent sharing of intent and awareness can shape teamwork fluency and trust between workers and robots.

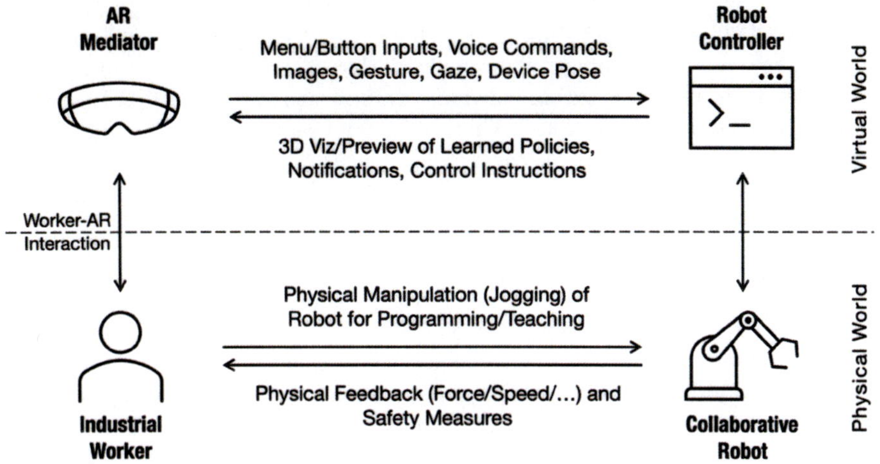

Fig. 7. Two-way communication of information and intent between workers and robots through an intelligent AR mediator.

5 Conclusions

The newer wave of industrial automation is not so much to replace workers but rather to increase precision, safety, and product quality [65]. Modern automation is about continuing to automate tasks that are dirty, dull, and dangerous, but preserving the ones that are "value-added" and often desirable parts of the jobs for human workers. Those kinds of value-added jobs are specifically the ones that are hard to automate, either because they require sophisticated and precise manipulation of physical objects that only a human is capable of or because they require complex reasoning and decision-making that machines are not capable of. Informed by the experiments and conceptual frameworks presented in Sects. 3 and 4, respectively, the author and his team assembled a panel of ten experts from major industrial, academia, and federal institutions in the U.S. and Europe to further illuminate the potentials and risks of industrial AR in the human-centered automation era.

The discussions were facilitated by four high-level questions. (1) How widespread do you think the adoption of AR technology in manufacturing will be in the next 5 years? Which firms would be best suited to adopt such technologies (*e.g.*, size, product type, capital/labor mix)? What impact might the adoption of AR technologies have on the skill requirements for specific job roles in assembly? To what degree can AR technologies be used to train the future manufacturing workforce? (2) What are the potential benefits and risks of AR for workplace-based learning on complex, career-spanning expertise in areas such as assembly and maintenance? Do you see other training techniques/technology alternatives on the horizon? (3) There is some evidence that overreliance of workers on AR can cause "brittleness" of knowledge [63], hinder learning, and deteriorate performance in adapting to novel situations. In your opinion, what are the impacts of AR on the ability of workers to learn new tasks in a way that enhances their flexibility in transferring their knowledge to new situations? (4) How can we interpret, predict, and guide the behavior of AR-supported assembly workers through adaptive scaffolding of instructions to the expertise level of individual workers, and immediate AR-based feedback on their actions and decisions? What are the implications for the design of future AR technologies?

This chapter is concluded by presenting seven key insights drawn from the panel discussion about the challenges and future trends in industrial AR:

1. AR can potentially be a disruptive assistive technology for manufacturing tasks that are not rote and require complex reasoning and decision-making; for example, inspection in regulated industries such as aerospace.
2. The acceptability of AR as a "tool" is likely to differ among incumbent and future workers and different demographics. The experiments presented in this paper only featured young and educated engineering students. Current AR technology may not be well received by more senior workers because the interfaces are not as intuitive as they should be for someone with little or no experience with AR or even with computers.
3. AR can increase the accessibility of manufacturing jobs to workers with different demographic characteristics by allowing for self-guided learning without the need for physical and real-time interaction with a trainer.
4. AR can create new opportunities for remote learning and assistance from larger, and possibly more diverse, pools of physically/temporally distant coworkers. It can also enable remote assistance and collaboration by allowing the on-site worker to share their experience with a remote expert and get immediate feedback with 3D visual cues.
5. The adoption of AR by companies will require rigorous justification through both proof-of-concept and economic cost-benefit analyses. It is important to educate companies on the potential impacts of AR on efficiency and productivity, the skills required for building, maintaining, and updating the content, the costs of software and hardware, and the acceptability of the technology among both incumbent and entry-level workforce.
6. Scalability must be regarded as a key criterion for the ideation and design of AR technologies. The marine engine manufacturer studied, for example, makes tens or hundreds of different variations of a given part family.
7. AR can be coupled with digital thread technologies to provide workers with part, process, and task information such as geometric dimensions and tolerances (GD&T), 3D annotations, material specifications, and process notes [66, 67] in real-time. AR can also leverage industrial Internet-of-Things (IoT) data to enable access to real-time machine data in semiautomated tasks such as robotic assembly or CNC machining.

Acknowledgement. This material is based upon work supported by the National Science Foundation under the Future of Work at the Human-Technology Frontier Grant No. 2128743. Any opinions, findings, and conclusions or recommendations expressed in this material are those of the author(s) and do not necessarily reflect the views of the National Science Foundation. I acknowledge the support of our expert panel and industry partners.

References

1. Brynjolfsson, E., McAfee, A.: The Second Machine Age: Work, Progress, and Prosperity in a Time of Brilliant Technologies. W W Norton & Co. (2014). Accessed 10 Mar 2021. https://psycnet.apa.org/record/2014-07087-000
2. NAM, "NAM Manufacturers' Outlook Survey (2020). https://www.nam.org/wp-content/uploads/2020/09/NAM-Outlook-Survey-Q3-2020.pdf. Accessed 17 Mar 2021
3. Houseman, S.N.: Understanding the Decline of U.S. Manufacturing Employment. Upjohn Institute Working Paper 18–287, Kalamazoo, MI (2018). https://doi.org/10.17848/wp18-287
4. Lund, S., et al.: Risk, Resilience, and Rebalancing in Global Value Chains. McKinsey & Company (2020). https://www.mckinsey.com/business-functions/operations/our-insights/risk-resilience-and-rebalancing-in-global-value-chains. Accessed 17 Mar 2021
5. SHRM, "Preparing for an Aging Workforce: Manufacturing Industry Report," Society for Human Resource Management (2015)
6. NAP, Building America's Skilled Technical Workforce (National Academies of Sciences, Engineering, and Medicine). National Academies Press (2017). https://doi.org/10.17226/23472
7. Giffi, C., Wellener, P., Dollar, B., Ashton Manolian, H., Monck, L., Moutray, C.: Deloitte and The Manufacturing Institute Skills Gap and Future of Work Study (2018). http://www.themanufacturinginstitute.org/~/media/E323C4D8F75A470E8C96D7A07F0A14FB/DI_2018_Deloitte_MFI_skills_gap_FoW_study.pdf
8. PTC, "Closing the Industrial Skills Gap with Industrial Augmented Reality," PTC (2019). https://www.ptc.com/en/resources/augmented-reality/ebook/closing-the-industrial-skills-gap-with-augmented-reality. Accessed 19 Feb 2020
9. Conway, S.: The Total Economic Impact TM Of PTC Vuforia: Cost Savings And Business Benefits Enabled By Industrial Augmented Reality (2019). https://www.ptc.com/en/resources/augmented-reality/report/forrester-total-economic-impact
10. Mizell, D.: Boeing's wire bundle assembly project. In: Barfield, W., Caudell, T. (eds.), Fundamentals of Wearable Computers and Augmented Reality. CRC Press, pp. 447–467 (2001)
11. Schwald, B., et al.: STARMATE: using augmented reality technology for computer guided maintenance of complex mechanical elements. In: Smith, B., Chiozza, E. (eds.), E-work and E-commerce-Novel Solutions and Practices for a Global Networked Economy, pp. 196–202 (2001)
12. Schwald, B., Laval, B.D.: An augmented reality system for training and assistance to maintenance in the industrial context. In: International Conference in Central Europe on Computer Graphics, Visualization and Computer Vision (WSCG), pp. 425–432 (2003)
13. Webel, S., Bockholt, U., Engelke, T., Gavish, N., Olbrich, M., Preusche, C.: An augmented reality training platform for assembly and maintenance skills. Robot. Auton. Syst. **61**(4), 398–403 (2013). https://doi.org/10.1016/j.robot.2012.09.013
14. Syberfeldt, A., Danielsson, O., Holm, M., Wang, L.: Visual assembling guidance using augmented reality. Procedia Manufacturing **1**, 98–109 (2015). https://doi.org/10.1016/j.promfg.2015.09.068

15. Holm, M., Danielsson, O., Syberfeldt, A., Moore, P., Wang, L.: Adaptive instructions to novice shop-floor operators using augmented reality. J. Ind. Prod. Eng. **34**(5), 362–374 (2017). https://doi.org/10.1080/21681015.2017.1320592
16. Marr, B.: The Amazing Ways Honeywell Is Using Virtual And Augmented Reality To Transfer Skills To Millennials. Forbes (2018). https://www.forbes.com/sites/bernardmarr/2018/03/07/the-amazing-ways-honeywell-is-using-virtual-and-augmented-reality-to-transfer-skills-to-millennials/#586ec524536a. Accessed 19 Feb 2020
17. Polladino, T.: Porsche Adopts Atheer's AR Platform to Connect Mechanics with Remote Experts « Next Reality. Next Reality (2017). https://next.reality.news/news/porsche-adopts-atheers-ar-platform-connect-mechanics-with-remote-experts-0181255/. Accessed 19 Feb 2020
18. O'Donnell, R.: How Mercedes-Benz uses augmented reality to train employees of all types. HRDive (2018). https://www.hrdive.com/news/how-mercedes-benz-uses-augmented-reality-to-train-employees-of-all-types/530425/. Accessed 19 Feb 2020
19. Vanneste, P., Huang, Y., Park, J.Y., Cornillie, F., Decloedt, B., van den Noortgate, W.: Cognitive support for assembly operations by means of augmented reality: an exploratory study. International Journal of Human Computer Studies **143**(October 2019), 102480 (2020). https://doi.org/10.1016/j.ijhcs.2020.102480
20. Lai, Z.H., Tao, W., Leu, M.C., Yin, Z.: Smart augmented reality instructional system for mechanical assembly towards worker-centered intelligent manufacturing. J. Manuf. Syst. **55**(February), 69–81 (2020). https://doi.org/10.1016/j.jmsy.2020.02.010
21. Arbeláez, J.C., Viganò, R., Osorio-Gómez, G.: Haptic augmented reality (HapticAR) for assembly guidance. Int. J. Interact. Des. Manuf. **13**(2), 673–687 (2019). https://doi.org/10.1007/s12008-019-00532-3
22. Erkoyuncu, J.A., del Amo, I.F., Dalle Mura, M., Roy, R., Dini, G.: Improving efficiency of industrial maintenance with context aware adaptive authoring in augmented reality. CIRP Annals - Manufacturing Technol. **66**(1), 465–468 (2017). https://doi.org/10.1016/j.cirp.2017.04.006
23. Siew, C.Y., Ong, S.K., Nee, A.Y.C.: A practical augmented reality-assisted maintenance system framework for adaptive user support. Robotics and Computer-Integrated Manufacturing **59**, 115–129 (2019). https://doi.org/10.1016/j.rcim.2019.03.010
24. Urbas, U., Vrabič, R., Vukašinović, N.: Displaying product manufacturing information in augmented reality for inspection. Procedia CIRP **81**, 832–837 (2019). https://doi.org/10.1016/j.procir.2019.03.208
25. Polvi, J., et al.: Handheld guides in inspection tasks: augmented reality versus picture. IEEE Trans. Visual Comput. Graphics **24**(7), 2118–2128 (2018). https://doi.org/10.1109/TVCG.2017.2709746
26. Runji, J.M., Lin, C.Y.: Markerless cooperative augmented reality-based smart manufacturing double-check system: case of safe PCBA inspection following automatic optical inspection. Robotics and Computer-Integrated Manufacturing **64**(February 2019), 101957 (2020). https://doi.org/10.1016/j.rcim.2020.101957
27. Zhu, J., Ong, S.K., Nee, A.Y.C.: A context-aware augmented reality assisted maintenance system. Int. J. Comput. Integr. Manuf. **28**(2), 213–225 (2015). https://doi.org/10.1080/0951192X.2013.874589
28. Angrisani, L., Arpaia, P., Esposito, A., Moccaldi, N.: A wearable brain-computer interface instrument for augmented reality-based inspection in industry 4.0. IEEE Trans. Instrum. Meas. **69**(4), 1530–1539 (2020). https://doi.org/10.1109/TIM.2019.2914712
29. Gattullo, M., Scurati, G.W., Fiorentino, M., Uva, A.E., Ferrise, F., Bordegoni, M.: Towards augmented reality manuals for industry 4.0: a methodology. Robotics and Computer-Integrated Manufacturing **56**(August 2018), 276–286 (2019). https://doi.org/10.1016/j.rcim.2018.10.001

30. Wang, T., et al.: CAPturAR: An augmented reality tool for authoring human-involved context-aware applications. In: UIST 2020 - Proceedings of the 33rd Annual ACM Symposium on User Interface Software and Technology, pp. 328–341 (2020). https://doi.org/10.1145/3379337.3415815
31. Cao, Y., et al.: GhostAR: a Time-space Editor for Embodied Authoring of Human-Robot Collaborative Task with Augmented Reality (2019). Accessed 06 Feb 2021. https://doi.org/10.1145/3332165.3347902
32. Wang, Y., Zhang, S., Wan, B., He, W., Bai, X.: Point cloud and visual feature-based tracking method for an augmented reality-aided mechanical assembly system. Int. J. Adv. Manuf. Technol. **99**(9–12), 2341–2352 (2018). https://doi.org/10.1007/s00170-018-2575-8
33. Wang, X., Ong, S.K., Nee, A.Y.C.: Multi-modal augmented-reality assembly guidance based on bare-hand interface. Adv. Eng. Inform. **30**(3), 406–421 (2016). https://doi.org/10.1016/j.aei.2016.05.004
34. Blattgerste, J., Strenge, B., Renner, P., Pfeiffer, T., Essig, K.: Comparing conventional and augmented reality instructions for manual assembly tasks. ACM International Conference Proceeding Series, Part F1285, pp. 75–82 (2017). https://doi.org/10.1145/3056540.3056547
35. Danielsson, O., Syberfeldt, A., Holm, M., Wang, L.: Operators perspective on augmented reality as a support tool in engine assembly. Procedia CIRP **72**, 45–50 (2018). https://doi.org/10.1016/j.procir.2018.03.153
36. Mourtzis, D., Vlachou, A., Zogopoulos, V.: Cloud-based augmented reality remote maintenance through shop-floor monitoring: a product-service system approach. J. Manuf. Sci. E. T. ASME **139**(6), 1–11 (2017). https://doi.org/10.1115/1.4035721
37. Hu, S.J., Zhu, X., Wang, H., Koren, Y.: Product variety and manufacturing complexity in assembly systems and supply chains. CIRP Ann. Manuf. Technol. **57**(1), 45–48 (2008). https://doi.org/10.1016/j.cirp.2008.03.138
38. Orfi, N., Terpenny, J., Sahin-Sariisik, A.: Harnessing product complexity: step 1establishing product complexity dimensions and indicators. Engineering Economist **56**(1), 59–79 (2011). https://doi.org/10.1080/0013791X.2010.549935
39. Bujak, K.R., Radu, I., Catrambone, R., MacIntyre, B., Zheng, R., Golubski, G.: A psychological perspective on augmented reality in the mathematics classroom. Comput. Educ. **68**, 536–544 (2013). https://doi.org/10.1016/j.compedu.2013.02.017
40. Zydney, J.M., Warner, Z.: Mobile apps for science learning: review of research. Comput. Educ. **94**, 1–17 (2016). https://doi.org/10.1016/j.compedu.2015.11.001
41. Ibáñez, M.B., Delgado-Kloos, C.: Augmented reality for STEM learning: a systematic review. Comput. Educ. **123**(April), 109–123 (2018). https://doi.org/10.1016/j.compedu.2018.05.002
42. Becerra-Fernandez, I., Sabherwal, R.: Knowledge Management: Systems and Processes. Routledge (2014)
43. Wang, X., Ong, S.K., Nee, A.Y.C.: A comprehensive survey of augmented reality assembly research. Advances in Manufacturing **4**(1), 1–22 (2016). https://doi.org/10.1007/s40436-015-0131-4
44. Fernández del Amo I.F., Erkoyuncu, J.A., Roy, R., Palmarini, R., Onoufriou, D.: A systematic review of augmented Reality content-related techniques for knowledge transfer in maintenance applications. Computers in Industry **103**, 47–71 (2018). https://doi.org/10.1016/j.compind.2018.08.007
45. Palmarini, R., Erkoyuncu, J.A., Roy, R., Torabmostaedi, H.: A systematic review of augmented reality applications in maintenance. Robotics and Computer-Integrated Manufacturing **49**(July 2017), 215–228 (2018). https://doi.org/10.1016/j.rcim.2017.06.002
46. Masood, T., Egger, J.: Augmented reality in support of Industry 4.0—implementation challenges and success factors. Robotics and Computer-Integrated Manufacturing **58**, 181–195 (2019). https://doi.org/10.1016/j.rcim.2019.02.003

47. Egger, J., Masood, T.: Augmented reality in support of intelligent manufacturing – a systematic literature review. Computers and Industrial Engineering, **140**. Elsevier Ltd, 106195 (2020). https://doi.org/10.1016/j.cie.2019.106195
48. Koumaditis, K., Venckute, S., Jensen, F.S., Chinello, F.: Immersive training: outcomes from small scale AR/VR pilot-studies. In: 26th IEEE Conference on Virtual Reality and 3D User Interfaces, VR 2019 - Proceedings, vol. 2019-January, pp. 1894–1898 (2019). https://doi.org/10.1109/VR44988.2019.9044162
49. Werrlich, S., Daniel, A., Ginger, A., Nguyen, P.A., Notni, G.: Comparing HMD-based and paper-based training. In: Proceedings of the 2018 IEEE International Symposium on Mixed and Augmented Reality, ISMAR 2018, pp. 134–142 (2019). https://doi.org/10.1109/ISMAR.2018.00046
50. Smith, E., McRae, K., Semple, G., Welsh, H., Evans, D., Blackwell, P.: Enhancing vocational training in the post-COVID era through mobile mixed reality. Sustainability **13**(11), 6144 (2021). https://doi.org/10.3390/SU13116144
51. Moghaddam, M., Wilson, N.C., Modestino, A.S., Jona, K., Marsella, S.C.: Exploring augmented reality for worker assistance versus training. Advanced Engineering Informatics **50**(April), 101410 (2021). https://doi.org/10.1016/j.aei.2021.101410
52. Masood, T., Egger, J.: Adopting augmented reality in the age of industrial digitalisation. Comput. Ind. **115**, 103112 (2020). https://doi.org/10.1016/J.COMPIND.2019.07.002
53. Werrlich, S., Nguyen, P.A., Notni, G.: Evaluating the training transfer of head-mounted display based training for assembly tasks. In: ACM International Conference Proceeding Series, pp. 297–302 (2018). https://doi.org/10.1145/3197768.3201564
54. Brice, D., Rafferty, K., McLoone, S.: AugmenTech: the usability evaluation of an AR system for maintenance in industry. Lecture Notes in Computer Science (including subseries Lecture Notes in Artificial Intelligence and Lecture Notes in Bioinformatics), vol. 12243 LNCS, pp. 284–303 (2020). https://doi.org/10.1007/978-3-030-58468-9_21
55. Wang, Z.B., Ong, S.K., Nee, A.Y.C.: Augmented reality aided interactive manual assembly design. The International Journal of Advanced Manufacturing Technol. **69**(5), 1311–1321 (2013). https://doi.org/10.1007/S00170-013-5091-X
56. Westerfield, G., Mitrovic, A., Billinghurst, M.: Intelligent augmented reality training for motherboard assembly. International Journal of Artificial Intelligence in Education **25**(1), 157–172 (2014). https://doi.org/10.1007/S40593-014-0032-X
57. Sahu, C.K., Young, C., Rai, R.: Artificial Intelligence (AI) in Augmented Reality (AR)-Assisted Manufacturing Applications: A Review (2020). https://doi.org/10.1080/00207543.2020.1859636
58. Noroozi, O., Kirschner, P.A., Biemans, H.J.A., Mulder, M.: Promoting argumentation competence: extending from first- to second-order scaffolding through adaptive fading. Educational Psychology Review **30**(1), 153–176 (2018). Springer New York LLC. https://doi.org/10.1007/s10648-017-9400-z
59. Cabello, V.M., Lohrmann, M.E.S.: Fading scaffolds in stem: supporting students' learning on explanations of natural phenomena. Adv. Intell. Syst. Comput. **596**, 350–360 (2018). https://doi.org/10.1007/978-3-319-60018-5_34
60. Belland, B.R., Walker, A.E., Kim, N.J., Lefler, M.: Synthesizing results from empirical research on computer-based scaffolding in STEM education: a meta-analysis. Rev. Educ. Res. **87**(2), 309–344 (2017). https://doi.org/10.3102/0034654316670999
61. Lin, T.C., Hsu, Y.S., Lin, S.S., Changlai, M.L., Yang, K.Y., Lai, T.L.: A review of empirical evidence on scaffolding for science education. Int. J. Sci. Math. Educ. **10**(2), 437–455 (2012). https://doi.org/10.1007/s10763-011-9322-z
62. Hart, S.G., Staveland, L.E.: Development of NASA-TLX (Task Load Index): results of empirical and theoretical research. Advances in Psychology **52**(C), 139–183 (1988). https://doi.org/10.1016/S0166-4115(08)62386-9

63. Radu, I.: Augmented reality in education: a meta-review and cross-media analysis. Personal and Ubiquitous Computing **18**(6), 1533–1543 (2014). https://doi.org/10.1007/s00779-013-0747-y
64. Shah, A.: Fluid Coordination of Human-Robot Teams Eytan Modiano Chairman, Department Commiltee on Graduate Theses. Massachusetts Institute of Technology, Cambridge (2011). Accessed 09 Jun 2021. https://dspace.mit.edu/handle/1721.1/63034
65. Autor, D., Mindel, D., Reynolds, E.: The Work of the Future: Building Better Jobs in an Age of Intelligent Machines. MIT Work of the Future (2020). https://workofthefuture.mit.edu/research-post/the-work-of-the-future-building-better-jobs-in-an-age-of-intelligent-machines/. Accessed 17 Mar 2021
66. Frechette, S.P., Jones, A.T., Fischer, B.R.: Strategy for testing conformance to geometric dimensioning & tolerancing standards. Procedia CIRP **10**, 211–215 (2013). https://doi.org/10.1016/j.procir.2013.08.033
67. Hedberg, T., Lubell, J., Fischer, L., Maggiano, L., Feeney, A.B.: Testing the digital thread in support of model-based manufacturing and inspection. J. Computing and Inf. Science in Eng. **16**(2) (2016). https://doi.org/10.1115/1.4032697

Printed Wearable Sensors for Robotics

Don Perera and Wenzhuo Wu(✉)

School of Industrial Engineering, Purdue University, West Lafayette, USA
wu966@purdue.edu

Abstract. In this chapter, we'll look at several tactile sensing and actuation mechanisms for robotics-based sensing applications. Piezoelectric, capacitive, and optical tactile sensing approaches are introduced under tactile sensing mechanisms, and their intriguing recent advancement is reviewed. Electrically responsive, thermally responsive, magnetically responsive, and photoresponsive robotic actuation techniques are described and their applications, including the most recent progress, are analyzed under robotic actuation mechanisms. Finally, the possibility for deploying robot-based solutions for preventing contagious diseases is discussed, as well as the application of robots in chemical sensing relevant to security, agriculture, and environmental protection.

1 Introduction

Additive manufacturing (AM), also known as three-dimensional (3D) printing, has a variety of uses in manufacturing, including the fabrication of physical prototypes and the production of end-use products. [1, 2] 3D printing is more cost effective than other traditional pricey prototyping and manufacturing technologies. In recent years, there has been a surge in interest in developing and improving the capabilities of 3D printing, which has resulted in a large growth in the number of 3D printable materials. The advancement of 3D printable smart and soft materials has aided in the evolution of robotics [3, 4]. These materials, when combined with printing techniques like fused deposition modeling, direct ink writing, selective laser sintering, and inkjet printing, have the potential to revolutionize wearable sensor technology for robotics applications based on mechanical and electrical actuation and sensing [2, 5].

The demand for robotics technologies to build new and more effective capabilities has continuously increased, thanks to breakthroughs in artificial intelligence and the ever-expanding Internet of Things (IoT). As a result of the increased research interest in this area, advancements in soft materials and additive manufacturing technologies have enabled the development of robots with sophisticated capabilities that are critical for a variety of applications, including manufacturing, manipulation, gripping, human–machine interaction, and locomotion. Furthermore, the use of compliant materials allows sensors to perform more complex jobs with greater flexibility and adaptability [4, 6].

This chapter examines the latest advancements in wearable sensing technology for robotic applications. We will discuss piezoelectric, piezoresistive, capacitive, and optical tactile sensors, biological and chemical sensors, as well as electrically, thermally, magnetically, and optically sensitive actuators, all of which are important in soft robotics.

C.-Y. Huang and S. W. Yoon (Eds.): ICPR1 2021, ACES 14, pp. 386–403, 2023.
https://doi.org/10.1007/978-3-031-44373-2_23

2 Tactile Sensing

The new generation of robots, e.g., soft robots, humanoids, social robots, medical robots, etc., are expected to perform work alongside and in conjunction with humans. Touch is an important sense for humans because it allows us to access contact factors, including form, surface roughness, stiffness, and warmth, and react appropriately. The sense of touch allows a robot to learn about real-world things. Tactile sensing is a critical component in the creation of the future generation of smarter robots [7]. Tactile sensors are devices that respond to touch force and can quantify interactions between the sensor and the surrounding environment [8]. Soft robotics relies heavily on these sensors. They allow robots to perceive physical interactions with their surroundings, gather data, and communicate that data in order to conduct desired tasks, including making precise motions and guiding robots through limited paths [9]. Where visual sensing is unavailable, tactile sensing plays a crucial role in ensuring safe interactions between robots and humans, objects, and the physical environment. Human-robot interaction, rehabilitation, and prosthetics could all benefit from improved tactile sensing capabilities in robots. Tactile sensing types include piezoelectric, piezoresistive, capacitive, and optical, while materials used to manufacture tactile sensors include composites, carbon nanotubes, conductive polymers, and gels [9, 10].

2.1 Piezoelectric Sensors

The piezoelectric effect is used in piezoelectric tactile sensors. The mechanical deformation of piezoelectric materials generates piezoelectric potential. The piezoelectric effect can be produced by the displacement of ions' centers in non-centrosymmetric crystal structures, such as lead zirconate titanate (PZT), or by materials that can be polarized by the alignment of dipole moments, such as polyvinylidene fluoride (PVDF) [11, 12]. The deformation in the piezoelectric material with a contact force of F generates a charge of +Q and -Q at the two electrodes. The charge induced which leads to a potential V across the tactile element is given by:

$$v = \frac{Q}{C} \approx \frac{dF}{C} = \frac{dt}{4\pi\varepsilon\varepsilon_r A}F \tag{1}$$

where d is the piezoelectric constant of the material, C is the static capacitance and ϵ_r is the relative permittivity [7]. Larger d/ϵ_r ratio of the material results in high sensitivity [7]. PZT and other inorganic materials have high d_{33} values, which is ideal for tactile sensor applications. The incorporation of lead and the high Young's modules, however, limit the application in wearable applications. Polymer-based materials, on the other hand, such as PVDF, have a high degree of flexibility, chemical stability, biocompatibility, and processability but a low d_{33}. As a result, researchers have looked into hybrid materials that have both inorganic and organic properties. Piezoelectric tactile sensors have good energy harvesting capabilities in addition to rapid response, great sensitivity, and low power consumption [11, 13]. Furthermore, piezoelectric transducers are pyroelectric, meaning they create charge when exposed to temperature fluctuations. Because of this feature, piezoelectric sensors can monitor a variety of factors, including temperature and force. When both piezoelectric and pyroelectric effects occur at the same time, however, it is difficult to distinguish between the responses [7]. Song et al, utilized a voltage mapping technique to distinguish between the pyroelectric and piezoelectric effect. In this research, the conjunction of Pyro-Piezoelectric effect for self-powered temperature and pressure was explored. Here, touching pressure and finger heat is applied to a "password keyboard" like application consisting of an Ag/BTO/Ag temperature-pressure sensor. Based on the human behavior of entering the password, representative voltage signals are mapped and analyzed. Researchers identified two distinctive inductive signals as the finger touches the sensing array. A sharp voltage peak was generated by the piezoelectric effect induced by the touching pressure and a gentle voltage peak was generated by the pyroelectric effect induced by the finger temperature [108].

To directly print PVDF into piezoelectric tactile sensors, Lee et al presented a poling-assisted fused deposition modeling approach (EPAM) [14]. To make the PVDF-based piezoelectric tactile sensors, filament extrusion was paired with a strong electric field between the nozzle tip and the substrate. The production of PVDF in phase, which is responsible for PVDF's piezoelectric capabilities, is triggered by a strong electrical field. This technique involves a mix of high-temperature drawing, electric poling under a strong electrical field, and high-pressure annealing. The researchers first measured an output current of 1.5 nA for a single printed layer and found that a larger electric field resulted in more piezoelectricity in the printed structures. This process was further improved by Kim et al. [15] to further increase the electrical field and allow the fabrication of multiple layers. In this experiment, the researchers fabricated pressure sensors using a hybrid material consisting of PVDF and $BaTiO_3$. It was observed that the highest amount of 55.91% β phase PVDF was obtained when the $BaTiO_3$ content was 15%, suggesting that hybrid materials are desirable for sensor fabrication.

Recently, triboelectric effect has been considered for tactile sensing [9, 16]. Contact electrification and electrostatic induction are used to generate the electric signal. Vertical-contact-separation mode, lateral-sliding mode, single-electrode mode, and free-standing-triboelectric-layer mode can all be used for tactile sensor applications, depending on device architecture [11]. Haque et al, [17] developed a triboelectric tactile sensor in vertical contact separation mode by utilizing direct ink writing. The sensor consists of three printed layers; PDMS layer with electrodes, Polyamide (PA) spring structure layer, and TangoBlack (TB) layer with electrodes. PA material was chosen for the spring mechanism since it provides structural integrity along with flexibility. Here, PDMS, combined with TB showed the highest triboelectric responses 300% higher average power output than nearest other pairs considered. The final device showed an output power of 60 μW when at an operating frequency is 5 Hz.

2.2 Piezoresistive Sensors

The working principle of piezoresistive tactile sensors is the variation of the resistivity of the material due to the applied mechanical stimulus [18]. The resistance of the sensing material changes depending on the contact force or contact pressure in piezoresistive tactile sensors. The two types of piezoresistive sensors are those whose resistance decreases with increasing pressure and those whose resistance increases with increasing pressure; the former are known as negative-type piezoresistive sensors (NPSs), while the latter are known as positive-type piezoresistive sensors (PTPSs) (PPSs) [19]. Semiconducting materials like silicon and germanium, as well as metals like nickel and platinum alloys, are commonly utilized to make piezoresistors and have a high piezoresistive response [20–26]. These materials, on the other hand, easily shatter when subjected to a substantial mechanical load. As a result, polymer-based materials and carbon-based materials including carbon nanotubes (CNTs), graphene, and (CNTs)-polymer composites are being investigated extensively for piezoresistive sensors.

Saeb et al developed a fused deposition modeling 3D printed stretchable sensor with a piezoresistive sensing material made of polylactic acid-graphene (PLA-G) conductive polymer composite (CPC) sandwiched between two stretchable thermoplastic polyurethane (TPU) structural layers for acquiring tactile feedback like pressure and bending angle [27]. Because of its large surface area and high conductivity, graphene is an excellent choice for tactile sensing. The completed device had a thickness of 1 mm and was stretchable and bendable. The device had a gauge factor of 550 and was particularly sensitive to bending produced strain. The system was able to distinguish between pressure and bending stimuli, with the minimum observable applied pressure being 292 Pa.

A 3D-printable, flexible, and conductive thermoplastic-based touch sensor was created by Christ et al [28]. The material was created using a win-screw extrusion method that combined polyurethane and multiwalled carbon nanotubes (TPU/MWCNT). This material was extruded into a filament with a constant diameter using a secondary mild single-screw extrusion procedure, and the filament was then printed using the FDS printing technique. MWNCTs increased the stiffness of TPU, which improved its printing capabilities. Under 100% applied strain, a significant gauge factor of 176 was reached, and cyclic loading revealed a high resistance-strain response. Tang et al. address the trade-off between sensitivity and measurement range by using a 3D printing approach to improve both sensitivity and sensing range via the positive piezoresistive effect [19]. CNTs and silicon nanoparticles (SiNPs) were used as conductive filler and rheology modifier, respectively, in the conversion of a viscoelastic silicone rubber solution to a printable gel ink. The new technology allowed for room-temperature direct printing of soft and porous composites (SPCs), with the sensitivity and sensing range of the sensors being adjusted by altering the conducting CNT and insulating SiNP content in the ink. Because the change in tunneling resistance was prominent during deformation, the sensors with low CNT concentration displayed a positive piezoresistive effect and demonstrated good sensitivity and a wide sensing range. Smart insoles and E-skin were developed by the researchers to demonstrate the sensor's possible applications.

2.3 Capacitive Sensors

Capacitive tactile sensors have two parallel plates with a dielectric layer between them, commonly silicone or air. The capacitive tactile sensor's working principle is that when objects approach or touch the sensor, the capacitance changes [11, 29]. The capacitance of a parallel-plate capacitor is given by:

$$C = 4\pi \varepsilon_r \varepsilon_0 \frac{A}{d} + C_f \tag{2}$$

where d, is the separation between the two parallel plates, A is the overlapping area of the electrodes, ϵ_r is the dielectric constant of the material between the two plates, ϵ_0 is the permittivity of a vacuum, and C_f is the contribution from the edges of the electrode which tends to store more charge than the electrode [7].

The distance between the parallel plates or the effective area of the sensor changes depending on the force applied to the sensor, resulting in a change in capacitance. The application of a normal force changes the distance, whereas the application of a tangential force changes the capacitive sensor's effective area, indicating that the capacitive tactile sensor can sense touch in both normal and tangential directions. These sensors, on the other hand, are unable to distinguish between the two directions. For measuring the applied force, the change in capacitance is translated to a change in voltage [7]. High sensitivity, compatibility with static force measurement, minimal power consumption, and a high spatial resolution are all features of these tactile sensors. However, as compared to other sensing mechanisms, these sensors are more vulnerable to noise and require more filtration to give correct signals [11]. Since capacitive tactile sensors are in high demand and have applications in artificial skin for robotics, non-planar surfaces of robotic grippers, and wearable textiles, capacitive sensing elements based on silicone and other polymer-based flexible materials have been extensively investigated over the last few decades [30].

The creation of a 3D printed soft capacitive sensor with a capacitance-to-digital converter chip on a PCB, entirely embedded within the 3D printed robot hand to deliver pressure-sensing and signal-processing activities was presented by Ntagios et al [31]. The fingers were 3D printed step by step with multiple layers of polylactic acid (PLA), thermoplastic polyurethane (TPU), and acrylonitrile butadiene styrene (ABS), while the capacitive sensors' conductive and dielectric layers, as well as the conductive tracks, were printed within the finger using a modified FDM 3D printer following each step of the finger printing. The dielectrics was a two-part rubber and commercially available flexible thermoplastic polyurethane (TPU), and the electrodes of the capacitive sensor were a silver paste, conductive polylactic acid (PLA) composite, and a graphite ink produced by the researchers. The sensors that were built and tested on the robotic hands in this experiment had a high sensitivity and could detect pressures as low as 1 kPa. Li and colleagues [32] created a sensor with novel coplanar designs that can be used in tactile and electrochemical sensing applications. The conductive electrodes were entirely immersed in polydimethylsiloxane (PDMS) matrix and the sensor was manufactured using a 3D printer. Because fringing capacitance contributions outside the coplanar surface are more prominent in coplanar capacitors and are sensitive to the dielectric permittivity of the surrounding medium, the authors of this experiment suggest that the unique coplanar design is more effective in tactile sensing applications than conventional parallel plate capacitors.

2.4 Optical Sensors

Optical tactile sensors function by analyzing variations in internal or output light [33]. Optical tactile sensors have several advantages over traditional tactile sensing systems, including the elimination of stray capacitances and thermal noises, the reduction of crosstalk, and the high spatial resolution of tactile imaging [11]. Furthermore, electromagnetic interference has little effect on optical tactile sensors, making them compatible with magnetic resonance imaging (MRI) [33]. Transparency, flexibility, and the capacity to distinguish multiple contacts are critical requirements for optical tactile sensors in robotics, electronic skin, and other practical applications such as touch screens [34, 35]. Soft marker-based optical tactile sensors and soft reflection-based optical tactile sensors are the two major forms of surface deformation identifications in optical tactile sensors [36]. The former measures the (lateral) shear deformation of the sensing surface. A good example of this method is the measurement of gel force [37], where the markers embedded in the supporting gel measure the shear deformation [38, 39]. When a pressure is applied, the reflection-reflection based optical sensor analyzes the normal depression left on the sensing surface. GelSight, which was first created by Johnson and Adelson [40] in 2009 and uses the surface shading of numerous internal lights to infer a depth map via photometric stereo, is an example of this technology. GelSight's ability to acquire material-independent microgeometry and act independently of the optical properties of the surface being touched is an important benefit of this technology [41].

Yuan et al. [41] used 3D printing to create a high-resolution GelSight robot tactile sensor for assessing geometry and force. This work produced a sensor that can monitor high-resolution geometry and infer local force and shear. A deformable elastomer piece serves as the sensor's contact medium, with an embedded camera capturing the elastomer's deformation and reconstructing the high-resolution 3D geometry. The authors also discussed the GelSights' ability to indicate other information, such as slip or incipient slip, and suggest that the figure print version of the sensor developed in this experiment can be successfully applied in robotic grippers, allowing the robot to perform a variety of complex perception and manipulation tasks.

Mechanoluminescence (ML) materials have emerged as prospective materials for electronic skin applications because they convert external mechanical energy into light emission without the use of electron or photon excitation. Qian et al. [42], for example, reported on the creation of printable Skin-Driven Mechanoluminescence devices using nano-doped matrix modification. By distributing rigid ZnS:M2+(Mn/Cu)@Al2O3 microparticles (ZMPs) into soft PDMS films and printing out flexible devices, the authors showed a flexible sensitive ML device. The resulting nanoparticle-doped matrix films achieved skin driving ML and had a sensitive and stable luminescence response to deformations.

3 Actuators

The advancement of actuation has been regarded as a critical component in the development of smarter robots. Actuation allows a robot to effectively distort its body and interact with its environment in order to perform a specific task, such as locomotion, manipulation, gripping, and human-machine contact [4]. Actuation can be carried out using a variety of stimuli, including electrical, thermal, magnetic, and photoresponse responses [43–48]. Traditional methods, such as pneumatic and hydraulic pressure, are still being studied for applications, but the need for pumps makes these systems unsuitable for soft robotics [6]. When it comes to the construction of soft actuators, material flexibility, adaptability, and reconfigurability are critical considerations. All actuating polymers, for example, must be able to demonstrate their shape reversibility [49]. Recently, materials such synthetic electroactive polymers, shape memory, fluids, shape memory alloys, liquid metals, hydrogels, 2D materials, and combinations of these materials have been explored for soft actuator applications [50–54]. Additive manufacturing techniques such as 3D printing, fused deposition, direct ink writing, selective laser sintering, inkjet printing, and stereolithography have been widely used in the fabrication of soft actuators due to advances in smart materials and the ability to process smart materials [2, 55–57]. In addition, 4D printing is defined as the process of creating a physical object by laying down successive layers of stimuli-responsive composite or multi-material with variable properties utilizing appropriate additive manufacturing technology. The 4D-printed object reacts to stimuli from the natural environment or through human intervention after it has been constructed, leading in a physical or chemical change of state over time [58–61].

3.1 Electrically Responsive Actuators

Actuators that are electrically responsive convert electrical energy into mechanical energy. Electrically sensitive materials, such as dielectric elastomers, ionic polymer–metal composites, and polyelectrolyte gel, have all been investigated for use in the manufacture of electrically responsive actuators. Electrically responsive actuators have the shortest response time when compared to other actuation approaches, and they are chosen because they can be immediately integrated with basic electronic devices and their actuation may be successfully controlled by programming [1, 62]. A dielectric actuator, which comprises of a soft dielectric sheet sandwiched between two compliant electrodes, is the most often used electrically responsive actuator. When an electric field is supplied across the electrodes, the electrostatic attraction of opposite charges on opposing electrodes and the repulsion of like charges on each electrode generate a stress on the film, causing it to contract in thickness and expand in area [1, 63]. The area expansion or thickness reduction of a soft flexible sheet is the actuation mechanism for all dielectric actuators. Electrically responsive actuators' performance is influenced by factors such as pre-strain, mechanical loads, actuator configuration, humidity, and temperature. Dielectric actuators, for example, have been noted as having limits due to the high electrical actuation voltage required and the requirement for pre-strain. Pre-strain-free dielectric actuators have been developed to address this issue [1]. Chortos et al [64] used 3D printing to create an interdigitated dielectric elastomer-based actuator. The in-plane contractile actuation modes of the actuators created in this work were achieved by 3D printing interdigitated electrodes using conductive elastomer ink. The electrodes were then encased in a chemically crosslinked polyurethane acrylate dielectric matrix that self-healed. The printed actuators had customizable mechanical characteristics and could withstand actuation strains of up to 9%. Wang et al. [1] used direct ink writing to create electrically controlled polyvinyl chloride (PVC) gel actuators. A gel sheet was layered between two copper electrodes to form the bending actuator. A jellyfish-like actuator with six arms was created to illustrate the actuation of the 3D-printed PVC gels. When an electric field is provided to the gel between two electrodes, the gel moves to the positively charged electrode, forcing the arms to bend upward. Due to the flexibility of the gel, the arms relax to their original positions when the voltage is removed. At a voltage of 1000 V, the bending angle reaches a maximum of 170°. Once the power is released, the jellyfish-like actuators can bend over 90 in less than 2 s and recover from 130 to 0 in 3 s.

Haghiashtiani et al. [43] created a hybrid material system that can be 3D printed as a soft dielectric elastomer actuator (DEA). The device was made in a unimorph arrangement, which resembles a cantilever with a rigid passive layer linked to the framework. The DEA device needed the stacking of many material layers, all of which were manufactured using DIW while carefully considering the inks' rheological properties, such as viscosity, yield stress, and viscoelastic moduli. While one end of the device was fixed to a rigid framework and the other end was allowed free to deflect, the performance of the printed device was tested under ramp-up electrical input, cyclic electrical loading, and payload masses. At a 5.44 kV applied electrical stimulus, the highest vertical tip displacement was 9.78 2.52 mm. Researchers believe that this device's self-sensing capabilities can be used to control soft robots in a closed-loop feedback loop without the use of additional optical or electromechanical sensors.

3.2 Thermally Responsive Actuators

Thermally responsive actuators convert thermal stimulus into mechanical energy, such as infrared (IR), near-infrared (NIR), thermal radiation, or Joule heating. Furthermore, these actuators can be stimulated and controlled locally or remotely using heat generated by lasers. Thermally responsive actuation is safer for biomedical and healthcare applications than other techniques that use electricity and UV light as stimuli [65]. These actuators, on the other hand, have a slower response time and are less efficient than other actuation technologies, such as electrically responsive actuators. The use of thinner films, better heat capacity materials, and higher power have all been identified as strategies to improve the performance of thermally sensitive actuators [3, 4]. Due to their high elastic deformation, low density, low cost, and ease of manufacture, materials such as Shape Memory Polymers (SMP), Shape Memory Alloys (SMA), and Liquid Crystal Elastomers are intensively investigated for the fabrication of thermally sensitive actuators. SMPs like polyurethane (PU) and thermoplastic polyurethane (TPU) can be programmed to remember a temporary shape and revert to it once a heat stimulus is provided. Thermally induced shape memory polymers convert from a transient deformed shape to a permanent shape by using the material's glass transition temperature (Tg). When the temperature rises over Tg, the polymer softens and rubberizes, allowing it to be deformed. Allowing the distorted polymer to cool below Tg enables the polymer to be kept in shape [66]. SMAs are made by combining different components like Fe-Mn-Si, Cu-Zn-Al, and Cu-Al-Ni. The qualities of the resulting alloy are determined by the materials used. These materials are distinguished by their reversible crystal structure, which allows them to deform and return to their original shape in response to temperature stimulation.

Cersoli et al. [66] used direct pellet extrusion to 3D print thermally induced shape memory polymers. The research combines the advantages of material extrusion and new materials to create smart 3D-printed parts with shape memory. To address the issue of manual SMP resetting, a hybrid material incorporating both SMA and SMP was employed to create a reversible actuated switch that was controlled by an external heat stimulus. In this experiment, the utilization of commercially available SMP pellets and the ability to print pieces straight from raw material without needing to tune the SMP pellets into a 3D printing filament is regarded as a benefit. The hybrid actuator successfully functioned as a thermal switch, allowing the actuation of an electronic circuit, and the final 3D printed AMP actuators had a shape recovery of over 96% and a shape retention of 90.7% across the heat cycles.

3.3 Magnetically Responsive Actuators

Magnetic particles are used as fillers in polymers, gels, papers, and fluids to create magnetically responsive actuators [67–69]. Magnetic actuation is appealing and widely investigated in the context of robotics because of its quick reaction, remote actuation, and shape quick reprogrammability. Furthermore, because magnetically sensitive actuators are contact-free, they can avoid the negative impacts of traditional techniques in applications such as medication administration, microsurgery, microfluidics, and internal body movements [3, 4, 70, 71]. The control of the magnetic field is the operational mechanism of magnetically responsive actuators. When a material containing magnetic particles is exposed to a magnetic field, the magnetic fillers align with the field, resulting in a variety of actuation modes including torque generation, deformation, elongation, and contraction, as well as bending. When the magnetic fillers in the material interact with the field spatial gradients, the actuation occurs [3, 4]. Magnetically sensitive actuators have the ability to generate movements in compact and enclosed spaces, which is a distinct advantage. The capacity of the magnetic field to penetrate numerous materials gives it this edge. Magnetically responsive actuators also have a fast response time when compared to other actuating methods. As a result, these actuators have been used in micropumps, swimmers, and crawlers, among other robotic applications [70–73]. However, external magnetic coils are huge and require a significant amount of power, despite the fact that the areas where the magnetic field and gradients are strong enough and adjustable are small [3, 4].

As previously stated, magnetically responsive actuation is being studied extensively for the purpose of producing robotic movements within the human body. Biomimetic soft swimmers have now acquired popularity as a treatment for thrombosis-related disorders [71]. These soft, small robots have narrow flagella-like microstructures and are inspired by flagella [74]. Khalil et al. [75] proposed the MagnetoSperm in 2014, a microrobot that navigates via weak magnetic fields. The micro robot had a sperm-like shape with one flexible tail that could propel by swinging the flexible tail in the magnetic field's direction. The magnetic dipole induced by the magnetic head, which featured a 200 nm cobalt-nickel layer, caused the tail to oscillate. The original design could only move in one direction, but enhancements were made in 2018 when dual tails were added. Hunter et al. [76] built a micro bio robot for cellular and chemical delivery applications based on this research, using 3D printing to fabricate microscale features. The micro robot's helical shape was achieved by pouring liquefied agar gel and iron oxide into a 3D printed mold. A three-axial Helmholtz coil arrangement generated the magnetic field, and the micro robot was magnetized along its long axis. The robot revolved along its long axis in a homogeneous rotating magnetic field and was able to propel forward in low Reynolds number settings. With two different concentrations of iron oxide, the magnetic response and robot motion were investigated, and the robot velocity increased as the external magnetic field frequency increased up to a critical frequency. The robot's speed slows after reaching the crucial frequency.

Hamilton et al. [77] created a flagella-based single particle ferromagnetic swimmer with a flexible tail and a rigid ferromagnetic head. The tails were constructed utilizing a 3D printed mold in order to generate the overall swimmer geometry. A Helmholtz coil device provided a consistent oscillating magnetic field that controlled the swimmer. With magnetic fields up to 3.5 mT, the frequency of the external magnetic field was changed between 30 and 170 Hz. The swimming behavior device is dependent on the length of the flagellum, the frequency of the applied field, the bending stiffness of the filament, and the fluid dynamic interactions of the tail, according to the researchers. After a critical the created devices was able to successfully propel in fluids with a range of Reynolds numbers and can be turned into a micropumping device by a change of reference frame, a quick decline in swimming performance was noticed. Ji et al. [48] devised a one-step multi-materials 3D printing approach for producing magnetically driven soft actuators that require no assembly. A flexible resin containing Fe_3O_4 nano particles was printed using the direct light processing (DLP) 3D printing technique.

3.4 Photoresponsive Actuators

Light-triggered actuators are a good choice for micro- and nanoscale applications because they can be controlled remotely, have a high degree of accuracy, and can be manipulated quickly. Photochromic molecules are capable of capturing optical impulses and translating them into diverse property changes. These molecules are employed in soft and flexible micro- and nanoscale actuators made of polymers, gels, and fluids [3, 4]. After being activated with light, photoresponsive polymers reversibly change their physical or chemical properties, such as shape, surface wettability, membrane potential, permeability, solubility, fluorescence, and phase-separation temperatures [3, 4, 78]. The bending, contraction, and swelling of the polymer are affected by several external parameters such as the wavelength and intensity of the light, as well as the time of exposure. Increasing the intensity of the light source or decreasing the thickness of the polymer film can usually improve the photochemical reaction in polymer films, and the deformations that occur during exposure can usually be reversed by applying heat, changing the wavelength of the light, or removing the light source [3, 78]. Liquid crystal polymer networked materials and carbon-based materials are the most commonly used photoresponsive actuators [4, 79, 80]. Because of their capacity to function in dry settings, LCNs are ideal for making soft actuators. Controlling the 3D arrangement of the molecular building blocks can also be used to design the deformation of polymer networks [80].

Ceamanos et al. [81] described 4D printing of LCN-based actuators with quick photoinduced mechanical response for robotic applications. The researchers developed an ink with an acrylate end-capped liquid crystalline polymer based on an azobenzene moiety. The printed free-standing film bends toward the UV light source when exposed to UV light illumination with 50 mW/cm^2 of UV light. When the UV light is turned off, the film relaxes in a matter of seconds while preserving some bending deformation. The ability of these films to conduct mechanical work was also studied by the researchers. A similar experiment was undertaken to raise a 1g weight when the film was illuminated with UV light at 50 mW/cm^2 and 200 mW/cm^2 to evaluate the lifting capabilities of the 4D printed films. The lighting of the LCE film with 50 mW/cm^2 UV light for 240s caused an 8 percent contraction, and the cessation of irradiation led in a quick relaxation of the LCE film to a new equilibrium length of 5% of the starting length within a few seconds. When the film was subjected to UV light with a power of 200 mW/cm^2 for 60 s, the film contracted and the 1 g weight was lifted over 4.8 mm. According to the researchers, light induced actuation includes the film's photothermal and photochemical capabilities, which may work together or separately depending on the actuation wavelength, light intensity, and pulse duration. Furthermore, the materials show good flexibility and the capacity to overcome brittleness, which is a common problem in thin film liquid crystal networks, and they have the potential to be used in micro-machines and robot-integrated elements.

4 Chemical Sensors

Robotic systems that mimic and surpass human sensory capabilities are critical for the advancement of numerous applications, as detailed in the previous sections. Security is another area that can gain tremendously from the evolution of robots with enhanced sensing skills. Traditional public health, environmental, and agricultural sensing methods have generally focused on physical factors such as pressure and temperature, and require detection of hazardous compounds in aqueous solutions, rendering them impracticable for dry-phase, real-time analysis [82]. Furthermore, autonomous trace-level hazard detection can prevent or reduce human exposure to toxic chemicals when operating in hazardous areas [82–84]. The most critical part of a chemical, biological, radiological, or nuclear (CBRN) disaster is time. Security personnel, such as law enforcement officers, and first responders, such as firefighters and emergency medical workers, must be able to make critical decisions quickly. To accomplish this, the development of field-deployable robotic systems capable of analyzing these hazards and delivering conclusions quickly is critical [85]. Robotic skins and glove-based wearable chemical sensors have recently been proposed as a feasible solution to this problem [86–90].

Amit et al. [91] created sensors to monitor organophosphate (OP) pesticides. The detection of OP is critical because it can endanger humans and animals and can be employed as a nerve agent in chemical warfare. The researchers created a flexible glove-based sensor device that can measure pressure and chemical signals at the time of application. When handling agricultural produce, the addition of pressure is critical to avoid hurting the robot or the object in contact as a result of uncontrolled force. The immobilized enzyme organophosphate hydrolase (OPH), which is specific for OP compounds, reacts with the OP analyte in solid form to operate the chemical sensor. Using a three-electrode electrochemical setup and square wave voltammetry, the p-nitrophenol product of the OPH reaction is determined (SWV). On disposable nitrile polymer gloves, the pressure and chemical sensors were created. Screen printed carbon, Ag/AgCl, and a protective insulator layer made of flexible, elastic adhesive were used to make the sensor. The three-electrode system used Ag/AgCl as the reference electrode, carbon ink as the counter electrode, and an insulating layer printed over Ag/AgCl interconnects as the dielectric separation. The hazardous chemical was detected via electrochemical analysis by swiping a contaminated surface with a carbon collecting pad, then pressing the collector pad against the printed measuring electrodes coated with electrolyte gel to complete the electrochemical cell. The data gained in this experiment, according to the researchers, has applications in security and will allow robotic fingertips to advance.

Barfidoht et al. [87] developed an electrochemical glove-based sensor that can detect fentanyl rapidly. Fentanyl is one of the most commonly abused opioids in the United States, responsible for thousands of deaths [92–94]. The flexible electrochemical sensors incorporated on the fingertips of the wearable glove-based sensor produced in this experiment are used to detect fentanyl on-site. The sensor was made with the help of screen printing. To finish the sensing electrode, a layer of Ag/AgCl was printed on the index finger of the nitrile glove as a reference electrode and a connecting pad, followed by a layer of carbon as a counter electrode and an insulator layer as a dielectric separation of the three electrode system. On the thumb was also printed a circular carbon pad for collecting drug leftovers. A glove-based sensor coupled to a SWV electrochemical analyser was used to conduct the test. First Swiping a contaminated glass slide with the thumb and combining the collection with the index finger covered in agarose gel, where the sensor is placed, completed the electrochemical cell. SWV was used to record the fentanyl oxidation with optimum parameters. The results collected by the portable electrochemical analyzer can be remotely transmitted for further study. The proposed sensor shows direct fentanyl oxidation in both liquid and powder forms, indicating that it could be used for point-of-need screening by first responders. Ciui et al. [90] created a chemical-flavor detecting robotic glove enabling fingertip detection of tastes in a variety of meals and beverages, including sweetness, sourness, and spiciness. Three printed fingertip sensing electrodes, driven by SWV and amperometric electrochemical methods, make up the screen-printed gadget. The carbon-based index finger detects sourness through ascorbic acid detection, the Prussian-blue modified enzyme-based biosensor printed on the middle finger detects sweetness through glucose detection, and the carbon-based ring finger detects spiciness through capsaicin molecules detection. The reference electrode and connections were printed with Ag/AgCl ink, similar to the prior trials, and the counter electrode was printed with a carbon-based ink. The researchers concluded that the chemically sensitive robotic technology with a novel sense of taste will pave the way for automated chemical sensing gear, with extensive implications for robotic sensing applications.

Yu et al. [82] constructed a mass-producible multimodal "M-Bot," an artificial intelligence-powered robotic sensing device. The flexible physicochemical sensor arrays were inkjet printed using custom-developed nanomaterial inks that could record electrophysiology, detect tactile perception, and sense a variety of hazardous compounds such as nitroaromatic explosives, insecticides, and nerve agents. The M-bot is made up of two stretchable inkjet-printed e-skin patches, e-skin-R and e-skin-H, that interact with the robot and human skin, respectively. Serial printing silver as interconnects and reference electrodes, carbon as counter electrode and temperature sensor, and polyimide (PI) as encapsulation were used to create E-skin-R. Target-specific nanoengineered sensing films were also used. For example, Pt-graphene was used to detect TNT. Before placing E-skin-H on human volunteers, silver interconnects were printed using DMP-2850, sealed with a layer of PDMS film, and an adhesive electrode gel was placed onto the electrodes. The e-skin-H captures electromyography (sEMG) signals from muscular contractions and decodes them using a machine learning model, while the e-skin-R does proximity sensing, tactile mapping, and temperature mapping. In addition, e-skin-R employs hydrogel-assisted electrochemical on-site hazardous substance sampling and analysis. Electrical stimulation of the human body via e-skin-H was used to perform real-time haptic and threat communication after detection. The sensor array made steady contact with human skin, allowing for accurate recording of neuromuscular activity, and it was an appealing technique for developing enhanced flexible and soft e-skins. This human-machine–interactive robotic sensing technology, according to the researchers, will open the way for novel wearable and robotic applications.

4.1 Robotics for Combating Infectious Diseases

Many countries and governments were taken off guard by the new coronavirus pandemic. Aside from the catastrophic death toll, the pandemic showed numerous areas, like as supply chain, public health, and communication, that need to be drastically improved to successfully tackle another global event of this scale and severity. Because the virus spreads by human-to-human contact, many countries have enacted lockdowns and strong social distancing measures, robotic-based solutions have been offered as a means of restricting the infection's transmission. Decontamination, patient management, telehealth, specimen collection, transportation, laboratory testing, reconnaissance, and home-based nursing were among the options offered [77, 95–103]. For example, in the context of disinfection, robots must meet stringent decontamination criteria to prevent disease transmission from robot to human, robot to robot, or robot to the environment. Traditional thermal imaging and vision-based sensing techniques were used in several of these systems. Although there is a scarcity of research on printed and wearable devices that focus specifically on infectious disease detection, it is critical that the scientific community concentrates its efforts on mass-producing robotic devices with these capabilities to advance the capabilities of the next generation of robots and better prepare our society for the next pandemic [104–106].

For example, Yu et al. [82] demonstrated the ability to monitor infectious biohazards such as SARS-CoV-2 without direct human exposure. The glove-based solution is a one-of-a-kind and game-changing innovation in the fight against infectious illnesses. A fabricated multiwalled carbon nanotube (CNT) electrode functionalized with antibodies specific to SARS-CoV-2 spike 1 protein was used to provide label-free SARS-CoV-2 viral detection. The SARS-CoV-2 S1 sensor revealed great selectivity over other viral proteins, as well as on-the-spot detection of dry phase protein.

Murphy et al. [107] evaluated 262 reports that appeared in various media and scientific papers between March and July 2020, describing 203 instances of employment of 104 robot models for the COVID-19 response. According to the authors, the largest category of robot applications is public safety, which includes law enforcement and emergency medical services. Clinical care was the second-largest category of robot applications discovered in the diagnosis and acute management of coronavirus patients. Robots with autonomous navigation capabilities were utilized for prescription and food dispensing, followed by telepresence robots for patient handling, robots that allow families to remotely visit patients, and inventory management robots. Expanding the usage of wearable robotics to include these duties to minimize physical encounters and manage infectious diseases through robotics-based solutions would require more technology advancements and study.

5 Conclusion

This chapter examines current advancements in the field of printed wearable sensors for robot applications. The advent of cost-effective printing techniques such as 3D and 4D printing, as well as the evolution of printable smart materials, have changed the study field relevant to the next generation of soft and rigid robots. It is clear from this research that wearable tactile sensing and robotic actuation research have progressed steadily over the previous few decades, and its applications have become widely used in daily life. Furthermore, multiple sensing solutions exist depending on the application and hybrid sensing technologies such as piezoelectric-piezoresistive tactile sensing and combined visual and tactile sensing have emerged as new developments in this field. It is expected that the growing demand for robot-based wearable solutions for on-the-spot chemical detection and analysis, as well as the identification of infectious diseases, will assist communities all over the world.

References

1. Wang, Z., et al.: 3D printing of electrically responsive PVC Gel actuators. ACS Applied Materials & Interfaces **13**(20), 24164–24172 (2021)
2. Wallin, T.J., Pikul, J., Shepherd, R.F.: 3D printing of soft robotic systems. Nature Reviews Mat. **3**(6), 84–100 (2018)
3. Hines, L., et al.: Soft actuators for small-scale robotics. Advanced Mat. **29**(13), 1603483 (2017)
4. El-Atab, N., et al.: Soft actuators for soft robotic applications: a review. Advanced Intelligent Syst. **2**(10), 2000128 (2020)
5. Zhan, S., et al.: 3D printing soft matters and applications: a review. Int. J. Molecular Sci. **23**(7), 3790 (2022)
6. Kim, J., et al.: Review of soft actuator materials. Int. J. Precision Eng. Manuf. **20**(12), 2221–2241 (2019)
7. Dahiya, R.S., Valle, M.: Robotic Tactile Sensing: Technologies and System. Springer (2013)
8. Gir˜ao, P.S., et al.: Tactile sensors for robotic applications. Measurement **46**(3), 1257– 1271 (2013)
9. Zhou, X., Lee, P.S.: Three-dimensional printing of tactile sensors for soft robotics. MRS Bulletin **46**(4), 330–336 (2021)
10. Xie, M., et al.: Flexible multifunctional sensors for wearable and robotic applications. Advanced Materials Technol. **4**(3), 1800626 (2019)
11. Wang, C., et al.: Tactile sensors for advanced intelligent systems. Advanced Intelligent Syst. **1**(8), 1900090 (2019)
12. Al-Handarish, Y., et al.: A survey of tactile-sensing systems and their applications in biomedical engineering. Advances in Materials Sci. Eng. 2020 (2020)

13. Senthil Kumar, K., Chen, P.-Y., Ren, H.: A review of printable flexible and stretchable tactile sensors. Research 2019 (2019)
14. Lee, C.B., Tarbutton, J.A.: Electric poling-assisted additive manufacturing process for PVDF polymer-based piezoelectric device applications. Smart Materials and Structures **23**(9), 095044 (2014)
15. Kim, H., et al.: Integrated 3D printing and corona poling process of PVDF piezoelectric films for pressure sensor application. Smart Materials and Struct. **26**(8), 085027 (2017)
16. Yao, G., et al.: Bioinspired triboelectric nanogenerators as self-powered electronic skin for robotic tactile sensing. Advanced Functional Materials **30**(6), 1907312 (2020)
17. Haque, R.I., et al.: Self-powered triboelectric touch sensor made of 3D printed materials. Nano Energy **52**, 54–62 (2018)
18. Stassi, S., et al.: Flexible tactile sensing based on piezoresistive composites: a review. Sensors **14**(3), 5296–5332 (2014)
19. Tang, Z., et al.: 3D printing of highly sensitive and large-measurement-range flexible pressure sensors with a positive piezoresistive effect. ACS Applied Materials & Interfaces **12**(25), 28669– 28680 (2020)
20. Singh, R., et al.: A silicon piezoresistive pressure sensor. In: Proceedings First IEEE International Workshop on Electronic Design, Test and Applications', pp. 181–184. IEEE (*2002*)
21. Zhou, G., et al.: A smart high accuracy silicon piezoresistive pressure sensor temperature compensation system. Sensors **14**(7), 12174–12190 (2014)
22. Shaby, S.M., Premi, M.S.G., Martin, B.: Enhancing the performance of MEMS piezoresistive pressure sensor using germanium nanowire. Procedia Materials Science **10**, 254– 262 (2015)
23. Gonzalez, P., et al.: A CMOS compatible polycrystalline silicon-germanium based piezoresistive pressure sensor. In: 2011 16th International Solid-State Sensors, Actuators and Microsystems Conference, pp. 1076–1079. IEEE (2011)
24. Wise, K.D., Angell, J.B., et al.: An IC piezoresistive pressure sensor for biomedical instrumentation. IEEE Trans. Biomedical Eng. **2**, 101–109 (1973)
25. Pang, C., et al.: An excellent platinum piezoresistive pressure sensor using adhesive bonding with SU-8. Advanced Materials Res. **60**, 68–73 (2009)
26. Wagner, S., et al.: Highly sensitive electromechanical piezoresistive pressure sensors based on large- area layered PtSe2 films. Nano letters **18**(6), 3738–3745 (2018)
27. Mousavi, S., et al.: An Ultrasensitive 3D Printed Tactile Sensor for Soft Robotics. In: *arXiv preprint* arXiv:1810.09236 (2018)
28. Christ, J.F., et al.: 3D printed highly elastic strain sensors of multiwalled carbon nanotube/thermoplastic polyurethane nanocomposites. Materials & Design **131**, 394–401 (2017)
29. Tang, Y., et al.: Recent advances of 4D printing technologies toward soft tactile sensors. Frontiers in Materials **8**, 658046 (2021)
30. Kapoor, A., et al.: Soft, flexible 3D printed fibers for capacitive tactile sensing. In: 2016 IEEE SENSORS, pp. 1–3 (2016). IEEE
31. Ntagios, M., et al.: Robotic hands with intrinsic tactile sensing via 3D printed soft pressure sensors. Advanced Intelligent Syst. **2**(6), 1900080 (2020)
32. Li, K., et al.: 3D printed stretchable capacitive sensors for highly sensitive tactile and electrochemical sensing. Nanotechnology **29**(18), 185501 (2018)
33. Chi, C., et al.: Recent progress in technologies for tactile sensors. Sensors **18**(4), 948 (2018)
34. Ramuz, M., et al.: Transparent, optical, pressure-sensitive artificial skin for large-area stretchable electronics. Advanced Mat. **24**(24), 3223–3227 (2012)
35. Yun, S., et al.: Polymer-waveguide-based flexible tactile sensor array for dynamic response. Advanced Mat. **26**(26), 4474–4480 (2014)

36. Lepora, N.F.: Soft biomimetic optical tactile sensing with the TacTip: a review. IEEE Sensors Journal (2021)
37. Kamiyama, K., et al.: Gelforce. In: ACM SIGGRAPH 2004 Emerging technologies, p. 5 (2004)
38. Kamiyama, K., et al.: Evaluation of a vision-based tactile sensor. In: IEEE International Conference on Robotics and Automation, 2004. Proceedings. ICRA'04, **2**, pp. 1542–1547 (2004). IEEE
39. Kamiyama, K., et al.: Vision-based sensor for real-time measuring of surface traction fields. IEEE Computer Graphics and Appl. **25**(1), 68–75 (2005)
40. Johnson, M.K., Adelson, E.H.: Retrographic sensing for the measurement of surface texture and shape. In: 2009 IEEE Conference on Computer Vision and Pattern Recognition, pp. 1070–1077 (2009). IEEE
41. Yuan, W., Dong, S., Adelson, E.H.: Gelsight: high-resolution robot tactile sensors for estimating geometry and force. Sensors **17**(12), 2762 (2017)
42. Qian, X., et al.: Printable Skin-driven mechanoluminescence devices via nanodoped matrix modification. Advanced Mat. **30**(25), 1800291 (2018)
43. Haghiashtiani, G., et al.: 3D printed electrically-driven soft actuators. Extreme Mechanics Letters **21**, 1–8 (2018)
44. Kwon, G.H., et al.: Biomimetic soft multifunctional miniature aquabots. Small **4**(12), 2148–2153 (2008)
45. Li, M., et al.: Soft actuators for real-world applications. Nature Reviews Materials, pp. 1– 15 (2021)
46. Cao, X., et al.: Review of soft linear actuator and the design of a dielectric elastomer linear actuator. Acta Mechanica Solida Sinica **32**(5), 566–579 (2019)
47. Wang, Y., et al.: Hierarchically structured self-healing actuators with superfast light-and magnetic- response. Advanced Functional Mat. **29**(50), 1906198 (2019)
48. Ji, Z., et al.: Multimaterials 3D printing for free assembly manufacturing of magnetic driving soft actuator. Advanced Mat. Interfaces **4**(22), 1700629 (2017)
49. Apsite, I., Salehi, S., Ionov, L.: Materials for smart soft actuator systems. Chemical Reviews **122**(1), 1349–1415 (2021)
50. Carrico, J.D., Tyler T., Leang, K.K.: A comprehensive review of select smart polymeric and gel actuators for soft mechatronics and robotics applications: fundamentals, freeform fabrication, and motion control. Int. J. Smart and Nano Mat. **8**(4), 144–213 (2017)
51. Martins, P., et al.: Polymer-based actuators: back to the future. Physical Chemistry Chemical Phys. **22**(27), 15163–15182 (2020)
52. Herath, M., et al.: Light activated shape memory polymers and composites: a review. European Polymer J. **136**, 109912 (2020)
53. Stoychev, G., Kirillova, A., Ionov, L.: Light-responsive shape-changing polymers. Advanced Optical Materials **7**(16), 1900067 (2019)
54. Kularatne, R.S., et al.: Liquid crystal elastomer actuators: synthesis, alignment, and applications. J. Polymer Sci. Part B: Polymer Physics **55**(5), 395–411 (2017)
55. Khoo, Z.X., et al.: 3D printing of smart materials: A review on recent progresses in 4D printing. Virtual and Physical Prototyping **10**(3), 103–122 (2015)
56. Ionov, L.: 3D microfabrication using stimuli-responsive self-folding polymer films. Polymer Rev. **53**(1), 92–107 (2013)
57. Zolfagharian, A., et al.: 3D printed soft parallel actuator. Smart Materials and Struct. **27**(4), 045019 (2018)
58. Pei, E.: 4D printing–revolution or fad? In: Assembly Automation (2014)
59. Nishiguchi, A., et al.: 4D printing of a light-driven soft actuator with programmed printing density. ACS Applied Mat. & Interfaces **12**(10), 12176–12185 (2020)

60. Zolfagharian, A., et al.: Topology-optimized 4D printing of a soft actuator. Acta Mechanica Solida Sinica **33**(3), 418–430 (2020)
61. Shiblee, M.D.N.I., et al.: 4D printing of shape-memory hydrogels for soft-robotic functions. Advanced Materials Technol. **4**(8), 1900071 (2019)
62. He, Q., et al.: Electrically controlled liquid crystal elastomer–based soft tubular actuator with multimodal actuation. Science Adv. **5**(10), eaax5746 (2019)
63. Brochu, P., Pei, Q.: Dielectric elastomers for actuators and artificial muscles. Electroactivity in polymeric materials. Springer, pp. 1–56 (2012)
64. Chortos, A., et al.: 3D printing of interdigitated dielectric elastomer actuators. Advanced Functional Materials **30**(1), 1907375 (2020)
65. Stroganov, V., et al.: Biodegradable self-folding polymer films with controlled thermo-triggered folding. Advanced Functional Materials **24**(27), 4357–4363 (2014)
66. Cersoli, T., et al.: 3D printed shape memory polymers produced via direct pellet extrusion. Micromachines **12**(1), 87 (2021)
67. Ogrin, F.Y., Petrov, P.G., Winlove, C.P.: Ferromagnetic microswimmers. Physical Review Letters **100**(21), 218102 (2008)
68. Cheang, U.K., Kim, M.J.: Self-assembly of robotic micro-and nanoswimmers using magnetic nanoparticles. J. Nanoparticle Research **17**(3), 1–11 (2015)
69. Chung, H.-J., Parsons, A.M., Zheng, L.: Magnetically controlled soft robotics utilizing elastomers and gels in actuation: a review. Advanced Intelligent Syst. **3**(3), 2000186 (2021)
70. Saren, A., Smith, A.R., Ullakko, K.: Integratable magnetic shape memory micropump for high-pressure, precision microfluidic applications. Microfluidics and Nanofluidics **22**(4), 1–10 (2018)
71. Pozhitkova, A.V., et al.: Reprogrammable soft swimmers for minimally invasive thrombus extraction. ACS Applied Materials & Interfaces (2022)
72. Doganay, S., et al.: A rotating permanent magnetic actuator for micropumping devices with magnetic nanofluids. J. Micromechanics and Microengineering **30**(7), 075012 (2020)
73. Fu, S., et al.: Biomimetic soft micro-swimmers: from actuation mechanisms to applications. Biomedical Microdevices **23**(1), 1–13 (2021)
74. Cicconofri, G., DeSimone, A.: Modelling biological and bio-inspired swimming at microscopic scales: recent results and perspectives. Comput. Fluids **179**, 799–805 (2019)
75. Khalil, I.S.M., et al.: MagnetoSperm: A microrobot that navigates using weak magnetic fields. Applied Physics Letters **104**(22), 223701 (2014)
76. Hunter, E.E,., et al.: Toward soft micro bio robots for cellular and chemical delivery. IEEE Robotics and Automation Letters **3**(3), 1592–1599 (2018)
77. Hamilton, J.K., et al.: Torque driven ferromagnetic swimmers. Physics of Fluids **30**(9), 092001 (2018)
78. Ercole, F.,, Davis, T.P., Evans, R.A.: Photo-responsive systems and biomaterials: photochromic polymers, light-triggered self-assembly, surface modification, fluorescence modulation and beyond. Polymer Chemistry **1**(1), 37–54 (2010)
79. Yu, L., et al.: Multifunctional liquid crystal polymer network soft actuators. J. Materials Chemistry A **8**(6), 3390–3396 (2020)
80. Pilz da Cunha, M., et al.: An untethered magnetic-and light-responsive rotary gripper: shedding light on photoresponsive liquid crystal actuators. Advanced Optical Materials **7**(7), 1801643 (2019)
81. Ceamanos, L., et al.: Four-dimensional printed liquid crystalline elastomer actuators with fast photoinduced mechanical response toward light-driven robotic functions. ACS Applied Materials & Interfaces **12**(39), 44195–44204 (2020)
82. Yu, Y., et al.: All-printed soft human-machine interface for robotic physicochemical sensing. Science Robotics **7**(67), eabn0495 (2022)

83. Ishida, H., Wada, Y., Matsukura, H.: Chemical sensing in robotic applications: a review. IEEE Sensors Journal **12**(11), 3163–3173 (2012)
84. Trevelyan, J., Hamel, W.R, Kang, S.-C.: Robotics in hazardous applications. In: Springer handbook of robotics. Springer, pp. 1521–1548 (2016)
85. Hubble, L.J., Wang, J.: Sensing at your fingertips: glove-based wearable chemical sensors. Electroanalysis **31**(3), 428–436 (2019)
86. Mishra, R.K., et al.: Wearable flexible and stretchable glove biosensor for on-site detection of organophosphorus chemical threats. ACS Sensors **2**(4), 553–561 (2017)
87. Barfidokht, A., et al.: Wearable electrochemical glove-based sensor for rapid and on-site detection of fentanyl. Sensors and Actuators B: Chemical **296**, 126422 (2019)
88. Bariya, M., et al.: Glove-based sensors for multimodal monitoring of natural sweat. Science Advances **6**(35), eabb8308 (2020)
89. Sempionatto, J.R., et al.: Wearable chemical sensors: emerging systems for on-body analytical chemistry. Analytical Chemistry **92**(1), 378–396 (2019)
90. Ciui, B., et al.: Chemical sensing at the robot fingertips: toward automated taste discrimination in food samples. ACS Sensors **3**(11), 2375–2384 (2018)
91. Amit, M., et al.: Point-of-use robotic sensors for simultaneous pressure detection and chemical analysis. Materials Horizons **6**(3), 604–611 (2019)
92. Torralva, R., Janowsky, : Noradrenergic mechanisms in fentanyl-mediated rapid death explain failure of naloxone in the opioid crisis. J. Pharmacology and Experimental Therapeutics **371**(2), 453–475 (2019)
93. Lutfy, K.: Opioid Crisis—An Emphasis on Fentanyl Analogs (2020)
94. Singer, J.A.: Stop Calling it an Opioid Crisis–it'sa Heroin and Fentanyl Crisis (2018)
95. Di Lallo, A., et al.: Medical robots for infectious diseases: lessons and challenges from the COVID-19 pandemic. IEEE Robotics & Automation Magazine **28**(1), 18–27 (2021)
96. Guettari, M., Gharbi, I., Hamza, S.: UVC disinfection robot. Environmental Science and Pollution Res. **28**(30), 40394–40399 (2021)
97. Javaid, M., et al.: Robotics applications in COVID-19: a review. J. Industrial Integration and Manage. **5**(04), 441–451 (2020)
98. Torres, F., Puente, S.T., Ubeda, A.: Assistance Robotics and Biosensors (2018)
99. Jat, D.S., Singh, C.: Artificial intelligence-enabled robotic drones for covid-19 outbreak. Intelligent Systems and Methods to Combat Covid-19, pp. 37–46 (2020). Springer
100. Porter, J., et al.: Society of Robotic Surgery Review: Recommendations Regarding the Risk of COVID-19 Transmission During Minimally Invasive Surgery (2020)
101. Standard, E.: Robots Offer a Contact-Free Way of Getting Swabbed for Coronavirus (2020)
102. Khan, H., et al.: Smart technologies driven approaches to tackle COVID-19 pandemic: a review. 3 Biotech **11**(2), 1–22 (2021)
103. Singh, J.: Doha Hamad airport launches coronavirus helmets and disinfection robots-simple flying. Simple Flying (2020)
104. Shen, Y., et al.: Robots under COVID-19 pandemic: a comprehensive survey. Ieee Access **9**, 1590–1615 (2020)
105. Gao, A., et al.: Progress in robotics for combating infectious diseases. Science Robotics **6**(52), eabf1462 (2021)
106. Yang, G.-Z., et al.: Combating COVID-19—The Role of Robotics in Managing Public Health And Infectious Diseases (2020)
107. Murphy, R.R., Gandudi, V.B.M., Adams, J.: Applications of robots for COVID-19 response. In: *arXiv preprint* arXiv:2008.06976 (2020)
108. Song, K., et al.: Conjuncted pyro-piezoelectric effect for self-powered simultaneous temperature and pressure sensing. Advanced Materials **31**(36), 1902831 (2019)

Soft Robotic Industrial Systems

Ramses V. Martinez[(✉)]

School of Industrial Engineering and Weldon School of Biomedical Engineering, Purdue University, Grissom Hall (GRIS) Rm. 284 315 N. Grant Street, West Lafayette, IN 47907-2023, USA
rmartinez@purdue.edu
https://engineering.purdue.edu/FlexiLab/

Abstract. Future industrial systems require humans and robots to work together safely and intuitively to optimize the production of goods. As a team, humans provide skills, experience, and knowledge, while robots offer physical assistance and take care of the performance of repetitive tasks at high speed. While this human-robot collaboration can minimize companies' response time to changes in supply or demand, several safety concerns need to be addressed before human operators can safely work near industrial robots. Soft robots, fabricated using elastic and compliant materials, have been developed to become intrinsically safe to human co-workers and to provide different manipulation approaches that exploit the deformability of the robot to enhance robotic dexterity while interacting with delicate and brittle materials. These soft robotic systems, however, required the development of new flexible sensors and control methods to achieve the accuracy of existing industrial robots. This chapter describes the development of soft robotic industrial systems. First, the design principles and manufacturing methods to create dexterous soft robots will be reviewed. Next, the main modeling and control methods used to implement soft robots in industrial environments will be described. Finally, current soft robotics challenges and emerging industrial applications for soft cyber-physical systems will be analyzed.

Keywords: Soft Robotics · Human-in-the-loop · Cyber-physical Systems · Human-Robot Interaction · Collaborative Robots · Manufacturing

1 Introduction

The field of robotics has significantly impacted the industrial and manufacturing sectors thanks to its capability to automate complex fabrication and handling processes, homogenizing and maximizing product quality. Unfortunately, most industrial environments and manufacturing plants are still in need of robotic systems able to operate safely in the vicinity of humans and adapt their manipulation strategies to new interactions with both rigid and delicate materials.

Recent trends in online product customization and the shortage of workers with the skills necessary to supervise and maintain robots have added even more pressure to current production systems, which struggle to meet the ever-expanding demand for products

C.-Y. Huang and S. W. Yoon (Eds.): ICPR1 2021, ACES 14, pp. 404–422, 2023.
https://doi.org/10.1007/978-3-031-44373-2_24

that meet the specific needs of the consumer. Manufacturing companies with *humans-in-the-loop*—human operators capable of supervising, maintaining, and re-programming robots during the production process—have demonstrated high flexibility in accommodating changes in demand and material supplies. Having a human-in-the-loop also allows companies to benefit from the rapid adaptability of dynamic automation processes, which could lead to a larger variety of product variants without significant cost increases. Unfortunately, current industrial robots are fabricated using hard components to increase fatigue resistance and minimize misalignment errors. These hard robots are commonly programmed to operate in highly structured environments that do not welcome the unpredictable motions of human workers. This incompatibility between hard robots and humans has negatively affected the adoption of robots in industrial production systems. To illustrate this, while one common prejudice against manufacturing robots is that they are invented to take away human jobs, the reality is that only between 5% and 40% of the manufacturing assembly tasks performed in industrial environments are carried out by robots [1]. This indicates that industrial systems still use robots as if they were individual machines that accomplish one isolated task, making robots highly dependent on humans to meet the production demands.

Therefore, human-robot collaboration is key to enhancing the flexibility and adaptability of industrial systems to varying constraints and changes in demand. Robotic collaborators, or *co-robots*, will also maximize the efficiency of human workers, allowing them to perform challenging manufacturing tasks. Additionally, the ubiquitous presence of co-robots in manufacturing plants could minimize hazards by eliminating the need for human workers to perform repetitive tasks and manipulate loads, reducing ergonomic risk factors. Furthermore, due to the relatively low cost of co-robots when compared with robotic systems with a high payload capacity, co-robotics could become a cost-effective solution for small and medium manufacturing companies aiming to maximize their competitiveness.

1.1 Soft Robotics: Towards Safe Human-Robot Collaboration

The potential benefits of human-robot collaboration led to the development of *collaborative robotics*. This research field explores multiple approaches to enhance human-robot interactivity so that the safety fences often used to separate human operators from working robots will no longer be necessary. Traditional industrial robots are robust, precise, and capable of automating complex tasks at high speed. Unfortunately, their rigid bodies often carry enough inertia to cause harm upon collision with humans, who, while doing their work, inadvertently intersect with their trajectories.

Several computer vision approaches have been developed to identify human intent and rapidly modify robotic motion to avoid unexpected obstacles and prevent crashes [2–4]. The advances in electronic sensors—sensors that encode physical parameters into electronic signals—made it possible for small and low-cost microcontrollers to monitor a large variety of environmental changes [5]. Moreover, the miniaturization of electronic sensors has facilitated their distribution along the length and end effectors of robotic arms commonly used in industrial environments [6, 7]. These sensors provide perception algorithms with the physical data necessary to identify the current configuration of the robotic arm (proprioception) and the contact with the objects to manipulate (exteroception) [8, 9]. Following this approach, robots equipped with force sensors across their structure and torque sensors in their joints have proven effectively operate in unknown environments [10]. Unfortunately, the high cost of these robotic systems makes it difficult for small and medium-sized companies to include robots in their production chain [4].

A novel approach to circumvent the safety problems inherent to traditional industrial robots is to depart from the use of rigid components in their fabrication and create *soft robots*. Soft robots are fabricated using soft and stretchable polymeric bodies that deform upon actuation generating a robotic motion that imitates the natural movements of soft-bodied animals like the octopus or the squid [11, 12]. Soft robots are inherently safe during human-robot interaction, as their actuation mechanism does not rely on rigid skeletons or hard moving parts. Their soft structure allows soft robots to deform upon accidental collisions with humans or fragile objects, absorbing the energy of the impact. Their intrinsic safety makes soft robots ideal as co-robots, demonstrating efficient interaction with other workers [13, 14], appropriate assistance in rehabilitation tasks [15, 16], and even provide elderly care [17, 18]. Moreover, the safety and continuum motion of soft robots has also generated considerable interest in the field of medicine, leading to the exploration of the use of soft robots as implantable artificial organs [19], surgical tools [20], and wearable assistive devices [21].

1.2 Enhancing Industrial Competitiveness Using Soft Robots

Pneumatics, hydraulics, and motor-tendon are the actuation mechanisms most commonly used in soft robotics as they all allow for the distribution of forces across the soft structure of the robot, which deforms accordingly. Upon actuation, the compliant surface of soft robots generates continuum robotic motion, which excels at gripping delicate objects, as the grasping forces exerted by the robot get uniformly distributed over the surface of the manipulated object. Dexterous soft robotic manipulation has many industrial applications, particularly in manipulating bendable and delicate objects [13].

Soft robots, due to their compliant structures, are often more resistant to damage than their rigid counterparts. For example, soft robotic tentacles, quadrupeds, and grippers have demonstrated impressive impact resistance by being able to operate normally after withstanding hammering, high-height falls, and even being run over by a car [22]. Furthermore, thanks to the chemical stability of elastomeric composites, soft robots also exhibit significant resistance to degradation during the manipulation of acids and other harsh chemicals.

Due to their deformability, impact resistance, and chemical stability, soft robots could reduce the existing operation constraints of industrial robots. For example, soft robots could help to depart from the common requirement of highly structured environments for industrial robots, allowing companies to become more dynamic by facilitating the rapid implementation of changes in the manufacturing of their products. Similarly, companies with more dynamic industrial environments and human workers benefiting from their safe collaborations with robots will significantly increase their competitiveness, leading to the next industrial revolution.

2 Soft Robotic Design Principles and Manufacturing Approaches

Nature has served roboticists as a source of Inspiration for the design of soft robots [23]. Taking plants and animals as an example of mechanical adaptability and multifunctionality, researchers found that the most critical parameters to control the motion of a soft robot are its design and the elasticity and viscoelasticity of the soft materials used in their fabrication [11]. Applying the latest advances in polymer science and design rules that combine bio-inspiration and architecture, the field of soft robotics has been able to provide a variety of robots and actuators capable of exhibiting excellent gripping abilities, on dry land and underwater maneuverability, and even flying capabilities [24].

2.1 Material Selection for Soft Robots

Silicone elastomers, soft polyurethanes, and rubber composites are the most common materials used in the fabrication of soft robots. During the development of a new soft robot, it is necessary to coordinate its desired motions and behaviors with the material properties of its soft body. Following this approach, if we are interested in the design of soft robotic actuators with high elastic hysteresis and toughness, we should choose polyurethanes and polyacrylates. Similarly, if we want to use electric fields to produce large strains in the soft robot, VHB and silicon-based dielectric elastomers actuators (DEA) should be our choice. Soft robots designed to interact with humans or brittle materials will benefit from the use of low-viscosity elastomers (such as silicone, latex, or natural rubber) or hydrogels in their fabrication. Finally, a critical criterion when choosing elastomeric materials is their resistance to fatigue. Upon fatigue testing (the performance of multiple actuation cycles at a certain temperature), soft robots fabricated with tougher materials outperform those fabricated with soft elastomers characterized by a non-linear elastic behavior. To enhance the toughness of soft robots without compromising their rigidity, a variety of elastomeric composites have been proposed [25]. These composites are fabricated by embedding fibers, textiles, and other bendable but tough materials into the elastomeric body of the soft robots. The embedded components of the composite prevent the overstretching of the elastomer, imparting these robots with high fatigue resistance and even crack and puncture resistance. Additionally, depending on the task that the soft robot will accomplish, we will have different work energy density requirements. Typically, soft robots performing high-energy tasks benefit from materials with high energy density or a high capacity to store actuation energy. Figure 1 shows that, while there is not a purely linear relationship between work energy density and stiffness (characterized by the Young's modulus), the lower the stiffness of a material, the more energy this material can store before it breaks.

The storing of elastic energy in their soft structure allows soft robots to achieve energy efficiencies greater than those of traditional rigid robots [26]. The capability of soft robots to store elastic energy depends on the storage modulus (G') and the low modulus (G") of the material used in the fabrication of the soft robot. The storage modulus can be defined as the ratio between the stress applied to the material and its strain. The low modulus of elastomeric materials can be understood as the amount of force required to stretch them. Figure 2, provides a comparison between the soft materials used in the fabrication of soft robots and soft organic materials such as fat, brain, or muscle tissue.

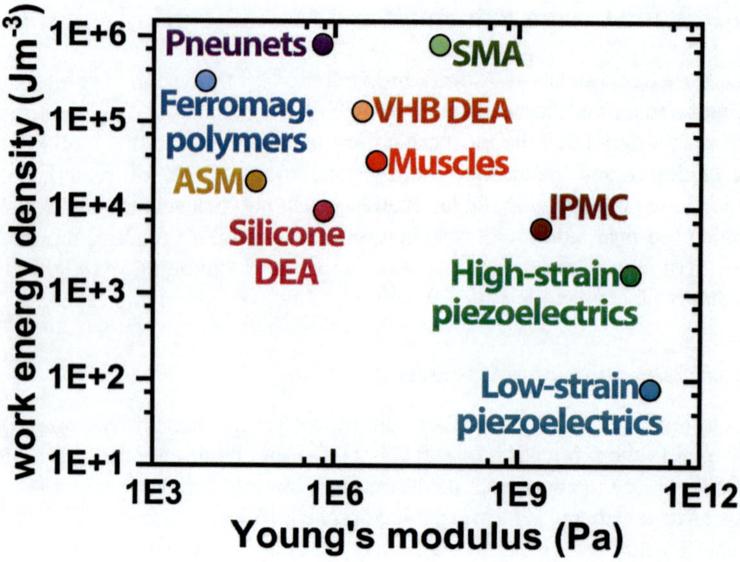

Fig. 1. Relationship between the work energy density and the Young's modulus of the most common soft robotic actuation methods. Actuation methods and materials: Pneunet—pneumatic networks; Ferromag.—ferromagnetic polymers; SMA—shape memory alloy; ASM—Architected Soft Machines; DEA—dielectric elastomer; IPMC—ionic polymer-metal composite.

Popular polymer choices in the construction of soft robots include polydimethylsiloxane (PDMS), Ecoflex (00-10 and 00-30), and Elastosil (M-4601), following bioinspiration from the low mechanical impedance of skin and other soft organic tissues.

2.2 Bio-Inspired Design of Soft Robots

A large variety of bio-inspired designs have been used to allow soft robots to mimic the actuation of natural organisms. Copying muscle distributions in vertebrates and invertebrates, several soft robots were able to mimic walking, gripping, and swimming motions [24]. The distribution of soft materials with different stiffness across their structure allowed soft robots to minimize stress concentration on their interface with their user or the object to manipulate.

The resulting motion of the soft robot upon actuation can be programmed by embedding strain-limiting layers into their structure. These strain-limiting materials (flexible but not stretchable meshes, fabrics, braided structures . . .) only redirect the deformation of the soft robot upon actuation in the desired directions of motion. The orientation of these strain-limiting materials allows them to mimic the complex 3D motions that include: contraction, extension, twisting, and bending [25].

2.3 Computer-Aided Optimization of Soft Robotic Actuators

Traditionally, roboticists used intuition, bio-inspiration, and other heuristic methods to create the design of the soft robot. Then, after the soft robot was created, its design was

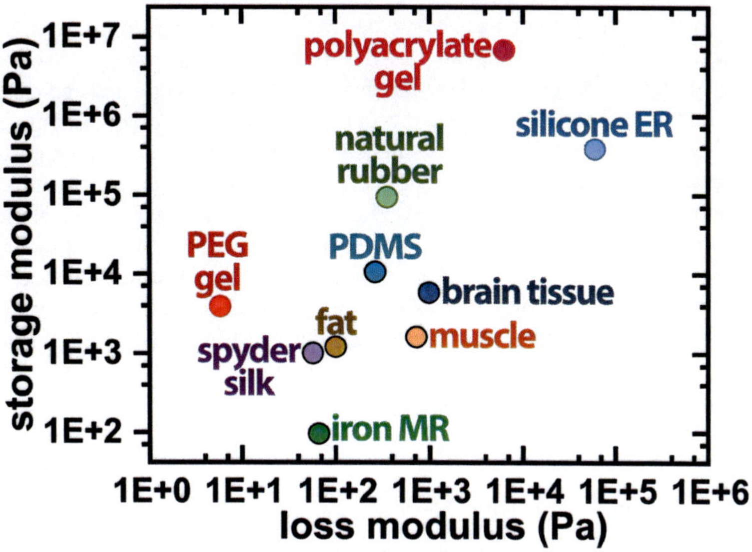

Fig. 2. Approximate dependence of the loss modulus on the storage modulus of several materials commonly used to fabricate soft robots. Materials: Polyethylene glycol (PEG) hydrogel, silicone-based electrorheological (ER), magnetorheological (MR) iron-based composites, silicone-based dielectric elastomers (DEA). Biological materials such as fat, muscle, and brain tissue are provided as a comparison.

improved by trial and error across multiple versions. These heuristic approaches often result in soft robots with motions or actuations that are too specific for a particular task, requiring the addition or removal of multiple components to achieve multifunctionality [27].

Computer-aided design (CAD) has allowed the automatic generation of soft robotic designs that can be evaluated, in simulations, using finite element analysis (FEA) and evolutionary algorithms. For example, Goswami *et al.* [28] demonstrated how, by downloading a CAD model of a human hand from the internet and painting over the regions that should bend upon actuation, a tessellation algorithm automatically transforms the CAD model into an architected soft machine (ASM) with the desired motion. This tessellation algorithm modifies the resolution of the original mesh of the CAD model so that, after a Voronoi algorithm is applied, new polygonal Voronoi cells of different sizes are created. A tendon cable between the tip of each finger and an individual servo motor placed on the wrist of these soft hand actuators allows for the actuation of each finger. The tensile forces distributed over the architecture of these soft machines upon tendon-based actuation lead to the bending of its cellular structure according to the relative size of the Voronoi cells [28]. Tessellation algorithms can be coupled with evolutionary soft robotic algorithms, which efficiently simulate sequential actuations such as gripping or walking and assign them a score in terms of their speed and energy efficiency. Following this approach, roboticists only need to create a "*seed*" design, which is automatically tessellated and simulated as part of a multi-step evolutionary process that is guided by the optimization of the performance of the desired motion and the minimization of the energy resources and time to complete it. At the end of this computer-based evolutionary process, the user obtains an optimal design that is ready to be fabricated and tested.

The automation of soft robotic designs not only saves a considerable amount of time but also allows researchers to discover counter-intuitive methods of actuation that do not occur in natural organisms. For example, Pal *et al.* [24] demonstrated how bi-directional twisting upon actuation can be achieved by exploiting the mechanical instabilities of architected soft machines. This motion makes the structure twist in one direction first and then in the opposite one during a single tendon pull, allowing the encoding of complex joint-like motions into soft robots [30]. Future evolutionary design systems will incorporate more specific tools to declare the constitutive equations of the soft structure of the robot and its desired kinematic motion and constraints, allowing the optimization process to reach even more realistic designs using a reduced number of evolutionary iterations.

2.4 Manufacturing Soft Robotic Actuators

Elastomeric materials often require mixing a polymeric base material with a curing agent. Curing times required to complete the polymerization reaction span from a few minutes to 48 h, depending on the type of elastomer. This curing time affects the strategy used to manufacture soft robotic actuators.

The relatively high viscosity and long curing times of soft elastomers and silicone rubbers have promoted the use of a soft lithographic method called "*replica molding*" to fabricate soft robotic actuators (see Figs. 6a and 7a). Replica molding consists of casting an elastomeric prepolymer—uncured mixture of the polymer base and its curing agent—over a patterned surface that serves as a mold. Prior to the molding process, the bubbles typically generated during the mixing of the base polymer with the curing agent need to be removed by placing the prepolymer in a desiccator at reduced pressure. The removal of the bubbles in the prepolymer is often performed after the prepolymer is cast on the mold, particularly if the mold has a complex geometry. Avoiding the trapping of air bubbles within the prepolymer or between the prepolymer and the mold improves the quality of the replica molding process and homogenizes the elastic properties of the resulting soft actuator. Some polymerization reactions allow for the speed-up of the molding process by placing the mold filled with prepolymer into an oven with temperatures ranging 60–120 °C [30]. The application of external temperature as a source of energy to promote the polymerization of the elastomer allows for significantly reducing the curing time (from 4 h to 30 min in the case of Ecoflex 00-10), improving manufacturing speed. However, thermal retraction might cause the shrinkage of the elastomer during the curing process, which affects the elastic properties of the soft actuator and requires the redesign of the mold to compensate for shrinkage.

2.5 3D and 4D Printing Methods for Soft Robotics

Recent advances in additive manufacturing have expanded the palette of polymers that can be deposited in small volumes during a layer-by-layer printing process to create complex 3D printed parts [31]. The compatibility of additive manufacturing with some elastomers, silicone rubbers, hydrogels, and other soft polymers has allowed the construction of complex 3D soft actuators composed of polymers with different stiffness and even polymers with flexible 2D and 3D electrical interconnects [32]. The latest advances in photopolymerization have allowed the miniaturization of flexible soft actuators to sizes of a few micrometers. When combined with magnetic actuation using global magnetic fields, the small size of photopolymerized soft robots enables new medical applications of this technology [33] (Fig. 3).

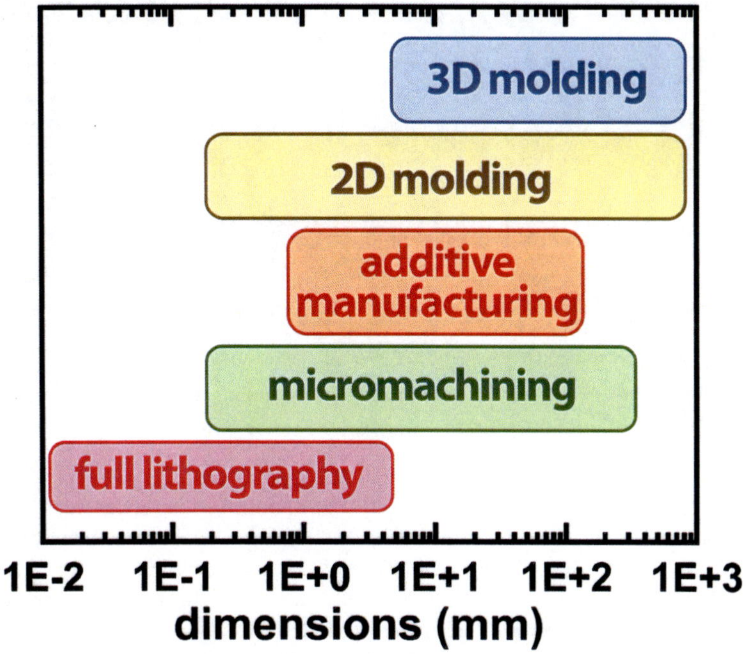

Fig. 3. Relationship between the work energy density and the Young's modulus of the most common soft robotic actuation methods. Materials: Pneunet—pneumatic networks; Ferromag.—ferromagnetic polymers; SMA—shape memory alloy; ASM—Architected Soft Machines; DEA—dielectric elastomer; IPMC—ionic polymer-metal composite.

Since photopolymerization leads to higher resolutions than standard fused deposition processes, a considerable interest in the field of material science has focused on the development of new photopolymers with different mechanical properties [34]. These research efforts have made flexible and even stretchable photopolymers available to stereolithographic 3D printers, which allow the fabrication of customized soft robotic actuators. Similarly, laser sintering 3D printing approaches have now different powdered polymers that can be sequentially sintered in a layer-by-layer approach to create soft robots. The possibility of rapidly manufacturing soft robots by simply 3D printing a CAD model maximizes the customizability of soft machines and enables their *in-situ* fabrication in remote locations to assist in disaster response situations [35].

Additive manufacturing also creates soft robots with more complex designs than those achievable by molding or machining methods (see Fig. 4). Moreover, by combining different 3Dprintable inks, soft robots benefit from graded stiffness and the possibility to encode the actuation of the soft robot in the 3D composition of its soft body [36]. Moreover, the layer-by-layer 3D printing process allows for the creation of cavities inside the robot that can be filled up with functional elements such as motors or sensors [37].

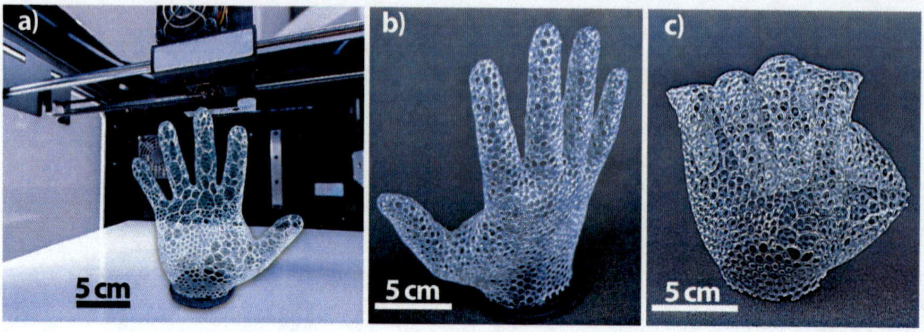

Fig. 4. 3D printing soft actuators. a) Architected soft robotic hand right after being 3D printed out of flexible polymer [29]. b) Image of the architected soft hand showing its Voronoi cellular structure, which reduces weight and makes the fingers to bend in a preferential direction upon tendon-based actuation. c) Soft robotic hand closed after the pulling the internal tendon wires (tied at the tip of each finger) from the wrist of the hand.

The adaptation of smart materials into additive manufacturing processes has allowed soft robotics to benefit from stimuli-sensitive materials capable of changing shape, chemical composition, and crystalline structure over time, creating new soft actuation approaches. Due to their time-varying properties, the use of these materials in additive manufacturing processes is called *4D printing*. Multimaterial 4D printing has demonstrated the creation of morphing soft robots capable of changing their shape upon changes in temperature or level of hydration. The recent application of gelatin and other protein-based compounds into 4D printing led to the development of completely biodegradable soft robots, opening a new alternative technology to minimize robotic waste [38].

3D and 4D printing approaches are still unable to achieve the vast range of mechanical properties accessible to molding manufacturing processes. Additionally, due to the need for supporting materials during the 3D printing or complex geometries with overhangs, 3D printed robots require the removal of those scaffolding materials before use. The removal of supporting material often requires machining or chemical etching, which have a deleterious effect on the soft structure of the soft actuator [31].

3 Soft Sensing Skins for Hard Robots

When compared with rigid robots used in industrial environments, soft robots are still far from achieving the robustness and reliability of rigid robotic arms. Industrial robotic applications requiring very high accuracy are, therefore, constrained to use commercially available hard robots, which have limited compliance and adaptability. Soft robotics have also explored the possibility to team up with hard robots through the design of soft sensing skins that can be mounted or wrapped around the hard robot to provide it with higher compliance and safer intractability.

Electronic skins (*e-skins*) are stretchable electronic circuits embedded in a polymeric slab. The flexibility of e-skins allows the electronic sensors to conform to the surface of the hard robot, providing it with an artificial tactile sense. Sensors for pressure, metallic contact, magnetic fields, and temperature can be easily embedded into elastomers for the development of e-skins. The combination of some of these sensors within the same e-skin has endowed hard robotic systems with a human-like tactile perception. For example, Pang *et al.* [39] developed a flexible and stretchable e-skin capable of providing YuMi robots with distributed exteroception across its surface. Such tactile perception, allows hard robots to identify collisions in unstructured environments, while the deformation of the soft skin dissipates some of the energy of the impact.

Following a bioinspired approach, e-skins are evolving into systems that provide information about the environment and reduce computational needs by automatically processing the tactile information sensed. Many biological organisms follow this approach too. For example, humans react to certain inputs with automatic muscular responses commonly known as *reflexes*. Similarly, thanks to the distribution of logic gates through the e-skin, hard robots do not require to constantly compute the distributed sensing of the e-skin and can simply rely on the computational power of the skin to decide when it is appropriate to influence the control system of the robot. This brain-free decision-making process is called *morphological computation*, a very active research field in soft robotics that aims to maximize self-adaptability. Soft robotics demonstrations of morphological computation include: smart grippers capable of automatically sorting objects by surface texture [40], morphing wings [41], and adaptive propulsion mechanisms for energy-efficient locomotion [42].

The development of *soft cyber-physical interfaces* has enabled natural interactions between operators and industrial robotic systems. These soft sensing interfaces can deform according to the action performed by the operator, providing intuitive haptic feedback [43]. Soft human-machine interfaces have demonstrated a more effective human control of industrial processes involving delicate materials and are rapidly finding new applications in the video game industry (with flexible and wearable sensors that can be used in augmented reality environments [44]) and the biomedical field (with the development of deformable instrumentation that ensures the safe interaction between surgeons and the tissue of the patient during surgery [20]).

4 Modeling Frameworks for Soft Robots and Soft Cyber-Physical Systems

The proliferation of soft robots and soft cyber-physical systems has been restrained by the lack of systematic frameworks capable of combining material properties, design constraints, and modeling tools. To address this limitation, several approaches have been explored to develop holistic methodologies capable of automating the design and modeling-based control of soft robots. To be effective these frameworks need to unite low-level material composition with high-level robotic design. Two main approaches are followed to model the actuation of deformable robots: Finite Element Analysis (FEA) and kinematic modeling.

4.1 Finite Element Analysis (FEA) of Soft Actuators

The combination of three-dimensional structural solvers and CAD designs led to the accurate simulation of the effect of loads on different architectures [45]. These 3D modeling tools usually divide the CAD model into a 3D mesh comprising a large number of **finite elements** of small size, whose position and deformation are provided by the solver. This Finite Element Analysis (FEA) leads to very accurate solutions, even when the soft robot has a very complex geometry or is exposed to multiple loads that induce highly non-linear deformations [46], see Fig. 5a, b.

Table 1 summarizes the most common FEA methods used to model soft robotic actuation. These methods use different formulations to model the non-linear and incompressible response of elastomers and silicone rubbers exposed to quasi-static loading forces. Most of these formulations are applicable to any elastic material so long their constitutive parameters are experimentally obtained by uniaxial tensile testing [12, 26].

Unfortunately, the accuracy of FEA is proportional to its mesh size and large numbers of finite elements require considerable computational power to be solved. Therefore, the distributed and continuum bending of soft robots leads to cumbersome and time-consuming finite element analysis to achieve accurate results. The computational burden of FEA hinders the use of finite element modeling for the development of real-time actuation predictors [12, 29, 46–54]

Table 1. Finite Elements Analysis (FEA) methods commonly used to model the elastic and viscoelastic behaviors of soft robotic actuators.

Software Used	Material Model	Refs.
MSC Marc	2nd-order Ogden	[42]
	3nd-order Ogden	[43]
	3nd-order Rivlin	[44]
Comsol Multiphysics	Hookean	[45]
	Neo-Hookean	[46]
	2nd-order Ogden	[47]
	Mooney-Rivlin	[4]
Abaqus	3th-order Ogden	[27]
	Yeoh	[48]
Ansys	Hookean	[49]
	Neo-Hookean	[50]
Own Development	4th-order Rivlin	[50]

4.2 Kinematic Modeling of Soft Actuators

To circumvent the limitations of FEA, several kinematic modeling tools have been developed to accelerate the simulation of soft robotic actuators. Kinematic modeling approaches aim to simplify the calculation of the non-linear deformations of soft actuators by representing their elastomeric body as an arrangement of beams and joints that describe different curvatures (see Fig. 5d). Industrial kinematic modeling techniques can be separated into two complementary mappings: i) a mapping that connects the actuator space with the configuration space (also known as shape prediction); ii) a mapping that connects the configuration space with the task space (also known as dexterity).

4.2.1 Piecewise Constant Curvature (PCC) Kinematic Modeling

One of the first kinematic modeling approaches used to predict the time-dependent curvature of soft robotic actuators was the piecewise constant curvature (PCC) approach [55]. The main advantage of the PCC method relies on its modularity, simplicity, and rapid calculation. In contrast with the complex integral functions used by FEA to compute the curvature variations across the different parts of the soft actuator, PCC is only applied to the bending actuators of the soft robot, and their curvature is described as a finite number of links describing a curve (see Fig. 5d). This simplification makes PCC much faster to compute than FEA but not as accurate. Comparing the results of the PCC kinematic modeling with experimental results allows for the optimization of the length of the links used to describe the curvature of the moving parts of the soft robot (typically limbs and fingers), achieving a compromise between the number of links and the accuracy of the simulation.

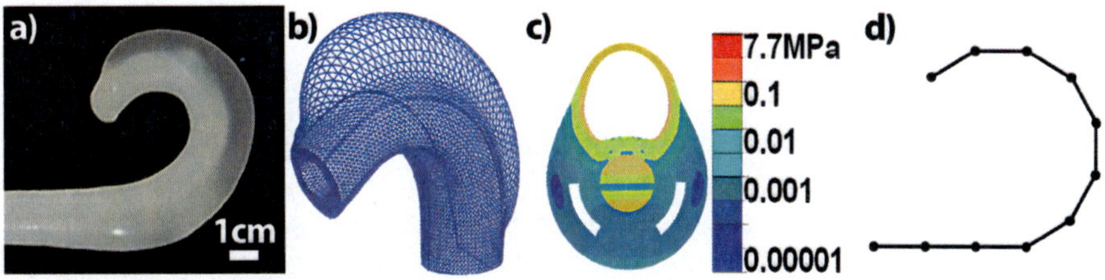

Fig. 5. Typical modeling techniques used to predict the motion of soft robotic actuators. a) Image of a soft pneumatic actuator exhibiting continuous bending upon the pressurization of one of its internal pneumatic channels. b) Finite element analysis (FEA) predicting the shape upon deformation of the actuator. c) FEA prediction of the crossectional deformation and distribution of stresses on the soft actuator upon pressurization. d) Prediction of the final shape of the actuator using a piecewise constant curvature (PCC) kinematic model.

4.2.2 Kinematic Modeling of Soft Actuators Using Artificial Neural Networks

The compromise between the number of interconnected links describing the soft actuator and the accuracy of the model makes kinematic modeling share limitations with FEA (remember that the accuracy of FEA depends on the number of finite elements used to model the soft actuator).

The rapid expansion of algorithms to interact with *big data* and rapidly identify trends and even predict sequences has been applied to soft robotic modeling. For example, to further accelerate kinematic modeling and enable the real-time motion prediction of soft robotic actuators, machine learning methods have been used to rapidly interpolate between accurate simulation and experimental conditions. Long-term, short-term, memory (LSTM) recurrent neural networks are commonly used for the kinematic modeling and control of soft robots. LSTM networks use optimization algorithms that need to be trained using a combination of experimental images of the soft actuator and their corresponding power inputs and environmental constraints. Then, after their time-consuming training, LSTMs can run quite rapidly and enable real-time prediction of the non-linearities of the kinetics and the kinematics of soft actuators.

Similarly, reinforced learning and neural networks of different depths have been demonstrated to accurately model the actuation of soft robotic actuators in real-time, which allow for the optimal utilization of soft robotic actuators during the performance of complex industrial tasks.

5 Industrial Advantages of Soft Robotic Actuators

The combination of flexible materials, stretchable sensing elements, an adequate robotic design, and fast modeling techniques enables the development of soft robotic actuators capable of outperforming their rigid counterparts in the performance of some industrial processes.

5.1 Dexterous Soft Robotic Manipulation

Soft robotic grippers are among the most desired soft robotic actuators. These purely elastomeric grippers are composed of different numbers of fingers, which bend in a continuous manner upon actuation (see Fig. 6b). The main advantage of soft robotic grippers when compared with rigid grippers is the fact that the bending actuation of the soft fingers translates into a conformal wrapping of the grasped object due to the lack of rigid skeletons of soft grippers [25]. This wrapping-based gripping strategy allows soft grippers to distribute the gripping force over the surface of the object, avoiding localized pressure points that can damage brittle objects.

Figure 6 shows a soft robotic gripper that mimics the bending motion of a human hand. This gripper has an internal pneumatic channel that, upon pressurization, makes the fingers bend over the palm side of the gripper, grasping objects of arbitrary shape with ease. The manipulation of delicate elements, such as eggs, is a highly desirable functionality expected in industrial environments (see Fig. 6b). This skill, thanks to the real-time modeling of the soft fingers using PCC, allows the control system of the soft robot to find the best strategy to grasp a target object according to its shape and texture. Note that, while traditional hard robots can be equipped with soft skins, their internal hard skeletons prevail and causes the pressure exerted over the gripped objects to be localized on the contact areas. Soft robotic grippers, therefore, outperform hard gripping actuation of the same size, as they are capable to provide a safe manipulation of the target object and even operate underwater for long periods (see Fig. 6c).

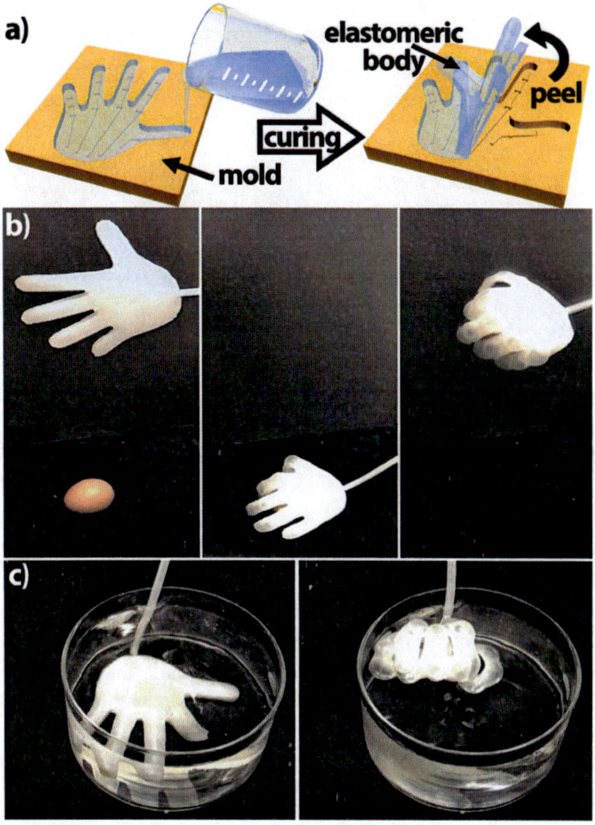

Fig. 6. Conformal manipulation of delicate objects using soft grippers. a) The soft gripper is designed in CAD and its negative is 3D printed and used as a mold for the liquid elastomer (prepolymer). The polymerization of the elastomer (curing) allows the simple peeling of the soft elastomeric body of the actuator (identical to the original CAD). b) After sealing the elastomeric body against a flexible strain-limiting layer (on the palm side), the actuator can deform continuously and wrap around the arbitrary shape of the object to grasp. Since there are no internal hard components in this actuator, the integrity of the egg is not compromised during the gripping process as the pressure is evenly distributed over the surface of the egg. c) The lack of internal circuitry makes this soft gripper to efficiently operate underwater.

5.2 High-Speed Soft Robots

When compared with industrial hard robots, soft robots exhibit relatively low actuation speeds. Their slow actuation is a consequence of the slow elastic recovery of their elastomeric body, as soft actuators require their constitutive material to roll back into their original shape after the actuation input dissipates (see Fig. 7c). While the time require for soft robots to actuate (response time) can be quite slow, their slow elastic recovery increases significantly their actuation cycle. This slow actuation cycle cannot compete with the reliable and accurate actuation of the fast electronic servo motors powering industrial hard robots. To mitigate this intrinsic (material-based) limitation of soft robots, several bioinspired approaches have aimed to exploit the fast movements exhibited by animals and plants in nature.

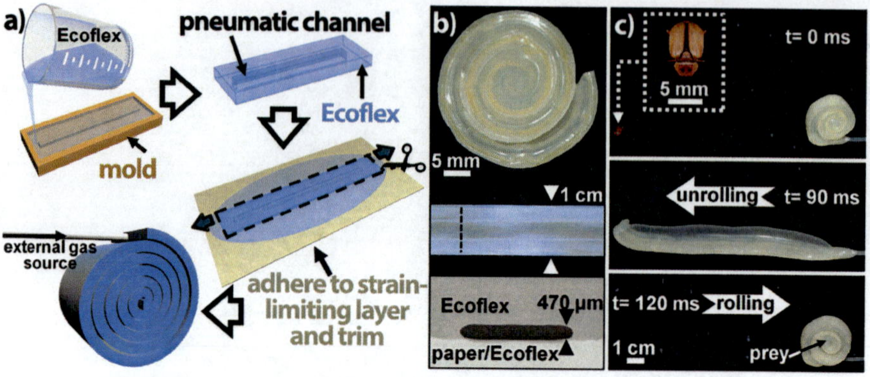

Fig. 7. High speed soft robotic actuation exploiting elastic energy storage. a) Fabrication of a simple soft robotic actuator with stored elastic energy. This actuator is manufactured using replica molding, then stretched (application of elastic energy), and then sealed in its stretched form against the strain-limiting layer that dictates its motion upon actuation. After this actuator expands upon pressurization, it can rapidly go back to its curled relaxed state using the elastic energy stored in its elastomeric body [27]. b) Images of the curled actuator (top), the longitudinal view of its internal pneumatic channel (middle), and the crossectional view of the pneumatic channel (bottom). c) Sequence of images showing how this soft actuator is able of catching an alive beetle, without killing it, in less than a quarter of a second.

For example, viper snakes can efficiently store energy in their muscles. This stored elastic energy can be suddenly released, on-demand, in order to generate the high-speed motion necessary to strike unsuspecting prey. The red kangaroo, as another example, stores elastic energy on the tendons of its legs. This elastic energy is sequentially stored and released during the jumping gait of the kangaroo, enhancing its locomotion energy efficiency. The ultimate example of elastic energy storage, however, can be found in grasshoppers and Venus flytraps, which have developed natural catapult mechanisms that enable impressive motion amplification [24].

Soft robots, alike the above-mentioned natural organisms, are capable of storing elastic energy into their structures and releasing that elastic energy during actuation or during their elastic recovery, which significantly enhances their actuation speed and energy efficiency, while maintaining their intrinsic interaction safety (see Fig. 7c).

6 Conclusions

The field of industrial robotics has recognized soft robotics as a promising approach to expand robotic motion, gain manipulation dexterity, and enhance safety in human-robot interactions. Soft robots benefit from the latest developments in biocompatible and biodegradable polymers, making soft robotics an ideal environmentally friendly approach to creating robots. Soft robots, however, require the development of new design frameworks capable of exploiting the mechanical responses of new polymers and elastomers and choosing the most appropriate architecture for the soft robot according to its final motion or function. While bioinspiration has served roboticists in the past as a heuristic method to guide the design of soft actuators, evolutionary algorithms have been demonstrated to significantly speed up the soft robotic design process by optimizing the design according to the desired functionality. Future evolutionary algorithms, empowered by artificial intelligence, will be able to automatically generate the design of the optimal soft actuators required to achieve different tasks, which will significantly expand the current use of soft robots.

Recent advances in material science and manufacturing processes have enabled the creation of soft actuators with a variety of mechanical properties and capacities to delicately manipulate different objects. In particular, recent advances in additive manufacturing have demonstrated that soft actuators do not require highly complex manufacturing systems (as hard robots do), but can be simply 3D printed using elastomeric inks. The deformable body of soft robots, however, can also be used to host electronic components and sensors, which expand the functionality of soft actuators. Furthermore, soft and compliant e-skins can be wrapped around hard robots to provide robots with artificial tactile sensing capabilities that enhance their manipulation of objects. Similarly, e-skins allow the creation of soft cyber-physical systems that serve as safe interfaces for human operators, who benefit from the intuitive haptic feedback provided by the e-skin. Moreover, the 3D molding of new soft materials with graded stiffness with multiple embedded electronic sensors will lead to the generation of soft robots with distributed sensing and control systems, which will allow for the creation of highly functional and realistic prosthetics and human-like robots. Artificial intelligence will also get a fundamental role in the prediction of the shape of the robot upon actuation, facilitating soft robotic control and optimizing human-robot interactions. The development of biodegradable polymers will lead to new applications of soft robotics in healthcare, including implantable robotic actuators assisting biological organs and a variety of biomedical instrumentation that will facilitate surgeries and enhance healthcare outcomes.

Industrial applications of soft robots are rapidly expanding, particularly in manufacturing plants requiring robots to work in close proximity to humans or manipulate brittle materials. New multi-finger grippers cappable to grasp objects of high complexity without crashing them will prove essential in tasks such as manufacturing and packaging. However, current research in soft robotics prioritizes getting soft robots to match the robustness and accuracy of hard industrial robots. Future industrial soft robots will be able to perform dexterous robotic manipulation at high-speed, which will decrease manufacturing times and cost. The possibility of soft actuators containing embedded logic gates along their structure and storing elastic energy in their limbs, allows soft robotics to explore capabilities such as embodied computation and energy harvesting, offering a promising path to the development of safer, eco-friendly, and energy-efficient industrial robots.

Declaration of Competing Interest. The author declares no competing financial interest.

References

1. Runge, G., Raatz, A.: A framework for the automated design and modelling of soft robotic systems. CIRP Ann. **66**(1), 9–12 (2017)
2. Hentout, A., Aouache, M., Maoudj, A., Akli, I.: Human– robot interaction in industrial collaborative robotics: a literature review of the decade 2008–2017. Adv. Robot. **33**(15–16), 764–799 (2019)
3. Berg, J., Lu, S.; Review of interfaces for industrial human-robot interaction. Curr. Robot. Reports **1**(2), 27–34 (2020)
4. Demir, K.A., Döven, G., Sezen, B.: Industry 5.0 and human-robot co-working. Procedia Compute. Sci. **158**, 688–695 (2019)
5. Shih, B., et al.: Electronic skins and machine learning for intelligent soft robots. Sci. Robot. **5**(41), eaaz9239 (2020)
6. Yogeswaran, N., et al.: New materials and advances in making electronic skin for interactive robots. Adv. Robot. **29**(21), 1359–1373 (2015)
7. Cheng, G., Dean-Leon, E., Bergner, F., Olvera, J.R.G., Leboutet, Q., Mittendorfer, P.: A comprehensive realization of robot skin: sensors, sensing, control, and applications. Proc. IEEE **107**(10), 2034–2051 (2019)
8. Rashid, F., Burns, D., Song, Y.S.: Sensing small interaction forces through proprioception. Sci. Reports **11**(1), 1–10 (2021)
9. Miki, T., Lee, J., Hwangbo, J., Wellhausen, L., Koltun, V., Hutter, M.: Learning robust perceptive locomotion for quadrupedal robots in the wild. Sci. Robot. **7**(62), eabk2822 (2022)
10. Lee, J., Hwangbo, J., Wellhausen, L., Koltun, V., Hutter, V.: Learning quadrupedal locomotion over challenging terrain. Sci. Robot. **5**(47), eabc5986 (2020)
11. Coyle, S., Majidi, C., LeDuc, P., Hsia, K.J.: Bio-inspired soft robotics: material selection, actuation, and design. Extreme Mech. Let. **22**, 51– 59 (2018)
12. Martinez, R.V., et al.: Robotic tentacles with threedimensional mobility based on flexible elastomers. Adv. Mater. **25**(2), 205–212 (2013)
13. Rossiter, J., Hauser, H.: Soft robotics—the next industrial revolution. IEEE Robot. Autom. Mag. **23**(3), 17–20 (2016)
14. Chen, S., Pang, Y., Cao, Y., Tan, X., Cao, C.: Soft robotic manipulation system capable of stiffness variation and dexterous operation for safe human–machine interactions. Adv. Mater. Technol. **6**(5), 2100084 (2021)
15. Polygerinos, P., Wang, Z., Galloway, K.C., Wood, R.J., Walsh, C.J.: Soft robotic glove for combined assistance and at-home rehabilitation. Robot. Auton. Syst. **73**, 135–143 (2015)
16. Awad, L.N., Esquenazi, A., Francisco, G.E., Nolan, K.J., Jayaraman, A.: The rewalk restore™ soft robotic exosuit: a multi-site clinical trial of the safety, reliability, and feasibility of exosuit-augmented post-stroke gait rehabilitation. J. Neuroeng. Rehabil. **17**(1), 1–11 (2020)
17. Dahl, T.S., Boulos, M.N.K.: Robots in health and social care: a complementary technology to home care and telehealthcare? Robotics **3**(1), 1–21 (2013)
18. Ansari, Y., Manti, M., Falotico, E., Mollard, Y., Cianchetti, M., Laschi, C.: Towards the development of a soft manipulator as an assistive robot for personal care of elderly people. Int. J. Adv. Robot. Syst. **14**(2), 1729881416687132 (2017)

19. Park, C., et al.: An organosynthetic dynamic heart model with enhanced biomimicry guided by cardiac diffusion tensor imaging. Sci. Robot. **5**(38), eaay9106 (2020)
20. Cianchetti, M., Laschi, C., Menciassi, A., Dario, P.: Biomedical applications of soft robotics. Nat. Rev. Mater. **3**(6), 143–153 (2018)
21. Koo, S.K.: Design factors and preferences in wearable soft robots for movement disabilities. Int. J. Cloth. Sci. Technol. (2018)
22. Martinez, R.V., Glavan, A.C., Keplinger, C., Oyetibo, A.I., Whitesides, G.M.: Soft actuators and robots that are resistant to mechanical damage. Adv. Function. Mater. **24**(20), 3003–3010 (2014)
23. Kovač, M.: The bioinspiration design paradigm: a perspective for soft robotics. Soft Rob. **1**(1), 28–37 (2014)
24. Pal, A., Restrepo, V., Goswami, D., Martinez, R.V.: Exploiting mechanical instabilities in soft robotics: control, sensing, and actuation. Adv. Mater. **33**(19), 2006939 (2021)
25. Martinez, R.V., Fish, C.R., Chen, X., Whitesides, G.M.: Elastomeric origami: programmable paper-elastomer composites as pneumatic actuators. Adv. Function. Mater. **22**(7), 1376–1384 (2012)
26. Pal, A., Goswami, D., Martinez, R.V.: Elastic energy storage enables rapid and programmable actuation in soft machines. Adv. Function. Mater. **30**(1), 1906603 (2020)
27. Kwok, S.W., et al.:. Magnetic assembly of soft robots with hard components. Adv. Function. Mater. **24**(15), 2180–2187 (2014)
28. Goswami, D., Liu, S., Pal, A., Silva, L.G., Martinez, R.V.: 3darchitected soft machines with topologically encoded motion. Adv. Function. Mater. **29**(24), 1808713 (2019)
29. Goswami, D., Zhang, Y., Liu, S., Abdalla, O.A., Zavattieri, P.D., Martinez, R.V.: Mechanical metamaterials with programmable compression-twist coupling. Smart Mater. Struct. **30**(1), 015005 (2020)
30. Schmitt, F., Piccin, O., Barbé, L., Bayle, B.: Soft robots manufacturing: a review. Front. Robot. AI **5**, 84 (2018)
31. Stano, G., Percoco, G.: Additive manufacturing aimed to soft robots fabrication: a review. Extreme Mech. Let. **42**, 101079 (2021)
32. Gul, J.Z., et al.: 3d printing for soft robotics–a review. Sci. Technol. Adv. Mater. **19**(1), 243–262 (2018)
33. Layani, M., Wang, X., Magdassi, S.: Novel materials for 3d printing by photopolymerization. Adv. Mater. **30**(41), 1706344 (2018)
34. Bagheri, A., Jin, J.: Photopolymerization in 3d printing. ACS Appl. Polymer Mater. **1**(4), 593–611 (2019)
35. Luo, M., et al.: Motion planning and iterative learning control of a modular soft robotic snake. Front. Robot. AI 191 (2020)
36. Skylar-Scott, M.A., Mueller, J., Visser, C.W., Lewis, J.A.: Voxelated soft matter via multimaterial multinozzle 3d printing. Nature **575**(7782), 330–335 (2019)
37. Valentine, A.D., et al.: Hybrid 3d printing of soft electronics. Adv. Mater. **29**(40), 1703817 (2017)
38. Baumgartner, M., et al. Resilient yet entirely degradable gelatin-based biogels for soft robots and electronics. Nat. Mater. **19**(10), 1102–1109 (2020)
39. Pang, G., et al.: Coboskin: soft robot skin with variable stiffness for safer human–robot collaboration. IEEE Trans. Industr. Electron. **68**(4), 3303–3314 (2020)
40. Roberts, P., Zadan, M., Majidi, C.: Soft tactile sensing skins for robotics. Curr. Robot. Reports **2**(3), 343–354 (2021)
41. Jenett, B., et al.: Digital morphing wing: active wing shaping concept using composite lattice-based cellular structures. Soft Robot. **4**(1), 33–48 (2017)
42. Shui, L., Zhu, L., Yang, Z., Liu, Y., Chen, X.: Energy efficiency of mobile soft robots. Soft Matter **13**(44), 8223–8233 (2017)

43. Dong, W., et al.: Soft human–machine interfaces: design, sensing and stimulation. Int. J. Intell. Robot. Appl. **2**(3), 313–338 (2018)
44. Jadhav, S., Majit, M.R.A., Shih, B., Schulze, J.P., Tolley, M.T.: Variable stiffness devices using fiber jamming for application in soft robotics and wearable haptics. Soft Robot. **9**(1), 173–186 (2022)
45. Grazioso, S., Di Gironimo, G., Siciliano, B.: A geometrically exact model for soft continuum robots: the finite element deformation space formulation. Soft Rob. **6**(6), 790–811 (2019)
46. Xavier, M.S., Fleming, A.J., Yong, Y.K.: Finite element modeling of soft fluidic actuators: overview and recent developments. Adv. Intell. Syst. **3**(2), 2000187 (2021)
47. Gorissen, B., Donose, R., Reynaerts, D., De Volder, M.: Flexible pneumatic micro-actuators: analysis and production. Procedia Eng. **25**, 681–684 (2011)
48. Dilibal, S., Sahin, H., Celik, Y.: Experimental and numerical analysis on the bending response of the geometrically gradient soft robotics actuator. Arch. Mech. **70**(5), 391–404 (2018)
49. Ding, L., et al.: Dynamic finite element modeling and simulation of soft robots. Chin. J. Mech. Eng.g **35**(1), 1–11 (2022)
50. Majidi, C.: Soft-matter engineering for soft robotics. Adv. Mater. Technol. **4**(2), 1800477 (2019)
51. Salem, L., Gat, A.D., Or, Y.: Fluid-driven traveling waves in soft robots. Soft Robot. (2022)
52. Hwang, Y., Paydar, O.H., Candler, R.N.: Pneumatic microfinger with balloon fins for linear motion using 3d printed molds. Sens. Actuat. A Phys. **234**, 65–71 (2015)
53. Ranzani, T., Cianchetti, M., Gerboni, G., Falco, I.D., Menciassi, A.: A soft modular manipulator for minimally invasive surgery: design and characterization of a single module. IEEE Trans. Robot. **32**(1), 187–200 (2016)
54. Gorissen, B., Reynaerts, D., Konishi, S., Yoshida, K., Kim, J.-W., De Volder, M.: Elastic inflatable actuators for soft robotic applications. Adv. Mater. **29**(43), 1604977 (2017)
55. Santina, C.D., Bicchi, A., Rus, D.: On an improved state parametrization for soft robots with piecewise constant curvature and its use in model based control. IEEE Robot. Autom. Let. **5**(2), 1001–1008 (2020)

Skill and Knowledge Sharing in Cyber-Augmented Collaborative Physical Work Systems with HUB-CI

Praditya Ajidarma and Shimon Y. Nof(✉)

PRISM Center, Purdue University, West Lafayette, IN, USA
{pajidarm,nof}@purdue.edu

Abstract. Recent advancements of manufacturing systems and supply networks towards Cyber-Augmented Collaborative Physical System (CCPS) necessitate skill and knowledge sharing model in a human-robot collaborative e-Work environment. Previous research on skill and knowledge sharing has mostly ignored the need to share knowledge with skills. It was also limited on how such helpful, collaborative augmentation can be enabled by the huge amount of data availability, and collaborative intelligence analytics. Such augmentation is further enabled by emerging, sophisticated computing resources, including machine learning, virtual/augmented reality, and hardware such as wearables, sensors, and IoT/IoS. This chapter aims to explore the state of the art of skill and knowledge sharing in manufacturing systems, and highlight the key areas and future research directions of the topic. A variety of case studies are also presented, particularly related to augmented reality and HUB-CI as the key enablers for skill and knowledge sharing.

Keywords: Augmented Reality · Collaborative Intelligence Augmentation · Cyber-Augmented Collaboration · Cyber Physical System · HUB-CI · Skill and Knowledge Sharing

Table of Acronyms and their definitions.

No	Abbreviation	Definition
1	AI	Artificial Intelligence
2	AR	Augmented Reality
3	ARS	Agricultural Robotics System
4	CAD	Computer-Aided Design
5	CC-Management	Cyber-Augmented Collaborative Management
6	CCPS	Cyber-Augmented Collaborative (e-Work with) Physical System

(*continued*)

C.-Y. Huang and S. W. Yoon (Eds.): ICPR1 2021, ACES 14, pp. 423–443, 2023.
https://doi.org/10.1007/978-3-031-44373-2_25

(*continued*)

No	Abbreviation	Definition
7	CCT	Collaborative Control Theory
8	CC-Work	Cyber-Augmented Collaborative Work
9	CNC	Computer Numerical Control
10	CPS	Cyber-Physical System
11	CRISP-DM	Cross-Industry Standard Process for Data Mining
12	CRP	Collaboration Requirement Planning
13	CTR	Collaborative Telerobotics
14	CSCD	Computer-Supported Collaborative Design
15	DCSP	Demand-Capacity Sharing Protocols
16	ERP	Enterprise resource planning
17	GUI	Graphical User Interface
18	HITL	Human In The Loop perspective
19	HITN	Human In The Network perspective
20	HRI	Human-Robot Interface
21	HUB-CI	HUB of Collaborative Intelligence, or a network of such HUBs
22	ICT	Information and Communications Technology
23	IoS	Internet of Services which leverage data from IoT/IIoT
24	IoT, IIoT	Internet of Thinks, also known as Industrial Internet of Things
25	TTC	Time to Complete (collaborative tasks executed by humans and robots)

1 Collaboration Automation in a Cyber-Augmented Collaborative Physical System (CCPS)

In the collaborative work and factories of the future (Moghaddam & Nof 2015, 2017) multiple systems, including humans as participants and as clients, are designed to work together, and cooperate towards accomplishing given objectives, collaboration and integration are necessary. With highly variable job and task requirements in modern work, and highly variable levels of preparedness of workers and robots under such conditions, a major concern of researchers of future work and factories has been: How to share skills and knowledge, as soon as needed, with workers and robots dynamically. CCT, the Collaborative Control Theory, aims to optimize such collaboration and integration. The purpose of this chapter is to address the sharing of skills and knowledge among the human participants.

What do we mean by skills and knowledge, and why it is necessary to share them online, just as needed and when they are needed? As an illustration, consider a baking work case: A baker needs *skills* of baking, e.g., best practice of ingredients preparation, measuring and mixing; and *knowledge*: details of the cake to bake, its ingredients, specifications of the mixer and oven available, and ongoing process status of the equipment and tools. While skill and knowledge sharing can be helpful to novice bakers, they become essential and mandatory for preparing and enabling human workers most effectively, when they work, or e-Work with a network of automation and robotics work agents.

Four layers of e-Work CPS, which we define as cyber-augmented collaborative work with physical systems, or for short, Cyber-Augmented Collaborative Physical Systems (CCPS), are defined as follows: 1. Cyber; 2. Physical items & systems; 3. Networking; and 4. CC-Work and CC-Management (CC: Cyber-Collaborative). Each layer is described as follows:

- **Cyber Layer**: An interdependent network of information systems infrastructures, including the Internet, telecommunications networks, computer systems, embedded processors, and controllers.
- **Physical Items and Systems**: The physical part of a CCPS includes sensors, actuators, radio-frequency modules for communication, and any other hardware to support CCPS function and provide the interface.
- **Networking**: Interconnectivity between computational elements (data repository, algorithms, AI) and computerized physical entities (CNC machines, robots, sensors) in the CCPS. Networking includes the IoT/IoS.
- **CC-Work and CC-Management**: Any task and management practices that are executed with cyber-augmented collaborative, or cyber-collaborative support means.

2 Skill in Cyber-collaborative Physical System

A widely acceptable taxonomy for objective assessment in the cognitive domain is termed Bloom's Taxonomy *(Bloom et al.* 1964*)*. The taxonomy is divided based on an ordinal scale of cognitive ability. The categories for the cognitive domain and illustrative action words for each level are presented in Table 1.

Different cognitive-based taxonomies have been developed following Bloom's Taxonomy. RECAP model (Imrie 1995) adopted Bloom's taxonomy and simplified it into a two-tier structure for both in-course and end-course assessments, related to two levels of learning, as presented in Table 2.

Table 1. The Definition and Action Words of Bloom's Taxonomy

No	Category	Definition	Illustrative Action Words
1	Knowledge	The ability to repeat information verbatim	to list; to state;
2	Comprehension	The ability to demonstrate understanding of terms and concepts	to explain; to interpret; to describe
3	Application	The ability to implement learned information to solve a problem	to calculate; to solve; to utilize; to execute
4	Analysis	The ability to dismantle a structure into its elements and formulate explanations based on a theory, or mathematical or logical models for a certain observed phenomenon	to derive; to explain; to interpret; to infer
5	Synthesis	The ability to create and combine elements with a high degree of novelty	to formulate; to make up; to design; to integrate
6	Evaluation	The ability to select and justify a set of selections from other alternatives	to determine; to select; to critique; to assess

Table 2. RECAP-based cognitive domain taxonomies

Tier	Skill Level	Bloom's Category	Assessment Level
1	Recall; Comprehension; Application	Knowledge; Comprehension; Application	Essential skills are assessed by objective-based, structured, short questions and answers survey
2	Problem-solving skills	Analysis; Synthesis; Evaluation	Advanced problem-solving skills are assessed by case-study questions and other criterion-referenced or norm-referenced assessments

For a cyber-physical manufacturing system, a taxonomy of job complexity, required skills, examples of role, and technical requirements has been developed (Krachtt 2019). Their taxonomy is presented in Table 3:

Table 3. Skill Taxonomy for a Cyber-Physical Manufacturing System

Job Complexity	Required Skill	Roles	Technical Knowledge Requirement
Entry Level	Resource management, social, and context skills	Order Pickers	AR devices, RFID, mobile ICT
		Sub Assemblers	ERP Systems, AR devices
		Data Clerks	Data Analytics in CPS
		Lift Operators	Material handling, RFID, Mobile ICT
		Manual Operator	AR Devices, RFID
Mid Level	Resource Management, Social, Content, Cognitive, and Technical skills	Robotics Operator	Automation, AI, AR devices, CPS, SMOs
		Materials Lead	Data Analytics in CPS, AI, AR, simulations
		Machine Operators	Automation, AR devices, SMOs, IIoT*, ICT
		Welder	Automation, IIoT*, AR, ICT
		Production Analyst	Data Analytics in CPS, SMOs, IIoT, CPS, ICT, simulations
Advanced Level	Resource Management, Social, Content, Cognitive, Technical, and Process, System skills	Automation Technician	Automation, AR devices, AI, CPS, SMOs, ICT
		Systems Tester	IIoT, IoT, AI, AR devices, CPS, ICT, SMOs
		Systems Integrator	CPS, IIoT*, AR devices, ICT, SMOs
		Machine Programmer	SMOs, IIoT*, CPS, ICT, big data

* We note that IIoT, Industrial Internet of Things, always requires an IoS, Internet of Services, that are designed to intelligently utilize the data and signals obtained by the IIoT; and international standards exist for both.

3 Case studies of Skill Sharing to Enable Collaboration Automation

3.1 AR-Enabled Skill and Knowledge Sharing

This section presents recent advances in augmented reality (AR), particularly on how it enables skill and knowledge sharing (see Acknowledgment). The first study (A. Villanueva et al. 2022), Collab-AR, is developed to facilitate and improve collaboration in Tangible AR (TAR) with a customized haptics feedback. In terms of time to complete the experiment: AR + Haptics ($M = 60.8$ min, $SD = 4.26$), Zoom + Physical Components ($M = 81.2$ min, $SD = 4.71$). The decrease in time (25.2%) was statistically significant between conditions ($p < 0.05$), due to the combined use of haptics and voice, as opposed to voice-only.

Another form of enabling technology is wearables. One study (Paredes et al. 2021) proposed a wearables taxonomy; a database of research, tutorials, aesthetic approaches, concepts, and patents; and CHIMERA, an online interface that provides visual and taxonomic connections to the growing database. The wearables taxonomy consists of categories, elements, and grouping types. There are 4 categories: function, fabrication, materials, and body zones; and 5 grouping types: research, tutorials, aesthetic approaches, concept designs, and patents. The database consists of 842 resources which are published between the year 2010 and 2020. CHIMERA is validated across three groups: 24 participants conducting a multidisciplinary design task, a group of wearable experts, and students in a wearable class. In instances where the CCPS requires a highly-specific form of collaborative skill and knowledge sharing, wearables customization becomes essential. One study (Paredes, Reddy, et al. 2021) developed FabHandWear as a device capable of creating customized, functional, and manufacturable hand wearables. The system allows a user to fabricate functional prototype of wearables without special machinery, clean rooms, or tools. The system is validated by conducting wearable devices development by inexperienced users. The participants reported a mean NASA TLX score of 47.5 ($SD = 15.083$), and a mean system usability score (SUS) of 70.42 ($SD = 16.61$), ensuring the FabHandWear's applicability.

In a skill and knowledge sharing instance, the primary role of AR is to augment the capabilities of a human in the loop. One notable research (T. Wang, Qian, He, Hu, et al. 2021), GesturAR, studied the taxonomy of human hand gestures as an input in AR, and processed into a hand interaction model which maps the gesture inputs to the reactions of the AR contents. The trigger-action AR allows visual programming and instantaneous results in AR. Five scenarios are developed to justify the proposed model: creating interactive objects, humanoid and robotic agents, augmenting in-door environment with tangible AR games, making immersive AR presentations, and interacting with entertaining virtual contents. The hand detection network accuracy and usability are evaluated as the performance metrics of the proposed design.

Skill and knowledge sharing also require object interaction and environmental manipulation. For instance, the integration of sensors, IoT devices, and human operators within a CCPS. One study (Chidambaram et al. 2021) proposed ProcessAR as an AR-based system capable of developing 2D/3D content that captures subject matter expert's (SMEs) environment object interactions in situ. ProcessAR locates and identifies different tools/objects through computer vision within the workspace when the author looks at them, and could be featured with 2D videos of detected objects and user-adaptive triggers. Compared to the baseline scenario, ProcessAR has a lower task time, better usability, particularly for novice users, and statistically significant reduction of the perceived workload both for expert and novice users.

In cases where object interaction and environmental manipulation occur on a physically small scale, one study (Adam et al. 2021) proposed a robust and multifunctional micromanipulation system with 3D micro-force sensing capabilities. In this system, multiple probes are actuated to achieve and simplify more complex manipulation tasks while providing force feedback to the user. A graphical user interface (GUI) was developed as a robust and comprehensive platform to intuitively control the entire system and its many capabilities. Furthermore, a VR system has been implemented to provide intuitive manipulation, and with the use of the force sensing probes, the user is able to select a maximum threshold force to keep the manipulation process safe. In order to validate its capabilities, several experiments were conducted: automatic contact detection, simple and complex caging applications (manipulation/assembly), and the test of VR capabilities. In terms of accuracy, caging manipulation has an error of 7.73% for polygonal parts and 8.78% for circular parts, in comparison with pushing application which has a 14.07% error.

Table 4 summarizes other projects related to AR-enabled skill and knowledge sharing for cyber-collaborative physical system:

Table 4. Summary of AR-enabled Skill and Knowledge Sharing in a CCPS

Title	Type of Augmented Reality	Metrics and Measurement	Features	Skill and Knowledge Modeling	Fields of Skill and Knowledge	Skill and Knowledge Sharing Instances
A Large-scale Annotated Mechanical Components Benchmark for Classification and Retrieval Tasks with Deep Neural Networks (Kim, Chi, et al. 2020)	Mechanical Components Benchmark (MCB) for annotating, defining, and benchmarking deep learning shape classifiers	mean accuracy over objects, average accuracy per class, F1-score and average precision (AP), and precision-recall curves	7 shape classification algorithms from point cloud, multi-view, and voxel grids 3D shape representations	The ability to view, annotate, classify and analyze the knowledge of mechanical components data	Computer vision	Could be implemented as AR-enabled knowledge-driven classification benchmark for mechanical parts
AdapTutAR: An Adaptive Tutoring System for Machine Tasks in Augmented Reality (G. Huang et al. 2021)	AR-based tutorial to better adapt to workers' diverse experiences and learning behaviors, with different levels of details (LoDs)	tutoring time, repeating times, testing time, and count of mistakes; user preference	Unity3D, backend server running web framework in Pythonbased on Tensorfow (v2.1) and SVM	Skills are detailed into step- enabled Avatar, animated component, step expectation, and subtask description	General; laser-cutting machine	Tutors perform tasks; tasks are decoded by AR; AR is equipped by novice operators

(*continued*)

Table 4. *(continued)*

Title	Type of Augmented Reality	Metrics and Measurement	Features	Skill and Knowledge Modeling	Fields of Skill and Knowledge	Skill and Knowledge Sharing Instances
First-Person View Hand Segmentation of Multi-Modal Hand Activity Video Dataset (Kim, Hu, et al. 2020)	Multi-modal video dataset generation based on hand thermal information	The mean Intersection over Union (IoU) between the two class based on manually-annotated labels	Modification of DeepLabV3 + with 3 modalities LWIR, RGB, and depth	Knowledge of left and right hands segmentation is based on "hands using tools" videos	Computer vision	Accurate and faster hand segmentation allows better hand tracking for operators
LightPaintAR: Assist Light Painting Photography with Augmented Reality (T. Wang, Qian, He, & Ramani 2021)	AR for spatial reference to enable precise light sources movement	user evaluation (SUS) on accuracy and overall experience	Hololens 2 spatial tracking function, Lume Cube LED, Canon EOS M6ii EF-M 11-22mm lens	The skill to light-paint the words "CHI 2021" using the LED light	General Motoric Skill	Could be implemented for vision-based light-signal detection
Object Synthesis by Learning Part Geometry with Surface and Volumetric Representations (Kim et al. 2021)	Part Geometry Network (PG-Net) to simulate realistic objects for a robust feature descriptor, object reconstruction, and classification	task convergence time, fitting time, and inference time; classification accuracy of PG-Net; reconstruction measures	TensorFlow deep learning framework on ModelNet datasets with a linear SVM for 3D classification benchmark	Knowledge is modeled as object synthesis based on AR-enabled multi-task and part geometry learning	3D object synthesis, and classification	Knowledge sharing could be implemented for AR-enabled CAD
RobotAR: An Augmented Reality Compatible Teleconsulting Robotics Toolkit for Augmented Makerspace Experiences (A. M. Villanueva et al. 2021)	AR for assessment, teaching, and learning	key competencies assessment and usability survey	phone-mounted robot platform; Unity 3D for the software	Skills are modeled as the ability to assemble electrical circuitry components kit	Electronics and circuitry	Skill sharing via AR and teleconsulting enables better students assessment and teaching
Towards modeling of human skilling for electrical circuitry using augmented reality applications (A. Villanueva et al. 2021)	AR-enabled assessment, teaching, and learning to implement an educational curriculum	the attainment of learning outcome of micro-skills	Micro-skills are aligned with the AR content using Q-matrix; they are classified into perceptual, cognitive, and motor types of skill	Skills are modeled as micro-skills, which are mapped into learning outcome	Electronics and circuitry	The model allows a feedback loop between the micro-skills, delivery method (full or partial), and learning outcomes attainment

(continued)

Table 4. (*continued*)

Title	Type of Augmented Reality	Metrics and Measurement	Features	Skill and Knowledge Modeling	Fields of Skill and Knowledge	Skill and Knowledge Sharing Instances
VRFromX: From Scanned Reality to Interactive Virtual Experience with Human-in-the-Loop (Ipsita et al. 2021)	Do-It-Yourself (DIY) platform to create interactive virtual experiences	time required to finish each task; System Usability Scale (SUS)	Unity Engine in C# pre-loaded back-end neural networks; object classification is achieved by a PointNet	Skills are the VR-based ability to retrieve object, model behavior of virtual objects, and interact (weld virtually)	virtual Metal Inert Gas (MIG) Welding simulator	VRFromX enables worker to train and simulate welding in-situ

3.2 HUB-Enabled Skill and Knowledge Sharing

Multi-agent skill and knowledge sharing becomes critical in work and factories of the future. With an increasing degree of automation, remote operation, maintenance, reorganization and reconfiguration become objectives of the human-automation-robot skill sharing augmentation initiative. This section reviews previous research on the usage of hubs for collaborative intelligence (HUB-CI) for enabling skill and knowledge sharing in a CCPS.

HUB-CI focuses on improving human collaboration through e-collaboration tools and services. It significantly enhances synthesis and integration of knowledge and discoveries, as well as their sharing and delivery in a timely manner (Seok & Nof 2011). Additionally, HUB-CI connects humans and robots for collaborative control of physical automation and assembly in manufacturing (Zhong & Nof 2013). Multiple HUB-CIs can operate in a hub-to hub and multi-hub collaborations involving multiple networks. Recent advances of HUB-CI aim to optimize information flow, based on the current activity, physiological state, attained information, and unique attributes of each worker. The design framework is presented in the following diagram (Fig. 1):

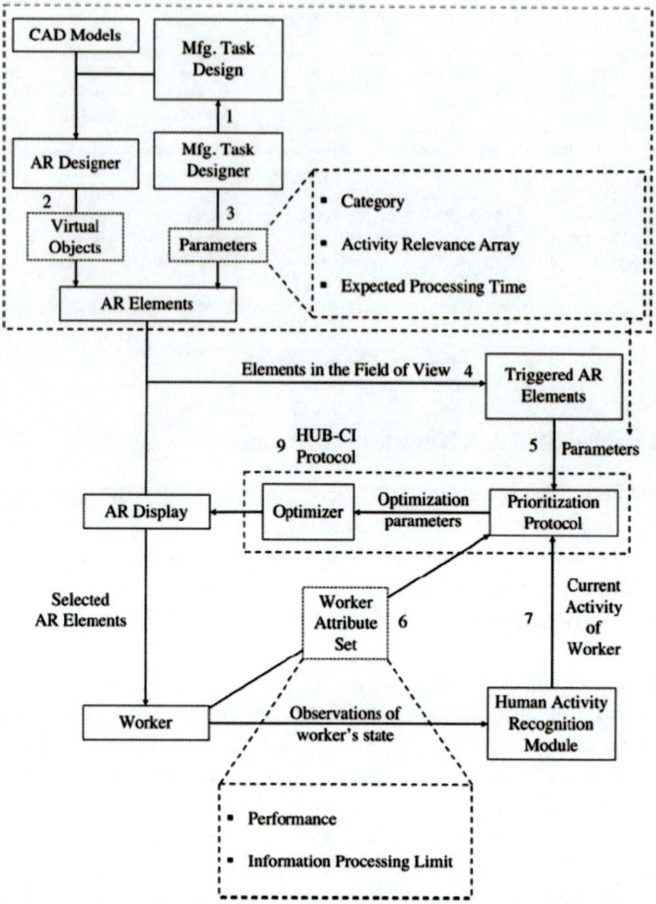

Fig. 1. AR design framework with HUB-CI protocol (Source: (Moghaddam & Nof 2022))

3.2.1 Collaborative Telerobotics for Product Design and Testing

A study in Collaborative Telerobotics (CTR) developed a model where humans (experienced and novices in the work tasks) and robot execute initial stages of skill and knowledge sharing based on a HUB-CI model (Zhong et al. 2013). Robot agents operating under collaboration protocols through the HUB-CI carry out their actions according to the aggregated command received. In turn, human agents acquire feedback from the robots from either a video stream, or 3D arrows which indicate the aggregated command (speed and direction) in spheres alpha-blended in the video. The conceptual model and mathematical formulation of the system are presented as follows:

Individual and collaborative experiments have been designed to evaluate the performance of CTR system implemented with HUB-CI model. Overall performance of the CTR system is determined by the time to complete (TTC) the given robotic task, the occurrence of conflict/error (CE) under each experiment, and its relationship to

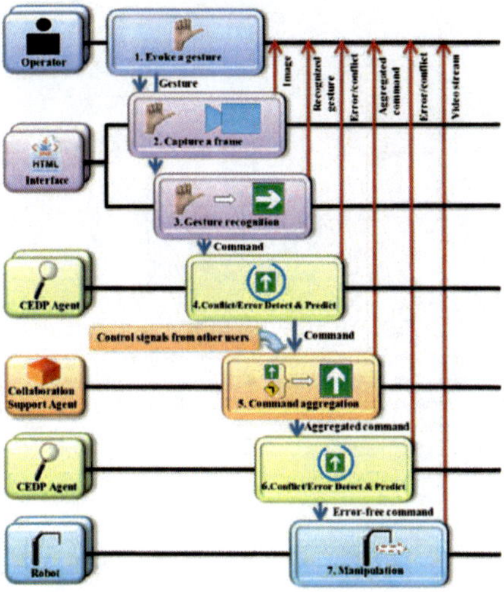

$$X_{ij}(t) = \begin{cases} 1, & \text{if operator } i \text{ issues } C_j \text{ at } t \\ 0, & \text{otherwise} \end{cases} \tag{1}$$

$$V(C_j, t) = \sum_{i=1}^{N} \left[W_{ij}(t) \cdot X_{ij}(t) \right] \tag{2}$$

$$\hat{C}(t) = \arg\max_{C_j} \left(V(C_j, t) \right) \tag{3}$$

$$E(\hat{C}(t)) = \begin{cases} 1, & \text{if } \hat{C}(t) \text{ is free of error} \\ 0, & \text{otherwise} \end{cases} \tag{4}$$

$$S(C_j, \hat{C}(t)) = \begin{cases} 1, & \text{if } C_j = \hat{C}(t) \\ 0, & \text{otherwise} \end{cases} \tag{5}$$

$$F_i(t) = \sum_{j=1}^{M} \left[X_{ij}(t) \cdot S(C_j, \hat{C}(t)) \right] \tag{6}$$

$$K_i(t) = \overline{XOR}(E(\hat{C}(t)), F_i(t)) \tag{7}$$

$$W_{ij}(t+1) = \frac{\sum_{\tau=t-p}^{t} K_i(\tau)}{\sum_{\tau=t-p}^{t} \sum_{j=1}^{M} X_{ij}(\tau)} \tag{8}$$

Fig. 2. Collaborative Telerobotics Sequence Diagram and Co-tolerance of Error and Conflict Algorithm (Source: (Zhong et al. 2013))

TTC. It is concluded that to achieve better performance, operators have to reduce errors and increase the frequency of error-free commands, as shown in Fig. 2. The intuitive and logical observation can be achieved by more effective skill and knowledge sharing augmentation, provided by HUB-CI.

3.2.2 Computer-Supported Collaborative, Integrated Life-Cycle Product Design

A second application of HUB-CI as an enabler for skill and knowledge sharing is described in a computer-supported collaborative design (CSCD) case study using CAD software (Zhong et al. 2014). The HUB-CI environment is hosted on a server which can be accessed via Internet, and offers the following elements and capabilities to support CSCD:

1. Defining the tasks and e-Work requirements;
2. Storage in an online database;
3. Collaborative coding and electronics CAD;
4. Structured Co-Insights Management as an environment used for the conceptual design of the physical product;
5. Capability to network, which is responsible for checking conflicts and errors throughout the development cycle of a new product;
6. Electronic CAD as the tool that supports the required software development and hardware design;
7. The physical development, testing, and validation, which are accomplished at the robotic prototyping cell;
8. The telerobotic cell, which is built as one of the service resources available to designers through the HUB-CI environment;

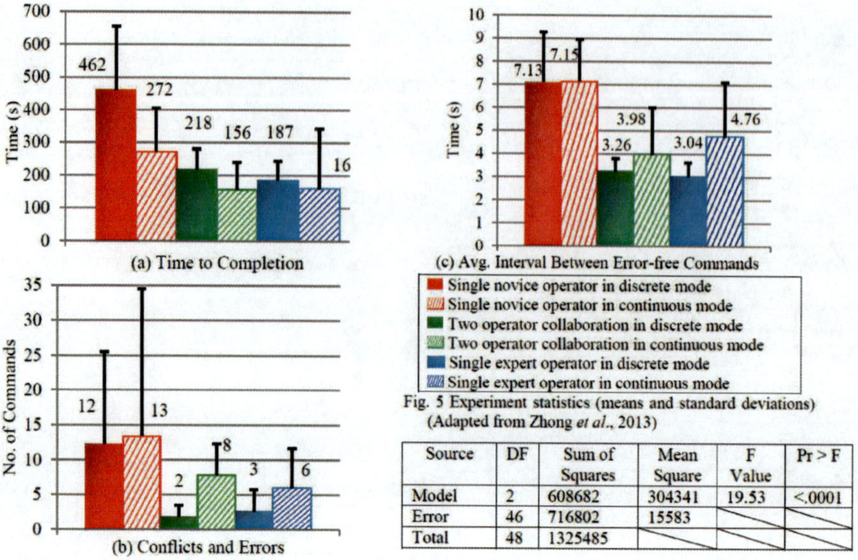

Source	DF	Sum of Squares	Mean Square	F Value	Pr > F
Model	2	608682	304341	19.53	<.0001
Error	46	716802	15583		
Total	48	1325485			

Fig. 3. Experimental Results in terms of e-Criteria and e-Measure of HUB-CI in CTR (Source: Zhong et al. (2013)

9. The designers working at the interaction tier of the HUB-CI environment;
10. The coordinate system representation of collaboration which indicates the multidimensionality of the collaboration space.

As a proof of concept, the study (Zhong et al. 2014) implemented a pilot system for collaborative design and prototyping based on HUBzero package. The distributed designers were asked to build a digital voltmeter from ten LEDs, ten resistors, a potentiometer and an Arduino controller. The output voltage of this voltmeter should be understandable by human as a form of knowledge intelligence. The result of this study has shown that an integrated system with HUB-CI can effectively provide the functionality required during the product development lifecycle.

3.2.3 Cyber-Physical Agricultural Robotic System

A third implementation of HUB-CI is within the field of precision farming, specifically in Agricultural Robotics System (ARS). Automation systems for greenhouses deal with tasks such as climate control, seedling production, spraying, and harvesting, however, few research projects have been conducted to optimize human-robot collaboration in the ARS. The HUB-CI for ARS (Nair et al. 2019) aims to develop an agricultural robotic system for early disease detection of pepper plants in greenhouses. The scope revolves around greenhouse monitoring, detection, and responding tasks, detailed in the system's architecture (Fig. 3 - Left) (Fig. 4).

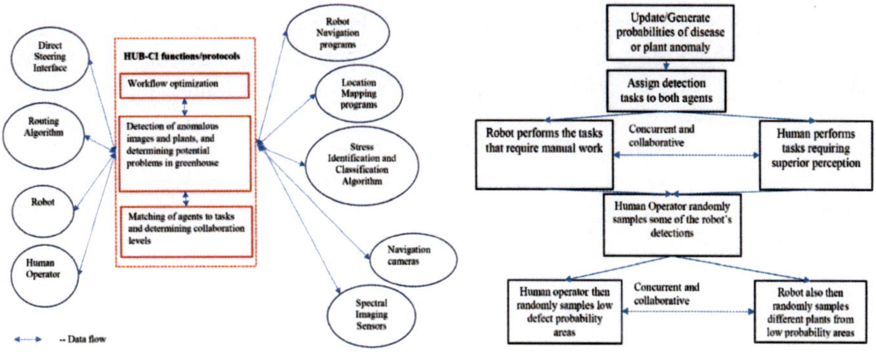

Fig. 4. HUB-CI model for Greenhouse monitoring (Left) and Workflow diagram for HUB-CI Collaboration Strategy (Right) (Source: (Nair et al. 2019))

The workflow is presented in the diagram shown in Fig. 3 (right). Specific CI tools developed for this purpose include: (1) spectral image segmentation for detecting and mapping to anomalies in growing pepper plants; (2) workflow/task administration protocols for managing/coordinating interactions between software, hardware, and human agents, engaged in the monitoring and detection, which would reliably lead to precise, responsive mitigation.

The study (Nair et al. 2019) experimented on how the HUB-CI improves human-robot skill sharing. Evidently, HUB CI yields significantly fewer errors and better early detection, improving the system efficiency by between 210% to 255% across 80 runs, compared to the system that does not implement decision support through HUB-CI. To simulate the remote operational nature of HUB-CI, commands were sent using Python and Robotic Operating System (ROS) programs via a Google Drive, between the PRISM lab in West Lafayette, Indiana, and the Volcani Institute agricultural robotic lab in Israel. Average lag time of remotely sent commands was 1.06 s across 2 different sets of runs of 30 min each. It is validated that HUB-CI yields significantly a higher quality of knowledge via collaborative workflow protocols, as indicated by fewer errors and better detection. The application enables precise monitoring for healthy growth of pepper plants in greenhouses.

3.2.4 Cyber-Collaborative Factory of the Future with Humans and Robots

The fourth case study (Dusadeerungsikul et al. 2019) is based on the implementation of collaboration requirement planning (CRP) for a HUB-CI within factories of the future. HUB-CI has been designed to comprise algorithms and protocols to improve the productivity and efficiency of a distributed system of networked agents via augmented collaboration. Multi-robot control in industry is a proven strategy of reducing production cost by having robots working faster and in parallel, with humans in the loop, leading to overall shorter processing time and higher flexibility.

The study (Dusadeerungsikul et al. 2019) developed and implemented two phases of CRP-H collaboration protocol: CRP-I (task assignment optimization) and CRP-II (agents schedule harmonization), These protocols are developed and validated in two test scenarios: A two-robot collaboration system with five tasks; and a two-robot-and-helper-robot collaboration system with 25 tasks. Simulation results indicate that under CRP-H, both operational cost and makespan of the production work are significantly reduced in both scenarios. The cost is slightly lower while average makespan of CRP-H is 45% less, compared to the baseline collaboration protocol scenario (Fig. 5).

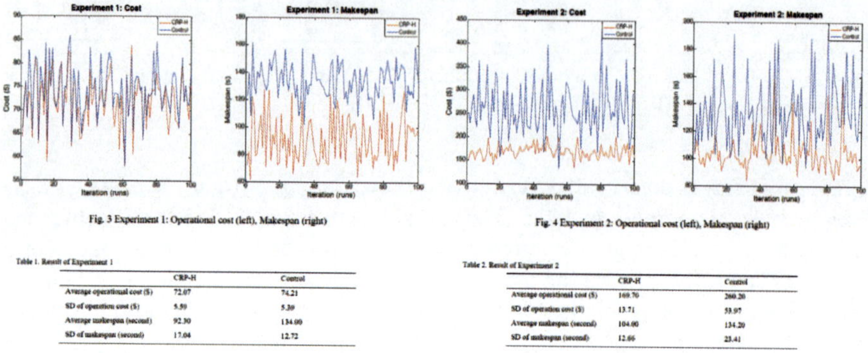

Table 1. Result of Experiment 1

	CRP-H	Control
Average operational cost ($)	72.07	74.21
SD of operation cost ($)	5.59	5.39
Average makespan (second)	92.30	134.00
SD of makespan (second)	17.04	12.72

Table 2. Result of Experiment 2

	CRP-H	Control
Average operational cost ($)	169.70	260.20
SD of operation cost ($)	13.71	53.97
Average makespan (second)	104.00	134.20
SD of makespan (second)	12.66	23.41

Fig. 5. Experimental Results of CRP-H (Source: Dusadeerungsikul et al. (2019))

It has been validated that the new CRP-H protocol delivers superior performance in terms of operational cost and makespan, when compared to a system logic that randomly assigns tasks to robots and instructs random scheduling. The better operational cost comes from CRP-I which optimally assigned tasks to robot(s). Moreover, makespan is minimized because of CRP-II which can updates schedule real-time from IoT/IoS devices' information.

3.3 Other Recently Researched Fields of Skill and Knowledge Sharing

Other recent research projects related to skill and knowledge sharing are summarized in Table 5:

Table 5. Summary of Other Related Research on Skill and Knowledge Sharing

No	Project Topic	Summary of the Approach	Relation to Skill and Knowledge Sharing
1	A novel social gamified collaboration platform enriched with shop-floor data and feedback for the improvement of the productivity, safety and engagement in factories (Lithoxoidou et al. 2020)	Gamified collaboration platform allows positive mood, engagement, and satisfaction, and increased human contact	Social enabler to skill sharing within a manufacturing enterprise
2	Affiliation/dissociation decision models in demand and capacity sharing collaborative network (Yoon & Nof 2011)	DCSP through affiliation and dissociation decision to ensure effective demand fulfilment through collaboration	Collaborative resource sharing between collaborative network of enterprises
3	Automated assembly skill acquisition and implementation through human demonstration (Gu et al. 2018)	Portable Assembly Demonstration (PAD) system to train robots for simple assembly tasks	Human-robot skill sharing protocols
4	Big data analytics-based fault prediction for shop floor scheduling (Ji & Wang 2017)	Big data analytics-based fault prediction model for shop floor scheduling	Knowledge sharing protocols for scheduling and maintenance operations
5	CausalWorld: A Robotic Manipulation Benchmark for Causal Structure and Transfer Learning (Ahmed et al. 2020)	benchmarking platform for reinforcement learning for pushing, picking, pick-and-place, and stacking	Simulation-based skill sharing protocols for robot arm manipulation
6	Collaborative capacity sharing among manufacturers on the same supply network horizontal layer for sustainable and balanced returns (Ahmed et al. 2020)	DCSP through horizontal (supplier) capacity sharing to ensure demand fulfilment through collaboration	Resource sharing between collaborative network of enterprises

(continued)

Table 5. (*continued*)

No	Project Topic	Summary of the Approach	Relation to Skill and Knowledge Sharing
7	Consultation length and no-show prediction for improving appointment scheduling efficiency at a cardiology clinic: A data analytics approach (Srinivas & Salah 2021)	CRISP-DM data analytics for appointment scheduling optimization	Knowledge sharing protocols for scheduling operations
8	Demand and capacity sharing decisions and protocols in a collaborative network of enterprises (Srinivas & Salah 2021)	DCSP through information sharing to ensure demand fulfilment through collaboration	Collaborative resource sharing between collaborative network of enterprises
9	Human-Robot Cross-Training: Computational Formulation, Modeling and Evaluation of a Human Team Training Strategy (Nikolaidis & Shah 2013)	The human-robot cross-training uses mutual adaptation process for learning fluency in joint-action	Human-robot skill sharing protocols
10	Human-Robot Teaming using Shared Mental Models (Nikolaidis & Shah 2012)	Theoretical model of SMM, how it plans, assesses, and promotes HRI	Human-robot knowledge sharing protocols
11	Improved Human-Robot Team Performance Using Chaski, A Human-Inspired Plan Execution System (Nikolaidis & Shah 2012)	A robot system capable of real-time workflow adaptation in a human-robot environment	Human-robot skill sharing protocols for a flexible collaborative workflow
12	Increasing Human Performance by Sharing Cognitive Load Using Brain-to-Brain Interface (Maksimenko et al. 2018)	Brain-to-Brain Interface allows workload-sharing and redistribution depending on current cognitive performance based on electrical brain activity	Human-to-human knowledge sharing protocols for a better teaming
13	Integrating representation learning and skill learning in a human-like intelligent agent (Li et al. 2015)	Deep feature learning SimStudent with transfer learning and feature focus to solve problems	Human-robot knowledge sharing protocols for a better tutoring system

(*continued*)

Table 5. (*continued*)

No	Project Topic	Summary of the Approach	Relation to Skill and Knowledge Sharing
14	Machine learning for predictive scheduling and resource allocation in large scale manufacturing systems (Morariu et al. 2020)	Cloud computing and machine learning for combined scheduling and maintenance optimization	Knowledge sharing protocols as an enabler of collaborative resource-sharing
15	Quantifying Task Similarity for Skill Generalisation in the Context of Human Motor Control (Sebastian et al. 2016)	Quantifying task similarity, learning, and transfer learning in motoric tasks	Skill sharing protocols in a sequential task assignment
16	Skill transfer support model based on deep learning (K.-J. Wang et al. 2021)	Skill transfer model aids new operator to execute tasks based on expert operators data, modeled with RNN and CNN	Human-to-human knowledge sharing protocols using machine learning
17	Towards Fully Autonomous Ultrasound Scanning Robot With Imitation Learning Based on Clinical Protocols (Y. Huang et al. 2021)	Imitation learning framework with One-Step Exploring (OSE) and Region of Attention (ROA) for Autonomous Ultrasound Scanning Robot	Human-robot skill sharing protocols for procedure-specified tasks
18	Virtual reality (VR) as a simulation modality for technical skills acquisition (Nassar et al. 2021)	VR as an enabler of skill acquisition and surgical simulation	Human-to-human skill sharing protocols for procedure-specified tasks

3.4 Emerging Research Challenges of Skill and Knowledge Sharing

3.4.1 Theoretical Research Challenges

Skill and knowledge sharing in production systems and supply networks is accomplished through the four layers of the cyber-collaborative physical system (CCPS), as illustrated in the figure below. Skill and knowledge sharing is preceded by a preliminary phase of skill and knowledge acquisition and documentation. The expert system developed in the first phase becomes the foundation of the next stage, the execution phase. In this second phase, the four layers of CCPS are streamlined and optimized to support the instance of skill and knowledge sharing. The outcome of skill and knowledge sharing is measured based on a set of performance metrics in the last stage, the evaluation phase (Fig. 6).

- Previous research has mainly focused on generating preliminary working systems of skill and knowledge sharing models. Some common developments include teaching

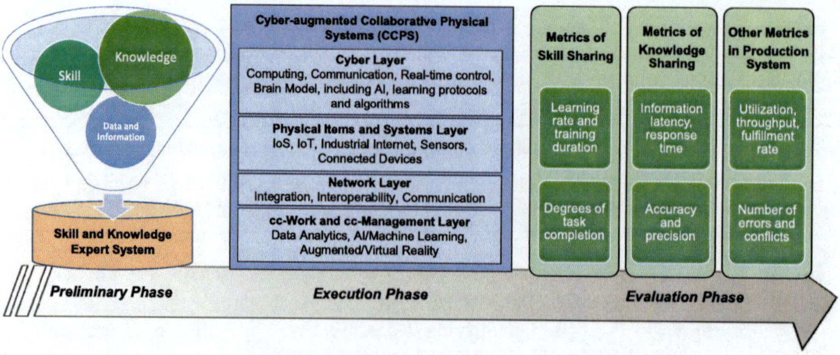

Fig. 6. The Framework of Skill and Knowledge Sharing in CCPS

robots to perform procedural tasks; guiding novice operators to execute a particular task using augmentation and wearables; and other applications in which the focus is put on developing a contextualized system, where instances of skill and knowledge sharing occur. The key finding of the previous studies is that skill and knowledge sharing occurs in various case studies, and is enabled by a wide range of tools, varying from machine learning, data analytics, industrial internet of things (IIoT) and virtual/augmented reality (VR/AR/XR).

- In terms of performance metrics, the effectivity of the working system is mostly measured by a usability survey, where human subjects are given a set of questionnaires to fill, and the options are formulated in Likert scale, or a similar scale. Despite their quantified nature, most of the surveys do not include objective assessments of the production system's or supply network's performance. Therefore, the framework defined here and shown in Fig. 6 above is able to measure the effectivity of skill and knowledge sharing into three different sub-metrics to deepen our understanding of the outcomes. The benefits of skill and knowledge sharing should be extended and related to general production systems and supply networks metrics, such as throughput, error reduction, conflict resolution, and on-time delivery.
- Previous research has only partially, not fully addressed the dynamic execution of skill and knowledge sharing under integrated, operational system conditions. Recent advances of automation have modeled human as an integral element in a smart manufacturing systems and supply networks, which is the human in the loop (HITL) perspective. As key decision making participants in the systems' network, human agents must be fully equipped and augmented with the necessary, timely skills and knowledge. The augmentation process should be streamlined so that the training and preparation duration is minimal. In cases where this time duration can be reduced, optimize and harmonized dynamic skill and knowledge sharing must be implemented.
- Further research is needed to address the three above emerging challenges in order to enable concurrent, optimized, and harmonized intelligence sharing in a skill and knowledge sharing CCPS. For this purpose, the HITL focus will have to expand to the NITN, Human in the network scale.

3.4.2 Future Research Plan

The emerging areas of research and major open questions about challenges concerning the field of Skill and Knowledge Sharing can be summarized in the following directions:

- *HUB-CI for information flow optimization*, which optimizes and harmonizes the collaborative intelligence of agents in a workflow by controlling data and information flow between them. This application of HUB-CI is particularly advantageous in cases where Augmented Reality (AR) and its variants are being used as tools for skill and knowledge sharing.
- *Learning protocols in the skill and knowledge sharing* (SaKS), which streamline the data exchange process for faster transmission. As the latency and coherence are maintained, SaKS can be organized and managed dynamically.
- *Machine learning-based ontology for SaKS taxonomy,* which provides an adaptive, interpretable definitions of basic concepts and the relationships between skill and knowledge. With the rising level of intelligence of computing resources, this subject will extend the classification that Bloom's Taxonomy and its derivatives provide, and improve researchers' understanding with contextual and iterative definitions of skill and knowledge.
- *Other collaborative augmentations*, which are related to physical wearables, augmented/extended reality and online, and real-time analytic systems based on the collaborative intelligence.

Several chapters in this book are already addressing these challenging directions.

Acknowledgment. This chapter was supported by the PRISM Center at Purdue University, and by NSF Grant #1839971: Pre-Skilling Workers, Understanding Labor Force Implications and Designing Future Factory Human-Robot Workflows Using Physical Simulation Platform. We also thank several colleagues who gave valuable comments to improve this study.

References

Adam, G., Chidambaram, S., Reddy, S.S., Ramani, K., Cappelleri, D.J.: Towards a comprehensive and robust micromanipulation system with force-sensing and VR capabilities. Micromachines **12**(7), 784 (2021). https://doi.org/10.3390/mi12070784

Ahmed, O., et al.: CausalWorld: A Robotic Manipulation Benchmark for Causal Structure and Transfer Learning (2020). https://doi.org/10.48550/arxiv.2010.04296

Bloom, B.S., of College, C., Examiners, U.: Taxonomy of Educational Objectives, vol. 2. Longmans, Green New York (1964)

Chidambaram, S., et al.: ProcessAR: an augmented reality-based tool to create in-situ procedural 2D/3D AR instructions. Design. Interact. Syst. Conf. **2021**, 234–249 (2021). https://doi.org/10.1145/3461778.3462126

Dusadeerungsikul, P.O., et al.: Collaboration requirement planning protocol for hub-ci in factories of the future. Procedia Manufac. **39**, 218–225 (2019). https://doi.org/10.1016/j.promfg.2020.01.327

Gu, Y., Sheng, W., Crick, C., Ou, Y.: Automated assembly skill acquisition and implementation through human demonstration. Robot. Auton. Syst. **99**, 1–16 (2018). https://doi.org/10.1016/j.robot.2017.10.002

Huang, G., et al.: AdapTutAR: an adaptive tutoring system for machine tasks in augmented reality. In: Proceedings of the 2021 CHI Conference on Human Factors in Computing Systems, pp. 1–15 (2021). https://doi.org/10.1145/3411764.3445283

Huang, Y., Xiao, W., Wang, C., Liu, H., Huang, R., Sun, Z.: Towards fully autonomous ultrasound scanning robot with imitation learning based on clinical protocols. IEEE Robot. Autom. Let. **6**(2), 3671–3678 (2021). https://doi.org/10.1109/LRA.2021.3064283

Imrie, B.W.: Assessment for learning: quality and taxonomies. Assess. Eval. High. Educ. **20**(2), 175–189 (1995). https://doi.org/10.1080/02602939508565719

Ipsita, A., et al.: VRFromX: from scanned reality to interactive virtual experience with human-in-the-loop. In: Extended Abstracts of the 2021 CHI Conference on Human Factors in Computing Systems, pp. 1–7 (2021). https://doi.org/10.1145/3411763.3451747

Ji, W., Wang, L.: Big data analytics based fault prediction for shop floor scheduling. J. Manuf. Syst. **43**, 187–194 (2017). https://doi.org/10.1016/j.jmsy.2017.03.008

Kim, S., Chi, H., Hu, X., Huang, Q., Ramani, K.: A Large-Scale Annotated Mechanical Components Benchmark for Classification and Retrieval Tasks with Deep Neural Networks, pp. 175–191 (2020). https://doi.org/10.1007/978-3-030-58523-5_11

Kim, S., Chi, H., Ramani, K.: Object synthesis by learning part geometry with surface and volumetric representations. Comput. Aided Des. **130**, 102932 (2021). https://doi.org/10.1016/j.cad.2020.102932

Kim, S., Hu, X., Vegesana, A., Ramani, K.: First-Person View Hand Segmentation of Multi-Modal Hand Activity Video Dataset. BMVC (2020)

Krachtt, N.: The workforce implications of Industry 4.0: manufacturing workforce strategies to enable enterprise transformation (2019)

Li, N., Matsuda, N., Cohen, W.W., Koedinger, K.R.: Integrating representation learning and skill learning in a human-like intelligent agent. Artific. Intell. **219**, 67–91 (2015). https://doi.org/10.1016/j.artint.2014.11.002

Lithoxoidou, E., et al.: A novel social gamified collaboration platform enriched with shop-floor data and feedback for the improvement of the productivity, safety and engagement in factories. Comput. Ind. Eng. **139**, 105691 (2020). https://doi.org/10.1016/j.cie.2019.02.005

Maksimenko, V.A., et al.: Increasing human performance by sharing cognitive load using brain-to-brain interface. Front. Neurosci. **12** (2018). https://doi.org/10.3389/fnins.2018.00949

Mohsen, M., Nof, S.Y.: The collaborative factory of the future. Int. J. Comput. Integrat. Manufac. published online 2015, 1–21 (2015). Printed **30**(1), 23–43 (2017). https://doi.org/10.1080/0951192X.2015.1066034

Mohsen, M., Nof, S.Y.: Collaborative service-component integration in cloud manufacturing. Int. J. Product. Res. published online 2017, 1–15 (2017).. Printed *56*(1–2), 677–691, 2018. https://doi.org/10.1080/00207543.2017.1374574

Mahdi, M., Nof, S.Y.: information flow optimization in augmented reality systems for production & manufacturing. In: 26th International Conference on Production Research. Curitiba, Brazil (2022)

Morariu, C., Morariu, O., Răileanu, S., Borangiu, T.: Machine learning for predictive scheduling and resource allocation in large scale manufacturing systems. Comput. Ind. **120**, 103244 (2020). https://doi.org/10.1016/j.compind.2020.103244

Nair, A.S., Bechar, A., Tao, Y., Nof, S.Y.: The HUB-CI model for telerobotics in greenhouse monitoring. Procedia Manufac. **39**, 414–421 (2019). https://doi.org/10.1016/j.promfg.2020.01.385

Nassar, A.K., Al-Manaseer, F., Knowlton, L.M., Tuma, F.: Virtual reality (VR) as a simulation modality for technical skills acquisition. Ann. Med. Surg. **71**, 102945 (2021). https://doi.org/10.1016/j.amsu.2021.102945

Nikolaidis, S., Shah, J.: Human-robot teaming using shared mental models. In: ACM/IEEE International Conference on Human Robot Interaction (HRI) (2012). https://interactive.mit.edu/human-robot-teaming-using-shared-mental-models

Nikolaidis, S., Shah, J.A.: Human-robot cross-training: Computational formulation, modeling and evaluation of a human team training strategy. In: 2013 8th ACM/IEEE International Conference on Human-Robot Interaction (HRI), pp. 33–40 (2013)

Paredes, L., et al.: CHIMERA. In: Proceedings of the ACM on Interactive, Mobile, Wearable and Ubiquitous Technologies, vol. 5(4), pp. 1–24 (2021). https://doi.org/10.1145/3494974

Sebastian, G., Fong, J., Crocher, V., Tan, Y., Oetomo, D., Mareels, I.: Quantifying task similarity for skill generalisation in the context of human motor control. In: 2016 14th International Conference on Control, Automation, Robotics and Vision (ICARCV), pp. 1–6 (2016). https://doi.org/10.1109/ICARCV.2016.7838705

Srinivas, S., Salah, H.: Consultation length and no-show prediction for improving appointment scheduling efficiency at a cardiology clinic: a data analytics approach. Int. J. Med. Inform. **145**, 104290 (2021). https://doi.org/10.1016/j.ijmedinf.2020.104290

Villanueva, A., et al.: Towards modeling of human skilling for electrical circuitry using augmented reality applications. Int. J. Educ. Technol. High. Educ. **18**(1), 39 (2021). https://doi.org/10.1186/s41239-021-00268-9

Villanueva, A.M., et al.: RobotAR: an augmented reality compatible teleconsulting robotics toolkit for augmented makerspace experiences. In: Proceedings of the 2021 CHI Conference on Human Factors in Computing Systems, pp. 1–13 (2021). https://doi.org/10.1145/3411764.3445726

Villanueva, A., Zhu, Z., Liu, Z., Wang, F., Chidambaram, S., Ramani, K.: ColabAR: a toolkit for remote collaboration in tangible augmented reality laboratories. Proc. ACM Hum. Comput. Interact. **6**(CSCW1), 1–22 (2022). https://doi.org/10.1145/3512928

Wang, K.-J., Rizqi, D.A., Nguyen, H.-P.: Skill transfer support model based on deep learning. J. Intell. Manuf. **32**(4), 1129–1146 (2021). https://doi.org/10.1007/s10845-020-01606-w

Wang, T., Qian, X., He, F., Hu, X., Cao, Y., Ramani, K.: GesturAR: an authoring system for creating freehand interactive augmented reality applications. In: The 34th Annual ACM Symposium on User Interface Software and Technology, pp. 552–567 (2021). https://doi.org/10.1145/3472749.3474769

Wang, T., Qian, X., He, F., Ramani, K.: LightPaintAR: assist light painting photography with augmented reality. In: Extended Abstracts of the 2021 CHI Conference on Human Factors in Computing Systems, pp. 1–6 (2021). https://doi.org/10.1145/3411763.3451672

Yoon, S.W., Nof, S.Y.: Affiliation/dissociation decision models in demand and capacity sharing collaborative network. Int. J. Prod. Econ. **130**(2), 135–143 (2011). https://doi.org/10.1016/j.ijpe.2010.10.002

Zhong, H., Nof, S.Y.: Collaborative design for assembly: the HUB-CI Model. In: 22nd International Conference on Production Research (ICPR) (2013)

Zhong, H., Wachs, J.P., Nof, S.Y.: HUB-CI model for collaborative telerobotics in manufacturing. IFAC Proc. Vol. **46**(7), 63–68 (2013). https://doi.org/10.3182/20130522-3-BR-4036.00059

Zhong, H., Wachs, J.P., Nof, S.Y.: Telerobot-enabled HUB-CI model for collaborative lifecycle management of design and prototyping. Comput. Ind. **65**(4), 550–562 (2014). https://doi.org/10.1016/j.compind.2013.12.011

Smart Agriculture and Agricultural Robotics: Review and Perspective

Avital Bechar[1] and Shimon Y. Nof[2](✉)

[1] Institute of Agriculture Engineering (IAE), Agriculture Research Organization (ARO), Bet Dagan, Israel
avital@volcani.agri.gov.il

[2] PRISM Center, and School of IE, Purdue University, West Lafayette, IN, USA
nof@purdue.edu

Abstract. The purpose of this chapter is to review the contribution of agricultural robotics to smart agriculture through the perspective of three contributing technology pillars: agricultural robotics; precision agriculture; and artificial intelligence. In this context, we describe contributions of recent research projects in agricultural robotics, their impacts on and prospects for smart agriculture and the next era in agriculture.

1 Introduction

In this chapter, we present the contributions of agricultural robotics to smart agriculture, along the three technology pillars of smart agriculture: agricultural robotics, precision agriculture, and artificial intelligence. We review the contributions to smart agriculture, and provide a perspective on the interaction and synergy of agricultural robots with the other two technology pillars. Throughout the chapter, we review a variety of research projects in the domain of agricultural robotics that are at the frontiers of smart agriculture.

Agriculture has been essential during all ages of human history and a foundation of human civilization, not only for feeding the increasing population, but also to serve other purposes, such as the production of medicines, fiber, and fuel (Alwis et al. 2022). With the new advances in science, technology and equipment, agrochemicals and genetically modified food have been introduced to agriculture with an aim to achieve a high yield while minimizing the labor cost.

During recent years, agriculture is undergoing the so called fourth revolution, by integrating Information and Communications Technologies (ICT) in traditional farming practices (Boursianis et al. 2022). Smart Agriculture (SA), also known as smart farming, smart irrigation and fertilization, climate smart farming, and smart pest control (Qureshi et al. 2022) is a management concept evolving from precision agriculture. It focuses on providing agriculture with the infrastructure to integrate advanced computing and automation technologies, such as big data, cloud computing, artificial intelligence (AI), robots, the internet of services (IoS) and the internet of things (IoT). The ultimate goal is increasing the quality and quantity of the crops while using fewer natural resources, achieving resilience, and optimizing use of human labor (Ayoub Shaikh et al. 2022).

C.-Y. Huang and S. W. Yoon (Eds.): ICPR1 2021, ACES 14, pp. 444–474, 2023.
https://doi.org/10.1007/978-3-031-44373-2_26

Smart agriculture is growing in importance due to the combination of several trends: the expanding global population, the increasing demand for higher crop yield, the need to use natural resources efficiently and sustainably, the increasing value and sophistication of ICT. SA aims to tackle three main objectives: i) sustainably increasing agricultural productivity and income; ii) adapting and building resilience to climate change and disruptions; and, iii) reducing and/or removing greenhouse gas emissions, where possible.

Smart agriculture is strongly related to three technology fields: 1) management information systems including artificial intelligence and machine learning models - planned systems for collecting, processing, storing, and disseminating data in the form needed to carry out a farm's operations and functions; 2) precision agriculture - Management of spatial and temporal variability to improve economic returns following the use of inputs and reduce environmental impact; and, 3) agricultural automation and robotics - the process of applying robotics, automatic control and artificial intelligence techniques at all levels of agricultural production. Figure 1 illustrates the three technology pillars of smart agriculture and the overlapping areas between them, indicating the synergy and relations between them.

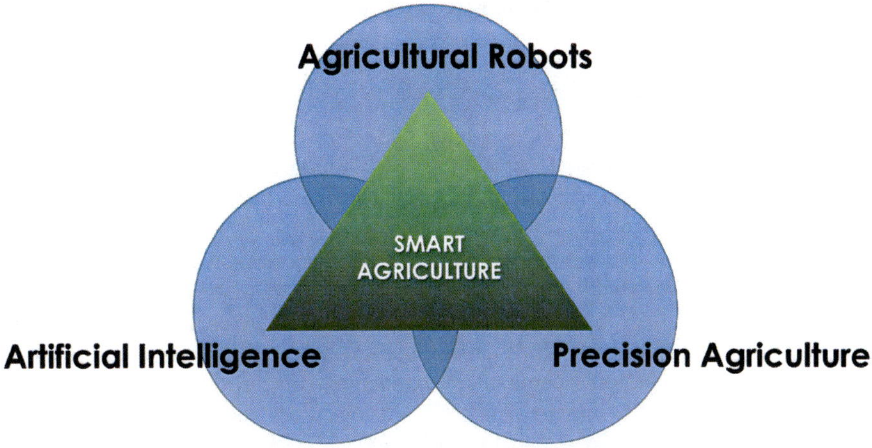

Fig. 1. The three technology pillars of smart agriculture.

Precision agriculture is a field in agriculture concentrating on selective decision making and planning based on the processing of detailed farm-timely information, knowledge and thoughtful expertise (Nair et al. 2021). It was first introduced 3–4 decades ago. Precision agriculture is designed to reduce, through technological means, the required amount of fertilizers and other chemicals, irrigation, fuel, manual work, lease and crop insurance payments (Mull 2013). In general, precision agriculture techniques and tools are designed to collect data, make good (optimal) decisions, and implement those decisions at relatively higher resolution (Becha 2021).

The developed techniques and research in precision agriculture were conducted to align with four main objectives: increase agricultural productivity; increase produce quality; reduce production costs; and reduce negative environmental impacts. Precision agriculture is the main beneficiary of the high variability that characterizes the agricultural domain: It aims to exploit the high variability using high resolution (up to an individual plant level) for data collection and decision-making, applying variate rate operations to increase the total plot revenue, and minimizing the total cost (Bechar 2021).

Robots are perceptive machines that can be programmed to perform specific tasks, make decisions, and act in real time. They are required in various fields that normally call for reductions in manpower and workload, and are best-suited for applications requiring repeatable accuracy and high yield under stable conditions (Holland and Nof 2007). However, they lack the capability to respond to ill-defined, unknown, changing, and unpredictable events (Moysiadis et al. 2020). Unlike industrial applications, which deal with often simple, repetitive, well-defined and predetermined tasks, agricultural applications of automation and robotics require relatively more advanced technologies to deal with complex and highly variable environments and produce (Nof 2009). The technical feasibility of agricultural robots for a variety of agricultural tasks has been widely validated. Nevertheless, despite the tremendous amount of research, commercial applications of robots in complex agricultural environments are not yet available (Urrea and Munoz 2015). Such applications of robotics in uncontrolled field environments are still in the developmental stages (Bac et al. 2013). The main limiting factors lie in production inefficiencies and lack of economic justification. Development of an agricultural robot must include the creation of sophisticated, intelligent algorithms for sensing, planning and controlling to cope with the difficult, unstructured and dynamic agricultural tasks (Bechar and Edan 2003).

Autonomous robots in real-world, dynamic and unstructured environments still yield inadequate results (Bechar 2010), because of inherent uncertainties, unknown operational settings and unpredictable environmental conditions. Inadequacies of sensor technologies further impair the capabilities of autonomous robotics. Therefore, the promise of automatic and efficient autonomous operations has fallen short of expectations in unstructured and complex environments. Complexity increases with involvement of natural objects, such as those encountered in medical and agricultural environments, because of the high variability in shape, texture, color, size, orientation and position of such objects (Bechar et al. 2009). In addition, the product being dealt with is of relatively of low cost; therefore, the cost of the automated system must be low in order for it to be economically justified. Also, the seasonal nature of agriculture makes it difficult to achieve the high utilization found in the manufacturing industries. The complex agricultural environment, combined with the intensive production, requires robust systems with short development time, at low cost (Nof 2009).

An important feature of artificial intelligence in smart agriculture is the ability to learn automatically from historical data and experiences (generally called 'machine learning'). Various learning methods and algorithms have been implemented in cyber physical systems, which facilitate continuous improvements, adaptations and learning from mistakes, as well as from success. Common applications of machine learning in cyber physical systems include, for example, fault detection (Sargolzaei et al. 2017), system security (Junejo and Goh 2016), pattern recognition or detection (Spezzano and Vinci 2015), predictive maintenance (Wu et al. 2018) and adaptive scheduling (Linard and Bueno 2016).

In agricultural cyber physical space (CPS), machine learning research (Airlangga and Liu 2019) has addressed several smart agriculture topics: image classification for plant recognition, plant disease detection using hyperspectral imaging (Wang et al. 2019), smart irrigation management (Goap et al. 2018), data mining and knowledge extraction (Dimitriadis and Goumopoulos 2008, Schuster et al. 2011), detection and prediction of biotic stresses in plants (Behmann et al. 2015, Wani and Ashtankar, 2017), crop yield evaluation (Finkelshtain et al. 2017), predicting environmental factors (Pandey et al., 2019, Taki et al. 2018) and automatic plant phenotyping (Yahata et al. 2017).

Future research could explore predictive maintenance, pattern detection, enhanced collaboration among agents (human or non-human agents) and system security, as related to agriculture.

In agriculture, the environment is unstructured and demands the motion of robots unlike that of machines in a factory or of vehicles in a parking lot (Canning et al. 2004). It changes in time and space, the environment conditions considered hostile and it requires mobile operation in 3D changing tracks. The terrain, vegetation, landscape, visibility, illumination and other atmospheric conditions are not well defined; they continuously vary, have inherent uncertainty, and generate unpredictable and dynamic situations (Bechar and Vigneault 2017). Complexity increases when dealing with natural objects, such as fruits and leaves, because of the high variability in shape, texture, color, size, orientation and position that in many cases cannot be determined a-priori.

From a robotic point of view, the world can be divided into four main domains, according to the structural characteristics of environments and objects (Bechar and Vigneault 2016): 1) the environment and the objects are structured; 2) the environment is unstructured and the objects are structured; 3) the environment is structured and the objects are unstructured; and, 4) the environment and the objects are unstructured. Each robotic area such as industry, medical, healthcare, etc. can be associated to one of the domains (Table 1). This classification illuminates the difference between the domains, their complexity and challenges. The agricultural domain is associated with the forth quadrant, in which none is structured and therefore, it is highly challenging to develop and commercialize robotics solutions. In such environments there are many situations in which autonomous robots fail due to the many unexpected events (Steinfeld 2004).

Table 1. The world domains from a robotic point of view, a variation on a Figure from Bechar and Vigneault (2016).

		Environment	
		Structured	Unstructured
Objects	Structured	Industrial/Service domains	Militar/Space/Underwater domains
	Unstructured	Medical/Social Domains	Agricultural Domain

Growing and production processes in agriculture are complex, diverse, human labor intensive and usually unique to each crop (Bechar and Vigneault 2017). The process type and components are influenced by many factors, including: the crop characteristics and requirements, the geographical/geological environment, climatic and meteorological conditions (Tremblay et al. 2011), market demands, customers' requirements, and the farmer's capabilities and means. The technology, equipment and means that are required for a specific agricultural task involving any given crop and environment will not necessarily be applicable to another crop or in a different environment. The wide variety of agricultural systems and their diversity worldwide make it difficult to generalize the application of automation and control (Schueller 2006) and therefor, more efficient agricultural practices are needed.

Until recently, research in the fields of agricultural robotics and smart agriculture evolved in parallel paths with very little interaction, relationship or reference between the two research fields. Development of an agricultural robot to perform a smart agricultural task must start with development of integrated approaches and operation concepts of both robotics and smart agriculture, and include creation of sophisticated, intelligent algorithms for sensing, planning and control, and decision-making algorithms to cope with the difficult, unstructured and dynamic environment and the unique nature of smart agriculture missions.

Extensive research has focused on the application of robots and intelligent automation systems to a large variety of field operations, and their technical feasibility has been widely demonstrated (Bac et al. 2014, Bechar and Vigneault 2016). Nevertheless, and in spite of the tremendous amount of robotic applications in the industry, very few robots are in operation in agricultural production (Xiang et al. 2014). Complexity increases when dealing with natural objects, such as fruits and leaves. This complexity is due to the high variability of many of the parameters that require robot response, many of which cannot be determined a-priori. In addition, agricultural robots work with live and fragile produce, making the tasks and features of agricultural applications quite different from industrial applications, which work with inanimate products. The main limiting factors lie in production inefficiencies and the lack of economic justification due to the very short period of potential utilization each year (Bechar and Vigneault 2016). Development of a feasible agricultural robot must include the creation of sophisticated intelligent algorithms for sensing, planning and controlling to cope with the difficult, unstructured and dynamic agricultural environment (Edan and Bechar 1998), or integrating a human operator (HO) into the system.

2 Appraisal and Investigation of the Importance of Agricultural Robots to Smart Agriculture and Analysis of Emerging Research Topics

Referring to the three leading characteristics of the agricultural domain: the variability level of the product; the structuredness level of the environment; and the systems' cost, as dimensions in a domination space (Fig. 2). Whereas the agricultural domain appears in the right lower area with high product variability, low structuredness level and low-cost demand. It reveals the gap that needs to be covered and the challenges of robotic systems for agriculture and for smart agriculture in particular, since robotics located on the other side of the domination space dealing usually with low variability of the product, high structuredness level of the environment and relatively high costs. The way to reduce the gap could be by developing concepts and approaches that are more suitable for smart agriculture operations, such as focusing on a specific task, integrating a human operator in the robotic system, simplifying the robotic systems by creating robot teams, and combinations of the above.

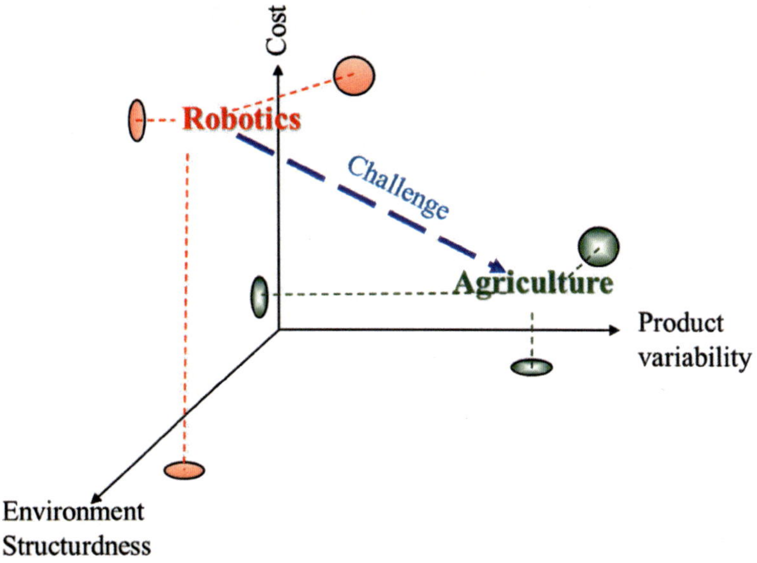

Fig. 2. The domination space of the three dimensions: the product variability level; the environment structuredness level; the cost. The blue line represents the gap robotics will need to cover and the challenges in this area (Bechar 2021).

The relative research effort related to the following areas: agriculture, robotics, precision agriculture (PA, including precision farming and precision irrigation), agricultural robotics (AR), smart agriculture (SA), artificial intelligence (AI), Agricultural Robots for Smart Agriculture (ARSA), Artificial Intelligence in Agriculture (AIA), Artificial Intelligence for Smart Agriculture (AISA) and Agricultural Robots and Artificial Intelligence for Smart Agriculture (ARAISA) in the past 5 years is summarized in Fig. 3. It is based on Peer-reviewed articles that were published since 2017 according to Scopus.

The number of annual peer-reviewed articles published from 2017 to 2021 is given in Fig. 4. All topics showed annual increase in the number of publications between the years 2017 to 2021 with one exception. The agricultural robotics topic showed a decrease of about 11% between years 2020 to 2021 (from 19,551 publications in 2020 to 17,269 publications in 2021). In the first three years (2017 – 2020) the annual number of publications in this topic increased, with the highest rate for all topics and all years in between the 2019 to 2020. From 3892 to 19,551 publications, above 500% increase, and probably the reduction in year 2021 compensate this high jump.

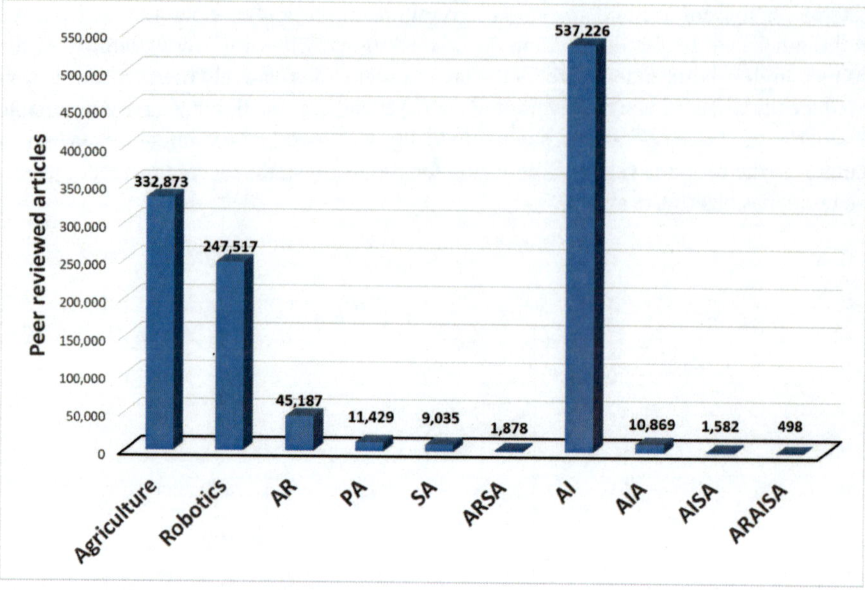

Fig. 3. Peer-reviewed articles on the main topics related to agricultural robotics for smart agriculture since year 2017 until September 2022 (Source: Scopus, accessed in October 2022). AR – Agricultural Robots, PA – Precision Agriculture, SA – Smart Agriculture, ARSA – Agricultural Robots for Smart Agriculture, AI – Artificial Intelligence, AIA - Artificial Intelligence in Agriculture, AISA - Artificial Intelligence for Smart Agriculture, and ARAISA - Agricultural Robots and Artificial Intelligence for Smart Agriculture.

The average annual increase in the number of articles in the research topics: agriculture and smart agriculture is 16% and 44% respectively, indicating that the research and development area of smart agriculture in evolving much faster than in agriculture in general. The average annual increase in the number of articles in agricultural robotics stands at 141% in comparison to only 16% in robotics. About 4.5% of the articles published on robotics in 2017 deal in agriculture, in comparison to 37% and 31% in 2020 and 2021, meaning that agriculture becomes a major topic in robotics. A similar situation is found in the AI topics: the average annual increase in the number of articles in AI in agriculture (AIA) stands at 65% in comparison to 32% in all AI. The analysis of smart agriculture topic with its subtopics shows that the annual number of articles published in the research topics of ARSA and AISA increased, on average, by 150% and 113% respectively, and that about 30% and 20% of the publication in smart agriculture during 2021 were related to robotics or AI. About 8% of the publications in smart agriculture during 2021 deal with integrating robotics and AI. In addition, the gap between the annual number of publications is reduced between 2017 to 2021 (Fig. 4) meaning that the topic of smart agriculture becomes relatively more popular and since it consists a field of precision agriculture, there is a shift from precision agriculture to smart agriculture.

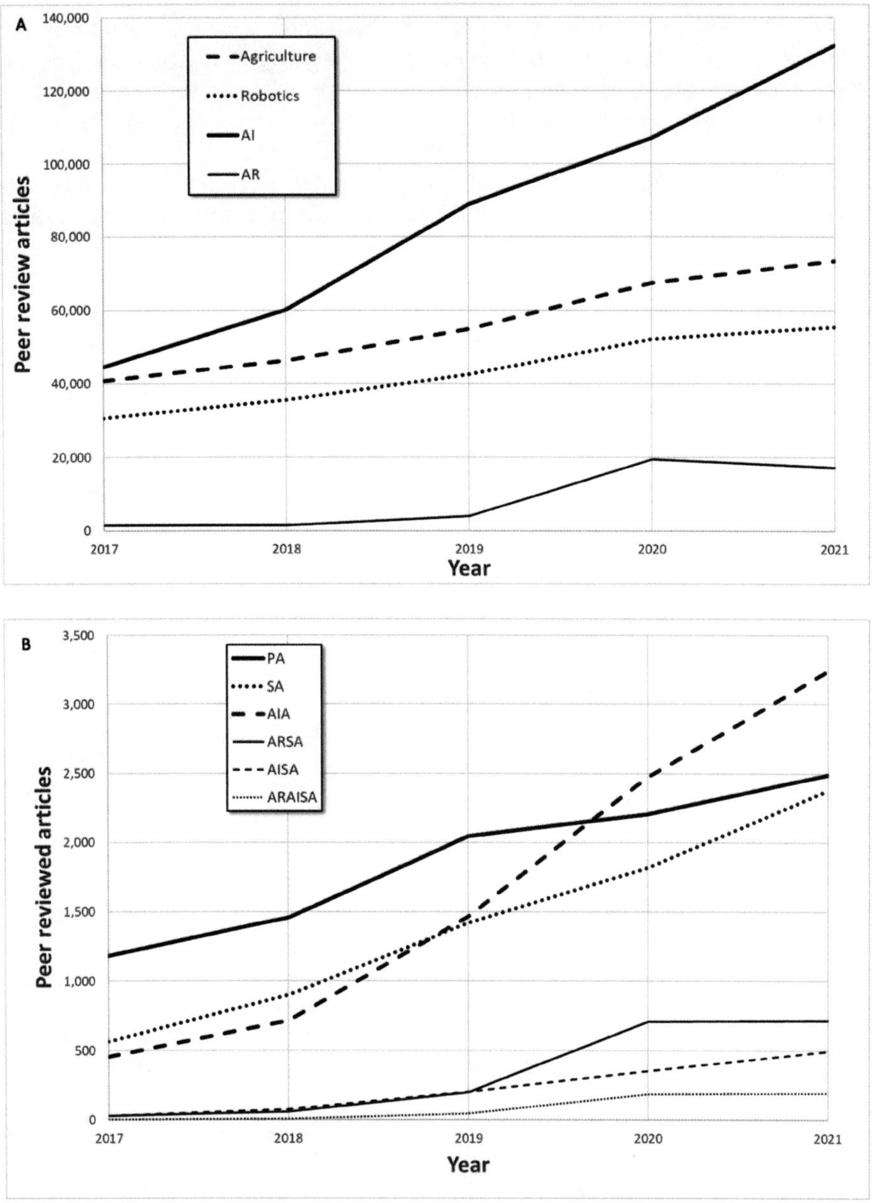

Fig. 4. Annual peer-reviewed articles on the main topics related to agricultural robotics for smart agriculture in 2017 to 2021 (Source: Scopus, accessed in October 2022). AR – Agricultural Robots, PA – Precision Agriculture, SA – Smart Agriculture, ARSA – Agricultural Robots for Smart Agriculture, AI – Artificial Intelligence, AIA - Artificial Intelligence in Agriculture, AISA - Artificial Intelligence for Smart Agriculture, and ARAISA - Agricultural Robots and Artificial Intelligence for Smart Agriculture.

Comparison of the annual number of peer-reviewed articles published in the past five years (comparing 2021 to 2017) reveals that the main emerging topics in research are agricultural robotics (AR), robotics for smart agriculture (ARSA), AI for smart agriculture (AISA) and the integration of robotics and AI for smart agriculture (ARAISA). AR, ARSA and AISA increased the annual number of publications over the past five years by 12.69-fold, 24.72-fold and 17.75-fold, respectively. The research effort dealing with the synergy between robotics and AI for smart agriculture increased by almost 48-fold (Fig. 5).

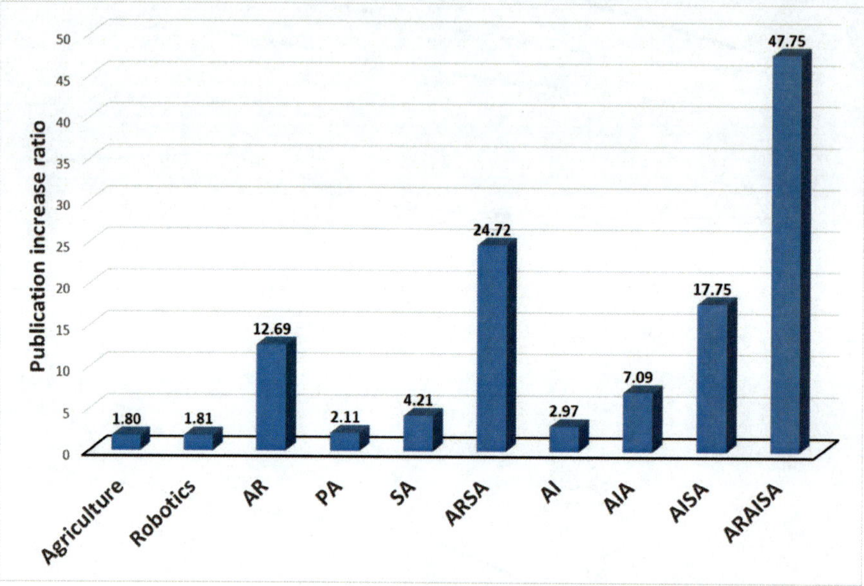

Fig. 5. Publication increase ratio between the annual peer-reviewed articles published in 2021 in comparison to 2017 on the main topics related to agricultural robotics for smart agriculture (Source: Scopus, accessed in October 2022). AR – Agricultural Robots, PA – Precision Agriculture, SA – Smart Agriculture, ARSA – Agricultural Robots for Smart Agriculture, AI – Artificial Intelligence, AIA - Artificial Intelligence in Agriculture, AISA - Artificial Intelligence for Smart Agriculture, and ARAISA - Agricultural Robots and Artificial Intelligence for Smart Agriculture.

3 Basic Terms, Guidelines, Principles and Conditions for the Synergy Between Agricultural Robots and Smart Agriculture

Much research was conducted on agricultural robotics in the past 40 years. Very few reached the commercialization stage. The main causes for incompletion were the extensive costs of the developed robots, inability to execute the required agricultural task, low robustness of the system, and its inability to successfully reproduce the same task in slightly different contexts, or to satisfy operational or economic aspects of the agricultural task. In addition, most approaches were imported from the industrial domain (Vidoni et al. 2015) that was not fit for the tasks at hand. All the effort conducted so far enabled the formulation of guidelines and definitions of the basic conditions required for development of agricultural robots (Benos et al. 2020), with modification to smart agriculture.

The classical operation process of agricultural robots consists of four general stages: 1) the robot senses and acquires raw data from and about the environment, task and/or its state, by using various sensors; 2) the robot processes and analyzes the data received from its sensors to generate reasoning and a perception of the environment, the task, or its state to a certain level of situation awareness; 3) the robot generates an operational plan based on its perception of the environment and state, or the task objectives; and 4) the robot executes the required actions included in the operational plan.

In performing an agricultural task in an unstructured environment, the robot must repeat these four steps continuously, since its state, the task status, and the environment are changing constantly. In fact, the limited ability of robotic systems to reason and plan in such environments results in poor global performance (Bechar et al. 2009), which makes this process appear ineffective.

The concept of 'Precision Collaboration' (Bechar et al. 2015) is the underlying aspect in all emerging trends in smart agriculture. The essence of this concept is that many often highly dispersed and distributed agents and resources are integrated to enable and accomplish the goals of smart agriculture collaboratively, with precision. It requires collaborative automation, the integration of robotics automation with collaborative control (including humans in the network), AI, and precision automation (Ajidarma and Nof 2021, Dusadeerungsikul and Nof 2021, Nof 2022, Sreeram and Nof 2021).

In complex systems and systems-of-systems, intelligent control techniques and systems are necessary for dynamic, real-time interpretation and guidance of the environment and the objects operating in it (Nof 2009). Many smart agriculture related projects have been undertaken that use the potential of technologies and concepts, such as Cloud computing, Internet of Things (IoT), Internet of Services (IoS), Cyber Physical System (CPS), robotic simulators with realistic motion simulations, cyber augmented collaborative control and Human–Robot Collaboration.

Robots for smart agricultural tasks are composed of numerous sub-systems and components that enable them to operate and perform their tasks. These sub-systems and devices deal with path planning, navigation, or guidance abilities (Carpio et al. 2020, Zaidner and Shapiro 2016), mobility, steering and control (Lipinski et al. 2016), sensing, manipulators or similar functional devices (Mann et al. 2014), end effectors, control and decision support systems to manage individual or simultaneous unexpected events, and some level of autonomy (van Henten et al. 2013). Robots for smart agriculture are generally designed to execute a specific agricultural task, such as specific spraying (Asaei et al. 2019), selective weeding (Wu et al. 2020), disease monitoring (Kerkech et al. 2020, Liang et al. 2020), selective pruning (Bechar et al. 2014a), etc. These various domain examples are considered as the 'main task' to be performed by the robotic system. To execute successfully the 'main task', the robotic system requires to perform several 'supporting tasks' considered as services, e.g., localization and navigation, detection of the object to treat, etc. Data, information and commands are transferred between the 'supporting tasks' and the 'main task'. Each 'supporting task' controls one or several sub-systems, components or devices, and a sub-system or a device may serve several 'supporting tasks' (Fig. 6). For instance, in developing a disease monitoring robot (Schor et al. 2016a), the 'main task' is disease monitoring, the robotic system needs to have the ability to perform the 'supporting tasks' of self-localization, trajectory planning, steering and navigating in the plot from its actual location to the next sampling location, collaborating with a human operator or interacting with a human presence and other robots or unexpected objects on the pathway, and modifying its trajectory planning as necessary. Nguyen et al. (2013) developed and implemented a framework for motion and hierarchical task planning for apple harvesting robot, Bechar et al. (2009) developed a methodology for melon detection by a human-robot system to be used by a melon harvesting robot, and Ceres et al. (1998) developed and implemented a framework for human integrated citrus harvesting robot. A framework for agricultural and forestry robots was developed by Hellstrom and Ringdahl (2013).

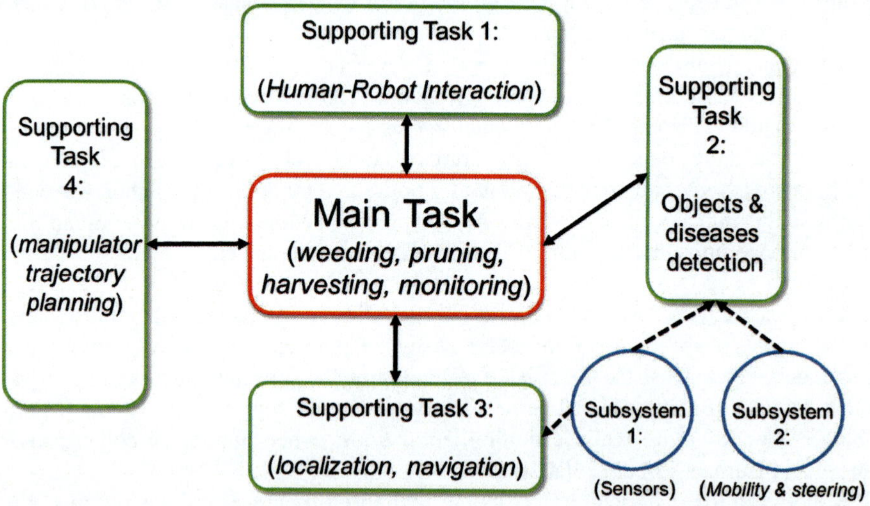

Fig. 6. Structure of task sub-systems in an agricultural robot. Solid arrows represent commands, data and information transfer; dashed arrows represent conceptual connections. In parentheses are examples for agricultural robot 'main tasks', 'supporting tasks' and subsystems (Bechar and Vigneault 2016).

Investigation of the agricultural task characteristics reveals that it can be classified into three levels, based on the task complexity. The task complexity is defined by the level of robot-plant interaction, while relatively higher level represents relatively higher complexity. A lower level of the robot-plant interaction represents no physical contact between the robot and the plant. In this level, the agricultural tasks are involved mainly in i) data collection using visual and other sensors, e.g., early detection of diseases and pests, abiotic stress diagnostics and anomalies identification (Sanchez et al. 2020, Freitas et al. 2020); ii) transportation of produce, materials and tools between different locations in the farm (Guzman et al. 2016); and, iii) material dispersion and deposition such as variable rate fertilization, selective and specific spraying, pollination (Bechar et al. 2008). The middle level of complexity requires a physical contact between the robot and the plant but no handling of produce, materials or plant parts. Typical tasks in this level are selective mechanical weeding (Tillett et al. 2008) which physically damage the weed but do not collect or handle it (Lati et al. 2021), seedling, fruit thinning and branch pruning that removes fruitlets and branches (Bechar et al. 2014b, Shoshan et al. 2022). The third level of robot-plant interaction and the most challenging one requires both physical contact between the robot and the plant, and handling of produce, materials or plant parts. Among the tasks at this level can be found fruit picking, harvesting of leaf crops, which require precise operation; decision making and handling the produce without impairing it or reducing its quality; transplanting of plants and trees, transferring of pots (with plants) in plant nurseries, etc.

In addition, since the main objective of an agricultural task is either to collect data, analyze it, make decision or act accordingly, smart agricultural tasks can be defined and classified according to the task's main objective from an agricultural robot perspective. The first type of smart agriculture 'main task' deals with data collection. Examples for smart agricultural tasks of this type are high spatial and temporal resolution monitoring of climate and environmental conditions, soil sampling (Lukowska et al. 2019, Schnug et al. 1998) for nutrients, pests and bacteria, visual and acoustic monitoring (Finkelshtain et al. 2017, Schor et al. 2016b) of anomalies, biotic and abiotic stresses (Wang et al. 2019), yield and plant condition. The second type is attributed to decision making, optimization and decision support processes. Representative smart agricultural tasks in this stage are irrigation management interfaces, classification task, and farm processes planning. The third task type relates to tasks that require action or physical performance such as specific spraying, transplanting and seeding (Gao et al. 2016, Bhimanpallewar and Narasingarao 2020), weed control (Wu et al. 2020, Raja et al. 2020), fruit picking and harvesting (Bloch et al. 2018).

Combining the above-mentioned classifications, the task complexity level and the task type, creates a task classification space (Fig. 7), that can be used to position a specific smart agricultural task and estimate the challenge level and the required research and development effort for designing and implementing a robot to perform that task (Fig. 7). Additional dimensions such as costs and limitation could be added to the space in order to better estimate the challenges.

The development and application of robots for smart agricultural tasks has to comply with the following five guidelines:

1. The farmer requirements must be considered first.
2. The agricultural task and its components must be feasible using the existing technology and the required complexity.
3. The required spatial and temporal resolution must be applicable by the robotic system and synchronized with other tasks in the process chain.
4. The cost of the robotic system solution must be lower than the expected revenue.
5. The robotic system developed must have an added value for the performance of the task or for other tasks in that process.

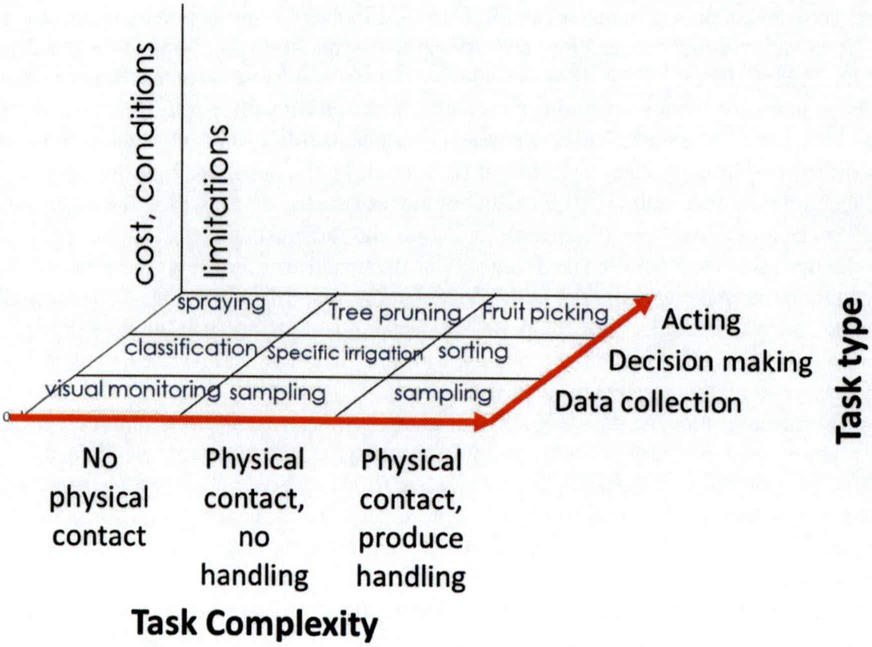

Fig. 7. The task classification space based on the task complexity and the task type. The location of several different tasks on this space can demonstrate the challenge level. The red arrows represent increasing challenge.

In most cases, the implementation of robots to perform a smart agricultural task is achievable if at least one of the following conditions is met:

i) The cost of utilizing robotics is lower than the cost of any concurrent methods.
ii) The usage of robotics enables increasing farm production capability, produce, profit, safety and survivability under competitive market conditions.
iii) The usage of robotics improves the quality and uniformity of the produce.
iv) The usage of robotics minimizes the uncertainty and variance in growing and production processes.
v) The usage of robotic systems enables the farmer to make decisions and act at higher temporal or spatial resolution in comparison to the concurrent system to achieve optimization in the growing and production stages in an equivalent manner to 'lean manufacturing' in the classic industry.
vi) The usage of robotic systems enables to increase the service or information quality.
vii) The robotic system is able to perform specific tasks that are defined as hazardous, or cannot be performed manually.

4 Simulations, Optimizations and Planners of an Agricultural Robotic System for Tasks in Smart Agriculture

As more robotic systems are being developed and implemented in the agricultural domain, it would be cost effective to simulate such systems during the development phase. Recently there have been a few research projects on simulating a robotic system for human–robot collaboration. A computational simulation environment named 'Simulation Environment for Precision Agriculture Tasks using Robot Fleets' (SEARFS) was developed to study and evaluate the execution of agricultural tasks that can be performed by an autonomous fleet of robots (Emmi et al. 2013). The environment is based on a mobile robot simulation tool that enables the performance, cooperation and interaction of a set of autonomous robots to be analyzed while simulating the execution of specific actions on a three-dimensional (3D) crop field. The SEARFS computational simulation environment is capable of simulating technological advances such as GPS, GIS, automatic control, in-field and remote sensing, and mobile computing, which will enable the evaluation of new algorithms derived from precision agriculture techniques and can contribute to smart agriculture. This environment was designed as an open source computer application. The SEARFS environment consists of four levels of configurations: 1) setting the simulation scene; 2) setting the mission parameters; 3) creating 3D virtual environment; and 4) executing the simulation.

A general method for the development of customized robot simulation and control system software with a robot operating system (ROS) was also developed by (Wang et al. 2016). The simulation designed in this research involves: a) a 3-D visualization model, created in URDF (unified robot description format) and viewed in Rviz to achieve motion planning with the MoveIt! software package, b) machine vision provided by a camera driver package in ROS to enable the use of tools for image processing, and 3-D point cloud analysis to reconstruct the environment to achieve accurate target locations, and c) communication protocols provided by ROS for serial, Modbus support of the communication system development. A tomato harvesting scenario was simulated using this methodology to demonstrate its features and effectiveness.

Automated processes in an uncertain and unstructured environment (such as agriculture) are challenged by changing peripheral requirements (Zhong et al. 2015). Addition of extra flexibility to the existing equipment to handle a larger range of tasks is a desirable solution, which can be offered, for example by Reconfigurable End-Effectors (REEs). A REE system has an adjustable structure to facilitate the adaptation of the end-effectors to various objects, therefore it is an intermediate solution between flexible and dedicated end-effectors (Zhong et al. 2015). In harvesting processes, the grasp quality is one of the most important factors for production quality, therefore research on effective design and control of reconfigurable end effectors is highly relevant. Use of multiple end effectors enables the robot to adapt directly to multiple agricultural functions as and when required. For effective REE operations, the asynchronous cooperation requirement planning (ACRP) framework was created to facilitate the design and control of REE. The ACRP provides a dynamic solution, extending from the planning facet of collaborative control theory (CCT) for designing (offline) and controlling (online) multi-agent collaborations. The ACRP methodology is illustrated in Fig. 8:

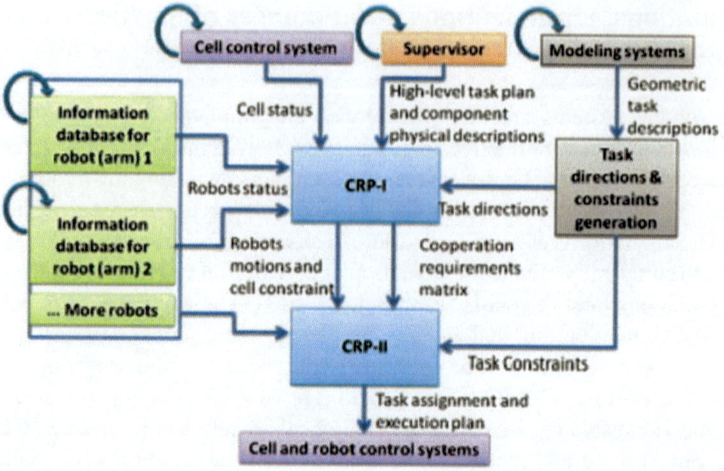

Fig. 8. Framework of Asynchronous Cooperation Requirement Planning (ACRP) (Courtesy of Zhong et al. 2015)

Robotic manipulators can perform a variety of agricultural tasks, many of them with precision. However, despite decades of research, few agricultural robots have been commercialized. One of the reasons for the lack of agricultural robots on the market today is their high cost and lack of precision enabling functions, which makes them unprofitable for farmers.

Bloch et al. (2017) from the Agricultural Research Organization in Israel designed robotic systems that are optimal for specific tasks. In the optimization process, the robot's performance is maximized while allowing it to perform the task. To achieve a reliable result, the actual field task must be described and modelled with sufficient precision. However, the complex and unstructured environment of agricultural tasks complicates the task description as well as the robot-design process.

The main goal was to characterize and analyze the environment of a given orchard and the required agricultural tasks, to understand their combined influence and interaction with the optimal design of a task-based robot for that orchard, and to determine the optimal robot base location (Bloch et al. 2018). This analysis allows the task description to be simplified by characteristics of the environment during simultaneous design of the robot and its environment.

To solve the robot-optimization problem for fruit picking in complex environments, a method was developed for characterizing the agricultural environment by fruit clustering and reaching cones. The method systematically reduces the complexity of the environments, thereby decreasing the number of calculations and providing a near-optimal solution. The method was approved and successfully applied to complex environments, solving the optimization problem in hours, rather than after weeks of calculations. The expected precision of the achieved solutions was 10% in the case examined.

In general, a set of tools and methodology was developed for analysis and design of the agricultural environment, together with optimal robot design. This methodology is novel in robot design, in particular in agriculture. It helps to improve the robot performance while designing low-cost robots affordable for farmers. The methods developed in this research are applied to apple and nectarine harvesting, although they can be used for robotic harvesting of any type of fruits, for other agricultural tasks, or in any area where the robot-environment design is used or is applicable.

5 Examples of Agricultural Robots' Projects for Smart Agriculture

5.1 Human–Robot Collaborative System for Selective Tree Pruning

Orchard pruning is a labor-intensive task that involves more than 25% of the labor costs. The main agricultural objectives of this task are to increase exposure to sunlight, control the tree shape, and remove unwanted branches. In most orchards, this task is performed once a year and up to 20% of the branches are removed selectively.

A human–robot collaborative system for selective tree pruning has been developed (Bechar et al. 2014b). The system consists of a Motoman manipulator, a color camera, a single-beam laser distance sensor, a human-machine interface (HMI) and a cutting tool based on a circular saw developed specifically for this task. The cutting tool, camera, and laser sensor are mounted on the manipulator's end-effector, and aligned parallel to each other (Fig. 9).

Fig. 9. Cutting tool design for tree pruning (Source: Agricultural Research Organization, Israel)

Experiments were established to examine the performance of the system under different conditions, human–robot collaboration methods and different trajectory types (Bechar et al. 2014b). A cutting tool was designed for pruning branches with a diameter of up to 26 mm at 45 degrees cutting angle. The saw diameter was determined to be 115 mm with a standard shaft diameter of 41 mm. An interface to connect the cutting tool to the robot's end effector was designed to minimize the total dimensions of the tool and to increase e robot dexterity. An average cycle time of 9.2 s was achieved when the human operator and robot perform simultaneously. The results also revealed that the average time required to determine the location and orientation of the cut was 2.51 s.

5.2 Robot for Automatic Melon Collection

Melon and watermelon harvesting require intensive manual labor. Machines with automatic robotic arms may replace personnel, especially in a simple routine that requires considerable physical effort. In this project, a human is involved but in a different way. Based on preliminary tests it was found that about 80% of the workers' time is invested in transferring the picked melons from the bed and only 20% in locating and disconnecting the ripe melons from the plant. Therefore, the task is conducted in two steps. In the first, the human passes in the field, detects the ripe melons, marks their locations and disconnects them from the plants with pruning shears. In the second steps, the robotic system passes and collects only the melons that were marked and harvested. A robotic arm system has been developed (Fig. 10) that can collect the melons automatically knowing their coordinates, while moving through the collection area. An electro-mechanical robotic arm system has been assembled that consists of a wheeled frame, cylindrical rails with end limit-switches, stepper motors with encoder for X- and Y-axis arm movement, a pneumatically operated robotic arm system for additional Y- and Z-axis movements, vacuum operated gripper, motor controllers and a PLC (Nair et al. 2021).

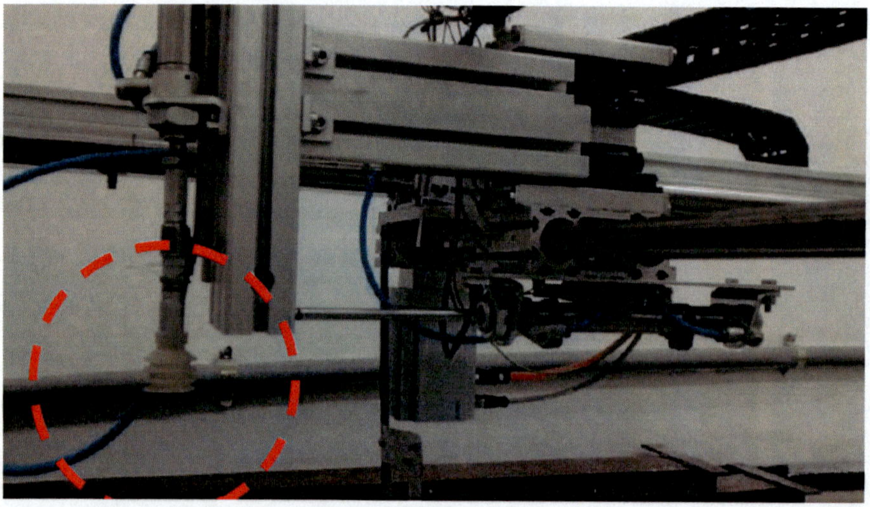

Fig. 10. A close-up of the melon picking robot and the robotic arm for melon picking (circled) (Source: Agricultural Research Organization, Israel).

A human-machine interface has been developed to enable operator intervention. A melon 'picking-up' simulator program has been created, capable of demonstrating the process of collecting melons by the robotic arm. For experimental applications, the melon collecting path optimization algorithm was used. The system was tested and succeeded in reaching up to seven target points in sequence with an accuracy of 84% (with a target-reaching error of 7–10 mm, collection time of 7–8 melons per min, at a distance of up to 4000 mm, with arm velocity of up to 800 mm/s and acceleration of up to 50 m/s^2).

In this project, an outdoor field multi-arm robot prototype was developed by Zion et al. (2014). The robot is towed by a tractor, marks the melons to be collected, assigns which arm will collect which melon, determines the collecting sequence, reaching above the melon, grasping it and lifting it from the ground (Fig. 11).

Fig. 11. field prototype of a multi-arm melon collector (Source: Agricultural Research Organization, Israel).

5.3 Development of a Selective Autonomous Sprayer for Greenhouses

The essential process of pest control and chemical application of nutrients is one of the most important processes in any agricultural production. Nevertheless, the application requires human resources; it is a time-consuming task and exposes the operators to the danger of contamination with hazardous chemicals. Integrating autonomous robots and machinery for agricultural tasks involving expensive labor, with tasks that are monotonous and hazardous, has accelerated recently. An autonomous robot is an alternative in many cases. This research focuses on the development of a navigation procedure for an autonomous sprayer in a greenhouse growing sweet peppers. An autonomous package was developed and installed on the sprayer. The autonomous sprayer includes a PC computer, controller, drivers and internal and external sensors, such as encoders, camera and LIDAR (Fig. 12).

Fig. 12. A selective autonomous sprayer for greenhouses (Source: Agricultural Research Organization, Israel).

5.4 Human–Robot Collaborative System for Early Detection of Diseases

Traditional agricultural management practices assume that fields growing crops have homogeneous properties (Steiner et al. 2008). In contrast, smart agriculture integrates different technologies, such as: sensors, information and management systems for adapting agricultural practices to variation within the field (Dong et al. 2013).

Crop yield is affected by different stresses, e.g. pests, diseases, weeds, environmental conditions, nutrition or water deficiencies, which can impair production. (Oerke and Dehne 2004) indicated that the impact of diseases, insects and weeds represents a potential annual loss of 40% of world food production. The occurrence of diseases depends on environmental factors and they often have a sporadic spatial distribution, therefore sensing techniques can be useful in identifying primary disease foci and distribution (Franke et al. 2009, Franke and Menz 2007). Sankaran et al. (2010) and Lee et al. (2010) suggested that detection and quantification of diseases with visible and infrared spectroscopy would be feasible. If a symptom or a disease can be detected by the naked eye, a sensor should also be able to record the stress symptoms (Stafford 2000).

Currently, disease detection and monitoring in greenhouses are conducted manually by an expert inspector and are limited because of the lack of human resources, sparse sampling, and large monitoring costs. Sampling intensity and resolution are low with about 20 arbitrarily locations sampled per hectare in a fixed pattern (the same locations are revisited) and each plot is monitored every 7–10 days. The plants are inspected for symptoms by an inspector crossing the greenhouse rows on foot. Thus, the inspector walks about 20 km per day, covering about 8 hectares, and a designated inspector is required for every 80 hectares. The limitations of the current inspection methods can lead to late detection and inability to mitigate a disease. As a preventive measure, repeated, large doses of pesticide are often applied even when symptoms are far below thresholds that require pesticide application. Moreover, pesticides are typically applied uniformly throughout the greenhouse, while disease distribution is typically centered in distinct locations. The result is additional pesticides usage, increased material cost, and adverse environmental effects.

In greenhouses, a current challenge is the early detection of stresses (potentially leading to diseases) and of other crop risks, to prevent the spread of uncontrolled disease and hence improve productivity and yield. Often, detection is too late even though there is enough knowledge on how to address specific stresses in plants, as soon as they are detected. Different biotic and abiotic stresses affect the expected potential crop yield. These stresses and other factors that limit yields must be detected as early as possible, such that appropriate and precise counter measures may be applied in a timely response. In the absence of an affordable and effective monitoring mechanism or system, the decisions taken by farmers could be wrong and might result in over- or under-application of pesticides, nutrients and water, often at unnecessary locations.

Robotic systems in greenhouses enable early detection and improved control of plant diseases. They are expected to increase yield, improve quality, reduce pesticide application, increase sustainability and reduce costs. Symptoms vary for each disease and crop, and each plant might suffer from multiple threats, thus, dedicated integrated disease detection systems and algorithms are required.

Automation of disease detection can alleviate current difficulties and lead to improvement in yield together with considerable reduction in pesticide use (Bock et al. 2010, Franke et al. 2009, Franke and Menz 2007). In addition to reduce production costs, this robotics solution will also lead to reduced exposure to pesticides of farm workers and inspectors, and increased sustainability. Plant diseases can affect various optical foliage characteristics, therefore disease detection can be based on different spectral ranges (Lee et al. 2010). Image processing of foliage light reflection has been applied to many different diseases and cultivars (Arnal Barbedo 2013, Veerendra et al. 2021). Methods based on fluorescence (Wetterich et al. 2016) or thermography (Oerke et al. 2011) can also be used for disease detection and have been extensively studied, but they are less relevant for a robotic detection system operating in the field because of cost, payload weight, or required setup. Mobile robotic manipulators with a combination of various sensing capabilities offer an automated solution suitable for early disease detection in greenhouses. There has been, however, little comprehensive research on the development of such integrated robotic disease detection systems for greenhouses, probably because the primary challenge of developing robust disease detection algorithms is still an open research challenge. Aerial platforms and ground mobile robotic platforms with fixed sensor configurations (Moshou et al. 2014) for disease detection have been tested for open field crops. Yet, in greenhouses, both solutions have inherent shortcomings. The maneuverability and flight duration of aerial systems within greenhouses is limited, navigation and disease locating cannot rely on GPS sensors because the structure can cause unpredictable errors, and therefore they lose their main outdoor advantage. In greenhouses, sensory position and adaptation of orientation can greatly improve detection, especially early detection where symptoms are typically centered on distinct locations. For fixed sensor installations, position and orientation adaptation are not possible. Moreover, in fixed configuration systems, the requirement for multiple disease detection can lead to a requirement for multiple detection positions and orientations, which tend to increase system complexity and hinder maneuverability.

To address this issue, a robotic disease detection system for greenhouse pepper plants was developed based on the concept of a mobile robotic manipulator (Schor et al. 2016a, 2017), which provides the required maneuverability and flexibility (Fig. 13). Prior to the above, no major system had been developed for disease detection for specialty crops in greenhouses that involved a mobile robotic manipulator.

The robotic disease detection system was developed holistically, i.e. system architecture, operation cycle, and detection algorithms for multiple threats to a pepper crop were developed in an integrated manner.

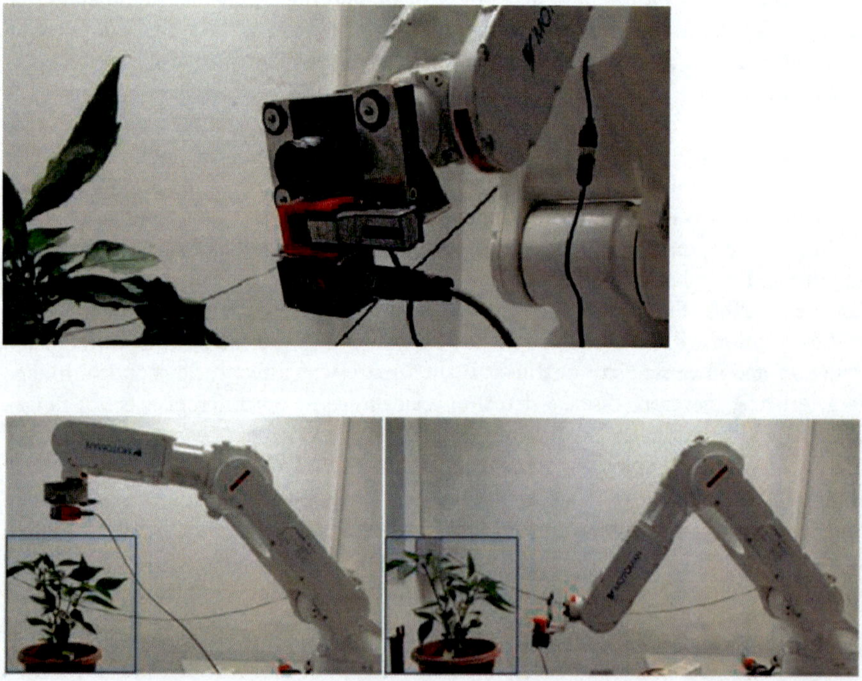

Fig. 13. The apparatus for disease detection for pepper plants (Source: Agricultural Research Organization, Israel).

The detection system comprises a mechanical structure, sensor suite, motion planning and disease detection algorithms. Visual spectrum imagery is used for motion planning and disease detection for fast, non-destructive and cost-effective operation. An algorithm based on principal component analysis (PCA) was developed for powdery mildew, and three algorithms were developed for tomato spotted wilt virus (TSWV) disease detection; one based on PCA, and two on the coefficient of variation (CV).

The algorithms were tested using images of healthy and infected plants taken from a greenhouse. For RGB-based detection of TSWV, PCA-based classification with leaf veins removed achieved the greatest classification accuracy (90%), and the accuracy of CV methods was also relatively high (85%, 87%). For powdery mildew (PM), the accuracy of pixel-level classification was relatively high (95.2%) while that of leaf condition classification was relatively low (64.3%), because leaf images came from the top of the leaf while disease symptoms start appearing from the bottom. The NIR-R-G-based detection produced inferior results for both diseases. The components of the system were integrated, and preliminary integration tests were done in a laboratory environment to verify that all system components would work together. The integrated system operated successfully for 110 consecutive minutes, with an average cycle time of 26.7 s for end-effector velocity of 15 mm/s and PCA-based detection algorithms.

Results are encouraging, because although the cycle time attained was slower than the calculated required baseline (Schor et al. 2017), the laboratory environment (comprising a conveyor belt, stationary sensor system, and black background for simplifying plant identification and background removal procedures) makes the disease detection task relatively easier and faster. Conducting a disease detection task in an unstructured environment such as a greenhouse will require more sophisticated algorithms for motion control, path planning and image processing because of a more complex environment that includes obstacles, background noises, illumination etc. As a result, the cycle time may end up being extended.

The robotic platform (mobile cart) was modified at ARO to improve the control and autonomous navigation, and to better suit the disease detection tasks in a greenhouse. The platform is equipped with a UR5 manipulator, a sensory system comprising two depth cameras, Kinect V2 and RealSense 435, and an RGB 1080p camera to create 3-D models and 2-D maps of the greenhouse. A real-time environment mapping application was developed and modified with the robot sensors while it moves in the environment and generates a 3-D model of it. A 3-D mapping experiment was conducted in the laboratory and in a pepper greenhouse at ARO (Fig. 14).

For the ‘human-in-the-loop’ tasks of the agricultural robot system, a HUB-CI (hub for collaborative intelligence) system was developed by the PRISM team at Purdue and the ARO team (Bechar et al. 2020). The objective: To enable effective and timely integration, and resulting collaboration tasks, by optimized exchange and leveraging of signals and information gathered in real-time from distributed components. The outcome of the HUB-CI is collaborative intelligence from the ARS networked components, thus enabling precision tasks (Nair et al. 2019). The following algorithms and protocols were developed by Dusadeerungsikul and Nof (2019): a) a collaboration (task administration) protocol and algorithms, to determine which image or case must be further reviewed by remote human operators or experts; b) an adaptive search collaboration

(a) (b)

Fig. 14. Three-D mapping of a pepper greenhouse (a); the robotic platform (b).

(task administration) protocol and algorithms, applying knowledge-based information; c) a routing collaboration (task administration) protocol and algorithms, creating and adapting a given tour for a mobile robot; d) detection-routing protocol and algorithms, based on a mechanism for remote disease detection algorithm to communicate with the routing algorithms; e) a manual control collaboration protocol providing mechanism and constraints for manual control of the robot system; and f) human-in-the-loop collaboration protocol - a mechanism for human operators to communicate with the search, detection, and routing algorithms.

The HUB-CI system has been designed as a virtual platform to integrate the streams of signals, data, and control logic from multiple participating agents, including cyber and human agents (Dusadeerungsikul et al. 2019, Moghaddam and Nof 2022, Seok and Nof 2011, Nair et al. 2019). It enables the cyber-collaborative protocols to make local control decisions based on global and local real-time information. A new type of HUB-CI prototype was developed and tested in the experiments. Unique features designed with the HUB-CI system include (Nair et al. 2019): i) planned collaboration between diverse users (farmers, engineers, pathology experts, etc.) of the agricultural robotic system in a HUB-CI environment, ii) collaborative semi-automated and manual control (remote and local) of agricultural robot, iii) learning-based filtering algorithm for spectral images taken off greenhouse plants, iv) collaborative decision making regarding the greenhouse system based on intelligent information sharing, v) scheduling and task administration of all cyber and human agents in the agricultural robotic system (ARS), and vi) adaptive search and routing collaboration protocols and algorithms, optimizing resources and time to perform monitoring and inspection tasks. Three experiments were conducted to examine the collaborative control of the system in action (Nair et al. 2021).

In all experiments, the robot at ARO site was controlled remotely from Purdue University. Two-way collaboration frames were developed: 1) an ad-hoc connection using TeamViewer, in which researchers at Purdue controlled the robot's computer directly; and 2) through a server using Dropbox.

For the disease detection algorithm, a machine-learning model was developed based on hyperspectral data. The hyperspectral imaging analysis can be divided into two research stages: 1) a classification algorithm needs to be developed based on full spectral information of healthy and diseased spots; 2) some key hyperspectral bands need to be selected specifically for real-time in-field detection. The Bio-Imaging and Machine Vision Lab, University of Maryland research group developed a new method of hyperspectral analysis named 'outlier removal auxiliary classifier generative adversarial nets' (OR-AC-GAN). The model uses full spectral information (395–1005 nm) to integrate the tasks of background removal, pixel-level spectral analysis and image-level plant classification. The model starts from generative adversarial nets (GAN) to learning the data distribution of different spectral classes. It can augment the training dataset online according to the data distribution and effectively remove the side effects of data outliers and imbalance on the dataset. This model can classify the one-dimensional spectral signal into different classes. Images were taken at ARO site using a specimen VNIR hyperspectral camera with a high-resolution, high-speed image acquisition GPU (NI PCIe-1427) installed on an i7-4770 CPU PC. The computer was equipped with the Specimen data recording software for hyperspectral images (HSI): Lumo Scanner. In the experiment, 54 independent test images of the TSWV disease database constructed, the model reached 96.25% prediction accuracy (92.59% sensitivity, 100% specificity) before visible symptoms appear (as early as 5 days after disease inoculation) using only 8 selected bands (Wang et al. 2019). Expert phytopathologists can detect the diseased plants only 15 days after disease inoculation. For pixel-level classification accuracy, the prediction of false positives in healthy plants was as low as 1.47%. The OR-AC-GAN is an all-in-one model meeting the first requirement of hyperspectral data analysis. The experiment proved that the augmented data, a 'by-product' of OR-AC-GAN could markedly improve the performance of existing band selection algorithms.

5.5 Multi-sensor Fault Tolerant Learning Algorithm in an Agricultural Robotic System

Ajidarma (2017) and Ajidarma and Nof (2021) aimed to develop a new fault tolerant interface design based on the collaborative control theory (CCT) principles of best matching, error prevention and conflict resolution (EPCR) for an agricultural robotic system. They developed a fault tolerant learning algorithm to process the data of moving sensors in an agricultural robotic system. The sensor data and actual state of the object were modelled as a function of error and rate of conflict. Two learning algorithms, adaptive learning algorithm (ALA) and cumulative learning algorithm (CLA) were developed and tested. This method involves collaboration with a human operator and an adaptive learning mechanism to minimize measurement and detection errors. It is a useful example of the precision collaboration concept (Bechar et al. 2015) discussed earlier in this chapter. This research addressed the problem of having an interface with fault tolerant sensor data processing in a collaborative agricultural robotic system where multiple sensors are mounted on a mobile robot, and a human operator performs supervisory functions (Nair et al. 2021).

6 Summary and Perspective

Research, developments and evaluations of robots to perform smart agricultural tasks are highly diverse in terms of objectives, structures, techniques, and components. In this context, it is difficult to compare different robots and transfer developed technology from one smart agricultural task to another. The limiting factors for the development of such systems are unique to each robotic system and smart agricultural task. In this chapter, an attempt to investigate characteristics of smart agricultural tasks and to study the interaction between robotics and smart agriculture was conducted.

Robots and intelligent automation systems are generally highly complex since they combine several different sub-systems that need to be orchestrated, integrated and correctly matched and synchronized to perform tasks optimally as a whole, and to successfully transfer, sift, and effectively utilize the required information. This collaborative integration needs to consider time delays, errors and conflicts, cycle times, and the characteristics of communication among all sub-systems. Agricultural robots are even more sophisticated since they must operate under unstructured agricultural environments without compromising productivity and work quality relative to concurrent methods. In that area, there has been considerable progress in the past few decades.

Research and development of robotic systems to perform smart agricultural tasks need to follow several steps. First, Investigation and study of the nature of tasks, processes, and their environments in relation to the variability of the leading parameters must be conducted in order to evaluate the feasibility of proposed solutions. Second, technologies and methodologies must be developed and modified to fit high variance cases, and overcome difficult challenges such as the continuously changing conditions, the significant variability of the produce and the environment, and hostile environmental conditions, such as vibration, dust, extreme temperatures, and humidity. Third, Identification of processes and tasks that can be 'robotized', and evaluation of the overall task complexity and the smart agriculture type. Forth, evaluation of the challenge level and the required research and development effort for such systems and such tasks. For high complexity tasks, high challenge level, or high research and development effort, possible solutions to overcoming this problem might be agronomical modifications, or a human integration, or both. Fifth, investigate whether the proposed and presented solution complies with the guidelines and conditions discussed in Sect. 3 of this chapter. Finally, sixth, agricultural robotic systems should be developed only for tasks and processes where alternative solutions, such as mechanical or lower-intelligence automation, cannot exist, or that robotics does not have a diminishing marginal utility relative to them.

The robots that are to be used for smart agricultural tasks must recognize and understand the physical properties of each specific object, and must be able to work under varied, uncertain environmental conditions in fields, or in controlled environments. Therefore, they need sensing systems that can work under variable conditions, as well as specialized manipulators and end-effectors. The environmental conditions are occasionally so severe with regard to high temperature, humidity, dust, wind, rain, and/or storm, that electrical circuit and material corrosion problems can be a major concern. These conditions must be taken into consideration when designing robotic systems for smart agriculture. In this sense, development and application of robots for smart agricultural tasks are expected to be an iterative process.

In this chapter, we presented the contribution of agricultural robotics to smart agriculture, which comprises the three technology pillars: agricultural robotics, precision, and artificial intelligence. This contribution is based on the close interaction and synergy of agricultural robots with the other two technological pillars.

First and foremost, agricultural robots are information technology (IT) platforms. Their ability to collect data of different types from multiple sources; filter, integrate and label data in real time and on site; overcome errors and conflicts; and control the way, location, time and duration of collecting data, make them a supreme instrument to practice smart agriculture and enable optimal and selective data collection. In addition, the performance of smart agricultural tasks by robots requires a massive collection of labelled data. Agricultural robots are the means for smart agriculture to evolve and to develop and examine new concepts and approaches. The characteristics of robots make them best suited for AI and machine learning models by nature, due to their intrinsic abilities, i.e., automatic and massive data labeling and optimal sensor pose. Reciprocally, AI increases agricultural robots' abilities and performance. Regarding precision agriculture, robots can improve the performance of site-specific operations and variable-rate applications, and are both the tool to perform precision agriculture and the means to expand its borders. The relations and interaction between agricultural robots to AI and precision agriculture in the sense of smart agriculture is presented in Fig. 15.

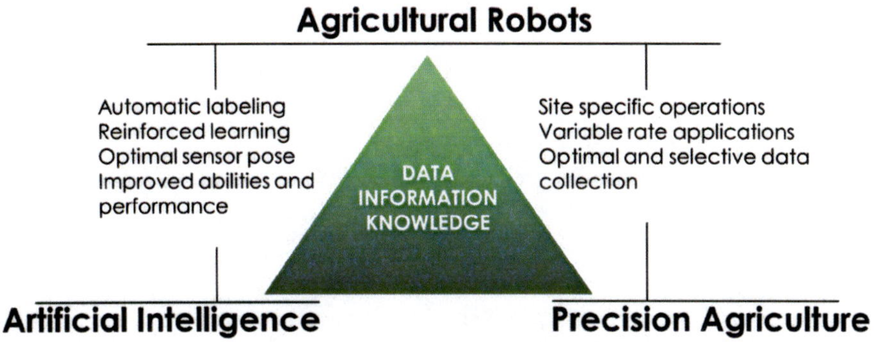

Fig. 15. The three technology pillars of smart agriculture and the interaction and synergy between them.

Emerging trends and future developments are planned and anticipated in all the above areas. Particular advantages can be expected by cyber-augmentation for further smart automation and autonomy, including cyber-augmented precision collaboration of stakeholder farmers and human–robot agents of smart agriculture. It is exciting to realize that we may be at the dawn of the next information era in agriculture.

References

Airlangga, G., Liu, A.: Initial machine learning framework development of agriculture cyber physical systems. J. Phys. Conf. Ser. (2019)

Ajidarma, P.: Multi-Sensor Fault Tolerant Learning Algorithm in an Agricultural Robotic System. M.Sc., Purdue University (2017)

Ajidarma, P., Nof, S.Y.: Collaborative detection and prevention of errors and conflicts in an agricultural robotic system. Stud. Inform. Control **30**, 19–28 (2021)

Alwis, S.D., Hou, Z., Zhang, Y., Na, M.H., Ofoghi, B., Sajjanhar, A.: A survey on smart farming data, applications and techniques. Comput. Indust. **138** (2022)

Arnal Barbedo, J.G.: Digital image processing techniques for detecting, quantifying and classifying plant diseases. SpringerPlus **2,** 1–12 (2013)

Asaei, H., Jafari, A., Loghavi, M.: Site-specific orchard sprayer equipped with machine vision for chemical usage management. Comput. Electron. Agric. **162**, 431–439 (2019)

Ayoub Shaikh, T., Rasool, T., Rasheed Lone, F.: Towards leveraging the role of machine learning and artificial intelligence in precision agriculture and smart farming. Comput. Electron. Agricul. **198** (2022)

Bac, C.W., Hemming, J., van Henten, E.J.: Robust pixel-based classification of obstacles for robotic harvesting of sweet-pepper. Comput. Electron. Agric. **96**, 148–162 (2013)

Bac, C.W., van Henten, E.J., Hemming, J., Edan, Y.: Harvesting robots for high-value crops: state-of-the-art review and challenges ahead. J. Field Robot. **31**, 888–911 (2014)

Bechar, A.: Robotics in horticultural field production. Stewart Postharvest Rev. **6**(3), 1–11 (2010). https://doi.org/10.2212/spr.2010.3.11

Bechar, A.: Agricultural robotics for precision agriculture tasks: concepts and principles. In: Bechar, A. (ed.) Innovation in Agricultural Robotics for Precision Agriculture. PPA, pp. 17–30. Springer, Cham (2021). https://doi.org/10.1007/978-3-030-77036-5_2

Bechar, A., et al.: Visual Servoing Methodology for Selective Tree Pruning by Human-Robot Collaborative System AgEng 2014. Zurich, Switzerland (2014a)

Bechar, A., et al.: Visual Servoing Methodology for Selective Tree Pruning by Human-Robot Collaborative System. The EurAgEng 2014 International Conference. Zurich, Switzerland. C0287 (2014b)

Bechar, A., Edan, Y.: Human-robot collaboration for improved target recognition of agricultural robots. Ind. Robot. **30**, 432–436 (2003)

Bechar, A., Gan-Mor, S., Ronen, B.: A method for increasing the electrostatic deposition of pollen and powder. J. Electrostat. **66**, 375–380 (2008)

Bechar, A., Meyer, J., Edan, Y.: An objective function to evaluate performance of human-robot collaboration in target recognition tasks. IEEE Trans. Syst. Man Cybern. Part C-Appl. Rev. **39**, 611–620 (2009)

Bechar, A., Nof, S., Tao, Y.: Final report: Development of a robotic inspection system for early identification and locating of biotic and abiotic stresses in greenhouse crops. BARD Research Project IS-4886-16 R (2020)

Bechar, A., Nof, S.Y., Wachs, J.P.: A review and framework of laser-based collaboration support. Annu. Rev. Control. **39**, 30–45 (2015)

Bechar, A., Vigneault, C.: Agricultural robots for field operations: concepts and components. Biosys. Eng. **149**, 94–111 (2016)

Bechar, A., Vigneault, C.: Agricultural robots for field operations. Part 2: operations and systems. Biosys. Eng. **153**, 110–128 (2017)

Behmann, J., Mahlein, A.K., Rumpf, T., Römer, C., Plümer, L.: A review of advanced machine learning methods for the detection of biotic stress in precision crop protection. Precision Agric. **16**, 239–260 (2015)

Benos, L., Bechar, A., Bochtis, D.: Safety and ergonomics in human-robot interactive agricultural operations. Biosys. Eng. **200**, 55–72 (2020)

Bhimanpallewar, R.N., Narasingarao, M.R.: AgriRobot: implementation and evaluation of an automatic robot for seeding and fertiliser microdosing in precision agriculture. Int. J. Agric. Resour. Gov. Ecol. **16**, 33–50 (2020)

Bloch, V., Bechar, A., Degani, A.: Development of an environment characterization methodology for optimal design of an agricultural robot. Ind. Robot. **44**, 94–103 (2017)

Bloch, V., Degani, A., Bechar, A.: A methodology of orchard architecture design for an optimal harvesting robot. Biosys. Eng. **166**, 126–137 (2018)

Bock, C.H., Poole, G.H., Parker, P.E., Gottwald, T.R.: Plant disease severity estimated visually, by digital photography and image analysis, and by hyperspectral imaging. Crit. Rev. Plant Sci. **29**, 59–107 (2010)

Boursianis, A.D., et al.: Internet of Things (IoT) and Agricultural Unmanned Aerial Vehicles (UAVs) in smart farming: a comprehensive review. Internet of Things (Netherlands) **18** (2022)

Canning, J.R., Edwards, D.B., Anderson, M.J.: Development of a fuzzy logic controller for autonomous forest path navigation. Trans. Asae **47**, 301–310 (2004)

Carpio, R.F., et al.: A Navigation architecture for ackermann vehicles in precision farming. IEEE Robot. Autom. Let. **5**, 1103–1110 (2020)

Ceres, R., Pons, J.L., Jiménez, A.R., Martín, J.M., Calderón, L.: Design and implementation of an aided fruit-harvesting robot (Agribot). Indust. Robot Int. J. **25**(5), 337–346 (1998). https://doi.org/10.1108/01439919810232440

Dimitriadis, S., Goumopoulos, C.: Applying machine learning to extract new knowledge in precision agriculture applications. Proceedings - 12th Pan-Hellenic Conference on Informatics, PCI 2008, pp. 100–104 (2008)

Dong, X., Vuran, M.C., Irmak, S.: Autonomous precision agriculture through integration of wireless underground sensor networks with center pivot irrigation systems. Ad Hoc Netw. **11**, 1975–1987 (2013)

Dusadeerungsikul, P.O., Nof, S.Y.: A collaborative control protocol for agricultural robot routing with online adaptation. Comput. Ind. Eng. **135**, 456–466 (2019)

Dusadeerungsikul, P.O., Nof, S.Y.: A cyber collaborative protocol for real-time communication and control in human-robot-sensor work. Int. J. Comput. Commun. Control **16**, 1–11 (2021)

Dusadeerungsikul, P.O., et al.: Collaboration requirement planning protocol for hub-Ci in factories of the future, pp. 218–225 (2019)

Edan, Y., Bechar, A.: Multi-purpose agricultural robot. In: The Sixth IASTED International Conference, Robotics And Manufacturing, pp. 205–212. 1998 Banff, Canada. (1998)

Emmi, L., Paredes-Madrid, L., Ribeiro, A., Pajares, G., Gonzalez-De-santos, P.: Fleets of robots for precision agriculture: a simulation environment. Ind. Robot. **40**, 41–58 (2013)

Finkelshtain, R., Bechar, A., Yovel, Y., Kósa, G.: Investigation and analysis of an ultrasonic sensor for specific yield assessment and greenhouse features identification. Precision Agric. **18**, 916–931 (2017)

Franke, J., Gebhardt, S., Menz, G., Helfrich, H.P.: Geostatistical analysis of the spatiotemporal dynamics of powdery mildew and leaf rust in wheat. Phytopathology **99**, 974–984 (2009)

Franke, J., Menz, G.: Multi-temporal wheat disease detection by multi-spectral remote sensing. Precision Agric. **8**, 161–172 (2007)

Freitas, H., Faical, B. S., Silva, A., Ueyama, J.: Use of UAVs for an efficient capsule distribution and smart path planning for biological pest control. Comput. Electron. Agric. **173** (2020)

Gao, G.H., Feng, T.X., Yang, H., Li, F.: Development and optimization of end-effector for extraction of potted anthurium seedlings during transplanting. Appl. Eng. Agric. **32**, 37–46 (2016)

Goap, A., Sharma, D., Shukla, A.K., Rama Krishna, C.: An IoT based smart irrigation management system using Machine learning and open source technologies. Comput. Electron. Agric. **155**, 41–49 (2018)

Guzman, R., Navarro, R., Beneto, M., Carbonell, D.: Robotnik-professional service robotics applications with ROS. In: Koubaa, A. (ed.) Robot Operating System (2016)

Hellstrom, T., Ringdahl, O.: A software framework for agricultural and forestry robots. Indust. Robot Int. J. **40**, 20–26 (2013)

Holland, S.W., Nof, S.Y.: Emerging Trends and Industry Needs. Wiley, Handbook of Industrial Robotics (2007)

Junejo, K.N., Goh, J.: Behaviour-based attack detection and classification in cyber physical systems using machine learning. In: CPSS 2016 - Proceedings of the 2nd ACM International Workshop on Cyber-Physical System Security, Co-located with Asia CCS 2016, pp. 34–43 (2016)

Kerkech, M., Hafiane, A., Canals, R.: Vine disease detection in UAV multispectral images using optimized image registration and deep learning segmentation approach. Comput. Electron. Agric. **174** (2020)

Lati, R.N., Rosenfeld, L., David, I.B., Bechar, A.: Power on! Low-energy electrophysical treatment is an effective new weed control approach. Pest Manag. Sci. **77**, 4138–4147 (2021)

Lee, W.S., Alchanatis, V., Yang, C., Hirafuji, M., Moshou, D., Li, C.: Sensing technologies for precision specialty crop production. Comput. Electron. Agric. **74**, 2–33 (2010)

Liang, H., He, J., Lei, J.J.: Monitoring of corn canopy blight disease based on UAV hyperspectral method. Spectrosc. Spect. Anal. **40**, 1965–1972 (2020)

Linard, A., Bueno, M.L.P.: Towards adaptive scheduling of maintenance for Cyber-Physical Systems. In: Margaria, T., Steffen, B. (eds.) ISoLA 2016. LNCS, vol. 9952, pp. 134–150. Springer, Cham (2016). https://doi.org/10.1007/978-3-319-47166-2_9

Lipinski, A.J., Markowski, P., Lipinski, S., Pyra, P.: Precision of tractor operations with soil cultivation implements using manual and automatic steering modes. Biosys. Eng. **145**, 22–28 (2016)

Lukowska, A., Tomaszuk, P., Dzierzek, K., Magnuszewski, L.: Soil sampling mobile platform for Agriculture 4.0 (2019)

Mann, M.P., Rubinstein, D., Shmulevich, I., Linker, R., Zion, B.: Motion planning of a mobile cartesian manipulator for optimal harvesting of 2-D crops. Trans. ASABE **57**, 283–295 (2014)

Moghaddam, M., Nof, S.Y.: Information flow optimization in augmented reality systems for production & manufacturing. In: Proceedings of ICPR-AR. Curitiba, Brazil (2022)

Moshou, D., Pantazi, X.-E., Kateris, D., Gravalos, I.: Water stress detection based on optical multisensor fusion with a least squares support vector machine classifier. Biosys. Eng. **117**, 15–22 (2014)

Moysiadis, V., Tsolakis, N., Katikaridis, D., Sorensen, C.G., Pearson, S., Bochtis, D.: Mobile robotics in agricultural operations: a narrative review on planning aspects. Appl. Sci. (Switzerland), **10** (2020)

Mulla, D.J.: Twenty five years of remote sensing in precision agriculture: Key advances and remaining knowledge gaps. Biosyst. Eng. **114,** 358–371 (2013)

Nair, A.S., Bechar, A., Tao, Y., Nof, S.Y.: The HUB-CI model for telerobotics in greenhouse monitoring. Procedia Manufac. **39**, 414–421 (2019)

Nair, A.S., Nof, S.Y., Bechar, A.: Emerging directions of precision agriculture and agricultural robotics. In: Bechar, A. (ed.) Innovation in Agricultural Robotics for Precision Agriculture. PPA, pp. 177–210. Springer, Cham (2021). https://doi.org/10.1007/978-3-030-77036-5_8

Nguyen, T.T., Kayacan, E., de Baedemaeker, J., Saeys, W.: Task and motion planning for apple harvesting robot*. IFAC Proc. Vol. **46**, 247–252 (2013)

Nof, S.Y. (ed.): Handbook of Automation: Springer (2009)

Nof, S.Y.: Automation: what it means to us around the world. In: Nof, S.Y. (ed.) Handbook of Automation. 2nd ed. Springer (2022)

Oerke, E.C., Dehne, H.W.: Safeguarding production—losses in major crops and the role of crop protection. Crop Prot. **23**, 275–285 (2004)

Oerke, E.C., Fröhling, P., Steiner, U.: Thermographic assessment of scab disease on apple leaves. Precision Agric. **12,** 699–715 (2011)

Pandey, A., Kumar, S., Tiwary, P., Das, S.K.: A hybrid classifier approach to multivariate sensor data for climate smart agriculture cyber-physical systems. ACM Int. Conf. Proc. Ser. 337–341 (2019)

Qureshi, T., Saeed, M., Ahsan, K., Malik, A.A., Muhammad, E.S., Touheed, N.: Smart agriculture for sustainable food security using Internet of Things (IoT). Wireless Commun. Mobile Comput. (2022)

Raja, R., Nguyen, T.T., Slaughter, D.C., Fennimore, S.A.: Real-time weed-crop classification and localisation technique for robotic weed control in lettuce. Biosys. Eng. **192**, 257–274 (2020)

Sanchez, L., Pant, S., Mandadi, K., Kurouski, D.: Raman spectroscopy vs quantitative polymerase chain reaction in early stage huanglongbing diagnostics. Sci. Reports **10** (2020)

Sankaran, S., Mishra, A., Ehsani, R., Davis, C.: A review of advanced techniques for detecting plant diseases. Comput. Electron. Agric. **72**, 1–13 (2010)

Sargolzaei, A., Crane, C.D., III, Abbaspour, A., Noei, S.: A machine learning approach for fault detection in vehicular cyber-physical systems. In: Proceedings - 2016 15th IEEE International Conference on Machine Learning and Applications, ICMLA 2016, pp. 636–640 (2017)

Schnug, E., Panten, K., Haneklaus, S.: Sampling and nutrient recommendations - the future. Commun. Soil Sci. Plant Anal. **29**, 1455–1462 (1998)

Schor, N., Bechar, A., Ignat, T., Dombrovsky, A., Elad, Y., Berman, S.: Robotic disease detection in greenhouses: combined detection of powdery mildew and tomato spotted wilt virus. IEEE Robot. Autom. Let. **1**, 354–360 (2016a)

Schor, N., Berman, S., Dombrovsky, A., Elad, Y., Ignat, T., Bechar, A.: Development of a robotic detection system for greenhouse pepper plant diseases. Prec. Agric. **18**(3), 394–409 (2016b). https://doi.org/10.1007/s11119-017-9503-z

Schor, N., Berman, S., Dombrovsky, A., Elad, Y., Ignat, T., Bechar, A.: Development of a robotic detection system for greenhouse pepper plant diseases. Precision Agric. **18**, 394–409 (2017)

Schueller, J.K.: CIGR Handbook of Agricultural Engineering, CIGR – The International Commission of Agricultural Engineering (2006)

Schuster, R., et al.: Multi-cue learning and visualization of unusual events. In: Proceedings of the IEEE International Conference on Computer Vision, pp. 1933–1940 (2011)

Seok, H., Nof, S.: The HUB-CI initiative for cultural, education and training, and healthcare networks. 21st IICPR. Stuttgart, Germany (2011)

Shoshan, T., Bechar, A., Cohen, Y., Sadowsky, A., Berman, S.: Segmentation and motion parameter estimation for robotic Medjoul-date thinning. Precision Agric. **23**, 514–537 (2022)

Spezzano, G., Vinci, A.: Pattern detection in Cyber-Physical Systems. Procedia Comput. Sci. **52**, 1016–1021 (2015). https://doi.org/10.1016/j.procs.2015.05.096

Sreeram, M., Nof, S.: Human-in-the-loop of cyber physical agricultural robotic systems. Int. J. Computers, Comm. Control **16** (2021)

Stafford, J.V.: Implementing precision agriculture in the 21st century. J. Agric. Eng. Res. **76**(3), 267–275 (2000). https://doi.org/10.1006/jaer.2000.0577

Steiner, U., Burling, K., Oerke, E.C.: Sensor use in plant protection. Gesunde Pflanzen **60**, 131–141 (2008)

Steinfeld, A.: Interface lessons for fully and semi-autonomous mobile robots. In: IEEE International Conference on Robotics and Automation,2752–2757 (2004)

Taki, M., Mehdizadeh, S.A., Rohani, A., Rahnama, M., Rahmati-Joneidabad, M.: Applied machine learning in greenhouse simulation; new application and analysis. Inform. Process. Agric. **5**(2), 253–268 (2018). https://doi.org/10.1016/j.inpa.2018.01.003

Tillett, N.D., Hague, T., Grundy, A.C., Dedousis, A.P.: Mechanical within-row weed control for transplanted crops using computer vision. Biosys. Eng. **99**, 171–178 (2008)

Tremblay, N., Fallon, E., Ziadi, N.: Sensing of crop nitrogen status: opportunities, tools, limitations, and supporting information requirements. HortTechnology **21**, 274–281 (2011)

Urrea, C., Munoz, J.: Path tracking of mobile robot in crops. J. Intell. Rob. Syst. **80**, 193–205 (2015)

van Henten, E.J., Bac, C.W., Hemming, J., Edan, Y.: Robotics in protected cultivation. IFAC Proc. Vol. **46**, 170–177 (2013)

Veerendra, G., Swaroop, R., Dattu, D.S., Jyothi, C.A., Singh, M.K.: Detecting plant Diseases, quantifying and classifying digital image processing techniques **51,** 837–841 (2021)

Vidoni, R., Bietresato, M., Gasparetto, A., Mazzetto, F.: Evaluation and stability comparison of different vehicle configurations for robotic agricultural operations on side-slopes. Biosys. Eng. **129**, 197–211 (2015)

Wang, D., et al.: Early detection of tomato spotted wilt virus by hyperspectral imaging and outlier removal auxiliary classifier generative adversarial nets (OR-AC-GAN). Sci. Reports **9** (2019)

Wang, Z., Gong, L., Chen, Q., Li, Y., Liu, C., Huang, Y.: Rapid developing the simulation and control systems for a multifunctional autonomous agricultural robot with ROS (2016)

Wani, H., Ashtankar, N.: An appropriate model predicting pest/diseases of crops using machine learning algorithms. In: 2017 4th International Conference on Advanced Computing and Communication Systems, ICACCS 2017 (2017)

Wetterich, C.B., De Oliveira Neves, R.F., Belasque, J., Marcassa, L.G.: Detection of citrus canker and Huanglongbing using fluorescence imaging spectroscopy and support vector machine technique. Appl. Opt. **55**, 400–407 (2016)

Wu, X., Aravecchia, S., Lottes, P., Stachniss, C., Pradalier, C.: Robotic weed control using automated weed and crop classification. J. Field Robot. **37**, 322–340 (2020)

Wu, Z., et al.: K-PdM: KPI-oriented machinery deterioration estimation framework for predictive maintenance using cluster-based hidden markov model. IEEE Access **6**, 41676–41687 (2018)

Xiang, R., Jiang, H., Ying, Y.: Recognition of clustered tomatoes based on binocular stereo vision. Comput. Electron. Agric. **106**, 75–90 (2014)

Yahata, S., et al.: A hybrid machine learning approach to automatic plant phenotyping for smart agriculture. In: Proceedings of the International Joint Conference on Neural Networks, 1787–1793 (2017)

Zaidner, G., Shapiro, A.: A novel data fusion algorithm for low-cost localisation and navigation of autonomous vineyard sprayer robots. Biosys. Eng. **146**, 133–148 (2016)

Zhong, H., Nof, S.Y., Berman, S.: Asynchronous cooperation requirement planning with reconfigurable end-effectors. Robot. Comput. Integrat. Manufac. **34**, 95–104 (2015)

Zion, B., Mann, M., Levin, D., Shilo, A., Rubinstein, D., Shmulevich, I.: Harvest-order planning for a multiarm robotic harvester. Comput. Electron. Agric. **103**, 75–81 (2014)

Correction to: Systems Collaboration and Integration

Chin-Yin Huang and Sang Won Yoon

Correction to:
C.-Y. Huang and S. W. Yoon (Eds.): *Systems Collaboration and Integration*, ACES 14,
https://doi.org/10.1007/978-3-031-44373-2

The original version of the book was inadvertently published with incorrect affiliation of the editor. The erratum chapters and the book have been updated with the changes.

The updated version of this book can be found at
https://doi.org/10.1007/978-3-031-44373-2

C.-Y. Huang and S. W. Yoon (Eds.): ICPR1 2021, ACES 14, p. C1, 2023.
https://doi.org/10.1007/978-3-031-44373-2_27

Author Index

C.-Y. Huang and S. W. Yoon (Eds.): ICPR1 2021, ACES 14, pp. 475–476, 2023.
https://doi.org/10.1007/978-3-031-44373-2